FIELD GUIDE TO THE
Birds of The Gambia
and Senegal

FIELD GUIDE TO THE

Birds of The Gambia
and Senegal

Clive Barlow and Tim Wacher

Illustrated by Tony Disley

CHRISTOPHER HELM
LONDON

First published 1997 by Christopher Helm, an imprint of A & C Black Publishers Ltd, 36 Soho Square, London W1D 3QY

www.acblack.com

Reprinted with amendments 1999, 2002
Reprinted in paperback 2005, 2007, 2010

ISBN: 978-0-7136-7549-8

A CIP catalogue record for this book is available from the British Library

Production and design by Fluke Art, Cornwall

This book is produced using paper that is made from wood grown in managed,
sustainable forests. It is natural, renewable and recyclable. The logging
and manufacturing processes conform to the environmental regulations
of the country of origin.

Printed in Hong Kong

10 9 8 7

CONTENTS

To my mother and to my late father
who have encouraged my interest in birds

Clive Barlow

To my parents, and to my friends in The Gambia

Tim Wacher

To my family, especially my parents,
whose guidance and support have been invaluable

Tony Disley

ACKNOWLEDGEMENTS

The production of a book such as this is only possible using information gathered by many people over many years. Inevitably, the great majority are not known personally to the authors, but the names listed in the bibliography give some indication of the scope of published work that has been consulted and which has acted as a cornerstone to researching and producing this book.

In addition, there have been many who have been more directly involved. In particular we would like to thank colleagues in The Gambia who between 1988 and 1993 assisted TW in compiling an up-to-date database on the distribution and sighting frequency of birds throughout The Gambia. This has been a major unpublished source on which much of the distributional and status information in this book is based. In particular, John Alder, Mathieu Chable, Gill Hirst, Rachel Jones, Donal Murray, Fernley Symons, and David and Karen Wheeler made a substantial effort over a sustained period. Their enthusiasm, coupled with meticulous systematic observation, played a significant role in extending our knowledge of the remoter regions and the often torrid wet season. They have thus provided a necessary counter to a creeping literature bias towards the coastal regions and the dry season, which tend to be well covered by migrant birders of Palearctic origin.

The Gambia Ornithological Society is also thanked for providing a focus for records submitted by the numerous visiting birdwatchers attracted to The Gambia over the years, many of whom contributed written records to the GOS. The names of all those whose information was used in the database are listed below with much appreciation: J. Alder, S. J. Aspinall, R. & C. Bertera, N. Borrow, C. F. Brooks, P. Bryant, A. Caspar, M. Chable, P. Chapman, D. Clifford, D. Clugston, A. Cole, P. Colston, M. & J. Cooke, T. Dean, A. del Nevo, M. Dryden, J. N. Dymond, M. Eccles, H. Ellenberg, G. Follows, C. Fox, A, Goodwin, J. & J. Geeson, L. Grandman, A. Gregory, M. Hall, N. J. Hallam, A. Harrop, P. V. Harvey, A. Hayward, C. Helm, R. Hibbert, G. Hirst, G. Hobson, A. Hunter, Y. Jaiteh, S. Jallow, R. M. Jones, A. Jones, M. Kennewell, R. Kersley, J. King, J. Klopfenstein, S. Lister, H. Marsh, E. Meek, B. Mines, D. Murray, J. Parrot, B. Perkins, G. Rainey, J. M. Randall, N. Riddiford, J. & L. Rimmer, M. Rolfe, L. Sangan, A. Shaw, R. & B. Shaw, J. Sly, W. F. Snow, R. & J. Spearman, G. Svahn, F. B. Symons, M. Taylor, C. Thomas, P. Tyler, T. Tyrberg, J. Verleysen, R. Wake, J. Walford, S. P. Warwick, W. Waters, R. Webzell, D. & K. Wheeler, C. White, C. Wilkins, K. Woodbridge, T. Wright. We are aware that the names of many others whose records were submitted prior to initiation of the database, or after work on it was concluded, do not appear. We trust they will understand the circumstances and accept our full acknowledgement that their contributions have been equally vital in developing ornithological awareness in The Gambia.

CB has been resident in the Gambia since 1985, and during this time has received the support and generosity of many people. He is particularly grateful to the following: the management and staff at the Atlantic Hotel in Banjul, who deserve very special thanks; Lloyd Burroughs and the staff at Kamino-Redcoat Freight for provision of a global courier service; colleagues at Tanji Birders for support in a variety of ways including supply of photographs and bird records; Nils Majgaard and PKP Architects, in particular for computer assistance.

For advice and support in The Gambia, CB would also like to thank the following: Bo Amilon, Ardy Sarge, Adama Bah and staff at The Bungalow Beach Hotel, Joost and Susanne de Batts, Foday Bojang (Director of the Department of Forestry and the Gambian-German Forestry Project), Eddie Brewer, the first Director of Dept. of Wildlife, Britannia Airways, the British Dragonfly Society, the Army Ornithological Society, Joseph Cann, John and Kay Carlin, Henry Carrol, Dr S. Conteh, Mike and Jen Cooke, Peter Cox, Isatou (Aisha) Davies, Nick Dennis, Marion Foon, The Gambia Experience, Gambia Hotel Association, Gambia Tours, Gamworks, Guardship, Gibou Janneh, Vernon C. Gilbert of WWF, Michael Gore, Robin Hellier, Alan and Avril Humphryes, Saikow Jallow, Sailou Jallow, Chief Sherriff Janneh, Malick Jeng, Momodou K.L and Ramou Jobe, Val and the late Reg Kersley, Ida King, Bob McKewan, Lamin Manneh, Paul Maroun, Nigel Marven and personnel of the BBC Natural History Unit, S. Lai M'Borge and Samba Fye and all the management and staff of Gamtours, Tamsir M'Bye, Francis Mendy, Catherine Mendy and Gambia National Library, the Ministry of Agriculture and Natural Resources, Saikou N'Jie, Omar N'Jie and the Ministry of External Affairs, Farma N'Jie and Kantong, Isatou N'day N'Jie and staff of The National Environment Agency, Dr Fred J.S Oldfield, Dr Samuel and Rachel Palmer, Hakan Persson, the late Graham Rainey, Major Mike and Janet Robjohn, Elizabeth Ross, Chris Rowles, Lamin Sammeteh, Ya Ya Sanyang, Pierre and Fumeké Sarr, the late Cherno Senghore, Aisha Tenneh Sillah, Peter Smith, Prince Abbey Sowe, Joyce Stavroulakis, Jim and Nancy Stone, Sunworld Holidays, Thomson Tour Operations, A. J. Todd,

Dr Omar Touray, Sarjo Touray and the staff of Tendaba Camp, Wandy Touray, Bernard Tréca of ORSTOM, Mrs Susan Waffa-Oggo and M.B.O Cham of the Ministry of Tourism, Stephen and Sonja Wilde, Richard and Dilys Wright.

In Europe, CB would like to thank the following: Phil Atkinson, Gwen Bonham, Nigel Cleere, Sainabou Drammeh at the Gambian High Commission in London, Gordon Gale, John and Dorothy Hook, Rob Hume, John and Alison Ovenden, Professor Chris Perrins, Dr Henri Quinque, RSPB (International Division), Iain Robertson, Bob Sharland, Dr John and Rosemary Tyler, the Senegalese High Commission in London, and Martin Woodcock.

Throughout TW's time in The Gambia, Bill and Judy Snow offered friendship, hospitality and good-humoured tolerance towards all matters ornithological. Their long accumulated knowledge of The Gambia provided context and valued perspective. Similarly, TW thanks many other friends at the International Trypanotolerance Centre. Dembo Touray and the cattle herding team showed him some of the places that have since become key birdwatching sites, and undertook the counting of displaying roadside bishops and other improbable activities with great good humour. It should not go without mention that, in similarly cooperative spirit, Dr Bakary Touray single-handedly brought about the identification of two new nightjar species for the Gambian list in the space of five minutes!

In London, TW thanks the Field Conservation and Consultancy division for providing office space while writing and working in the Zoological Society of London's library.

At the Natural History Museum in Tring, Peter Colston and Robert Prys-Jones facilitated inspection of a wide range of material in their collection, and Tony Parker did likewise at the Liverpool Museum.

For reading and correcting text drafts we thank Dr David Porter, and Professor Robert B. Payne. For assistance with matters concerning sound recordings we thank Dr Claude Chappuis, Dr Gerard Morel, John Hammick and Pat Sellar.

We are very grateful to Dr. A. Camara, Director of the Department of Parks and Wildlife Management (DPWM) in The Gambia and his entire staff for cooperation in the preparation of this book. Paul Murphy, Research and Development Officer, supplied the information cited about current conservation activity in The Gambia. We hope the book will assist them in securing the important long term objectives of the DPWM in The Gambia.

We are especially grateful to Christopher Helm for giving us the opportunity to write this book, and for his patience and commitment to the project throughout. Jim Flegg cheerfully undertook the initial editing of large portions of the text, and Nigel Redman provided substantial and dedicated editorial and ornithological expertise in the editing of the final version, which has resulted in a great improvement of the text. Responsibility for any factual errors, however, rests with the authors alone. Sally Young at Pica Press kept us in communication with each other when we were all in different continents, and Marc Dando and Julie Reynolds of Fluke Art exercised their considerable production and design skills to transform the raw text and artwork into a handsome book.

Finally, but most important of all, we note that the small but growing cadre of Gambian birdwatchers are beginning to add rapidly to knowledge of Gambian ornithology. We hope that they will find this book a source of encouragement, and use it to extend and improve our knowledge still further. It must also be recorded that the Gambian people as a whole, through their characteristically cheerful and welcoming response to birders and all visitors, provide the cornerstone which makes The Gambia the very enjoyable and special place that it is.

INTRODUCTION

GEOGRAPHY

The Gambia, at close to 11,000 square kilometres in area, is one of the world's smallest nation states. The unusual, elongated east-west boundary lines of the country are defined by the course of the Gambia River, which along with the Senegal and Niger Rivers is one of three great West African rivers all rising in the Fouta Djallon mountain range of Guinea.

The Gambia follows the meandering course of the river over its last 400 kilometres, cutting through the flat laterite plates of the West African shield to the Atlantic Ocean. The landscape is of gently undulating plains with a few laterite ridges cut by old and dry meanders. There are no hills to speak of and the entire country lies below 100m. The total length of the country is around 330km, and it averages only some 25-30kms in width over much of this length.

Except for its relatively short marine coastline, The Gambia is entirely surrounded by Senegal. At about 200,000 sq km, Senegal is about 20 times the size of The Gambia and penetrates some 500km inland from the Atlantic Ocean. Lying between 12°30'N and 16°30'N, much of the northern area of Senegal lies within a comparatively arid belt, with the Senegal River defining the northern and eastern boundaries and the Casamance River forming a complex of mudflat and lagoon areas in the south. The arid northern plain, or Ferlo, is wide and flat. Within Senegal the Gambia River is extensively engorged along much of its length, with numerous shallow sandbanks and rock snags which are largely absent downstream in The Gambia. In the extreme south-east there are gently rising hills, notably at Mount Asserik in Niokolo Koba National Park, which offers emotive views to far horizons of the West African interior.

The region of Senegal lying to the south of The Gambia, the Casamance, receives the highest overall rainfall of the region and vegetation growth is correspondingly rich. Its former dense forests are now considerably reduced and their future is by no means assured, with even long established national parks such as Basse-Casamance suffering substantial damage in recent years. Nevertheless the region continues to support a rich avifauna, with several forest specialists not seen further north.

The two large southern rivers, the Gambia and Casamance, create very large estuarine inlets. The Gambia River is salty or brackish from its mouth for some 200km inland, changing to fresh water somewhere in the vicinity of the ferry crossing at Farafenni through to Kaur. The precise extent of saltwater intrusion varies with season, and the volume of freshwater flow coming down the river. Tidal movements throughout the length of the river are extensive, and even the freshwater sections are rocked up and down on a daily basis, pushed by the great weight of saltwater ebbing and flowing in the funnel-shaped lower section.

CLIMATE AND VEGETATION

At these latitudes the climatic system is overwhelmingly dominated by the inter-tropical convergence zone (ITCZ). The ITCZ is created on the African landmass where the sun lies directly overhead at midday. The intense heating at this point causes air to rise, drawing convergent winds from north and south. The warm rising air gains moisture and is associated with massive convection clouds and rainfall. The tilt of the earth's axis relative to the plane of its orbit around the sun causes this massive weather formation to process in a cyclical seasonal passage north and south of the equator through each calendar year. Its arrival over The Gambia and Senegal, typically in late May or June is preceded by a build-up of oppressive heat and humidity over the parched land. Fierce thunderstorms may break as the weather fronts move over and rain typically persists on and off through June and July, peaking in August, as the airmass system halts and slowly turns back south with progression of the earth's orbit.

Overall rainfall averages in excess of 1000mm but the seasonal distribution in a single wet season has large ecological consequences for birds and all life, and most notably the vital farming activities on which much of the local economy depends. The 'wet season' falls during the northern hemisphere summer to autumn, and in the text is defined as July to October in reference primarily to the state of the vegetation.

Mean monthly temperatures do not vary greatly during the year (23°C January, 27°C October) but accompanying changes in humidity, coupled with high afternoon maximum temperatures in the late dry season and wet season, lead to some torrid days at that time of year, while comparatively cool breezes occasionally lead to surprisingly chilly evenings in January and February. These effects are most exaggerated in the inland divisions with some of the most extreme temperatures, in excess of 42°C, most likely to be recorded from URD in The Gambia and Niokolo Koba in Senegal in the late dry season.

In some years hot dusty air from the Sahara may move south and engulf the region in a choking yellow-white suspension of minute dust particles; this is the harmattan. In a bad harmattan the dust may hang in the air for several days, with visibility slowly pulsing in and out, and coating everything in a layer of dust.

Changes in vegetation of all habitats not surprisingly follow the rainfall cycle. Much of the natural vegetation undergoes a burst of new growth with the onset of the rains. Small lilies and wildflowers push quickly through the burnt soils, only to be overtaken by a mass of dense grasses and herbs which form a tangled scrub layer throughout the woodlands by the end of the rains. The clumps of giant *Andropogon* grasses, so characteristic of roadside and field margins as well as woodland, keep on growing into the early dry season before producing seed heads 2-2.5m above the ground. The freshwater swamps swell and flood with the increased flow down the river. Only the large Winterthorn Acacia *Feidherbia albida*, prominent in fields with canopies all tilted toward the south, does not conform. It has the unusual habit of casting all its leaves in the wet season, looking grey and lifeless at a time when all else is lush and growing.

The farmers' crops complete growth after the end of the rains, in the 'early dry season', and harvesting is completed around December-January. At this time, farmers are hard at work in the fields separating ground nuts from the chaff, and bundles of sorghum and millet are tied up and stacked out of the way of domestic stock on roofs and storage frames. The tangled mass of vegetation in woodlands and the bush becomes parched and the great stands of giant grass dry to form expansive mats of brittle tinder.

The 'late dry season' (March to May) is a time of bare fields, tinder-dry vegetation and sometimes frequent bush fires. The fires are sometimes accidental, sometimes deliberate, but probably nearly always man-made. This fire pressure, together with intense collection of wood for fuel results in relentless thinning of the woodlands, a problem now beginning to be addressed through community forest programmes as well as the establishment of a network of Forest Parks. The village N'dama cattle and other livestock are free to wander and forage where they can. Often they cluster, patiently waiting for mangos to fall, or search beneath wild figs and winterthorn, now perversely green when all else is dry and depositing large nutritious seed pods which are eagerly sought by the hungry animals.

MAJOR HABITATS

In terms of broad Pan-African vegetation zones, The Gambia and Casamance lie in a transition zone along the interface between the forest-savanna mosaics of the moister Guinean zone and the drier Sudanian woodlands. The last forest islands are found in southernmost Gambia, with one or two very small patches north of the Gambia River marking the northern limit. Sudanian woodlands dominate much of The Gambia and inland parts of Senegal, including Niokolo Koba, whilst increasing aridity northwards into Senegal results in the scattered dwarf shrub and acacia plains of the north.

Within this broad scheme there is a range of important specific habitat types critical to a variety of bird species. A brief synopsis of these follows.

Marine System

Offshore from The Gambia and Senegal, the marine environment is dominated by the shallow continental shelf, which here continues for 40km before dropping to oceanic depths. Only Dakar, situated out on a long peninsula from the mainland, lies close to the drop-off. Numbers of pelagic seabirds offshore peak during periods of passage, notably March-April and September-October. In the shallower waters inshore intensive fisheries provide abundant feeding opportunities for seabirds. Gannets, shearwaters petrels and skuas all gather in these waters, and increased seawatching activity in recent years, especially off Dakar, is already beginning to yield many new discoveries.

Coastal Shoreline

The greater part of the shoreline is dominated by long shelving sandy beaches. Western Reef Herons stalk the surf, Pied Kingfishers hover above, whilst Sanderlings and Ruddy Turnstones are very numerous for much of the dry season.

The beaches are punctuated by occasional fishing villages where gulls take full advantage of fish processing operations on the beach. At sand spits formed opposite bends in the coastline, semi-permanent roosts of gulls, terns and waders find rest when not feeding at high tide. Rarities in these gatherings might include Audouin's Gull and Kelp Gull. Similar gatherings form opposite the few freshwater outlets along the beaches; here White-fronted Plover and Giant Kingfisher might also be encountered. In a few places lateritic outcrops form low cliffs overlooking the sea, providing lookout points for gannets and skuas offshore.

Estuary and Mangroves

Both the Gambia and Casamance rivers are associated with very large and important systems of mudflats and mangroves, as is the Delta du Saloum. On the Gambia River this extends inland for nearly 200km. Two mangrove types dominate; the smaller more grey-green 'White' Mangrove *Avicennia nitida* pushes aerial roots up through the mud from below, while the much taller, darker 'Red' Mangrove *Rhizophora racemosa* props itself out of the water on branching stilt-like stems. Red Mangroves form the long dark profile lining the skyline along the banks of the main river through most of its lower sections. They also create the leaning galleries over narrower tidal tributaries.

The mudflats as everywhere are rich in invertebrates - annelid worms, molluscs and crustaceans, and the mangrove roots provide a haven for myriads of small fish. At low tide a host of Palearctic wader species, herons, pelicans, African Spoonbills, Sacred Ibises and Yellow-billed Storks all stalk the flats. In quieter creeks Goliath Herons stand watch while the occasional African Finfoot sneaks between the mangrove stems. The Mouse-brown Sunbird is an active resident strongly associated with the mangroves, but less expected forest species such as Common Wattle-eye may also take advantage of the shady and sheltered tangles as well as migrants such as Subalpine Warbler. In mid-river stretches especially, the mangroves may conceal several secretive rarities, notably Pel's Fishing Owl and White-backed Night Heron.

Freshwater Riverbanks

Much of the freshwater sections of the banks of the Gambia River are covered in a dense thorny wall of evergreen vegetation. This conceals thicket-loving species such as Oriole Warbler and White-crowned Robin Chat, while river-loving Swamp Flycatchers perch low over slow flowing water. In taller stands Violet Turaco, Western Banded Snake Eagle and African Fish Eagle are characteristic species. Some of the first records of the poorly known population of Adamawa Turtle Dove found in this part of West Africa came from this humid lowland habitat. Weavers, especially Yellow-backed Weaver, form huge colonies on low branches overhanging the water, and on some islets in the river, mixed breeding colonies of small herons cram themselves amongst the dense thorny vegetation; Black-crowned Night Heron, Black Egret, Squacco Heron and Cattle Egret all squabble cheek by jowl in the late rains and early dry season.

There remain in Central River Division of The Gambia a few isolated fragments of freshwater swamp forest, most significantly in Gambia River National Park. These are dominated by the massive vertical trunks of the tall broad-leaved tree *Mitragyna stipula*, while groves of the huge but beautifully arched fronds of the raffia palm *Raphia hookeri* add a classically tropical appearance. In the humid interior of these forests, small freshwater channels flood in from the main river; Square-tailed Drongo and Yellow-breasted Apalis occupy mid-layers of the canopy, Buff-spotted Woodpeckers work low stems in the densest clumps of lianas, whilst the brilliant Crimson Seed-cracker constructs domed nests of fern fronds among the twisted *Pandanus*.

Other Wetlands

During the rains and for variable times afterwards, many parts of the flat low-lying areas bordering the river are flooded, creating extensive swamps. The cores of these swamps are in many cases maintained throughout the dry season by tidal movements of freshwater flushing gently in and out from the main river. The local name for these swamps is 'banto faros'; they form the basis of the expanding Gambian rice-growing industry. Both rice-growing areas and the natural swamps play host to an abundance of birds. The elegant Black Egret, Winding Cisticola, Yellow-crowned Bishop and African Jacana are often prominent; Allen's Gallinule and Greater Painted-snipe hide in the rank vegetation and Eurasian Marsh Harriers quarter overhead. Numerous mixed hirundines swoop low for insects and as the air warms in the morning, these are good open places to view a diversity of raptors rising on the thermals.

Some parts of CRD support extensive reedbeds. Little Bitterns may accumulate in good numbers, and Greater Swamp Warblers chortle unseen. Much probably remains to be learnt about overwintering and resident smaller species here. Open sandflats with pools of freshwater nearby provide wet season refuges for the sought-after Egyptian Plover which visits downstream locations in The Gambia when its sandbank nesting sites further upriver in Senegal are flooded.

Villages, Farmland and Fallow Land

On higher ground extensive farms have been cleared from the woodland surrounding villages creating fields where in the wet season crops such as ground-nut, sorghum and millet are grown. Characteristically, farmers leave trees that are useful to them. Baobabs *Adansonia digitata*, figs *Ficus* spp. and winterthorn, frequently colonised by White-billed Buffalo Weaver, are prominent among them. Cultivated mango, cashew and citrus usually grow closer to the village compounds, sometimes maintained as orchards; these are a regular site for Olivaceous Warbler.

The village compounds themselves are used by Grey-headed Sparrows, while the ubiquitous Red-billed Firefinch and its accompanying brood parasite, the Village Indigobird, also make use of convenient nest sites close to the grain stores. In inland areas White-rumped Seedeaters are inconspicuous but regular in tall baobabs overlooking villages. Red-rumped and Mosque Swallows cruise around the thatched roofs and may nest under the eaves. Screaming parties of Little Swifts wheeling above the villages at dusk are a regular feature of rural life.

In the growing season, tall stands of millet are dotted with brilliant Northern Red Bishops, while elaborate tangles of strings with dangling tin cans are rigged from central platforms in the fields where they are shaken by small children to discourage the hordes of raiding weavers.

Not all fields are used every year and at the margins of farms fallow land develops varying stages of regenerating scrub and woodland. These fallow areas can attract an array of species providing, for example, abundant song posts for Whistling Cisticola, and look-out posts for gangs of Yellow-billed Shrikes or a solitary hunting Shikra. At dusk these are also places to find Long-tailed Nightjar, while White-faced Scops Owl and Pearl-spotted Owlet are usually first detected by their calls.

Hotel Gardens

Hotel gardens constitute a relatively minor, variable and artificial habitat. However, many of the hotel gardens in The Gambia and Senegal provide an unusually vivid first introduction to African birds for visitors from all over the world.

The ubiquitous intense sky-blue and red of the tiny cordon-bleus and firefinches are frequent subjects of enquiry, while the noisy and industrious Village Weavers, some bright black and yellow, others dowdier, almost impose their presence at poolside palm trees. Brilliant metallic sunbirds take full advantage of flowering hibiscus and other decorative shrubs introduced for tropical effect. Parties of metallic blue starlings strut across lawns, especially the swaggering and easily recognised Long-tailed Glossy Starling.

In the larger gardens black and scarlet Yellow-crowned Gonoleks, normally an elusive denizen of dense thickets, have become bold, hopping blatantly in the open. Where planting and irrigation are well established a progression of incoming species has been recorded, providing some of the best opportunities to view and photograph such thicket-loving specialists as the White-crowned Robin-Chat. Gardens that retain a few fig trees can turn up a host of species normally associated with woodlands. Even at the edge of Banjul, a small, well wooded and carefully managed hotel garden has attracted some highly unexpected species such as Yellow-breasted Apalis, and provided temporary shelter for genuine rarities such as African Cuckoo-Hawk.

These observations give insight into the dependency of some species on the maintenance of mosaic habitat patches. Gardens demonstrate that small protected oases do make a difference to the birds while at the same time offering opportunity to admire their beauty at close range.

Guinea Savanna

The southern margins of The Gambia and coastal Casamance were probably more extensively wooded than they are today. Tall emergent specimens of Grey Gingerplum *Parinari excelsa* would have spread above a dense mid-storey of a multitude of lush broad-leaved species. The mid-layers would have been draped with various creepers such as the spiky climbing Rattan Palm and leafier species. Throughout, the woodlands were dotted with straight-trunked Oil Palms *Elais guineensis*, except at the coast where the tall Rhun Palm *Borassus aethiopicus* dominated, as it still does at Bijilo Forest Park.

Today significant patches of this habitat can be found in protected locations such as Basse-Casamance, with a tiny remnant at Abuko and one or two other spots in The Gambia. It is here that a handful of true forest specialist species hang on. They constitute a comparatively impoverished subset of the broader forest-adapted avifauna found further to the south, perhaps comprising those species better able to cope with habitat fragmentation and the constraints of living in small patches of habitat. Among these are the African Pied Hornbill, Grey-headed Bristlebill and Little Greenbul, which can be found in comparatively tiny remnant thickets. The latter is one of the main subjects of recent genetic studies in West Africa which have demonstrated both mobility between patches and the existence of strong natural selective forces on patch-living populations. A few other species such as White-spotted Flufftail and Western Little Sparrowhawk are reported now only from the largest forest thickets in The Gambia and are probably entirely dependent on the protection of one or two locations.

Much of the former Guinea Savanna is nowadays more open. Field clearance has removed all but the numerous Oil Palms, kept for harvesting its leaves and palm wine. Figs and the attractively fronded Locust Bean *Parkia biglobosa* are also frequently spared. In places secondary growth creates denser vegetation, but seldom attains the height or structure of the original. The avifauna is perhaps less specialist, but still rich in interest and beauty; Blue-bellied Rollers, African Golden Oriole and Swallow-tailed Bee-eater create splashes of brilliant colour, while parties of Brown and Blackcap Babblers, Green Wood Hoopoes and Long-tailed Glossy Starlings add to the list of noisy and social species associated with this habitat. The intense sociability is significant, since many of these gregarious species have been found to be based on family groups using a system of cooperative assistance at the nest in which relatives, frequently the non-breeding offspring of a primary breeding pair, help raise subsequent offspring by nest building, bringing food and giving protection from predators. Species known to use this strategy are indicated in the text, but much remains to be learnt.

Sudan Savanna

A broad belt of central southern and south-eastern Senegal including the very large Niokolo Koba National Park, plus most of the Gambia north and east of the Bintang Bolon, lies within the Sudanian savanna belt. Where tall red termite mounds indicate iron hard lateritic soils, the dominant natural vegetation is a dry woodland, of moderate height, composed from a rich mix of small tree species, notably *Acacia macrostachya* with various *Terminalia* and *Combretum* species A few taller emergents are present such as the Ironwood *Prosopis africana*, Red Silk Cotton *Bombax buonopozense* and Rosewood *Pterocarpus erinaceus*. Typically a proportion of these persist after dying as gaunt, grey standing trunks with crudely broken branches. These provide favoured perches for many savanna birds, from tiny Rufous Cisticolas to immense Martial Eagles. Abyssinian Rollers perched along the roadside, en-trance unsuspecting observers with flashes of astonishing sky-blue and inky dark blue as they rise to their tumbling display flights. Mixed feeding parties of White-shouldered Black Tit, Senegal Batis, Green-backed Eremomela and Brown-backed Woodpecker work through bushes and shrubs. More secretive species such as Chestnut-crowned Sparrow-Weaver, Brown-rumped Bunting and Bronze-winged Courser can also be found. From September to December Exclamatory Paradise Whydahs come to attract mates at song posts beside roadside quarry waterholes where their Red-winged Pytilia hosts come to drink together with many other small waxbills and finches.

In areas of deeper soil very large West African Mahogany *Khaya senegalensis* and bigger stands of Rosewood create taller woodland. Here rarities include White-breasted Cuckoo-Shrike and Red-headed Weavers, while onion-shaped nests of Vitelline Masked Weavers are suspended decoratively from thorny *Acacia macrostachya* trees. Mixed flocks of glossy starlings attracted to termite emergences at the advent of the rains provide interesting identification challenges.

Dry Sahel of Northern Senegal

Over a large area of central and northern Senegal the vegetation becomes sparser with scattered shrubs including acacias and *Balanites aegyptiaca*. Bird species characteristic of this habitat include Black-crowned Sparrow-Lark, Sennar Penduline Tit, African Collared Dove, Desert Cisticola, espe-cially in grassy areas, and Grey-backed Eremomela. Rarities include Golden Nightjar and Little Grey Woodpecker.

In Ferlo district the Ostrich bred until forty years ago. Today it is all but exterminated, with only sporadic sightings of individuals over the last ten years. A number of migrants, typically breeding in the southern Mediterranean region, are found in the subdesert plains during the Palearctic winter, with numbers varying according to rainfall patterns and plant growth further north. Most of these are also seen sporadically in The Gambia. Examples include Desert, Isabelline and Black-eared Wheatears, Isabelline Shrike and Spectacled Warbler.

THE SENEGAMBIAN AVIFAUNA

The Senegambian region boasts a rich avifauna of some 660 species, about a third of which are migrants, largely from the Palearctic. Of this total about 540 species have been recorded from The Gambia alone, a country only one twentieth the size of Senegal. A high altitude satellite photograph of The Gambia and Senegal, tuned to indicate the distribution of vegetation and water, gives a compelling indication of why the region supports such a diverse avifauna. Senegambia lies at an ecological cross-roads between the Atlantic Ocean in the west and the African continent in the east, and between the arid Sahel/Sahara in the north and the humid tropics in the south, thus supporting a wide range of habitats.

Many Palearctic migrants, both landbirds and waterbirds, heading south to avoid the northern winter are guided along the western continental seaboard, crossing arid terrain until they arrive at the lower reaches of the Senegal River and encountering the first substantial continuous vegetation cover as they near The Gambia. The earliest arrive in the mid-rains, with major passages arriving as the rains finish. It is a time when the land is at its most productive and they find habitats catering to many specialist preferences; open ground and woodland, and freshwater and saline wetlands are all found within a small area.

This fine mosaic of habitats lying along a major ecotone between the arid north and the wetter, more productive south also provides a home for many resident African species. For most of these our region marks the northernmost and absolute limit of their ranges as well as their closest approach to Europe. Likewise, the regular seasonal passage of the inter-tropical convergence zone brings about such a burst of productivity that many intra-African migrants follow it north to breed in the wet season, further enlarging the avifauna.

It is because of this convergence of distinct regions that small but carefully located reserves can offer such rich bird watching prospects and make an effective contribution to conservation. Tanji Bird Reserve is a newly established example of such a reserve. It is located at one of the westernmost points of the African continent, close to the interface between the Afrotropical and Palearctic avifaunal zones. It is in line with a major north-south Palearctic-African migration route at a point where habitat diversity suddenly becomes richer. Lying beside the sea, its habitats comprise remnant forest with patches of scrub and orchard-like land on old dunes, bisected by a brackish estuary.

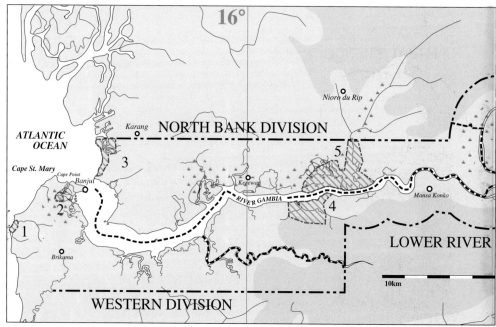

Map of The Gambia showing location of protected areas. 1. Tanji Bird Reserve; 2. Abuko Nature Reserve; 3.

Palearctic migrants, intra-African migrants and resident African species can all find sustenance in this small area; more than 300 species have been recorded within its 600 ha. boundary.

It is commonplace, even fashionable, to proclaim doom for the future, citing ever expanding human populations and heedless development as certain threats to all biodiversity everywhere. These are not hollow worries, and vigilance and effort are more necessary now than ever before. But there is cause for some optimism in the growing awareness at all levels across the globe that environmental protection, properly planned and instigated, is not a luxury, but a necessary and contributing part of any worthwhile development plan.

The Gambia and Senegal are showing their commitment to this long term view through the establishment of protected areas, often in the face of seemingly more pressing short term economic problems. It is for all those involved in this process, in the first place government officials and local citizens increasingly affected by habitat loss, to ensure that this process is seen to create human and economic benefits through protection of biodiversity. It is our hope that this book will contribute by helping to catalyse interest and support for broad-based conservation initiatives in the local region and thereby encourage international funding agencies, non-governmental organisations and visiting birdwatchers to support the national institutions in this mission.

PROTECTED AREAS IN THE GAMBIA

Being a small country with a rapidly expanding population, there is considerable pressure on the natural resources in The Gambia to cater for their growing needs. After decades of decline, and in some cases extinction, of the populations of many large mammals, The Gambia has endeavoured to halt this trend by a combination of establishing a protected area network and public awareness. The first protected area established was Abuko Nature Reserve in 1968, and over the years this has seen a great many visitors. In 1977 a Wildlife Act was passed and this is to be reviewed soon to take account of changes in the intervening years. The Department of Parks and Wildlife Management (DPWM) now has six protected areas under its care, totalling 39,768 ha (see Tables 1 and 2, and map below). It is intended to extend the network to include representative samples of every major habitat within the country, aiming to cover a minimum of 5% of the total land area.

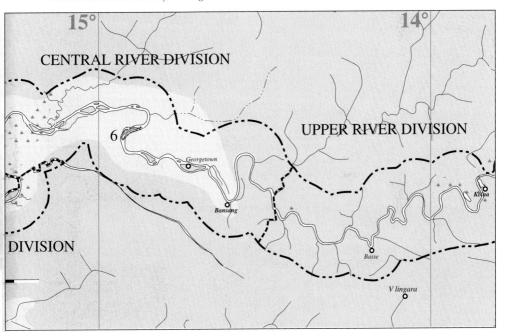

Niumi National Park; 4. Kiang West National Park; 5. Bao-bolon Wetland Reserve; 6. Gambia River National Park.

Table 1. National Parks in The Gambia

Site	Date gazetted	Location	Area (hectares)
Gambia River N. P. (Baboon Island)	1978	CRD	585
Niumi N. P.	1986	NBD	4,940
Kiang West N. P.	1987	LRD	11,526

Table 2. National Reserves in The Gambia

Site	Date gazetted	Location	Area (hectares)
Abuko Nature Reserve	1968	WD	105
Tanji Bird Reserve	1993	WD	612
Bao-bolon Wetland Reserve	not yet gazetted	NBD	22,000

WD = Western Division, LRD = Lower River Division, NBD = North Bank Division, CRD = Central River Division.

These protected areas (with the exception of the Gambia River National Park which is the site of a chimpanzee rehabilitation project) are open to the public throughout the year for a moderate fee of D30 per adult (about £2 or $3). Many protected areas are still being developed but this is a slow process due partly to limitations in financial resources, but also because the protected area system is attempting to integrate its conservation objectives with the sustainable utilisation of natural resources amongst local communities. In a country with limited resources this approach is essential for the long term survival of its natural heritage. Tourism plays a considerable role in the Gambian economy although much of it is centred on the coastal strip. The current emphasis on development of ecotourism will assist in the dispersal of income while placing a tangible value on the wild flora and fauna at the local level.

The Gambia is currently drawing up a National Biological Diversity Study and Action Plan with assistance from IUCN. This study will enable the country to focus on where it stands and its long term aims in relation to the conservation and wise use of its biodiversity. As a signatory to the Ramsar Convention on Wetlands of International Importance, an assessment of two further areas for designation as Ramsar sites is currently being made to bring the Gambian total to three sites. Plans are being developed for a national inventory of sites of high ecological value, including wetlands, with recognition being given to the importance of maintaining corridors and adequate buffer zones. Although the resources of DPWM are limited, the motivation and dynamism present in The Gambia contributes in moving towards a sustainable and ecologically balanced approach to development.

PROTECTED AREAS IN SENEGAL

Senegal currently maintains six nationals parks, together with a number of faunal and special reserves. The National Parks Directorate is the responsibility of the Ministry of Tourism and Nature Protection, while areas other than national parks are administered by the Direction des Eaux, Forêts et Chasse. The most important of these sites, which are frequently referred to in the text, are tabulated below (Tables 3 and 4) with an indication of their size and location. They are mapped on the front endpapers.

Table 3. National Parks in Senegal.

Site	Location	Area (sq.km.)
Oiseaux du Djoudj	North-west Senegal	160
Langue de Barbarie	Coastal North Senegal	20
Iles de la Madeleine	Islands off Dakar	5
Delta du Saloum	Coastal Central Senegal	760
Basse-Casamance	South-west Senegal	50
Niokolo Koba	South-east Senegal	9,130

Table 4. Major Faunal and Special Reserves in Senegal.

Site	Location	Area (sq.km.)
Ndiael	North-west Senegal	466
North Ferlo	North-east Senegal	4,870
South Ferlo	North-central Senegal	6,337
Kalissaye	Coastal South Senegal	16

The Senegal River valley in the far north represents a narrow green belt across the otherwise dry sahelian steppes of the Ferlo districts, terminating in the rich delta area at Djoudj. Further south the coastline is broken only at the basin of the Saloum delta, the Gambia River and the Casamance River, with the Dakar peninsula acting as a significant site for observing both pelagic species and migrants. The heavily wooded habitats of the east and south-east are best represented in the Parc National de Niokolo Koba, where the main strongholds of large ungulates and carnivores in our area are still to be found. The distinctive habitats of Casamance to the south of The Gambia include the best examples of intact Guinean forest-savanna, notably in the Parc National de Basse-Casamance.

Using This Book

AIMS AND SCOPE OF THIS GUIDE

This field guide aims to assist anyone in The Gambia and Senegal with the identification of any bird species listed up to 1997. The focus is on The Gambia, but full information on the avifauna of surrounding Senegal is presented because of the unusual circumstances of the local geography. All residents, all intra-African migrants and all Palearctic migrants are covered in the text – over 660 species in total. Illustrations depict all species except vagrants and accidentals; the majority of these omissions from the plates are Palearctic vagrants, many of which are known only from one or two records, plus a handful of North American, Asiatic and extralimital African species, also mostly recorded on very few occasions.

NOMENCLATURE

In recent years there has been a gathering momentum towards a much needed, unified system of common English names for African birds. Readers will find several instances in this guide of nomenclatural difficulties created by the proliferation of English names associated with particular species. In other examples the same English name applies to different species in different regions of Africa.

Now for the first time major new works are becoming available that treat the continental avifauna as a unified whole, and introducing necessary changes with the aim of removing such ambiguities. The policy of this guide has been to follow this trend with respect to the common name used in English as well as using these sources for scientific nomenclature and taxonomy. Hence we have in the main followed nomenclatural usage in the major series *The Birds of Africa*, Volumes I-V. Dowsett and Forbes-Watson (1993) has also been used as a major source.

The process of establishing a unified nomenclature is still in progress in the sense that while many changes are now accepted, not all are agreed and further changes are to be expected. We cannot and do not claim to have found definitive names in every case, but we hope to have minimised the occasions when names given here will be subject to further revision, thereby assisting a gradual acceptance of a unified system.

With the obvious need for reliable cross reference to other literature for common as well as scientific names, it remains essential to retain a list of 'other names' where these apply. In West Africa this need is particularly acute, since the long standing field guide for the whole region (Serle and Morel 1977) uses some names of distinctly West African bias (e. g. West African River Eagle for African Fish Eagle), which have been made redundant in more recent pan-African publications. Nevertheless that volume, now twenty years old, has played a crucial role in the development of West African ornithological knowledge and is very well known to many people. Therefore, although we do not use the same English names for the reasons given, we make a particular point of including all names used in Serle and Morel under 'Other names' where these now differ.

For French common names we have followed the usage of Morel & Morel (1990) as well as giving other French names where alternatives are listed in Dowsett and Forbes-Watson (1993) and *The Birds of Africa*, Volumes I-V.

PLATES AND ILLUSTRATIONS

The primary reference point in any field guide is the plate section, where field observation can be compared directly with the painted images. The 48 colour plates cover almost 570 species. Within the text, line drawings have been used to illustrate additional species as well as highlight certain identification features. Captions opposite the plates draw attention to key points, including on occasions useful features of habitat and behaviour relevant to identification. The page numbers for the corresponding text accounts are also given.

SPECIES ACCOUNTS

Identification

An 'Identification' section gives details of size (metric) and indicates names of local subspecies where appropriate. In most cases certain characteristic features of the bird, whether of appearance or behaviour, are then singled out; fuller descriptions of plumages for male, female and immature, plus soft parts coloration follows.

A 'Similar species' section draws attention to possible confusion species, facilitating cross-reference between species accounts and highlighting key points of separation. It follows that in many cases features mentioned in this section are frequently useful to have in mind in the field, when brief views may leave little time, and attention must be quickly focused on the critical details.

In the texts covering large raptors individual sections are devoted to flight characteristics to aid overhead identification; these sections are matched in the illustrations by provision of separate plates devoted to comparison of raptors in flight. Accounts of many of the weaver species also give separate attention to nest identification, since these are unusually prominent and distinctive for several members of this group.

Habits

This section is used to provide a range of supplementary information useful for identification. It includes a necessarily brief summary of ecology and behaviour, indicating for example whether the species is typically social or solitary, features of habitat preference and sometimes aspects of how the species uses its chosen habitat. Habitat types used here are explained in the introduction. These details often provide strong circumstantial support to the identification process and can also be used to predict the typical situations in which the species might be found.

It might be noted here that because The Gambia is a remarkably narrow country alongside a major river, and because variable habitats form a tight mosaic, a commonsense awareness of scale is useful. For example it is commonplace to see waterbirds flying over forest in The Gambia and even sometimes vice versa. Likewise there is a sense in which almost everywhere is 'near water'. In general text references to habitat preference are specific, but these circumstances mean that field observers can sometimes expect to see some species from a wider range of habitats than those listed. Similar effects apply to the rich habitat mosaics of southern Senegal, but this is obviously less of a problem in the less well covered expanses of the east and north.

Voice

A good knowledge of bird songs and calls is one of the most valuable, if hard to acquire, skills in field ornithology. It can transform perception of the avian community and the relative abundance of species in it.

In order to provide an impression of the vocalisations of the region's birds, we have drawn on descriptions from a wide range of the literature, combined with published recorded material (notably Chappuis 1974-1985) and our own field observations and recordings.

The weakness of the English language as a device to convey the complexity of bird calls is widely acknowledged. We have done our best to combine language with onomatopoeic constructions, some already published, others our own, to give basic descriptions in which we try to highlight structure and tonal characteristics of the calls. In many instances, for example the cisticolas and some forest birds, we believe this will be a useful and significant aid to identification; but inevitably in some cases imagination and a generous ear will be needed, and real recordings are in general much to be preferred.

Status and Distribution

This section integrates the most recent published information about status and distribution for the Senegambian avifauna. It combines spatial, seasonal and abundance information, and methods for organising these different aspects are described here.

For Senegal our principal source has been Morel & Morel (1990) who give distributions based on one-degree squares. For The Gambia we have used Gore (1990) in conjunction with Jensen & Kirkeby (1980), both updated by comparison with our database of almost 60,000 records covering all observed species across a 10km x 10km grid and collected primarily between 1987 and 1994, mostly in 1990-92.

It is important to note here that knowledge about status and distribution of Senegambian bird species is far from complete and there is still great scope for careful and systematic observation from all parts of the region. It is very relevant that while some interior areas, notably north of the river in The Gambia and the far interior of Senegal, are known to be less well covered than the coastal region, new discoveries of recent years, even of African resident species, have been as likely to come from the well known areas near the coast as from the interior.

11

Abundance. Statements about abundance are based on the definitions used in *Birds of Africa* Vol. 1. i.e.

> Vagrant – only a few records (probably extralimital)
> Rare – seen once in several years
> Uncommon – 10 or fewer seen in a year
> Frequent – quite often seen or heard, but some effort needed to locate it
> Common – 1-10 seen daily
> Abundant – 10-100 seen daily
> Very Abundant – more than 100 daily in preferred habitat

These aim to reflect relative abundance an observer might expect. In nearly all cases commonsense demands that these definitions have been heavily qualified and elaborated to reflect restrictions of habitat or other constraints.

Season. The local seasonal pattern is defined in the introductory section on climate. In the text three seasons are mainly used; Wet Season (July to October), Early Dry Season (November to February) and Late Dry Season (March to June). The ecological basis of this is the single wet season followed by a long dry season which for convenience can be divided into two equal periods reflecting different stages of drying vegetation and water supply (see introduction).

Occasional use of spring, summer, autumn and winter in connection with Palearctic migrants refers to the four seasons of the northern hemisphere. They are defined approximately as Spring (March to May), Summer (June to August), Autumn (September to November) and Winter (December to February). Inclusion of two parallel seasonal reference systems is justified because the migrant species involved are moving between two climatic zones and their movements are best described in context. Indeed migration is probably a device arising as one possible response to manipulate different seasons in different places in the quest for the best available food supply and breeding opportunity.

Location. The basic process and sources for determining species distributions was the same as that for Status. In The Gambia we integrated major recent published sources with our own observations to report species' presence or absence according to the administrative divisions of the country. These are tabulated below together with their acronyms as used in the text for short-hand reference.

Table 5. Administrative Divisions of The Gambia

Administrative Division	Region	River salinity	Woodland type	Wetland type
Western Division	WD	Estuarine	Moist	Saltflat & creek
Lower River Division	LRD	Estuarine	Dry	Saltflat & creek
North Bank Division	NBD	Estuarine	Dry	Saltflat & creek
Central River Division	CRD	Freshwater	Dry	Flooded swamps
Upper River Division	URD	Freshwater	Dry	Oxbow swamps

The Administrative Divisions are useful in this context for The Gambia because they are similarly sized units, all characterised by a fine scale of intermixed habitat types, but coinciding with changes in a series of broader ecological features. A further practical advantage is that their borders are usually signposted at the roadside as you travel around the country.

The potential for the broad river estuary to act as a barrier, especially to forest and woodland birds, is manifest, validating a distinction between Western Division and North Bank Division in the west. Lower River Division lies on the south bank of the river, between Bintang Bolon and Pakali Ba. It marks the start of the drier Sudanian woodland vegetation. Central River and Upper River Divisions both lie astride the freshwater sections of the river, with the latter marking sections where the river is partially engorged, as can be seen from Basse eastwards.

For Senegal a more extensive area reference system is appropriate for a larger country with much of it characterised by larger blocks of uniform habitat, and for which contemporary distributional information is available at a coarser scale of resolution. The main areas referred to in the Status and Distribution sections for Senegal can be found on the front endpaper map. In many cases circumstances allow simplification in the text by referring to extensive regions relative to The Gambia (e.g. 'in Senegal found only to the south of The Gambia'). This is possible because The Gambia itself lies at a major ecotonal changeover, with the result that many Senegalese bird distributions separate along the latitude of The Gambia.

Breeding

The prime sources for information on breeding have been integrated from a wide range of the published literature, with more recent opportunistic observations added to illustrate the kind of material that contributes to building up the life history profiles of these species.

The primary emphasis is on seasonality, since this has in many cases strong implications for identification issues, notably plumage phases and behavioural visibility.

In some cases basic details of nest structure and location are also added where these are known and of interest. Breeding information for long-distance migrants and vagrants is not supplied where breeding takes place extralimitally.

Of necessity much of the information is brief because breeding biology remains comparatively poorly known for many species. This is another area where much remains to be learned and where patient observers can make valuable contributions.

Additional Notes

In a few cases notes are given to explain particular points of interest or reasons for some of the choices in presentation of information.

Nomenclature and taxonomy in particular are areas notorious for differences of opinion and in some cases we have added a brief note to indicate such a background.

It is perhaps a danger that the traditional field guide format dealing species by species, especially when reinforced by the tidy field birder's tick list, can too easily create an illusion of species standing as neatly defined, equally differentiated genetic units in the environment. Modern biology is now replete with examples indicating how this is not so and the West African avifauna contributes a fair share of these; the recent history of the Yellow-legged Gull, Crimson Seed-cracker and the indigobirds are notable examples, whilst paradise flycatchers are likely to provide another.

Species vary extensively and while many are clearly distinct, in numerous other cases boundaries and divisions between breeding units can be more opaque than suspected or, alternatively, sharper, deeper and far more subtle than expected.

We hope that users of this book will find the information supplied about the existence of such problems interesting, a salient reminder of these complexities, and a clear stimulus to sharpen their own observation and enjoyment while watching the birds involved.

PLUMAGE TOPOGRAPHY

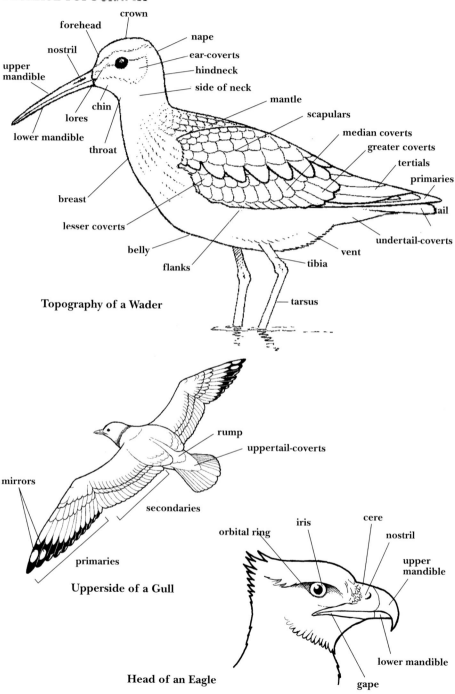

Topography of a Wader

crown
forehead
nape
nostril
ear-coverts
upper mandible
hindneck
side of neck
mantle
chin
scapulars
lores
median coverts
lower mandible
greater coverts
throat
tertials
primaries
breast
tail
lesser coverts
undertail-coverts
belly
vent
flanks
tibia
tarsus

Upperside of a Gull

rump
uppertail-coverts
mirrors
secondaries
primaries

Head of an Eagle

iris
cere
orbital ring
nostril
upper mandible
lower mandible
gape

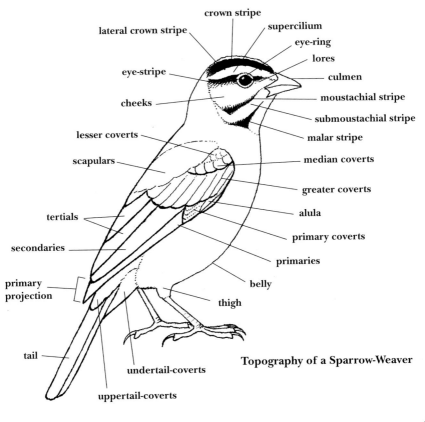

crown stripe

lateral crown stripe

supercilium

eye-ring

lores

eye-stripe

culmen

cheeks

moustachial stripe

submoustachial stripe

lesser coverts

malar stripe

scapulars

median coverts

greater coverts

alula

tertials

primary coverts

secondaries

primaries

primary
projection

belly

thigh

tail

Topography of a Sparrow-Weaver

undertail-coverts

uppertail-coverts

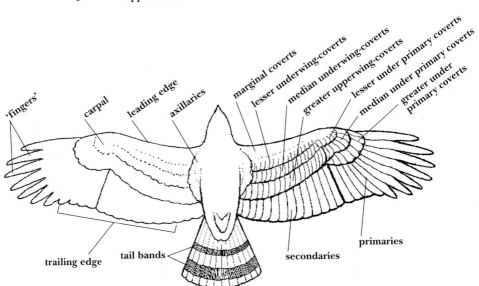

'fingers'

carpal

leading edge

axillaries

marginal coverts

lesser underwing-coverts

median underwing-coverts

greater upperwing-coverts

lesser under primary coverts

median under primary coverts

greater under
primary coverts

primaries

trailing edge

tail bands

secondaries

Underside of an Eagle

PLATE 1: GREBES, SHEARWATERS AND PETRELS

1. LITTLE GREBE *Tachybaptus ruficollis* (not to scale)   Page 116
Locally common on freshwater ponds and lakes. Nomadic with seasonal and irruptive increases. **1a Adult breeding**. Russet neck. Gape greenish-yellow. **1b Adult non-breeding**. Dull olive-brown. **1c Flight**. Pale indistinct wing panel.

2. CORY'S SHEARWATER *Calonectris diomedea*   Page 113
Off northern Senegal, April-November. Strongly built, very white belly. Atlantic race *C. d. borealis* has dark-tipped yellow bill and lacks capped appearance. **2a Upperside**. **2b Underside**. Cape Verde race, *C. (d.) edwardsii*, often considered a full species, is smaller, slimmer, longer-tailed and has a dark-capped appearance with a thin grey bill.

3. SOOTY SHEARWATER *Puffinus griseus*   Page 114
Off northern Senegal. All-dark above, silvery-white patches on underwing, slim bill blackish. **3a Upperside**. **3b Underside**.

4. GREAT SHEARWATER *Puffinus gravis*   Page 114
Off northern Senegal. Ashy-brown above, strong build, pale below with dark belly smudge, dark cap, pale nape band, white horseshoe at tail base. **4a Upperside**. **4b Underside**.

5. LITTLE SHEARWATER *Puffinus assimilis*   Page 114
Off northern Senegal. Like small Manx, with white above eye, and projecting feet. Note flight with bursts of rapid wing beats and short glides. **5a Upperside**. **5b Underside**.

6. MANX SHEARWATER *Puffinus puffinus*   Page 114
Uncommon to rare off Senegal and The Gambia. Well-defined black above and white below. Black reaches below eye. **6a Upperside**. **6b Underside**.

7. BULWER'S PETREL *Bulweria bulwerii*   Page 113
Off northern Senegal. Small-medium dark petrel with shearwater-like flight, tapered tail long and pointed, wedged when fanned. **7a Upperside**. Pale band on wing. **7b Underside**. All-dark.

8. LEACH'S STORM-PETREL *Oceanodroma leucorhoa*   Page 116
Uncommon to rare off Senegal and The Gambia. No feet projection. **8a Upperside**. Pale wing bands, rump with central dividing line. **8b Underside**.

9. MADEIRAN STORM-PETREL *Oceanodroma castro*   Page 116
Rare off Senegal. Tail less forked than Leach's, no central rump line, white rump not as extensive as Wilson's. **9a Upperside**. No pale wing band. **9b Underside**.

10. WHITE-FACED STORM-PETREL *Pelagodroma marina*   Page 115
Off northern Senegal. Pale, grey-brown, patterned face. **10a Upperside**. **10b Underside**.

11. WILSON'S STORM-PETREL *Oceanites oceanicus*   Page 115
Recorded off Senegal and The Gambia, April-September. Extensive white rump, squarish tail, rounded wings, long leg projection. **11a Upperside**. **11b Underside**. **11c Pattering over water**.

12. BRITISH STORM-PETREL *Hydrobates pelagicus*   Page 115
Commonest storm-petrel, regularly seen from land. Obvious white band on underwing, tail blunt and square, no leg projection. **12a Upperside**. **12b Underside**. **12c Pattering over water**.

13. FEA'S PETREL *Pterodroma feae*   Page 113
No modern Senegambian information. Ashy-grey, partial breast-band, dark stubby bill, wedged-shaped tail. **3a Upperside**. **13b Underside**.

PLATE 2: TROPICBIRD, GANNETS, PELICANS, FRIGATEBIRD, HAMERKOP, CORMORANTS, BITTERNS, DARTER

1. RED-BILLED TROPICBIRD *Phaethon aethereus*　　　　　Page 117
Mainly off Senegal. Wedge-shaped tail, flies high, red bill, very long streamers in breeding plumage (not illustrated). **Immature.** Bill yellow could be tipped black, finely barred back.

2. BROWN BOOBY *Sula leucogaster*　　　　　Page 118
Rare in coastal Senegambian waters. **2a Adult male.** Uniform chocolate brown above, white waistcoat below. **2b Immature.** Uniform brown above, buffy below.

3. NORTHERN GANNET *Sula bassana*　　　　　Page 117
Sometimes common in coastal waters. Adult (not illustrated) mostly white, washed yellow. **3a First-winter.** Dark brownish-grey above covered in pale spots. **3b Second-winter.** Becoming whiter.

4. GREAT WHITE PELICAN *Pelecanus onocrotalus*　　　　　Page 120
Widely distributed. Larger, whiter, less numerous than Pink-backed. **4a Adult.** Flushed pale pink in breeding season. **4b Flight.** Soaring underside. Strong black and white contrast on underwings.

5. PINK-BACKED PELICAN *Pelecanus rufescens*　　　　　Page 120
Widely distributed. Grey, with pinkish back in flight. **5a Adult.** Greyish plumage with nuchal crest. **5b Flight.** Soaring underside. Not strongly contrasting; dirty grey.

6. MAGNIFICENT FRIGATEBIRD *Fregata magnificens*　　　　　Page 121
Vagrant. Deeply forked tail, long narrow wings, long hooked bill, very agile. **Immature.** White below. Very variable.

7. HAMERKOP *Scopus umbretta*　　　　　Page 128
Common in most aquatic habitats. **Adult.** All-brown, heavy bill, shaggy crest.

8. GREAT CORMORANT *Phalacrocorax carbo*　　　　　Page 118
Locally common at coast, less common inland. Clean-cut white and black below. **Adult breeding.** White thighs. Immature (not illustrated) browner above, all grey-white below.

9. LONG-TAILED CORMORANT *Phalacrocorax africanus*　　　　　Page 119
Common throughout. Weaker head and neck than Great. Mixed roosts in mangroves. **9a Adult breeding.** All-dark, crested, ruby-red eye. **9b Immature.** Duller above, dirty greyish-white below.

10. WHITE-CRESTED TIGER HERON *Tigriornis leucolophus*　　　　　Page 122
Rarely seen. Dense mangroves. **Adult male.** Barring well spaced; female much darker, nearly black. White nape plume shorter on immature.

11. EURASIAN BITTERN *Botaurus stellaris*　　　　　Page 121
Very rare Senegambia. Reedbeds, brackish swamps, paddies. **Adult.** In flight: golden-brown plumage, trailing long legs.

12. LITTLE BITTERN *Ixobrychus minutus*　　　　　Page 121
Uncommon to frequent. Very small. Freshwater, reedbeds. Large creamy patch on dark wing. Resident and Palearctic migrant populations. **12a Adult male** African *I. m. payesii.* **12b Adult male** Palearctic *I. m. minutus.* **12c Adult female** African *I. m. payesii.*

13. DWARF BITTERN *Ixobrychus sturmii*　　　　　Page 122
Rare to frequent. Freshwater, paddies, seasonal swamps; most likely in rains and just after. **13a Adult.** Dark slaty-grey above. **13b Immature.** Back barred, not spotted.

14. AFRICAN DARTER *Anhinga rufa*　　　　　Page 119
Seasonally common. Fresh and saline waters, most wetlands, commonest near mangroves. **14a Adult male.** Drying out. Long snake-like neck and dagger-like bill. **14b Immature.** Paler sandy-buff.

PLATE 3: HERONS AND EGRETS

1. WHITE-BACKED NIGHT HERON *Gorsachius leuconotus*   Page 123
Local. Mangroves and freshwater courses. Highly nocturnal. Furtive. Huge eye and white spectacles. **1a Adult breeding**. White back not always seen. **1b Immature**. Brown, less spots than greyer Black-crowned.

2. BLACK-CROWNED NIGHT HERON *Nycticorax nycticorax*   Page 123
Common in suitable habitat throughout. Sometimes active in day. **2a Adult breeding**. Black and white plumage. Long neck plumes. **2b Immature**. Grey-brown, heavily spotted, streaked below.

3. CATTLE EGRET *Bubulcus ibis*   Page 124
Abundant throughout. Stocky, heavy-jowled, yellow bill. **3a Adult non-breeding**. Mostly white. **3b Adult breeding in flight**. Tawny-brown and white. Brighter bare parts.

4. SQUACCO HERON *Ardeola ralloides*   Page 124
Common near water throughout. **4a Adult non-breeding**. Sandy-biscuit colour, bittern-like. **4b Adult breeding**. Strikingly white in flight with brown saddle. Cobalt-blue bill with black tip.

5. STRIATED HERON *Butorides striatus*   Page 124
Common throughout, most abundant in rains: mangroves and swamps. Erectile bushy crest. Good climber. Often approachable. **5a Adult**. Back blackish-green. **5b Immature**. Spotted wings.

6. BLACK EGRET *Egretta ardesiaca*   Page 125
Common, sometimes large numbers. Marshes, paddies, swamps. Entirely black. **6a Adult**. Shaggy crest, thin-faced, rapacious appearance. **6b**. Diagnostic 'umbrella' hunting position.

7. INTERMEDIATE EGRET *Egretta intermedia*   Page 126
Entirely white. Shorter-necked than Great White. Gape line does not extend beyond eye. **7a Adult non-breeding**. Bill all yellow. **7b Adult breeding**. Bill red, tipped orange.

8. WESTERN REEF HERON *Egretta gularis*   Page 125
Common, often in good numbers. Beaches, mangroves, swamps. Blue-black, white throat. Heavier bill than Little Egret. **8a Adult breeding**. Typical dark morph. **8b Adult breeding**. Typical white morph, much less common.

9. LITTLE EGRET *Egretta garzetta*   Page 126
Common throughout. **9a Adult non-breeding**. Dagger-like jet black bill. Dark morph (not illustrated) not known in our area. **9b Adult breeding**. Face skin orange.

10. GREAT WHITE EGRET *Egretta alba*   Page 126
Common throughout. Gape extends behind eye. **10a Adult non-breeding**. Bill yellow. **10b Adult breeding**. Bill darkens from tip to base.

11. BLACK-HEADED HERON *Ardea melanocephala*   Page 127
Common throughout, more in dry months. Often in dry areas. **11a Adult**. Snake-like neck clearly black and white. **11b Adult flight**. Distinct underwing contrast. **11c Immature**. Neck pattern less clear.

12. GREY HERON *Ardea cinerea*   Page 127
Common throughout, usually near water. Head, face, throat mostly white; underwing more uniform grey than Black-headed Heron. **Adult** Palearctic nominate race. African race (not illustrated) paler.

13. GOLIATH HERON *Ardea goliath*   Page 128
Local throughout, usually singly. Mangrove creeks, river edges. Shy. **Adult** Huge, powerful bill, immense broad grey wings in flight.

14. PURPLE HERON *Ardea pururea*   Page 127
Found throughout all wetlands. Secretive. **14a Adult**. Slim, willowy neck, slender bill. **14b Immature**. Brown and duller.

6b

PLATE 4: STORKS, SPOONBILLS, CRANE AND FLAMINGOS

1. WHITE STORK *Ciconia ciconia* — Page 130
Very rare in The Gambia, more frequent in northern Senegal. **1a Adult.** Straight red bill and legs. All-white with black flight feathers. **1b Underside** soaring; tail white.

2. BLACK STORK *Ciconia nigra* — Page 129
Very rare in The Gambia, more frequent in northern Senegal. Wetlands. **2a Adult.** Legs, feet, bill red. **2b Underside** soaring. Thin elegant neck, black back and rump. White belly and axillaries.

3. ABDIM'S STORK *Ciconia abdimii* — Page 130
Very rare in The Gambia. Regular in eastern Senegal. Recalls small Black Stork with blue face. **3a Adult.** Shortish pink legs, joints redder. **3b Underside** soaring. More white on leading edge of underwing than Black.

4. WOOLLY-NECKED STORK *Ciconia episcopus* — Page 130
Seasonally and locally very common. Mostly dark with white neck. **4a Adult.** Dark face mask. **4b Underside** soaring. Underwing all dark, undertail-coverts longer than tail.

5. MARABOU STORK *Leptoptilos crumeniferus* — Page 131
Frequent to common. Massive bill and gular sac. **5a Adult.** Bald head, white legs. **5b Underside** soaring. Often mixes with large vultures.

6. YELLOW-BILLED STORK *Mycteria ibis* — Page 129
Found throughout, all seasons. Wetlands. Often with pelicans. **6a Adult.** Downcurved yellow bill, bare red face. Immature (not illustrated) brown-grey. **6b Underside** soaring. Black tail.

7. AFRICAN OPENBILL STORK *Anastomus lamelligerus* — Page 129
Rare in Senegal. **Adult.** All-dark. Vociferous. Gap in mandibles.

8. SADDLE-BILLED STORK *Ephippiorhynchus senegalensis* — Page 131
Rare, any season. Very tall, debonair. **Adult** Black and white with colourful bill. White primaries prominent in flight.

9. AFRICAN SPOONBILL *Platalea alba* — Page 133
Common in wetlands throughout, largest numbers in dry season. **9a Adult.** Red-faced, no prominent upper mandible ridges. Red legs. **9b Immature.** More feathered face than European.

10. EUROPEAN SPOONBILL *Platalea leucorodia* — Page 133
Rare passage migrant; commoner northern Senegal. **10a Adult.** No bare red face, with black lores and yellow flush on neck. Black bill tipped yellow, with ridges. **10b Immature.** Bill paler.

11. BLACK CROWNED CRANE *Balearica pavonina* — Page 175
Widespread but local. Wetlands. Gregarious. Vociferous. **Adult.** Unique crest. Large pale wing patch in flight.

12. LESSER FLAMINGO *Phoenicopterus minor* — Page 134
Local in The Gambia. Numbers fluctuate in northern Senegal. Blotchy red coverts in flight. **12a & b Adult.** Smaller, pinker than Greater. Bill uniform coral-red. Immature (not illustrated) grey-brown.

13. GREATER FLAMINGO *Phoenicopterus ruber* — Page 133
Small numbers, locally common, all seasons, river and coast. Breeds in large numbers in northern Senegal. Uniform scarlet wing-coverts in flight. **13a & b Adult.** Taller than Lesser. Some birds very white. Bill pink tipped black. Immature (not illustrated) grey bill tipped black.

PLATE 5: DUCKS, GEESE AND IBISES

1. EGYPTIAN GOOSE *Alopochen aegyptiacus* Page 136
Scarce on freshwater parts of river and fields. Commoner northern Senegal. **Adult**. Oval eye patches and belly patch chestnut. Striking white wing patch contrasts with black primaries in flight.

2. KNOB-BILLED DUCK *Sarkidiornis melanotos* Page 136
Common and widespread. Goose-like, speckled head and neck. **2a Adult male**. Protuberance on bill. **2b Adult female**. Smaller than male. **2c Immature**. Duller than adult.

3. SPUR-WINGED GOOSE *Plectropterus gambensis* Page 136
Common and widespread, most during rains. **3a Adult**. Male larger than female, knob on forehead. **3b Immature**. Smaller, duller and brownish.

4. FULVOUS WHISTLING DUCK *Dendrocygna bicolor* Page 135
Uncommon in The Gambia, mostly in dry season. Commoner in northern Senegal. Black rump, pale V-shape on uppertail-coverts in flight. **Adult**. Golden-brown with creamy flank stripes, no white face.

5. WHITE-FACED WHISTLING DUCK *Dendrocygna viduata* Page 135
Common to very abundant, mostly freshwater habitats. All seasons. Long neck and legs. **5a Adult**. Dark chocolate brown. Face white. Barred flanks. **5b Immature**. Duller, face buff-brown.

6. TUFTED DUCK *Aythya fuligula* Page 139
Infrequent Senegambia in dry season. **6a Male eclipse**. Black and white, partial crest. Female (not illustrated) brown; both sexes show bright yellow eye. **6b Speculum**. Less bright than Ferruginous.

7. FERRUGINOUS DUCK *Aythya nyroca* Page 139
Rare Senegambia in dry season. All-dark, white vent. **7a Male eclipse**. White eye. Female (not illustrated) eye brown. **7b Speculum**. Brighter than Tufted.

8. NORTHERN SHOVELER *Anas clypeata* Page 138
Annual dry season visitor. Commoner in northern Senegal. **8a Male** Green head and chestnut flanks. **8b Female**. Brown, spatulate bill. **8c Speculum**. Large blue shoulder patch.

9. GARGANEY *Anas querquedula* Page 138
Commonest Palearctic duck in our area. Favours freshwater. **9a Male**. White stripe over eye to nape. **9b Female**. Dark eye stripe, pale patch at bill base. **9c Speculum**. Dull green bordered white.

10. COMMON TEAL *Anas crecca* Page 137
Uncommon dry season visitor. Commoner in northern Senegal. **10a Male**. Green eye patch, pale rear end with dark border. **10b Female**. Pale face with darker crown. **10c Speculum**. Black and green.

11. NORTHERN PINTAIL *Anas acuta* Page 138
Dry season visitor, all divisions. Abundant northern Senegal. **11a Male** Elegant, long pointed tail. **11b Female**. Brown, slim, tapered tail. **11c Speculum**. Green tones, white trailing edge.

12. AFRICAN PYGMY GOOSE *Nettapus auritus* Page 137
Local, seasonally frequent to common. Freshwater with dense surface vegetation. Small, pretty. Sexes differ. **12a Male**. White face, jade ear-coverts. **12b Female**. Duller, smudgy markings.

13. WHITE-BACKED DUCK *Thalassornis leuconotus* Page 135
Northern Senegal only. Freshwater with dense surface vegetation. **Adult**. Mottled rufous-brown. Speckled bill. Large white crescent-shaped spot between eye and bill extends to tip of chin.

14. SACRED IBIS *Threskiornis aethiopicus* Page 132
Widespread in wetlands. **Adult** Naked head black, body white. White wings bordered black in flight.

15. GLOSSY IBIS *Plegadis falcinellus* Page 132
Uncommon in dry season inland. Commoner in northern Senegal. Dark curlew-like. Trails legs in flight. **Adult non-breeding**. Head and neck streaked whitish. Bronzy-green sheen in breeding plumage.

16. HADADA IBIS *Bostrychia hagedash* Page 132
Frequent to common in all seasons; riverside and swamps inland. **Adult**. Heavy-bodied, white malar stripe.

24

PLATE 6: OSPREY, HARRIER-HAWK, VULTURES AND CROWS

1. OSPREY *Pandion haliaetus* Page 161
Common Palearctic migrant. Records every month, October-March peak. Coast, mangroves and river. **Adult.** Small-headed. Brown and white.

2. EGYPTIAN VULTURE *Neophron percnopterus* Page 145
Infrequent. Dishevelled head. Weak bill, yellow face. **2a Adult.** Creamy-white, black flight feathers. **2b Immature.** Greenish face, long nape feathers.

3. AFRICAN HARRIER-HAWK *Polyboroides typus* Page 149
Common, widespread; favours wetter places. Long bare legs, small head, bare face changes colour, weak crest. **3a Adult.** Grey, large wing spots, barred below, **3b Immature.** Brown, darker wings, tail bars. Grey-blue facial skin. Intermediate plumages common.

4. PALM-NUT VULTURE *Gypohierax angolensis* Page 144
Common throughout. **4a Adult.** All-white underparts, wings black and white; eagle-like. Bare red skin around eye. **4b Immature.** All-brown. Intermediate plumages common.

5. BROWN-NECKED RAVEN *Corvus ruficollis* Page 347
Rare in The Gambia, beaches, tips, road verges. Regular northern Senegal. **Adult.** All-dark, brown nape if seen well. Slimmer, narrower-winged than Pied in flight.

6. PIED CROW *Corvus albus* Page 346
Very common on coast, fewer inland. **Adult.** Black and white.

7. HOODED VULTURE *Necrosyrtes monachus* Page 145
Very numerous throughout. Urban tips, abattoirs, beaches, bush. **Adult.** All-brown. Small head, weak bill, indistinct hood. Immature (not illustrated) has dark down (not feathers) on back of head.

8. RÜPPELL'S GRIFFON VULTURE *Gyps rueppellii* Page 146
Few at coast, increasing inland. **8a Adult.** Larger than White-backed. Heavily scaled white, horn bill. **8b Immature.** Bill pale, darker ruff than shorter-necked White-backed.

9. EUROPEAN GRIFFON VULTURE *Gyps fulvus* Page 146
Uncommon palearctic migrant in Senegambia. Palearctic migrant. Ginger-khaki tones. Stands taller than Rüppell's but neck similar. **9a Adult.** Very sandy. **9b Immature.** Streakier, darker.

10. WHITE-BACKED VULTURE *Gyps africanus* Page 145
Common and widespread. Bill dark at all ages. **10a Adult.** White back obvious on take off. **10b Immature.** No white on back. Less tall than Rüppell's.

11. WHITE-HEADED VULTURE *Trigonoceps occipitalis* Page 147
Uncommon in Senegambia. **11a Adult.** Flat-topped head, always pink bill with blue cere. White down on head. **11b Immature.** Colourful bill, diagnostic head shape.

12. LAPPET-FACED VULTURE *Torgos tracheliotus* Page 147
Small numbers in The Gambia. Widespread Senegal. **12a Adult.** Massive bill, square head. Prominent ridged wattles. **12b Immature.** Ghosts adult but with whitish head down. Brown thigh feathering.

PLATE 7: EAGLES AND BUZZARDS

1. MARTIAL EAGLE *Polemaetus bellicosus*   Page 161
Rare resident. Enormous. Flat head, short crest. Wings reach well down tail. May hover. **1a Adult**. Dark brown and white. Sparingly blotched. **1b Immature**. Greyish, crest prominent.

2. CROWNED EAGLE *Stephanoaetus coronatus*   Page 160
Very rare. Gallery forest. Hunts primates through trees. Wings much shorter than tail. **2a Adult**. Blue-black. Double-crested. **2b Immature**. Much paler. Conspicuous barred tail.

3. AFRICAN FISH EAGLE *Haliaeetus vocifer*   Page 144
Common. Wetlands throughout. Very vocal. **3a Adult**. White head-cape, chestnut belly, all-dark wings. **3b Immature**. Dark brown and white. Appears rather vulturine.

4. TAWNY EAGLE *Aquila rapax*   Page 157
Widespread. All divisions, all months. Plumage variable. No obvious crest. Strong, wide gape to under eye. **Adult**. Dark morph. Pale morph (not illustrated) less frequent.

5. WAHLBERG'S EAGLE *Aquila wahlbergi*   Page 158
Widespread. All divisions, all months. Smaller-faced than Tawny. Slight crest. **Adult**. Dark morph. Pale morph (not illustrated) less frequent.

6. AFRICAN HAWK EAGLE *Hieraaetus spilogaster*   Page 158
Frequent throughout. Dry woodland with tall trees. No crest. Feathered legs. Note: Palearctic Bonelli's Eagle reported in northern Senegal since 1990, November-April (see text). **6a Adult**. Female more heavily streaked. **6b Immature**. Tawny-brown above, rufous-orange below.

7. AYRES'S HAWK EAGLE *Hieraaetus ayresii*   Page 159
Very rare, few records. Smaller than African Hawk Eagle. **Adult**. Small crest. White leading edge 'landing-lights' on wing. Heavily streaked below. Immature (not illustrated) is buff-streaked below.

8. BOOTED EAGLE *Hieraaetus pennatus*   Page 159
Frequent throughout dry months. Palearctic visitor. Rounded tail. White 'landing-light' shoulder edges. **8a Adult**. Pale morph. **8b Adult**. Dark morph. Intermediates occur.

9. COMMON BUZZARD *Buteo buteo*   Page 156
Rare Palearctic vagrant. Bare lower legs, rounded head, barrel-chested. Race *vulpinus* is more rusty. **Adult**. Plumages very variable. See text for Long-legged Buzzard.

10. RED-NECKED BUZZARD *Buteo auguralis*   Page 157
Rare intra-African migrant. Movements rain-related. Rusty-brown and white. **10a Adult**. Chestnut-brown uppertail. **10b Immature**. White chest, with rusty blobs. Tail barred.

11. LONG-CRESTED EAGLE *Lophaetus occipitalis*   Page 160
Frequent to common throughout, all months. Floppy crest. **Adult**. Looks all black at distance. Immature (not illustrated) is scaly.

PLATE 8: EAGLES, BUZZARDS, KITES, BAT HAWK AND CUCKOO-FALCON

1. BATELEUR *Terathopius ecaudatus*   Page 149
Common to frequent. Colourful. Wing-tips extend beyond very short tail. **1a Adult female.** Male (not illustrated) has all-black secondaries, not grey. **1b Immature.** Brown.

2. BROWN SNAKE EAGLE *Circaetus cinereus*   Page 148
Uncommon on coast, frequent inland. Dry woodlands. **Adult.** All-dark brown, bonnet-shaped head, sturdy neck, bare legs, staring yellow eyes. Immature (not illustrated) is fringed white.

3. WESTERN BANDED SNAKE EAGLE *Circaetus cinerascens*   Page 149
Local to frequent in well established woodland or riverine forest. Grey. Large-headed. **3a Adult.** White tail band often obscured, tip grubby. Pale shoulder line. **3b Immature.** Creamy-white becoming greyer, tail band acquired gradually.

4. SHORT-TOED EAGLE *Circaetus gallicus*   Page 148
Palearctic *gallicus* is dry season visitor November-April, African *beaudouini* is less frequent. Dark hood clean-cut from very pale barred and blotched underparts. Grey-brown back. Plumages extremely variable. **4a Adult** *C. g. gallicus*. Short chest bars incomplete. **4b Adult** *C. g. beaudouini*. African migrant. Narrow chest bars complete. Creamy immature (not illustrated) has no tail band.

5. RED KITE *Milvus milvus*   Page 143
Rare Palearctic vagrant. Colourful, reddish-chestnut. Head whitish-rufous. Very agile, elastic wings, strongly patterned. **Adult flight.** Deeply forked tail. Pale wing patches.

6. BLACK KITE *Milvus migrans*   Page 143
Very common throughout, all habitats, all seasons. Brown plumage. Tail forked. Agile. **6a Adult** *M. m. parasitus.* Yellow bill, paler head. **6b Adult** *M. m. migrans.* Black bill.

7. BAT HAWK *Macheiramphus alcinus*   Page 141
Rare, few modern records. Mangroves, dry woodland. Predates bats. Active dusk. **Adult.** Dark falcon-like, small crest.

8. EUROPEAN HONEY BUZZARD *Pernis apivorus*   Page 141
Palearctic migrant, infrequent Senegambia. Typically forages in or near forest. **Adult.** Size of Black Kite. No crest. Small head. Note tail pattern. Yellow eye.

9. GRASSHOPPER BUZZARD *Butastur rufipennis*   Page 155
Seasonally common in The Gambia. Widespread Senegal. **Adult.** Colourful plumage with red and brown, streaky. Orange-ringed golden eye. Stoops falcon-like. Immature (not illustrated) has head foxy-brown with dark moustaches.

10. BLACK-SHOULDERED KITE *Elanus caeruleus*   Page 142
Common throughout. Fallow land and open country. Hovers. Short tail, pointed wings. **10a Adult.** Bandit mask, black epaulettes. **10b Adult.** Hovering. Immature (not illustrated) is scaly.

11. AFRICAN SWALLOW-TAILED KITE *Chelictinia riocourii*   Page 142
Infrequent, extensive arid zones. Graceful, tern-like, very aerial. Gregarious. **11a Adult.** Deeply forked tail. **11b Adult flight.** Black patch on underwing sometimes absent. Immature (not illustrated) is shorter tailed and scaly.

12. AFRICAN CUCKOO-HAWK *Aviceda cuculoides*   Page 140
Rare. Records all seasons. Tall trees with lianas, along watercourses. **12a Adult.** Recalls large cuckoo with small crest. Pot-bellied profile. Secretive. **12b Immature.** White below, dark streaks and blotches.

PLATE 9: HARRIERS, ACCIPITERS AND LIZARD BUZZARD

1. MONTAGU'S HARRIER *Circus pygargus*   Page 150
Widespread, locally common November-March. Low quartering flight. **1a Adult male.** Blue-grey. Black bar along upperwing. Extensive black wing-tips. **1b Adult female.** Dark trailing edge to all primaries as well as secondaries. Well-defined barring on secondaries. **1c Juvenile.** Weaker face pattern; no pale collar.

2. PALLID HARRIER *Circus macrourus*   Page 150
Infrequent to rare October-May. Lighter wing action than Montagu's. **2a Adult male.** Black wedge on wing-tips. **2b Adult female.** Almost no black trailing edge to inner primaries. **2c Juvenile.** Strong face pattern. Well-defined blackish mark below eye with buffy collar below.

3. EURASIAN MARSH HARRIER *Circus aeruginosus*   Page 151
Common Palearctic migrant October-March. Wetlands, mangroves, reedbeds. Characteristic flight pattern; tail not forked. **3a Adult male.** Two-tone greys and browns. **3b Adult female.** Immature similar. Chocolate, creamy head and forewing.

4. GREAT SPARROWHAWK *Accipiter melanoleucus*   Page 155
Rarely observed in coastal forest, dry months. **4a Adult** Strong. Black and white plumage. Bare lower legs. White flank spots. White nape patch. Male smaller. **4b Immature male.** Hawk-eagle immatures have legs all feathered.

5. DARK CHANTING GOSHAWK *Melierax metabates*   Page 151
Common roadside hawk. No wing spots or throat stripe, red legs, barred below. **5a Adult.** Pale rump finely barred, wings rounded. **5b Immature.** Grey-brown. Pale supercilium. Striped throat, barred below.

6. GABAR GOSHAWK *Micronisus gabar*   Page 151
Uncommon at coast; widespread inland. Dry woodlands, riverside habitats. Like small Dark Chanting Goshawk. A melanistic form (not illustrated) is regularly reported. It is all-black with a paler, dark-barred tail. **6a Adult.** Bars grey-brown. Plain white rump. **6b Immature.** Pale brown, white rump.

7. AFRICAN GOSHAWK *Accipiter tachiro*   Page 152
Regular in coastal forests. Strong, plump. Prominent rufous bars, denser on flanks. Up to three ovoid white tail-spots. Sits in shadows. **7a Adult.** Dove-grey head, darker back. Long hind claw. **7b Immature.** Liver-brown or black above, white below, partial chest bars and blotches.

8. SHIKRA *Accipiter badius*   Page 153
Throughout, in most habitats, all divisions. Pale grey back. Delicate pink barring below. Rump grey, tail centrally unbarred. Considerable size variation. Weakish legs and feet. **8a Adult male.** Bars uniform pink at distance. **8b Adult female.** Larger size. **8c Immature female.** Brown above, pale nape patches. Entirely blotched below.

9. OVAMBO SPARROWHAWK *Accipiter ovampensis*   Page 154
Uncommon, in western Gambia. Underside entirely dark-barred including throat. White shaft streaks on tail. **9a Adult.** Long, slim gait. **9b Immature female.** Conspicuous supercilium.

10. WESTERN LITTLE SPARROWHAWK *Accipiter erythropus*   Page 153
Very rarely seen, forest dweller. Diminutive, dove-sized. Very dark above, pinkish-white below, uniform pink at distance. **10a Adult male.** Pink flanks. Looks capped in flight. **10b Adult female.** More barred. Immature (not illustrated) has bars and spots.

11. LIZARD BUZZARD *Kaupifalco monogrammicus*   Page 155
Common throughout, all seasons. Sexes alike. **Adult.** Prominent black mesial stripe. Grey rump. Single white band on black tail.

PLATE 10: FALCONS

1. PEREGRINE FALCON *Falco peregrinus* Page 166
Uncommon, widespread, often near wader haunts; dry season. Large, thickset, heavy, but considerable size variation. No rufous on head. **1a Adult** (nominate). Thick moustachial. Densely barred. **1b Juvenile.** Heavily streaked. No rufous on crown. **1c Adult flight.** Triangular pointed wings, short tapered tail.

2. LANNER FALCON *Falco biarmicus* Page 165
Frequent, all divisons, all months. Race *erlangeri* in northern Senegal. Large. Slimmer, longer-tailed, less blue-grey than scarcer Peregrine. **2a Adult.** Thin moustachial, rufous crown. Pink-buff below, some spotting. **2b Juvenile.** Pale rufous crown. Streaked not barred below. **2c Adult flight.** Longer wings and tail than Peregrine. Blunter wing-tips.

3. BARBARY FALCON *Falco pelegrinoides* Page 166
Sporadic dry season sightings, northern Senegal. Desert dweller. Smaller and paler than most Peregrines. **Adult.** Fine diffuse barring below. Rufous in centre of crown. Broad subterminal tail bar eliminates Lanner.

4. AFRICAN HOBBY *Falco cuvierii* Page 165
Frequent, all divisions, all seasons. Restricted in northern Senegal. Small, slim, dark above. **4a Adult.** Orange-red below, fine streaks. **4b Juvenile.** Paler below, heavily streaked.

5. EURASIAN HOBBY *Falco subbuteo* Page 165
Uncommon Palearctic visitor. Occasional dry season records. Small, slim. Heavily streaked with paler ground colour below. **5a Adult.** Russet trousers. **5b Juvenile.** Pale crown markings, browner-backed than adult.

6. RED-NECKED FALCON *Falco chicquera* Page 164
Local, near rhun palms and well established trees. Crown and nape warm rufous, face white. Broad subterminal tail bar. **6a Adult.** Pale grey back, prominent neat barring below. **6b Juvenile.** Darker above, irregular barring below.

7. FOX KESTREL *Falco alopex* Page 163
Rare vagrant in Senegambia. Rocky areas. Long wings, long tail. **7a Adult.** Deep foxy-red. Black primaries. **7b Adult in flight.** Slim, tail looks longer than body.

8. GREY KESTREL *Falco ardosiaceus* Page 164
Common throughout. All grey with large head. Very fast direct flight. **Adult.** Yellow spectacles. Legs and feet yellow. Immature (not illustrated) is darker with sparse black streaking below; facial skin whitish.

9. LESSER KESTREL *Falco naumanni* Page 163
Frequent Palearctic passage migrant, mostly inland. Gregarious. Attends bush fires. White claws. Slightly elongated central tail feathers. **9a Adult male.** Plain rufous mantle. Very pale below, lacks moustachial. **9b Adult female.** Weak moustachial streak.

10. COMMON KESTREL *Falco tinnunculus* Page 163
Frequent to locally common Palearctic migrant throughout. November-April. Open habitats. Mostly singly. Black claws. Hovers. **10a Adult male.** Spotted rufous back. Slight moustachial streak. **10b Adult female.** Stronger moustachial than Lesser Kestrel.

PLATE 11: OVERHEAD RAPTORS: VULTURES, EAGLES, OSPREY AND HARRIER-HAWK

1. EUROPEAN GRIFFON VULTURE *Gyps fulvus*   Page 146
Subadult. Very large. Sandy-khaki body, well-defined contrast with all-dark primaries, secondaries and shortish tail. Immatures paler with more contrast.

2. WHITE-BACKED VULTURE *Gyps africanus*   Page 145
2a Adult. Extensive white underwing-coverts contrast with darker flight feathers. White centre of back clearly visible from above. **2b Immature.** More mottled, contrast insignificant.

3. RÜPPELL'S GRIFFON VULTURE *Gyps rueppellii*   Page 146
3a Adult. Underwing-coverts black-brown, obvious white frontal bar and two complete rows of white spots across coverts. Scaly above. **3b Immature.** Less well-defined than adult.

4. WHITE-HEADED VULTURE *Trigonoceps occipitalis*   Page 147
4a Adult. Black at front, white or pale at rear. **4b Immature.** Distinct thin white line along middle of underwing.

5. LAPPET-FACED VULTURE *Torgos tracheliotus*   Page 147
Adult. Huge. All-dark wings, narrow white band along leading edge, jagged trailing edge, shortish broad diamond-shaped tail. White thighs, square head.

6. PALM-NUT VULTURE *Gypohierax angolensis*   Page 144
6a Adult. Underparts all-white, shortish, rounded tail black tipped white. Broad rounded white wings, black trailing edge to translucent primaries. **6b Immature.** Brownish, eagle-like but shorter tail.

7. HOODED VULTURE *Necrosyrtes monachus*   Page 145
Adult. Nondescript, weak head and bill. Forewing darkest. Tail square, with rounded edges.

8. EGYPTIAN VULTURE *Neophron percnopterus*   Page 145
8a Adult Creamy-white wing-coverts and tail contrast with blackish flight feathers. Tail strikingly wedge-shaped. **8b Immature.** All brown, becoming adult through a series of mixed plumages.

9. MARTIAL EAGLE *Polemaetus bellicosus*   Page 161
9a Adult. Immense. Broad, long wings, short tail. Clean-cut bib effect of brown upper breast contrasts with spotted white belly. **9b Immature.** Very pale, grey tail barred as adult.

10. CROWNED EAGLE *Stephanoaetus coronatus*   Page 160
10a Adult. Broad banded wings with rufous forewing; long banded tail. **10b Immature.** Underwing-coverts light rufous.

11. OSPREY *Pandion haliaetus*   Page 161
Adult. Very white, streaky breast-band. Blackish carpal patches and wing band. Flies with wings bowed.

12. TAWNY EAGLE *Aquila rapax*   Page 157
12a Adult. Variable. Broad tail rounded when fanned. No prominent contrasting plumage features. Seven well spaced 'fingers'. **12b Immature.** More contrast, sturdy head and neck. Feathered legs.

13. WAHLBERG'S EAGLE *Aquila wahlbergi*   Page 158
13a Adult dark phase. **13b Adult** pale phase. Slim body; flat, horizontally held wings. Long, narrow tail.

14. BOOTED EAGLE *Hieraaetus pennatus*   Page 159
14a Adult dark phase above. **14b Adult** dark phase below. **14c Adult** pale phase. Both phases show pale area on innermost three primaries, and small white shoulder patches on front of wing.

15. AFRICAN FISH EAGLE *Haliaeetus vocifer*   Page 144
15a Adult. White head and tail, chestnut body and wing-coverts, rest of wing black. **15b Immature.** Mostly whitish heavily blotched and streaked blackish-brown.

16. AFRICAN HARRIER-HAWK *Polyboroides typus*   Page 149
16a Adult. Weak head, bare legs. Broad wings, underside finely barred, broad black trailing edge. Broad white band across tail. **16b Immature.** Underwing-coverts rufous-brown. Tail with up to five dark bands.

PLATE 12: OVERHEAD RAPTORS: KITES, BUZZARDS, HAWKS AND EAGLES

1. BLACK KITE *Milvus migrans* Page 143
Adult. Yellow-billed race *M. m. parasitus*. Variable dark chocolate brown, suffused rufous with black streaking. Tail fork less prominent when spread.

2. GRASSHOPPER BUZZARD *Butastur rufipennis* Page 155
Adult Obvious ginger-red on underside of dark-bordered underwing. Wings slightly pointed.

3. COMMON BUZZARD *Buteo buteo* Page 156
Adult. Very variable. Barrel-chested, warm brown, pale chest band, upper breast darker. Barred whitish underwing, darker carpal patch, dark trailing edge. Tail barred, broad terminal bar.

4. RED-NECKED BUZZARD *Buteo augularis* Page 157
Adult. Dark rusty head and chest contrasts with white below. Underwing whitish, dark carpal patches. Unbarred chestnut tail with dark subterminal band.

5. AYRES'S HAWK-EAGLE *Hieraaetus ayresii* Page 159
5a & b Adult. Heavily streaked body. Shortish wings heavily barred and spotted, 'head-lamps' on leading edge. Upperwing dark. Undertail with 3 or 4 bands, tipped black. **5c Immature**. Strongly barred underwing, with pale rufous underwing-coverts.

6. AFRICAN HAWK EAGLE *Hieraaetus spilogaster* Page 158
6a Adult. Black and white pattern; rounded wings. Body white, some black streaks. Broad black terminal tail band. **6b Immature**. Tawny-brown above, orange-rufous below with streaks.

7. BAT HAWK *Macheiramphus alcinus* Page 141
Adult. Wings long narrow, sharply pointed, broad-based. Silhouette recalls bulky dark falcon. Flight languid.

8. BATELEUR *Terathopius ecaudatus* Page 149
8a Adult male. Upswept wings, almost tailless. Rolling flight. Female (not illustrated) has prominent all-white underwing. Feet project beyond tail. **8b Immature**. All brown. Tail slightly longer than adult.

9. SHORT-TOED EAGLE *Circaetus gallicus* Page 148
Variable. Chocolate hood contrasts with pale body on typical forms. No dark carpal patch. Densest marking on forewing. Square tail with 3 or 4 narrowish dark bands. **9a Palearctic** *C. g. gallicus*. Often incomplete barring. **9b African** *C. g. beaudouini*. Complete barring; darker above than nominate *gallicus*.

10. WESTERN BANDED SNAKE EAGLE *Circaetus cinerascens* Page 149
'Front-heavy' appearance. Heavy grey body and head, thick neck. **10a Adult**. Single broad thick white bar on pale-tipped shortish tail. **10b Immature**. Very creamy, tail bar visible.

11. BROWN SNAKE EAGLE *Circaetus cinereus* Page 148
Dark brown. Big head, thick neck. **11a Adult**. Underwing largely unbarred pearly white, contrasts with brown body and forewing. Narrow tail bars equally spaced, whitish narrow tip. **11b Immature**. Blotchy, generally ghosts adult.

12. AFRICAN CUCKOO-HAWK *Aviceda cuculoides* Page 140
Adult. Longish wings and tail. Grey head. Underwing-coverts densely barred chestnut on white. Undertail has 3 bars, with pale tip. Flight languid. Immature (not illustrated) has blotchy belly.

13. LONG-CRESTED EAGLE *Lophaetus occipitalis* Page 160
Adult. At distance looks all-black with two large white wing patches. On closer views shows more white in wing and discernible crest.

PLATE 13: OVERHEAD RAPTORS: ACCIPITERS AND FALCONS

1. AFRICAN GOSHAWK *Accipiter tachiro* Page 152
1a Adult male. Chequered underwing, longish narrow tail when closed, broad when flared. Rufous-toned body, reddest on flanks, pale throat. **1b Immature female**. Dark spots and blotches below, partial chest bars.

2. GREAT SPARROWHAWK *Accipiter melanoleucus* Page 155
2a Adult male. White below, black-marked flanks. Undertail-coverts very white. Short wings:long tail ratio. **2b Immature female**. Rufous below with some long narrow blackish streaks.

3. WESTERN LITTLE SPARROWHAWK *Accipiter erythropus* Page 153
3a Male. Very small. Dark above. Pale rump, tail spots. **3b Female**. Better defined bars below. Underwing-coverts grey, fine chestnut barring. Wing-tips blackish-grey.

4. SHIKRA *Accipiter badius* Page 153
Russet underwing barring, tips edged blackish. Belly with very narrow pinkish bars. **4a Male**. Smallish. Undertail with 5-6 narrow bars. **4b Immature female**. Body blotched with rich brown.

5. GABAR GOSHAWK *Micronisus gabar* Page 151
5a Adult. Plain grey bib, barred below and across wings, long tail. Prominent white rump. **5b Immature female**. Throat and chin vertically streaked soft brown with brown barring on underparts. Pale supercilium.

6. DARK CHANTING GOSHAWK *Melierax metabates* Page 151
6a Adult. Slender largish body, grey bib and barred belly. Black primary tips. **6b Immature female**. All brown-grey, neck and upper breast streaky, rest of underparts with irregular bars but not clear.

7. OVAMBO SPARROWHAWK *Accipiter ovampensis* Page 154
7a Adult. Small-headed. Entirely narrowly dark-barred below, no grey bib. Underwing barred black and white. **7b Immature female**. Pale rufous below, flanks barred, distinct white supercilium, dark patch behind eye.

8. LIZARD BUZZARD *Kaupifalco monogrammicus* Page 155
Adult. Thickset. Large head. Black vertical stripe on white throat. Neatly barred belly.

9. RED-NECKED FALCON *Falco chicquera* Page 164
Adult. Rich chestnut crown and nape, all neatly barred below.

10. GREY KESTREL *Falco ardosiaceus* Page 164
Adult. Uniform slate-grey with some fine markings; yellow spectacles, large head.

11. RED-FOOTED FALCON *Falco vespertinus* Page 164
Gregarious. **Adult female**. Rich buff body and forewings. Crown rufous, legs red. See text for male.

12. LESSER KESTREL *Falco naumanni* Page 163
Subtly rounded tail in both sexes. Highly gregarious. **12a Male**. Whitish underwings, black wing-tips. No moustachial. **12b Female**. Paler than female Common Kestrel, with reduced barring.

13. COMMON KESTREL *Falco tinnunculus* Page 163
Adult female. Tail barred dark, broad subterminal band. Underwing barred and spotted grey.

14. LANNER FALCON *Falco biarmicus* Page 165
Juvenile. Heavily streaked below. Dark band across underwing-coverts. Dull rufous crown and nape.

15. BARBARY FALCON *Falco pelegrinoides* Page 166
Adult. Chunky and compact. Strongly washed rufous. Underwing paler than Peregrine.

16. PEREGRINE FALCON *Falco peregrinus* Page 166
Juvenile. Face as adult. Heavy body buffy, very heavily streaked black.

17. EURASIAN HOBBY *Falco subbuteo* Page 165
Adult. Smallish, sickle-shaped wings, below buff and pale rufous, prominently streaked blackish. Black moustachial contrasts with whitish face.

18. AFRICAN HOBBY *Falco cuvieri* Page 165
Adult. Smallish. Sickle-shaped wings; body and forewings rich uniform dark rufous.

PLATE 14: FRANCOLINS, QUAIL, STONE PARTRIDGE, GUINEAFOWL, BUTTON-QUAILS, AND SANDGROUSE

1. DOUBLE-SPURRED FRANCOLIN *Francolinus bicalcaratus* Page 169
Very common throughout. **Adult**. Strong head and face markings. Prominent supercilium. Bold teardrop chestnut and black spots on breast. Legs and bill olive-green. Black line from bill to eye.

2. WHITE-THROATED FRANCOLIN *Francolinus albogularis* Page 168
Uncommon and local. Small francolin. **Adult**. Underparts plain buff, some streaks on male, more barred on female. Chestnut wings. White chin and throat.

3. AHANTA FRANCOLIN *Francolinus ahantensis* Page 168
A few pairs in coastal forest thickets e.g. Abuko/Bijolo. Secretive, forages in shaded leaf-litter. **Adult**. Dark above, milky coffee below with pale feather margins. Legs and bill pale orange.

4. STONE PARTRIDGE *Ptilopachus petrosus* Page 167
Widespread throughout wooded country. Bantam-like, with cocked tail. Shy. Noisy at dawn and dusk. **Adult**. Dark, contrasting pale belly patch. Legs dark red.

5. HELMETED GUINEAFOWL *Numida meleagris* Page 169
Frequent to common. Only guineafowl in our area. Shy. Roosts off ground sometimes in mangroves. **Adult** *N. m. galeata*. Casque reduced on immature (not illustrated).

6. LITTLE BUTTON-QUAIL *Turnix sylvatica* Page 170
Infrequent. All seasons. Grassland. Very small and quail-like. Face not striped. Crouches. **6a Adult**. Heart-shaped dots on sides. **6b Flight**. Creamy coverts contrast with rest of buff-grey wing.

7. QUAIL-PLOVER *Ortyxelos meiffrenii* Page 170
Rare. Short grassy areas. Dumpy, diminutive courser-like stance. Fluttery lark-like flight. **7a Adult**. Creamy-white supercilium. Spotted breast. **7b Flight**. White wing-patch, trailing edge and shoulders contrasting with black primaries.

8. COMMON QUAIL *Coturnix coturnix* Page 167
Infrequently seen Palearctic migrant. Furtive. Grassy areas, dried swamps. **8a Adult**. Strong facial markings, weaker on female. **8b Adult flight**. Streaky back. Uniform wings, no panels.

9. CHESTNUT-BELLIED SANDGROUSE *Pterocles exustus* Page 219
Few Gambian records, common northern Senegal. Dark belly patch all plumages. Adults 'pin-tailed'. Uniform dark underwing in flight. **9a Adult male**. Plain face. Single dark chest band. **9b Adult female**. Head streaky, narrow chest band weak.

10. FOUR-BANDED SANDGROUSE *Pterocles quadricinctus* Page 220
Common, locally abundant. Drier areas. Short square tail. Pot-bellied in flight. Grey underwing contrasts with black tips. Visits waterholes at dusk. **10a Adult male**. Black and white face pattern. **10b Adult female**. Plain face. No breast bands.

PLATE 15: RAILS, CRAKES, GALLINULES, JACANA, PAINTED SNIPE AND FINFOOT

1. AFRICAN WATER RAIL *Rallus caerulescens* Page 172
Vagrant. Reedbeds. **Adult**. Very long, slightly decurved red bill. Zebra-striped flanks. Undertail white. Immature (not illustrated) is duller.

2. AFRICAN CRAKE *Crex egregia* Page 171
Frequent in freshwater swamps and reedbeds throughout; peaks in rains. Slight pale supercilium. Bold black and white barring well spaced. **Adult**. Bill grey, purplish-red base. Mottled black streaks above.

3. SPOTTED CRAKE *Porzana porzana* Page 173
Uncommon, mainly Senegal. Extensive barring and spots below. Streaked and flecked above. **Adult**. Stubby yellow bill, base orange-red. Creamy-white ventral area.

4. LITTLE CRAKE *Porzana parva* Page 172
Palearctic migrant, mainly Senegal. Long wings. **Adult**. Bill short, horn-yellow darker at tip, red spot at base. Brownish-slate narrowly barred white ventral area.

5. BAILLON'S CRAKE *Porzana pusilla* Page 172
Palearctic vagrant northern Senegal. Black-bordered white streaks and spots above. **Adult**. Bill greenish-yellow, darker at tip. Barred black and white ventral area.

6. EURASIAN COOT *Fulica atra* Page 175
Vagrant Gambia, locally abundant northern Senegal. **Adult**. Black. Short white bill and frontal shield.

7. COMMON MOORHEN *Gallinula chloropus* Page 174
Seasonal, widespread in suitable wetland habitat. Swims well. White undertail. White lateral line. **7a Adult**. Red frontal shield and bill, tipped yellow. **7b Immature**. Brownish-grey toned.

8. LESSER MOORHEN *Gallinula angulata* Page 174
Uncommon, rains associated. Nomadic. Black Crake-sized. Shy. **8a Adult**. Small red patch on bright yellow bill. **8b Immature**. Brown-yellow bill.

9. ALLEN'S GALLINULE *Porphyrio alleni* Page 174
Uncommon. Marshes and paddies. **9a Adult**. Ventral area white. Breeding male frontal shield blue, lime-green on female. **9b Immature**. Sandy-brown, buff undertail-coverts.

10. BLACK CRAKE *Amaurornis flavirostris* Page 173
Commonest crake. Widespread in freshwater, sometimes brackish wetlands. All inky-black. **10a Adult**. Yellowish bill. Fleshy-red legs. **10b Immature**. Dull. Bill greenish-black, legs dark.

11. WHITE-SPOTTED FLUFFTAIL *Sarothrura pulchra* Page 171
Very few pairs in coastal forest thickets e.g. Abuko. Sedentary. Often heard, rarely seen. Very small. Entirely terrestrial. **11a Male**. Polka-dotted. **11b Female**. Narrow buffy barring.

12. PURPLE SWAMPHEN *Porphyrio porphyrio* Page 173
Local. Bogs, swamps. Large. Undertail white. **Adult**. Massive bill and frontal shield.

13. GREATER PAINTED-SNIPE *Rostratula benghalensis* Page 179
Common in appropriate wetland. Needs cover. Crepuscular. White 'harness' band. Large eye with white mark behind. Legs dangle in weak flight. Bobbing walk. **13a Male**. Drabber. **13b Female**. Striking.

14. AFRICAN JACANA *Actophilornis africanus* Page 179
Common throughout in freshwater with surface vegetation, rice paddies. **14a Adult**. Frontal shield blue in breeding plumage. **14b Immature**. Duller, smaller. Lacks shield.

15. AFRICAN FINFOOT *Podica senegalensis* Page 176
Local, sporadic. Furtive. Dense mangroves, under tangled river banks. Dagger-like bill red. Feet lobed. **15a Male**. Dark. **15b Female**. Browner. Striking face pattern.

PLATE 16: BUSTARDS, COURSERS, PRATINCOLE AND THICK-KNEES

1. BLACK-BELLIED BUSTARD
Eupodotis melanogaster   Page 178
Uncommon to locally frequent. **1a Adult male**. Black line behind eye. **1b Male flight**. Expansive white areas. **1c Adult female**. Back heavily speckled black. **1d Female flight**. Dark underwing contrasts with pale belly.

2. WHITE-BELLIED BUSTARD
Eupodotis senegalensis   Page 178
Rare. No prominent black markings on back. **2a Adult male**. Blue neck. **2b Male flight**. Pale panel on primaries. **2c Adult female**. Back uniform. **2d Female flight**. Recalls male, paler neck and head.

3. SAVILE'S BUSTARD *Eupodotis savilei*   Page 177
Recently discovered in The Gambia but well known in northern Senegal. Open scrubby bush, farmland. Small size; some black markings above. Distinctive call. **3a Adult male**. No black line behind eye. Rufous-crested in display. **3b Adult female**. Black belly patch.

4. EGYPTIAN PLOVER *Pluvianus aegyptius*   Page 183
Locally common June-February in The Gambia; breeds Senegal. **4a Adult**. Stunningly patterned in black, white, blue-grey, and peach. Confiding. Gregarious. **4b Adult flight**. Flies low over water.

5. BRONZE-WINGED COURSER *Rhinoptilus chalcopterus*   Page 184
Locally frequent. Dry woodland savanna with trails and tracks. Nocturnal. Tall, plover-like, no crest. **Adult**. Long purplish-red legs. Iridescent violet wing-tips. Large bulbous eye. Pied face.

6. CREAM-COLOURED COURSER *Cursorius cursor*   Page 184
Rare Palearctic migrant. Saltpans, farmland. **Adult**. Blue-grey crown. No black belly patch.

7. TEMMINCK'S COURSER *Cursorius temminckii*   Page 184
Commonest courser. Mostly dry season. Fallow, burnt ground. Very active. **Adult**. Black belly patch. Rusty-chestnut cap. Immature (not illustrated) spangled buff.

8. COLLARED PRATINCOLE *Glareola pratincola*   Page 185
Palearctic and African migrant, mostly inland; wetlands. Highly gregarious. **8a Adult**. Black bib-ring. **8b Immature**. Speckled buff. Partial or no neck band. **8c Adult flight**. Brown, tern-like.

9. SPOTTED THICK-KNEE *Burhinus capensis*   Page 183
Locally frequent. Dry woodland savanna, emergent forest, grassland. Nocturnal. **Adult**. Heavily spotted. No contrasting wing panel. Conspicuous wing spots in flight.

10. SENEGAL THICK-KNEE *Burhinus senegalensis*   Page 182
Commonest thick-knee, widespread in wetter habitats. Very vocal when disturbed. **Adult**. One greyish wing panel bordered black above.

11. WATER THICK-KNEE *Burhinus vermiculatus*   Page 182
Southern Senegal only. Estuaries, rivers systems. **Adult**. Base of bill green. Conspicuous grey wing panel with a white line above, lacking lower black band.

12. STONE CURLEW *Burhinus oedicnemus*   Page 182
Palearctic migrant to northern Senegal. Arid places. **Adult**. Broad grey wing panel with narrow upper white band bordered above and below by narrow black bands.

PLATE 17: PLOVERS

1. SENEGAL PLOVER *Vanellus lugubris* Page 191
Scarce, near the coast; mainly Senegal. Near cattle or game. Drier places. Quick courser-like dashes.
Adult. Plain head, no wattles or crest. White forehead. Narrow black chest band.

2. BLACK-HEADED PLOVER *Vanellus tectus* Page 190
Very common in suitable dry habitat. Small parties rest in shade by day. Nocturnal. **Adult.** Earthy-brown
above. Pointed black crest. Very small red wattles. Staring golden eye.

3. SPUR-WINGED PLOVER *Vanellus spinosus* Page 190
Very common throughout. Usually near water. Noisy and irritable. **Adult.** Mainly black below, large white
face patches, black capped. Nominal crest lies flat. Wing spurs mostly hidden.

4. WATTLED PLOVER *Vanellus senegallus* Page 189
Common throughout. Prefers watersides. **Adult.** Long yellow legs. Yellow wattles, red at base.

5. WHITE-CROWNED PLOVER *Vanellus albiceps* Page 190
Local, mostly July-November in The Gambia; breeds Senegal. Riverbanks and flooded plains. Black area
on closed wing. **5a Adult.** Beard-like wattles. **5b Adult flight.** Startling colour transformation in flight.

6. NORTHERN LAPWING *Vanellus vanellus* Page 191
Vagrant from Palearctic. More frequent northern Senegal. Most likely sandbars, coastal flats. **Adult non-
breeding.** Wispy upswept crest. Green-black above, white below. Undertail cinnamon. Broad rounded wings.

7. GREY PLOVER *Pluvialis squatarola* Page 189
Palearctic migrant; all months. Commonest September-March. Muddy places, coastal beaches, mangroves.
Black axillaries in flight. **Adult non-breeding.** Greyish. Transitional plumage commonplace (see text).

8. FORBES'S PLOVER *Charadrius forbesi* Page 187
Rare Senegambia. Dry places. No white forecrown. Broad white stripe from eye to nape. **8a Adult breeding.** Double chest band, lower broadest. **8b Immature.** Weaker markings; buffy fringes.

9. LITTLE RINGED PLOVER *Charadrius dubius* Page 186
Palearctic migrant; all seasons. Widespread, October-March. **9a Adult non-breeding.** Yellow eye-ring. **9b
Flight.** Upperwing uniformly brown. No wing-bar. **9c Immature.** Sandier, eye-ring partial.

10. RINGED PLOVER *Charadrius hiaticula* Page 186
Palearctic migrant all months. October-March peak. **10a Adult non-breeding.** Strong face mask from March
onwards. **10b Flight.** White wing-bar. **10c Immature.** Whiter on head than Little Ringed.

11. KENTISH PLOVER *Charadrius alexandrius* Page 187
Palearctic migrant and regional breeder. Wings fall equal to or just beyond tail. Finer-billed than White-
fronted. Legs usually darkish. **11a Adult male breeding.** Partial breast bands black. **11b Adult female.**
Partial breast bars brown. **11c Immature.** White collar.

12. WHITE-FRONTED PLOVER *Charadrius marginatus* Page 187
Local resident breeder along coast, rare inland. Gingery tones, most paler than Kentish and heavier-
billed. Wing tips fall just short of tail. Legs usually pale. Neck collar
buffy. **12a Adult male breeding.** Legs reddish-buff. **12b Adult female.** Head bar browner. **12c & d Immature.** Greenish-olive legs.

13. KITTLITZ'S PLOVER *Charadrius pecuarius* Page 186
Uncommon. Muddy saltpans, marshy edges, playing fields. Tallish,
sandy-yellow, no breast band. **13a Adult male breeding.** Dark crown
band reaches eye. **13b Adult female.** White neck collar and supercilium connect. **13c Immature.** Strongly dappled, long-legged.

14. GREATER SAND PLOVER *Charadrius leschenaultii* Page 188
Vagrant northern Senegal. Much bigger than Ringed Plover. **Adult
non-breeding.** Heavy robust bill. Long-legged. Underwing pale.

6

PLATE 18: CURLEWS, GODWITS AND SANDPIPERS

1. EURASIAN CURLEW *Numenius arquata* — Page 198
Seasonally common Palearctic migrant on the coast. Largest wader. **Adult** *N. a. orientalis*. Plain crown, pale brown, enormous downcurved bill (longer on female). Whiter underwing than nominate. No seasonal plumage change.

2. WHIMBREL *Numenius phaeopus* — Page 198
Very common Palearctic migrant. Beaches, mangroves, inland. **Adult.** Streaky brown, decurved bill, tip kinked, bold head markings, central crown stripe. Seven-note flight call. No seasonal plumage change.

3. BLACK-TAILED GODWIT *Limosa limosa* — Page 197
Common to locally very abundant Palearctic migrant. Muddy environs, paddies, saltmarsh. Straight pink-based bill. Stands tall. Large flocks. **3a Adult breeding.** Head and neck chestnut. **3b Adult non-breeding.** Uniform grey-brown back. **3c Adult non-breeding in flight.** Broad white wing-bar, bold black bar on pure white tail, both absent on Bar-tailed.

4. BAR-TAILED GODWIT *Limosa lapponica* — Page 197
Common to locally abundant Palearctic migrant. Subtly upswept bill. **4a Adult breeding.** Underparts all chestnut, no bars. **4b Adult non-breeding.** Streaky back. **4c Adult non-breeding in flight.** Plain wings, narrowly barred tail.

5. COMMON GREENSHANK *Tringa nebularia* — Page 199
Common Palearctic migrant. All wetland types. **Adult non-breeding.** White face. Broad-based bill subtly uptilted. Legs greenish. Calls *tew-tew-tew*.

6. MARSH SANDPIPER *Tringa stagnatilis* — Page 199
Locally common Palearctic migrant. Favours wide open water. **Adult non-breeding.** Graceful, fine black bill. Dainty feeder. Spindly legs.

7. COMMON SANDPIPER *Actitis hypoleucos* — Page 201
Common Palearctic migrant. All wetland types. **Adult non-breeding.** White shoulder mark. Short legs, low gait. Constantly bobs and wags body wagtail-like.

8. GREEN SANDPIPER *Tringa ochropus* — Page 200
Frequent to common Palearctic migrant. Watercourses, ditches, prefers sheltered places. Insignificant spotting on back. **8a Adult non-breeding.** Dark uniform olive-green back, streaky breast, very white below. Weak supercilium. **8b Adult upperside in flight.** Dark back contrasting with white rump, bold tail bars.

9. WOOD SANDPIPER *Tringa glareola* — Page 200
Common, sometimes abundant, Palearctic migrant throughout. Near cover, most wetland types. **Adult non-breeding.** Well spotted back. Leggier than Green Sandpiper. Strong supercilium.

10. SPOTTED REDSHANK *Tringa erythropus* — Page 198
Uncommon Palearctic migrant. Bill long and straight. **Adult non-breeding.** Dark line through eye, strong white supercilium. Graceful. Ashy-grey. Legs orange-red. Breeding plumage (not illustrated) is sooty black, spotted white.

11. COMMON REDSHANK *Tringa totanus* — Page 199
Very common Palearctic migrant. All wetlands, beaches. **Adult non-breeding.** Mottled brown above, whitish faintly streaked below. Eye-ring. Red legs, orange-yellow on immature.

PLATE 19: MISCELLANEOUS WADERS

1. EURASIAN OYSTERCATCHER *Haematopus ostralegus*   Page 180
Coastal beaches, lagoons, near mangroves. All months, peaks October-April. **Adult non-breeding.** Black and white, chunky stature, long sturdy orange bill. Shortish red legs.

2. PIED AVOCET *Recurvirostra avosetta*   Page 181
Frequent to common. All seasons, peaks October-March. Muddy places. Fine upswept bill. Characteristic feeding manner sweeping surface of water. **Adult.** Graceful. Lead-blue legs.

3. BLACK-WINGED STILT *Himantopus himantopus*   Page 181
Common to locally abundant, all divisions, all months, peaks October-April. Elegant. Extremely long pink legs. Fine straight bill. **3a Adult female.** All-white head. **3b First-winter** Variable dusky markings.

4. RUFF *Philomachus pugnax*   Page 195
Frequent to locally abundant, all seasons, peaks November-March. Shortish bill. Neck thick, long when stretched. **4a Male partial breeding,** but most are like female. **4b Female non-breeding.** Smaller than male.

5. GREY PHALAROPE *Phalaropus fulicaria*   Page 202
Rare. Bill broad, thickish. Sits high on water. Spins. **Adult non-breeding.** Dark mark through eye.

6. JACK SNIPE *Lymnocryptes minimus*   Page 196
Rare to uncommon, few records November-February. Very furtive. **Adult.** Small snipe, medium-length straight bill. Double pale supercilium.

7. COMMON SNIPE *Gallinago gallinago*   Page 196
Most frequent snipe. October-March, all divisions. Most wetlands with cover. **Adult.** Very long straight bill.

8. GREAT SNIPE *Gallinago media*   Page 197
Rare to uncommon; marshy ground, rice fields, inland. **Adult.** Bulky, deep-chested body. Barred belly. Long stout bill.

9. RUDDY TURNSTONE *Arenaria interpres*   Page 201
Common to very abundant at coast, some inland upriver. All months, peaks October-March. **Adult.** Smallish, thickset. Short legs orange. Tortoiseshell back, white below.

10. RED KNOT *Calidris canutus*   Page 192
Frequent in small numbers on the coast, all seasons. **Adult non-breeding.** Stout straight bill. Greyish back. White supercilium.

11. TEMMINCK'S STINT *Calidris temminckii*   Page 193
Rare to uncommon **Adult non-breeding.** Diffuse partial breast band. Legs, feet, toes greenish. Plain back. Bobs when feeding.

12. LITTLE STINT *Calidris minuta*   Page 193
Frequent to very abundant, all divisions, all months. Shallow open muddy or marshy places. Highly gregarious. **Adult non-breeding.** Indistinct breast markings. Dark legs.

13. SANDERLING *Calidris alba*   Page 192
Common to very abundant on the coast, some inland upriver. All months. Rapid feeder. Short black bill. **13a Adult partial breeding.** Infrequent, admixed rufous. **13b Adult non-breeding.** Greyish-white. Black legs.

14. DUNLIN *Calidris alpina*   Page 194
Frequent to locally common in small numbers. All months. Dumpy, smaller than Red Knot. Subtly downcurved bill, less than Curlew Sandpiper. **14a Adult partial breeding.** Patchy black belly. **14b Adult non-breeding.** Weak supercilium. Greyish-brown above.

15. CURLEW SANDPIPER *Calidris ferruginea*   Page 194
Common to locally very abundant. All months. Bill longish, evenly decurved. Taller than less common Dunlin. **15a Adult partial breeding.** Blotchy orange. **15b Adult non-breeding.** Long white supercilium.

PLATE 20: SKUAS AND GULLS

1. POMARINE SKUA *Stercorarius pomarinus*   Page 203
Locally common October-March. Plumages very variable. Bulky body, thick, pale-based bill. **1a Adult non-breeding**, pale phase **1b Adult**, dark phase. Tail projection long and spoon-shaped, often absent. **1c Juvenile**. Most regular type seen. Barred rump/uppertail-coverts more black and white than buffier Arctic. Shape, jizz important.

2. ARCTIC SKUA *Stercorarius parasiticus*   Page 203
Locally common October-March. Plumages very variable. Slimmer, lighter than Pomarine, more tern-like. **2a Adult non-breeding**, pale phase. **2b Adult**, dark phase. Tail extensions spiky, often absent. **2c Juvenile**. Most regular type seen. Buffy tones, not pale-bellied. Shape, jizz important. (Any lightweight, small-headed, pale-bellied skua, see Long-tailed Skua text). **2d Adult breeding** (pale phase) chasing tern.

3. GREAT SKUA *Catharacta skua*   Page 204
Rare off northern Senegal. **Adult**. Huge, all-brown, large white patches in outer wings, wedge-shaped tail short. Body barrel-shaped. See text for distinction from South Polar Skua.

4. AUDOUIN'S GULL *Larus audouinii*   Page 209
Locally frequent migrant on coast, November-March. Pearly-grey and white gull, heavyish angular tri-coloured bill. **4a Adult breeding**. White head, coral-red bill, subterminal black band with yellow tip, legs grey. **4b Second-winter**. Primaries all black, sloping head. Grey legs are best distinction.

5. MEDITERRANEAN GULL *Larus melanocephalus*   Page 204
Rare Palearctic migrant, coast, November-February. Very white-winged, thick scarlet, bulbous, droop-tipped bill (beware Nearctic vagrants, see text). **5a Adult breeding**. Jet black hood extends well down neck, thick broken white eye-crescents. **5b Adult non-breeding**. White primary tips. **5c Second-winter**. Black marks on white wing-tips.

6. LITTLE GULL *Larus minutus*   Page 206
Rare Palearctic vagrant. Diminutive, graceful tern-like flight, rounded wings, short legs. Smallest gull. Immature (not illustrated) has dark zig-zag band on wings, see text. **6a Adult non-breeding**. Wings pale above, dark underwings. Dark hood when breeding. **6b Second-winter**. Dark primary tips.

7. BLACK-HEADED GULL *Larus ridibundus*   Page 207
Common all seasons, most November-April. Most likely dark-hooded gull. **7a Adult breeding**. Chocolate hood, nape and neck white. **7b Adult non-breeding**. Dark smudge behind eye. **7c Flight** Underwing whitish and pale grey, white leading edge.

8. GREY-HEADED GULL *Larus cirrocephalus*   Page 207
Abundant on coast, some upriver. Commonest gull. **8a Adult breeding**. Pale lavender-grey hood. **8b Second-winter**. Darker grey back than Black-headed, pink bill dark-tipped. **8c Flight** Underwing dark grey, black wing-tips, prominent white mirrors.

9. SLENDER-BILLED GULL *Larus genei*   Page 208
Common to abundant on coast, peaks January-July. Very white gull, long bill and sloping forehead. **9a Adult breeding**. Pure white head, rosy tints below. **9b First-winter**. Smudgy ear spot, bill orange, legs pale. **9c Flight**. Arched backed, protracted neck, long bill, underwing recalls Black-headed Gull.

PLATE 21: GULLS AND TERNS

1. COMMON GULL *Larus canus*   Page 210
Rare vagrant from Palearctic. Gentle looks. Medium-sized. Weak-looking greenish bill, may be ringed at tip. **Adult non-breeding.** Faint streaks on neat, round head.

2. KELP GULL *Larus dominicanus*   Page 210
Small numbers on the coast, e.g. Tanji Bird Reserve. Largest, darkest-backed gull. **Adult.** Heavier-billed than Lesser Black-backed Gull. Shorter wing projection. Olive legs.

3. YELLOW-LEGGED GULL *Larus cachinnans*   Page 211
Palearctic migrant. Likely October-March. Medium-grey, paler then Lesser Black-backed Gull. **3a Adult.** Head unstreaked. Larger than Lesser Black-backed. **3b Second-winter.** Develops mid-grey saddle. Clean-cut tail band.

4. LESSER BLACK-BACKED GULL *Larus fuscus*   Page 211
Common to very abundant Palearctic migrant, peaks October-March. Upperparts slate on *graellsii* to black on *fuscus*. **4a Adult non-breeding.** Streaky head. **4b Second-winter.** Becoming darker with age. **4c First-winter.** Browner tertials than immature Yellow-legged which is overall paler grey-brown, with less pronounced dark tail band.

5. CASPIAN TERN *Sterna caspia*   Page 213
Abundant most months on coast, frequent upriver. Largest tern, bigger than many gulls. Huge scarlet red bill, big head. **5a Adult breeding.** Glossy black cap, short cut-off crest. **5b Adult non-breeding.** Head grey with streaks, sometimes white. **5c Juvenile.** Scaly, bill paler. **5d Flight.** Strong powerful, thick neck. Dark wedge on primaries.

6. ROYAL TERN *Sterna maxima*   Page 213
Abundant most months on coast, common upriver. Commonest big tern. Less heavy than Caspian. Dagger-like bill (size varies). **6a Adult breeding.** Slick black crown, shaggy short nape feathers, bill very orange. **6b Adult non-breeding.** Bill more yellow-orange. White forecrown. **6c Juvenile.** Bill yellower. Scalloped. **6d Flight.** White tail and rump.

7. LESSER CRESTED TERN *Sterna bengalensis*   Page 214
Frequent to common on coast, much reduced May-September. Bill yellow or orange, thinner, straighter, less droopy than Royal. Body structure comparable to Sandwich. **7a Adult breeding.** Crest shaggier, more extensive than Royal. **7b Adult non-breeding.** Bill yellow-orange. **7c Juvenile.** Yellow bill, blackish shoulder bar. **7d Flight.** Grey rump and tail.

8. SANDWICH TERN *Sterna sandvicensis*   Page 214
Common to very abundant migrant on coast from Palearctic, all months, peaks November-March. Medium-large tern. Pale yellow-white tip on narrow pointed all-black bill. **8a Adult breeding.** Shaggy crest. **8b Adult non-breeding.** Note bill colour. **8c Juvenile.** Mottled. May lack pale bill-tip. **8d Flight.** Medium-length forked tail white, rump white.

PLATE 22: TERNS AND SKIMMER

1. GULL-BILLED TERN *Gelochelidon nilotica*   Page 212
Frequent at coast, less so upriver, all months, peak September-March. All-black, stubby gull-like bill. Very pale. Forages by quartering, often alone. **1a Adult non-breeding**. Slight gonys separates immature Sandwich. Head sometimes all white. **1b Adult breeding**, in flight. Glossy black cap to nape. No streamers.

2. ROSEATE TERN *Sterna dougallii*   Page 214
Rare to uncommon coastal Senegambia July/September/April. More pelagic than other terns. Passage migrant from Palearctic. Very pale, whiter than Arctic and Common Terns. Broad translucent area along secondaries and inner primaries. **2a Adult non-breeding**. Slender, spiky bill. Tail projecting clearly beyond wings. **2b Juvenile**. Forehead not as clean white as Arctic and Common. Beware immature Sandwich. **2c Adult breeding**, in flight. Very long streamers. Rosy tints below.

3. ARCTIC TERN *Sterna paradisaea*   Page 215
Occasionally reported Palearctic passage migrant. Coastal, most August-September. Very short-legged, short-necked. Darker underwing than Roseate. **3a Adult non-breeding**. Short legs. **3b Juvenile**. Less conspicuous dusky shoulder than Common and Roseate. **3c Adult breeding**, in flight. Streamers longer than Common, shorter than Roseate.

4. COMMON TERN *Sterna hirundo*   Page 215
Frequent to abundant on coast November-July, reduced numbers in rains. Palearctic visitor and regional breeder. Darker shoulder bar than Arctic. Only inner webs of outer primaries translucent. **4a Adult non-breeding**. Longer legs than Arctic. **4b Juvenile**. More white above gape than Arctic. **4c Adult breeding**, in flight. Streamers shorter than Arctic; blacker leading edge.

5. WHISKERED TERN *Chlidonias hybridus*   Page 217
Uncommon to frequent on coast and inland. Sporadic visitor from Palearctic. Singly, small parties. Heaviest-bodied, longest-billed marsh-tern. **5a Adult breeding**. Dark cap. White face stripe. Lead-grey body. **5b Adult non-breeding**. No black patch behind eye, no dark shoulder mark. **5c Juvenile**. Brown, scalloped back. **5d Adult non-breeding**, in flight. Uniform grey back and tail.

6. WHITE-WINGED BLACK TERN *Chlidonias leucopterus*   Page 217
Frequent to abundant Palearctic passage migrant. All seasons, most March-May. Transitional plumages common. **6a Adult non-breeding**. Eye patch more than Whiskered, less than Black. **6b Juvenile**. Shorter bill than Black Tern. **6c Adult breeding**, in flight. Jet black underwing-coverts contrasts with silvery-white flight feathers. **6d Adult non-breeding**, in flight. No black smudge on side of breast.

7. BLACK TERN *Chlidonias niger*   Page 217
Abundant to very abundant. All seasons, most June-October. Transitional plumages common. **7a Adult non-breeding**. Eye patch more pronounced than White-winged Black. **7b Juvenile**. Dark smudge on sides of breast. **7c Adult breeding**, in flight. Jet black body contrasts with dove-grey underwings. **7d Adult non-breeding**, in flight. Black smudge on side of breast diagnostic.

8. LITTLE TERN *Sterna albifrons*   Page 216
Frequent to common on coast or upriver. Palearctic migrant and regional breeder. Smallest tern. **8a Adult non-breeding**. Extensive white forehead, black streak from eye to nape. **8b Juvenile**. Black shoulder. No obvious patch beyond eye. **8c Adult breeding**, in flight. White narrow V mark over eye.

9. AFRICAN SKIMMER *Rynchops flavirostris*   Page 218
Uncommon to rare, coast or upriver. Breeds Senegal. Large black and white tern-like bird with curious bill structure. **9a Adult**. Different mandible lengths. Immature (not illustrated) mottled. **9b Adult fishing**. Trawls with V-postured wings.

PLATE 23: PIGEONS AND DOVES

1. LAUGHING DOVE *Streptopelia senegalensis*   Page 225
Common throughout, mostly in pairs. Smallish. **Adult.** Rusty above with blue-grey in wing. Necklace on foreneck, hindneck plain.

2. EUROPEAN TURTLE DOVE *Streptopelia turtur*   Page 224
Palearctic migrant. Gregarious; vast flocks inland. **Adult.** Heavily mottled, chequered wings. Black and white patch on side of hindneck.

3. BLUE-SPOTTED WOOD DOVE *Turtur afer*   Page 221
Common in well wooded areas, mostly coastal. **3a Adult.** Earthy-brown tones. More demarcated blue cap than Black-billed Wood Dove. Colour of wing spots difficult to see. Red bill, pale-tipped. **3b Flight.** Chestnut axillaries, barred back.

4. BLACK-BILLED WOOD DOVE *Turtur abyssinicus*   Page 222
Common and widespread throughout. Greyer plumage. **Adult.** Head less blue than Blue-spotted. Black bill, black lore line.

5. SPECKLED PIGEON *Columba guinea*   Page 222
Common throughout. **5a Adult.** Extensive bare red skin patches. Heavy looking. Wings peppered with white spots. **5b Flight.** Pale rump.

6. BRUCE'S GREEN PIGEON *Treron waalia*   Page 221
Common inland. Wide use of habitats including dry woodlands and mangroves. **Adult.** Belly lime-yellow. Head ashy-grey.

7. AFRICAN GREEN PIGEON *Treron calva*   Page 220
Frequent to common, mainly near coast. Flocks to fruiting figs. **Adult.** Fluffy white feathering on thighs. Uniform green underparts.

8. NAMAQUA DOVE *Oena capensis*   Page 222
Frequent to seasonally very abundant throughout, largest numbers December-January. Only dove with long thin tail, sometimes absent. Underwing rufous. **8a Adult male.** Black mask. **8b Adult female.** No mask.

9. TAMBOURINE DOVE *Turtur tympanistria*   Page 221
Casamance only. Forest streams. **Adult male.** Chocolate above, white below. Female greyer below.

10. RED-EYED DOVE *Streptopelia semitorquata*   Page 223
Common to very abundant throughout. Most habitats. Largish. Black collar indistinctly fringed white. Pale face and crown contrasts with darkish brown-grey nondescript body. **10a Adult.** Narrow bare skin around eye claret-red. **10b Immature.** Scalloped. **10c Tail upside.** No white.

11. AFRICAN MOURNING DOVE *Streptopelia decipiens*   Page 223
Common to abundant. Associated with water. Smaller, paler grey than Red-eyed Dove. Black collar clearly fringed white. **11a Adult.** Eye piercing, yellow with red orbital ring. **11b Tail upside.** Outer feathers broadly tipped white.

12. AFRICAN COLLARED DOVE *Streptopelia roseogrisea*   Page 224
Uncommon in The Gambia. Commoner in northern Senegal. Desert edges. Gregarious. **12a Adult.** Very pale. **12b Flight.** Much paler than Vinaceous, especially underwing. **12c Tail upside.** Much white.

13. VINACEOUS DOVE *Streptopelia vinacea*   Page 224
Abundant throughout most habitats. Smallest black-collared dove. **13a Adult.** Earthy grey-brown, tinged vinous below. **13b Flight.** Underwing dark. **13c Tail upside.** Broadly tipped white with black base.

14. ADAMAWA TURTLE DOVE *Streptopelia hypopyrrha*   Page 225
First reported 1990. Very local, riverbank forest and mango plantations. **Adult.** Very dark earth-brown above strongly scalloped, pink-cinnamon belly. White sides of face contrast with grey head. Solid black half-collar.

14

PLATE 24: CUCKOOS, COUCALS AND PIAPIAC

1. PIAPIAC *Ptilostomus afer* Page 346
Common throughout. Gregarious. **1a Adult**. All black. Long stiff tail (no white), pale patch in open wing. Long gangly legs. Black bill. **1b Immature**. Pink bill.

2. YELLOWBILL *Ceuthmochares aereus* Page 233
Rare to uncommon in a few coastal forests. Blue-grey. Squirrel-like movements, curious calls. **Adult**. Large yellow bill. Immature (not illustrated) has dusky bill.

3. GREAT SPOTTED CUCKOO *Clamator glandarius* Page 230
Frequent to uncommon, all seasons, peaks June-July and November-March. Palearctic and African populations. Wooded areas, drainage channels. Long-tailed, spotted-winged in all plumages. Parasitises Pied Crow. **3a Adult**. Bushy greyish crest. **3b Immature**. Black head, chestnut primaries.

4. BLACK COUCAL *Centropus grillii* Page 234
Very locally frequent to common in rains, well inland. Rank grassland with trees. **4a Adult**. Black below, rufous-brown above. **4b Immature**. Heavily barred and streaked.

5. SENEGAL COUCAL *Centropus senegalensis* Page 235
Common to abundant throughout. Most habitats. Creeps on ground. **5a Adult**. Creamy-white below, black head. Eye ruby-red. **5b Immature**. Duskier than adult with delicate barring.

6. BLACK-THROATED COUCAL *Centropus leucogaster* Page 234
Casamance forest only. Large. Shy. Sluggish flight. **Adult**. Black chest, white lower breast.

7. RED-CHESTED CUCKOO *Cuculus solitarius* Page 230
Uncommon to frequent in the rains. **Adult**. Chestnut-red breast-band. Female bars denser.

8. LEVAILLANT'S CUCKOO *Clamator levaillantii* Page 229
Common throughout. May-November. Well wooded areas. Spiky black crest, black wings, single white primary bar. **8a Adult**. Streaky throat. **8b Flight**. White terminal tail spots.

9. JACOBIN CUCKOO *Clamator jacobinus* Page 229
Rare to uncommon. Dry woodland, grassy marsh. **Adult**. Pure white below. Crested. Slim. Smaller than Levaillant's. Immature (not illustrated) is buffy below, not streaky.

10. AFRICAN CUCKOO *Cuculus gularis* Page 232
Frequent to common May-November. Commonest grey-backed cuckoo. Hawk-like. **10a & b Adult**. Yellowish-orange eye, prominent bare pale orange skin sometimes extends from gape to around eye. Bill mainly yellow tipped black,

11. COMMON CUCKOO *Cuculus canorus* Page 231
Rare. Best known in northern Senegal. Palearctic migrant. Normally silent. **11a & b Adult**. Eye yellow with neat orange-yellow orbital ring. Bill slighter than African, blackish-horn with yellow base.

12. KLAAS'S CUCKOO *Chrysococcyx klaas* Page 232
Common in the rains, present year-round in coastal thickets. **12a Adult male**. Single white mark behind eye. Clean white outer tail. Green half-collar patches. **12b Adult female**. Variable, ghosts male. Immature (not illustrated) see text.

13. AFRICAN EMERALD CUCKOO *Chrysococcyx cupreus* Page 232
Rare in The Gambia and Casamance forest, in the rains. **13a Adult male**. Emerald-green above, sulphur-yellow below. **13b Adult female**. Very heavily barred.

14. DIEDERIK CUCKOO *Chrysococcyx caprius* Page 233
Common in the rains, absent dry season. **14a Adult male**. Well marked head, blotchy wings. Spotted white outer tail. Moustachial streak. **14b Adult female**. Lacks white crown mark. **14c Immature**. Coral-red bill.

PLATE 25: OWLS

1. VERREAUX'S EAGLE OWL *Bubo lacteus*   Page 237
Local throughout. Tall trees in thickets, dry savanna and riverside habitats. Huge. **Adult.** Grey-brown above, milky-grey below. Pink eyelids. Immature (not illustrated) lacks tufts.

2. AFRICAN WOOD OWL *Strix woodfordii*   Page 239
Rare. A few pairs in coastal forest thickets. **Adult.** Medium-sized. Brown, heavily barred and spotted. No ear-tufts. Dark eyes.

3. SPOTTED EAGLE OWL *Bubo africanus*   Page 237
Uncommon; dry savanna woodland. **Adult.** Large. Grey-brown. Horned. Eyes brown-black.

4. AFRICAN SCOPS OWL *Otus senegalensis*   Page 236
Frequent to common throughout. Well-wooded country. Small. **4a Adult.** Grey plumage, lichen-like. Tufts sometimes relaxed. Grey and brown forms. **4b Adult.** Body erect posture. See text for rare European Scops Owl.

5. BARN OWL *Tyto alba*   Page 235
Common throughout in diverse habitats. **Adult.** Golden-buff, white belly speckled dark, sometimes darker. Screeches. No ear-tufts. Heart-shaped face.

6. WHITE-FACED SCOPS OWL *Otus leucotis*   Page 236
Common to frequent throughout. Favours acacias. Small. **6a Adult.** Distinctive white facial disc. Staring orange eyes. Prominent tufts. White lines down back. **6b Adult.** Body erect posture.

7. PEARL-SPOTTED OWLET *Glaucidium perlatum*   Page 238
Common to frequent throughout. Wooded areas. Active by day. Mobile tail. **Adult.** Chocolate coloration. Two 'eye-spots' on nape when head turned.

8. SHORT-EARED OWL *Asio flammeus*   Page 239
Rare in The Gambia (inland) and coastal northern Senegal, January-March. **8a Adult.** Pale sandy-tawny, heavily streaked. Lemon-yellow eyes. Small ear-tufts. **8b Flight.** Pale underwing with conspicuous carpal patches.

9. MARSH OWL *Asio capensis*   Page 239
Very local in The Gambia and Casamance. Small migrant population to the coast, August-December. **9a Adult.** Uniform brown. Pale buffy facial discs, darker around eyes. **9b Flight.** Contrasting buffy patches on upperwing.

10. PEL'S FISHING OWL *Scotopelia peli*   Page 238
A few sites along rivers. Fresh and saline waters. Dense tall riparian forest and mangroves. **Adult.** Large, awesome. Ginger coloration unique. Bonnet-headed. Large liquid eyes.

PLATE 26: NIGHTJARS

1. EUROPEAN NIGHTJAR *Caprimulgus europaeus* Page 243

Palearctic migrant, October-February, but no recent records. Large, darker than Red-necked. **1a Adult male**. Dead-leaf pattern. Heavily streaked grey. **1b & c Male**, wing and tail. White spots on outer three primaries, on both webs. White outer tail, quarter of length. **1d & e Female**, wing and tail: no white. **1f Flight silhouette**. Male.

2. RED-NECKED NIGHTJAR *Caprimulgus ruficollis* Page 242

Palearctic migrant. **2a Adult male**. Large. Big-headed, broad rusty hind collar. Flight hawk-like. **2b & c Male**. Wing spots (more prominent on outer webs) and tail patches white. **2d & e Female**. Wing spots and tail patches buffy.

3. LONG-TAILED NIGHTJAR *Caprimulgus climacurus* Page 240

Locally common all divisions, all months. **3a Adult male breeding**. Slim, small-bodied, elongated. Tail sometimes broken. Three white or buffy areas in open wing, including trailing edge. Female shorter-tailed. **3b Flight silhouette**.

4. PENNANT-WINGED NIGHTJAR *Macrodipteryx vexillarius* Page 244

Unverified, breeds south of equator. Non-breeding male (not illustrated) has large white bar on blackish wing. **4a Female**. Widely spaced, rusty and dark wing barring. Recalls Standard-winged but larger. **4b Flight silhouette**. Female. Largish, all dark. No white. Compare wing-tip shape to 5d.

5. STANDARD-WINGED NIGHTJAR *Macrodipteryx longipennis* Page 243

Locally common throughout, much reduced March-May. Most frequent nightjar after Long-tailed. No white in tail or wing in either sex. **5a Adult male breeding**. Unmistakable. Various stages of standard-emergence, November-February. **5b Female**. Short, square tail. Narrow tawny collar. **5c Male flight silhouette**. Unique trailing bats. **5d Female flight silhouette**. Smallish.

6. BLACK-SHOULDERED NIGHTJAR *Caprimulgus nigriscapularis* Page 241

Status unverified. Forest edges. **6a Adult male**. Rich tawny-brown. Spots on five primaries. Tail rufous barred black. Blackish-brown 'shoulder' line. **6b & c Male**. Wing spots and tail patches white. **6d & e Female**. Wing spots and tail patches buffy.

7. SWAMP NIGHTJAR *Caprimulgus natalensis* Page 240

Status uncertain. Freshwater swamps. **7a Adult male**. Squat, short-tailed, dusky. Spotted appearance. **7b & c Male**. Wing spots and tail patches extensive, white. **7d & e Female**. Wing spots and tail patches buffy.

8. PLAIN NIGHTJAR *Caprimulgus inornatus* Page 241

Erratic visitor to Senegambia, variable ground colour. **8a Adult male, pale form** (richer tawny-brown individuals frequent). Pin-head speckles, especially on crown. **8b & c Male**. White wing bar across darkish outer primaries, outer two tail feathers a third white. **8d & e Female**. Wing spots tawny-buff. Tail diffusely barred. See text for Egyptian Nightjar.

5a

PLATE 27: SWIFTS, MOUSEBIRD, WOOD HOOPOES AND HOOPOE

1. WHITE-RUMPED SWIFT *Apus caffer* — Page 246
Local in rains. **Adult upperside**. White rump, forked tail. Sleek.

2. ALPINE SWIFT *Apus melba* — Page 246
Rare, Palearctic migrant to Senegambia. Big and impressive. Languid flight action. **Adult underside**. White throat and breast separated by brown band.

3. PALLID SWIFT *Apus pallidus* — Page 245
Palearctic passage migrant. Gregarious. **3a Adult underside**. Milky-brown. Body colour similar to Palm Swift. Extensive pale throat patch. Blunter wing tips than Common with which it mixes. **3b Adult upperside**. Darker outer primaries contrast with paler wing.

4. COMMON SWIFT *Apus apus* — Page 245
Palearctic passage migrant. Gregarious. **4a Adult underside**. Sooty with pale throat. Tail forked when opened. **4b Adult upperside**. Uniform. Immature (not illustrated) is very dark with contrasting white throat.

5. LITTLE SWIFT *Apus affinis* — Page 246
Common throughout, seasonally abundant. Noisy flocks. Commonest white-rumped swift. **5a Adult upperside**. Short square tail. square white rump. Uniformly dark below. **5b Adult underside**. White throat patch.

6. MOTTLED SPINETAIL *Telacanthura ussheri* — Page 244
Frequent to locally common all divisions. Over forests and baobab plantations. Bat-like fluttering flight. Silent. **6a Adult underside**. Diagnostic cylindrical white body-band at rear. **6b Adult upperside**. Notch in secondaries.

7. AFRICAN PALM SWIFT *Cypsiurus parvus* — Page 245
Common to abundant throughout, in vicinity of palms. **7a Adult underside**. Tapered, frail, pencil-thin body. Long tail. Mouse-brown. **7b Flight silhouette**.

8. BLUE-NAPED MOUSEBIRD *Urocolius macrourus* — Page 247
Locally common northern Senegal. **Adult**. Short-cropped crest, long tapered stiff tail. Sociable. Low, direct flight. Facial skin red. Blue nape patch absent in immature.

9. BLACK WOOD HOOPOE *Rhinopomastus aterrimus* — Page 257
Frequent to locally common. Dry woodlands, thicket edges, locust-bean trees. Joins mixed feeding parties. Black Flycatcher-size. **Adult**. Black, violet sheen.White wing-bar in flight.

10. GREEN WOOD HOOPOE *Phoeniculus purpureus* — Page 257
Common to locally abundant throughout. Diverse habitats. Noisy, sociable groups. Clambers. Probes on trunks and ground. **10a Adult**. Long tapered tail edged white, white in wings. Red bill. Female shorter-billed than male. **10b Immature**. Black bill.

11. HOOPOE *Upupa epops* — Page 258
Frequent all divisions, all seasons. Palearctic and local populations. Open habitats. Striking orange, black and white bird. Erectile fan-like crest. See text for separation of races. **11a Adult**. Flattened crest. **11b Flight**. Crest raised.

PLATE 28: KINGFISHERS AND ROLLERS

1. GIANT KINGFISHER *Megaceryle maxima*   Page 250
Local. Coastal inlets; streams, ponds. Shaggy crest. **1a Adult male.** Breast
coppery-brown. **1b Adult female.** Breast white densely marked black.

2. WOODLAND KINGFISHER *Halcyon senegalensis*   Page 248
Rains associated, common. Woodlands. **Adult.** Mantle and scapulars all blue
and no black wedge behind eye. Largely insectivorous. Bicoloured bill.

3. BLUE-BREASTED KINGFISHER *Halcyon malimbica*   Page 248
Frequent to locally abundant. Usually by water. All divisions, all months. **Adult.**
Closed wing extensively black. Black mask behind eye. Ominvorous. Bicoloured
bill, upswept lower mandible 'shoveled'.

4. STRIPED KINGFISHER *Halcyon chelicuti*   Page 248
Common in all divsions. Open wooded country. Avoids water. **Adult.** Dull, unglossy. Turquoise
rump. Sexes differ on underwing pattern.

5. GREY-HEADED KINGFISIIER *Halcyon leucocephala*   Page 247
Frequent to seasonally common, recorded all seasons. Mixed woodland near water preferred. **Adult.** Belly warm chestnut. Immature
(not illustrated) lacks chestnut below.

6. PIED KINGFISHER *Ceryle rudis*   Page 251
Common to abundant in variety of habitats throughout. Sociable. Black
and white. Forages by hovering or diving from perch. **6a Adult male.** Double breast-bands. **6b Adult female.** Single broken band.

7. AFRICAN PYGMY KINGFISHER *Ceyx picta*   Page 249
Common to frequent resident in forest-thickets; more during rains throughout.
Insectivorous. **Adult.** Eyebrow orange. Flat head no crest.

8. MALACHITE KINGFISHER *Alcedo cristata*   Page 249
Widespread in aquatic places. Wet season increase. **8a Adult.** Eyebrow blue. Crest green-blue and
black, long, erectile. **8b Immature.** Black bill. Mantle and wing-coverts blackish spangled blue with
white markings.

9. SHINING-BLUE KINGFISHER *Alcedo quadribrachys*   Page 250
Rare. Wet season sightings most years July-September. Aquatic. **Adult.** Stunning luminous
purple-blue above, dark chestnut below. Long shiny black bill. Uncrested.

10. BROAD-BILLED ROLLER *Eurystomus glaucurus*   Page 256
Common to locally abundant in rains, widespread. Much reduced in dry season. Aerial feeder, most active
dawn and dusk. **Adult.** Wide yellow bill. Chocolate back, lilac front.

11. BLUE-BELLIED ROLLER *Coracias cyanogaster*   Page 255
Common throughout. Favours damp places with oil-palms. **Adult.** Dark blue, buff head and breast. Tail
wires. Turquoise and dark blue on wings.

12. RUFOUS-CROWNED ROLLER *Coracias naevia*   Page 255
Common in drier places throughout. **Adult.** Pinkish-brown flecked white. Broad white supercilium. White
nape patch. Square-tailed. Immature (not illustrated) is less pink.

13. ABYSSINIAN ROLLER *Coracias abyssiniica*   Page 255
Common to abundant, all divisions, dry months. Open country. Long tail streamers often absent in rains.
Adult. Purple-blue (not black) flight feathers in all plumages. Whiter-faced than infrequent European Roller.

14. EUROPEAN ROLLER *Coracias garrulus*   Page 256
Uncommon to rare Palearctic migrant in Senegambia. No streamers in any plumage. Thickset. Usually
singly. **Adult.** Face bluer than Abyssinian. Black (not purple-blue) primaries.

70

PLATE 29: BEE-EATERS, PARROTS AND TURACOS

1. SWALLOW-TAILED BEE-EATER *Merops hirundineus*   Page 251
Locally common. Sedentary. Near thickets, well wooded dry savanna. Singles or family groups. **1a Adult.**
Forked tail. **1b Immature.** Dull. Bluish tail, blunter than adult.

2. NORTHERN CARMINE BEE-EATER *Merops nubicus*   Page 254
Frequent to common. Near river, mangroves (not coastal), dry grassland. Large, long-tailed. **Adult.** Red
and blue plumage. Black mask. Facial hood greenish-blue.

3. BLUE-CHEEKED BEE-EATER *Merops persicus*   Page 253
Seasonally common, mainly coastal districts near mangroves, mostly in dry months. No gorget. Green
below. **3a Adult,** with full tail extensions, often absent. Immature (not illustrated) is duller, scaly, with
short tail. **3b Flight underside.** Coppery-russet underwing.

4. EUROPEAN BEE-EATER *Merops apiaster*   Page 253
Seasonal Palearctic migrant, records all months, peaks November-January. Favours dry savanna woodlands. **4a Adult.** Black gorget, sulphur-yellow throat. Turquoise-blue below. Short tail extensions. **4b Flight
underside.** Broad black trailing edge.

5. LITTLE BEE-EATER *Merops pusillus*   Page 251
Most widespread bee-eater. Confiding. No tail streamers. **5a Adult.** Yellow throat, black gorget, cinnamon
upper breast. **5b Immature.** Duller, no gorget.

6. WHITE-THROATED BEE-EATER *Merops albicollis*   Page 252
Small numbers any divison, all seasons. Distinctive pied head and face. **Adult.** Very long streamers are
often absent. Immature (not illustrated) has yellow wash on face.

7. RED-THROATED BEE-EATER *Merops bullocki*   Page 252
Common in all seasons, well inland. Highly gregarious. **Adult.** Scarlet-red throat, buffy breast and belly.
No tail streamers.

8. LITTLE GREEN BEE-EATER *Merops orientalis*   Page 253
Locally common, well inland. **8a Adult.** Small and slim with long streamers. Glossy lime-green. **8b Immature.**
Streamers lacking or emerging.

9. ROSE-RINGED PARAKEET *Psittacula krameri*   Page 227
Common. Sedentary pairs or mobile flocks. Large hooked bill. **Adult male.** All-green with black chin and
pink collar. Female (not illustrated) lacks collar.

10. SENEGAL PARROT *Poicephalus senegalus*   Page 226
Common. **Adult.** Yellow underparts. Short tail. Duller immature (not illustrated) has browner head.

11. BROWN-NECKED PARROT *Poicephalus robustus*   Page 226
Locally frequent. Dry woodland, mangrove forest. Large top-heavy head and bill. Pairs or small parties.
Unapproachable. **11a Adult male.** Silvery-grey head. **11b Adult female.** Red forehead.

12. VIOLET TURACO *Musophaga violacea*   Page 228
Frequent to locally common. Forests, drainage channels with big fruit trees, visits mangroves. **12a Adult.**
Yellow shield over red bill. **12b Flight.** Stunning crimson discs in dark wings.

13. WESTERN GREY PLANTAIN-EATER *Crinifer piscator*   Page 228
Common in most habitats. **Adult.** Grey, streaky, shaggy erectile crest. Large conical yellow-green bill.
Immature (not illustrated) has black woolly head.

14. GREEN TURACO *Tauraco persa*   Page 227
Local in The Gambia and southern Senegal; coastal forests. **14a Adult.** Grass green. Red in wings. Immature
(not illustrated) very dull, crest small. **14b Flight.** Arrow-headed crest, weaker-necked than Violet Turaco.

PLATE 30: TINKERBIRDS, BARBETS AND HORNBILLS

1. YELLOW-RUMPED TINKERBIRD *Pogoniulus bilineatus*   Page 261
Local. Near forest. Calls *tonk-tonk-tonk-tonk* delivered in short phrases with pauses. Small. **Adult.** Well-defined black and white facial markings.

2. YELLOW-FRONTED TINKERBIRD *Pogoniulus chrysoconus*   Page 262
Common all divisons. Wooded habitats. Calls repetitious *tink-tink-tink-tink* with about 100 notes per minute. Very small. **Adult.** Yellow-golden circular crown spot

3. RED-RUMPED TINKERBIRD *Pogoniulus atroflavus*   Page 261
Casamance coastal forest. Calls lower pitched than Yellow-fronted. A frequent *oop... oop... oop* in long series, or nasal *onk... onk... onk,* also trills *peet-peet-peet.* **Adult.** Scarlet-red rump.

4. YELLOW-THROATED TINKERBIRD *Pogoniulus subsulphureus*   Page 261
Casamance forest. Calls *pyop*, much faster in tempo, sharper quality, urgent delivery. Could be likened to a rapid burst of the popping call of Yellow-fronted. **Adult.** Yellow wash on underparts.

5. HAIRY-BREASTED BARBET *Tricholaema hirsuta*   Page 262
Casamance in coastal forest. Canopy. Calls a low, slow series of *oork* and a rapid *poop-poop-poop-poop*, recalling wood-dove. **Adult.** Black hair-like tips to lower throat.

6. BEARDED BARBET *Lybius dubius*   Page 263
Common throughout. Well wooded areas. Pairs. Heavy and direct flight. **Adult.** Mainly black and red. White patch on back. Tufted bill massive.

7. VIEILLOT'S BARBET *Lybius vieilloti*   Page 262
Locally common throughout. Dry savanna, farmland with trees. Trumpets out-of-time duet *whu-oupe whu-oupe.* **Adult.** Red head. Sulphur stripe up back.

8. AFRICAN PIED HORNBILL *Tockus fasciatus*   Page 259
Common in coastal forest thickets, fewer inland. **Adult.** Black with neat white belly. Black tail, only outer rectrices tipped white. Cream-yellow and black bill, modest casque-ridge, smaller and blacker in female.

9. RED-BILLED HORNBILL *Tockus erythrorhynchus*   Page 258
Very common throughout. Open woodland and dry savanna. **9a Adult male.** Larger mandibles than female. **9b Adult female.**

10. AFRICAN GREY HORNBILL *Tockus nasutus*   Page 259
Common throughout. Most habitats with tall trees. **10a Adult male.** Bill mainly black. **10b Adult female.** Dark red mandibles.

11. PIPING HORNBILL *Ceratogymna fistulator*   Page 260
Casamance, local; thick forest. **Adult.** Tail black, broadly tipped white (except central feathers). Heavily ridged bill, reduced casque.

12. BLACK-AND-WHITE-CASQUED HORNBILL
Ceratogymna subcylindricus   Page 260
Rare; Gambia only. **12a Adult male.** Vast size. Half black, half white wing; huge casqued bill. Tail white, centrally black. **12b Adult female.** Reduced casque.

13. YELLOW-CASQUED WATTLED HORNBILL
Ceratogymna elata   Page 260
Small population in Casamance forest. All-black, distinctive head, face, neck patterns. Tail white centrally black. **13a Adult male.** Black bushy crest. **13b Adult female.** Crest rufous. Immature similar.

14. ABYSSINIAN GROUND HORNBILL *Bucorvus abyssinicus* Page 258
Widespread throughout. Huge black walking bird. Roosts off ground. **14a Adult male.** Pink throat skin. **14b Adult female.** Blue throat skin. **14c Flight.** Prominent white primaries in both sexes.

13a

13b

PLATE 31: WOODPECKERS, HONEYGUIDES, WRYNECK AND SPOTTED CREEPER

1. GREY WOODPECKER *Dendropicos goertae* Page 268
Common all divisions. Most frequent woodpecker, using most habitats. Plain green back, plain grey face and underparts. Faint barring on wings. **1a Adult male.** Scarlet hindcrown. **1b Adult female.** All-grey head.

2. CARDINAL WOODPECKER *Dendropicos fuscescens* Page 268
Frequent to common all divisions but not regularly seen. Well wooded areas. Sparrow-sized. Green back faintly barred, yellow-buff underparts streaked brown. **2a Adult male.** Forehead brown, hindcrown to nape scarlet. **2b Adult female.** Crown and nape brownish-black.

3. LITTLE GREEN WOODPECKER *Campethera maculosa* Page 267
No Gambian records. Few sites in Senegal. Forest edges, overgrown streams. Small. Only woodpecker in region with green (not brown) back and barred (not streaked) underparts. **3a Adult male.** Red crown and nape. **3b Adult female.** Red nape patch absent.

4. FINE-SPOTTED WOODPECKER *Campethera punctuligera* Page 266
Common all divisions, regularly seen. Fond of palms and termite mounds. Tiny black dots on chin, throat, neck sides. **4a Adult male.** Forehead to crown, and strong moustache all crimson. **4b Adult female.** Crown bicoloured. Malar area peppered with dots.

5. GOLDEN-TAILED WOODPECKER *Campethera abingoni* Page 266
Rare. Darker than Fine-spotted. Streaked instead of spots below. **5a Adult male.** Uniform red crown. **5b Adult female.** Crown bicoloured. Malar area streaky.

6. BUFF-SPOTTED WOODPECKER *Campethera nivosa* Page 267
Frequent to common in forest islands, e.g. Abuko. Dark. Neat rounded head. Underparts spotted becoming barred on belly. Chisel-like bill. **Adult male.** Red nape patch, absent on female (not illustrated).

7. BROWN-BACKED WOODPECKER *Picoides obsoletus* Page 268
Frequent in dry savanna woodland. Small. Well-defined facial markings, white and brown streaked (not red) rump, no yellow in tail. **7a Adult male.** Nape scarlet. **7b Adult female.** Nape brown.

8. LITTLE GREY WOODPECKER *Dendropicos elachus* Page 267
Rare. Subdesert with acacias. Tinkerbird-sized. Conspicuous red rump, pale facial markings, greyish-brown back. Bleached appearance. **8a Adult male.** Hindcrown–nape red. **8b Adult female.** Head brown.

9. GREATER HONEYGUIDE *Indicator indicator* Page 264
Common to frequent throughout, dry savanna woodland, thickets, mangroves. Bulbul-sized. White outertail. **9a Adult male.** Grey with black throat and white ear patches. **9b Adult female.** Plainer without face pattern. **9c Immature.** Uniform brown back, pale yellow underparts.

10. LESSER HONEYGUIDE *Indicator minor* Page 265
Locally frequent in well wooded places. Active. Flycatches. **10a Adult.** Streaky olive-green back. Immature lacks loral/moustachial marks. **10b Flight.** White outer tail tipped dark.

11. SPOTTED HONEYGUIDE *Indicator maculatus* Page 264
Rare. Thickset. Greenbul-sized. **Adult.** Olive-brown above, yellowish-white spotting below. Conical bill.

12. CASSIN'S HONEYBIRD *Prodotiscus insignis* Page 263
One record northern Senegal. Forest associated. Small. Flycatcher-warbler habits. Flicks/flares immaculate white-edged tail. **Adult.** Greenish. White flanks. Sharp bill.

13. EURASIAN WRYNECK *Jynx torquilla* Page 265
Palearctic migrant, sandy wooded areas November–April. Feeds on ground. Undulating flight, flicks wings. **Adult.** Cryptic plumage. Small chisel bill.

14. SPOTTED CREEPER *Salpornis spilonotus* Page 330
Rare. Open woodland, recently burnt patches. **Adult.** Long decurved bill. Peppered or barred with black and white. Immature (not illustrated) has markings less defined.

PLATE 32: LARKS AND PIPITS

1. SINGING BUSH LARK *Mirafra cantillans* Page 270
Local migrant; inland dry grassy country. **1a Adult.** Streaked sandy-buff above, buff supercilium. Underparts white with gorget of buff streaks. White outer tail feathers, rufous wings conspicuous in flight. **1b Immature.** More boldly marked with 'arrowheads' on back. Robust bill as adult.

2. FLAPPET LARK *Mirafra rufocinnamomea* Page 270
Widespread but scarce; grassy plains with scattered bushes. **Adult.** Upperparts sandy-buff with darker streaking, greyer on mantle. Underparts tawny-buff, streaked across breast. Wings cinnamon, outer tail feathers rufous-buff. Whirring wingclaps a feature of high overhead flight display.

3. RUFOUS-RUMPED LARK *Pinarocorys erythropygia* Page 271
Vagrant; intra-African migrant associated with Sahelian conditions. **Adult.** Upperparts streaked sandy-buff, russet on rump. Underparts whitish, dark streaks on breast.

4. SHORT-TOED LARK *Calandrella brachydactyla* Page 271
Vagrant from Palearctic. Dry and sparse grassland. **Adult.** Upperparts pale sandy-buff with darker markings, crown chestnut with dark streaks. Underparts whitish-buff, often with greyish collar patch. Folded wing-tips covered by tertials.

5. BLACK-CROWNED SPARROW-LARK *Eremopterix nigriceps* Page 273
Northern Senegal. Grassland and light scrub to semi-desert. Crouches. **5a Adult male.** Pale grey-buff upperparts; black crown and underparts, white forehead and cheeks. Short-tailed. Pale finch-like bill. **5b Adult female.** Speckled sandy-buff above, with plain buff supercilium, whitish underparts.

6. CHESTNUT-BACKED SPARROW-LARK *Eremopterix leucotis* Page 272
Widely distributed, reasonably common resident; dry grassland and open bush. **6a Adult male.** Head and underparts black, white patches on cheeks and nape. Back rich chestnut. Pale finch-like bill. **6b Adult female.** Head and underparts pale buff, heavily streaked darker; white collar around nape.Mottled back.

7. CRESTED LARK *Galerida cristata* Page 272
Common resident, absent inland; busy roadsides, dunes, beaches. **7a Adult.** Upperparts mottled with brown and buff, underparts pale buff, lightly streaked on breast. Pale buff supercilium, long spiky crest, normally erect. Pale chestnut outer tail feathers. **7b Immature.** Shorter crest normally held flat.

8. SUN LARK *Galerida modesta* Page 272
Well inland. Small groups in rains. **8a Adult.** Mottled sandy-buff above, streaked below. Conspicuous supercilium extending to nape. Crest readily seen but not always raised. **8b Immature.** As adult but darker, more cinnamon-buff in flight feathers, with pale buff edgings.

9. RED-THROATED PIPIT *Anthus cervinus* Page 282
Regular Palearctic migrant to open ground. Often near water. **9a Adult non-breeding.** Upperparts streaked black on buff; streaked rump not easily seen. Underparts pale-buff, heavily streaked black. Buff supercilium. **9b Adult breeding.** Face and throat chestnut-red; transitional plumages frequent.

10. PLAIN-BACKED PIPIT *Anthus leucophrys* Page 282
Widespread resident but not common; short grassland, often near water. **Adult.** Large, long-legged, pale sandy-olive above, only faintly marked. Long whitish supercilium. Pale buff below, spotted faintly on breast.

11. TAWNY PIPIT *Anthus campestris* Page 281
Scarce Palearctic migrant to open arid areas and stubble fields. **11a Adult.** Pale, relatively long-tailed; only faintly marked above, pale yellow-buff below. **11b First-winter.** More streaked on throat and breast.

12. TREE PIPIT *Anthus trivialis* Page 282
Most frequent Palearctic pipit; open areas with scattered trees. **Adult.** Olive above with brown markings. Dark streaks on yellow-buff throat, paling to plain white belly; streaks lighter on flanks; plain rump.

PLATE 33: SWALLOWS AND MARTINS

1. FANTI SAW-WING *Psalidoprocne obscura* Page 273
Well wooded sites throughout; increases in rains. Only all-dark swallow in our area. **1a & b Adult male.**
Long deeply forked tail. Streamerless birds frequent. Female (not illustrated) has shorter tail.

2. RED-RUMPED SWALLOW *Hirundo daurica* Page 276
Common resident, increases inland. **2a & b Adult.** Elongated swallow with long streamers. Purple-black
back, wings and crown, chestnut cheeks and rump, whitish underparts. Black undertail-coverts diagnostic.

3. PIED-WINGED SWALLOW *Hirundo leucosoma* Page 278
Widespread, in low numbers in lightly wooded country, often near water. Forages low. **3a & b Adult.**
Martin-like with shallowly forked tail. Purplish-black above, white patch at base of upperwing.

4. WIRE-TAILED SWALLOW *Hirundo smithii* Page 277
Widespread near water. **4a & b Adult.** Bright metallic blue-black above, warm chestnut cap, all-white chin
and throat. Very long, slender wire streamers, often broken or difficult to see.

5. LESSER STRIPED SWALLOW *Hirundo abyssinica* Page 276
Rare in The Gambia, more regular in Niokolo Koba. Savanna woodland. **5a & b Adult.** Slender, only
swallow in our area with very heavily streaked underparts. Rufous rump and head.

6. MOSQUE SWALLOW *Hirundo senegalensis* Page 275
Common throughout in all seasons. Largest swallow. **6a, b & c Adult.** Dark blue-black above; dark cap with
rufous cheeks; rump, nape and underparts chestnut, paler on breast. Rufous undertail.

7. RUFOUS-CHESTED SWALLOW *Hirundo semirufa* Page 275
Mostly inland and in the rains. Savanna woodlands. **7a, b & c Adult.** Dark cap extends down to cheeks.
Richer chestnut underparts than Red-rumped and Mosque Swallows. Rufous undertail.

8. BARN SWALLOW *Hirundo rustica* Page 279
Paleartic migrant in low numbers. **8a & b Adult.** Long tail streamers. Blue-black above, chestnut face,
complete blue-black breast-band. **8c Immature.** Paler, duller, shorter streamers; brownish breast-band.

9. RED-CHESTED SWALLOW *Hirundo lucida* Page 278
Abundant resident throughout. **9a & b Adult.** Shorter tail streamers than Barn Swallow. More chestnut on
throat with narrower breast-band, often incomplete. Whiter below. **9c Immature.** Paler and duller.

10. GREY-RUMPED SWALLOW *Pseudhirundo griseopyga* Page 275
Rarely recored. Wooded savanna near water. **Adult.** Glossy blue-black back, wings and tail. Long tail
streamers. Crown, nape and rump grey; underparts white.

11. ETHIOPIAN SWALLOW *Hirundo aethiopica* Page 278
Niokolo Koba. **Adult.** Chestnut confined to a small forehead patch. Breast-band narrow and incomplete.

12. COMMON HOUSE MARTIN *Delichon urbica* Page 279
Palearctic migrant, sometimes in large flocks. **Adult non-breeding.** Tail notched; whitish rump.

13. AFRICAN ROCK MARTIN *Hirundo fuligula* Page 276
Very rare. **Adult.** Sandy-brown paler on throat. Tail square, white marks near tips.

14. CRAG MARTIN *Hirundo rupestris* Page 277
Very rare. **Adult.** Large brown martin, paler on upper breast. Dark underwing-coverts;
white markings near tips of slightly notched tail.

15. BROWN-THROATED MARTIN *Riparia paludicola* Page 274
Rarely recorded. Likely near water with other hirundines. **Adult.** Size of Sand Martin
but underparts buffer, pale towards undertail; lacks breast-band.

16. SAND MARTIN *Riparia riparia* Page 274
Common to abundant Palearctic migrant. Wetlands. **Adult.** Sandy-brown above, white
below with clean-cut, brownish pectoral band. Shallow, forked tail.

17. BANDED MARTIN *Riparia cincta* Page 274
Rare. **Adult.** Large, more robust than similar Sand Martin, with white underwing-coverts.

17

80

PLATE 34: ORIOLES, CUCKOO-SHRIKES, DRONGOS, WAGTAILS AND LONGCLAW

1. AFRICAN GOLDEN ORIOLE *Oriolus auratus* — Page 338
Widespread, common in well wooded areas. **1a Adult male.** Golden-yellow and black; much gold in wings. **1b Adult female.** Yellowish-olive above, whitish below. Broad yellow fringes to wing feathers. Dark eyepatch extends behind eye.

2. EUROPEAN GOLDEN ORIOLE *Oriolus oriolus* — Page 338
Rare Palearctic migrant. **2a Adult male.** Brilliant gold and black; wings mainly black, gold in tail confined to tip. **2b Adult female.** Greenish above, yellow on flanks; breast and belly whitish with black streaks. Yellow in tail confined to tip.

3. WHITE-BREASTED CUCKOO-SHRIKE *Coracina pectoralis* — Page 283
Uncommon local resident. Tall leafy trees. **Adult male.** Upperparts pale grey, underparts white. Conspicuous white orbital ring; powerful black, slightly-hooked bill.

4. RED-SHOULDERED CUCKOO-SHRIKE *Coracina phoenicea* — Page 283
Frequent, secretive, in well wooded areas. **4a Adult male.** Uniformly glossy black with conspicuous red shoulder patch. **4b Adult female.** Olive-green above, bright yellow fringes on wings; underparts white with dark barring.

5. FORK-TAILED DRONGO *Dicrurus adsimilis* — Page 345
Widespread and common. **5a Adult.** Uniformly black, long deeply forked tail, heavy head and bill. **5b Underside of wing.** Underwing shows pale flight feathers.

6. SQUARE-TAILED DRONGO *Dicrurus ludwigii* — Page 345
Localised resident in some forest pockets. **6a Adult.** Smaller than Fork-tailed Drongo, shorter tail only slightly notched. **6b Underside of wing.** Underwing uniformly black in flight.

7. AFRICAN PIED WAGTAIL *Motacilla aguimp* — Page 281
Rare on upper Gambia River, regular Niokolo Koba. **Adult.** Clean-cut, black above, white below; black collar, bold white supercilium, much white in wings.

8. YELLOW WAGTAIL *Motacilla flava* — Page 279
Common, often numerous, passage migrant on open ground and wetlands. Distinctive tail movements and shrill call. Breeding males all have olive back and yellow underparts. Females often subdued. **8a M. f. flava, non-breeding male.** Olive above, whitish below, yellower on belly. **8b M. f. flava, breeding male.** Blue head with bold white supercilium. **8c M. f. thunbergi, breeding male.** Slate-grey hood, bordered white on cheeks and throat, lacks supercilium. **8d M. f. iberiae, breeding male.** Blue-grey hood, slight white supercilium, white throat. **8e M. f. flavissima, breeding male.** Pale greenish crown and cheeks, bold yellow supercilium.

9. GREY WAGTAIL *Motacilla cinerea* — Page 280
Very rare Palearctic migrant to aquatic habitats. Wags tail vigorously. **Adult non-breeding.** Grey above, white supercilium, dark wings. Underparts white, some yellow on breast, distinctive lemon yellow undertail. Longer-tailed than Yellow Wagtail.

10. WHITE WAGTAIL *Motacilla alba* — Page 280
Common and widespread Palearctic migrant. **10a Adult non-breeding.** Pearl-grey crown and mantle, white underparts with black breast-band. Wings and tail dark grey and white. **10b Adult breeding.** Black crown and nape, white cheeks, black throat and bib.

11. YELLOW-THROATED LONGCLAW *Macronyx croceus* — Page 283
Uncommon local resident. Open wetland areas. Large. **Adult.** Streaky brown above. Underparts strikingly golden-yellow with black gorget. Pale corners to tail.

PLATE 35: BULBULS, GREENBULS, ORIOLE-WARBLER AND BABBLERS

1. COMMON BULBUL *Pycnonotus barbatus*   Page 288
Ubiquitous, often abundant. **Adult.** Upperparts and breast dull brown, darker on face; slightly crested. Very vocal.

2. YELLOW-THROATED LEAFLOVE *Chlorocichla flavicollis*   Page 285
Locally common along well vegetated streams, riverine forest, coastal mangroves. Social groups. **Adult.** Overall olive-brown, throat brilliant yellow.

3. SLENDER-BILLED GREENBUL *Andropadus gracilirostris*   Page 285
Casamance forest. **Adult.** Grey-brown head, reddish eye, rest of upperparts olive-brown. Underparts grey-buff, shaded yellow; throat whitish. Bill slender.

4. LITTLE GREENBUL *Andropadus virens*   Page 284
Locally common resident in coastal forest. **Adult.** Nondescript, often scruffy. Olive-green, paler on underparts, rufous-brown tinge on wings and tail.

5. YELLOW-WHISKERED GREENBUL *Andropadus latirostris*   Page 285
Casamance forest. **Adult.** Olive-grey, paler on underparts, wings and tail dark rufous-brown. Distinctive fluffy yellow moustachial stripes.

6. LEAFLOVE *Pyrrhurus scandens*   Page 286
Scarce resident in some forests. **Adult.** Grey head, rufous-brown back and tail. White throat, whitish-buff underparts.

7. WHITE-THROATED GREENBUL *Phyllastrephus albigularis*   Page 287
Casamance forest. **Adult.** Grey head, olive back, pale yellow belly.

8. GREY-HEADED BRISTLEBILL *Bleda canicapilla*   Page 287
Forest resident, in understorey. **Adult.** Pale olive-green above, rich yellow below from throat to vent; neat pale throat patch. Grey head and nape. Pale spots on outer tail feathers when tail is fanned.

9. RED-TAILED GREENBUL *Criniger calurus*   Page 288
Casamance forest. **Adult.** Olive above, yellow below. Grey head, blue skin around eye. Fluffy white throat.

10. SWAMP PALM GREENBUL *Thescelocichla leucopleura*   Page 286
Swamp forest. **Adult.** Rich brown above with speckled head; broadly white-tipped tail. Underparts buff-brown across breast with paler streaks.

11. WESTERN NICATOR *Nicator chloris*   Page 344
Rare forest dweller. **Adult.** Olive above with bold yellow spots on wings; yellowish cheeks. Underparts greyish-white. Robust bill.

12. ORIOLE WARBLER *Hypergerus atriceps*   Page 314
Resident, usually in densely vegetated watery places throughout. Slim. **Adult.** Black silvery-fringed head, upperparts olive-green; underparts bright yellow. Strident duets.

13. CAPUCHIN BABBLER *Phyllanthus atripennis*   Page 329
Rare forest dweller in deep shade. **Adult.** Body and wings dark reddish-brown, tail blackish. Head, nape and breast white with grey scaling. Pale bill.

14. BLACKCAP BABBLER *Turdoides reinwardtii*   Page 328
Widespread, common resident, prefers thicker vegetation. Garrulous, social groups. **Adult.** Black hood, grey back and tail, blackish wings, prominent pale eye. Throat white, rest of underparts whitish speckled pale grey.

15. BROWN BABBLER *Turdoides plebejus*   Page 328
Widespread, common resident. Often in sparser habitats. Typical babbler, gregarious. **Adult.** Orange eye. Brownish above, grey-brown and pale-flecked head and breast.

PLATE 36: WHEATEARS AND CHATS

1. WHINCHAT *Saxicola rubetra*   Page 292
Common Palearctic migrant. Open land with scattered bushes. Typical tail-flicking stance. **1a Male.** Streaked dark brown upperparts, white supercilium, dark cheeks, white in wings. Underparts variably buff to orange. **1b Female.** Paler and buffier, but with obvious supercilium.

2. NORTHERN WHEATEAR *Oenanthe oenanthe*   Page 292
Fairly common Palearctic migrant, on open ground. **2a Breeding female.** Grey back, black wings, distinctive black and white tail; buff underparts. **2b Breeding male** (not to scale). As female but with black cheek patch, bolder white supercilium, richer breast. **2c Non-breeding male** (not to scale). Pale, washed-out version of breeding plumage; buff feather fringes. **2d First-winter.** Pale and buff, but note tail pattern. **2e. Tail pattern.**

3. BLACK-EARED WHEATEAR *Oenanthe hispanica*   Page 293
Rare. Open ground with trees. **3a Breeding female.** Rich buff upperparts, paler underparts, white throat. Wings blackish. Note tail pattern. **3b & c Breeding male** (not to scale). Pale buff above, whitish below. Two morphs: Black-eared form (3c) and Black-throated form (3b). **3d Non-breeding male** (not to scale). Scaly buff feather fringes. **3e Non-breeding female.** Buff above, whitish below. **3f Tail pattern.**

4. DESERT WHEATEAR *Oenanthe deserti*   Page 293
Rare. Arid open ground. Black tail in all plumages. **4a Breeding female.** As Black-eared but paler with poorly defined cheek patch. **4b Breeding male** (not to scale). Paler, with more extensive black on throat and shoulders than Black-eared. **4c Non-breeding male** (not to scale). Pale feather fringes give scaly appearance to throat and wings. **4d Non-breeding female.** Very pale. **4e Tail pattern.**

5. ISABELLINE WHEATEAR *Oenanthe isabellina*   Page 293
Rare. Arid open ground. Sexes alike. **5a Non-breeding adult.** Bulky, upright stance; rather uniform, pale buff plumage. Broad pale edges to wing feathers. Broad terminal tail-bar. **5b Tail pattern.**

6. BLACKSTART *Cercomela melanura*   Page 294
Vagrant. Arid rocky areas. **Adult.** Drab grey-brown chat with distinctive black rump and tail.

7. FAMILIAR CHAT *Cercomela familiaris*   Page 294
Arid rocky areas, south-east Senegal. **7a Adult.** Drab brown chat, paler below. Rump and tail chestnut with darker tail centre and tips. **7b Tail pattern.**

8. MOCKING CLIFF-CHAT *Thamnolaea cinnamomeiventris*   Page 295
Open rocky areas, Senegal. **Adult.** Back, wings, breast and tail black; small white patch on wing. Rump and belly chestnut. No white line between breast and belly in local form.

9. WHITE-FRONTED BLACK CHAT *Myrmecocichla albifrons*   Page 295
Local resident. Dry savanna woodland. **9a Male.** Uniformly dark slate-black apart from white forehead. **9b Female.** Lacks white forehead.

10. NORTHERN ANTEATER CHAT *Myrmecocichla aethiops*   Page 294
Thinly but widely distributed, mainly north of the Gambia River. **Adult.** Uniformly dark; in fresh plumage shows pale feather fringes. White patches in wings obvious in flight.

PLATE 37: CHATS, ILLADOPSES AND THRUSHES

1. FIRE-CRESTED ALETHE *Alethe castanea*   Page 290
Forest undergrowth, Casamance. **Adult.** Upperparts brown, tail darker with white tips. Cheeks grey, crown orange. Underparts white.

2. RUFOUS-WINGED ILLADOPSIS *Illadopsis rufescens*   Page 327
Forest undergrowth, Casamance. **2a & b Adult.** Dull brown above, rufous-edged flight feathers. Throat white, diffuse partial breast-band, grey-buff on flanks, rest of underparts whitish. Legs whitish-flesh.

3. BROWN ILLADOPSIS *Illadopsis fulvescens*   Page 327
Forest undergrowth, Casamance. **Adult.** Upperparts brown, cheeks grey, throat white, Underparts pale buff-brown. Legs blue-grey.

4. PUVEL'S ILLADOPSIS *Illadopsis puveli*   Page 327
Forest undergrowth. **4a & b Adult.** Similar to Rufous-winged, but lacking rufous on flight feathers. Very white below; diffuse peachy-buff breast-band.

5. COMMON NIGHTINGALE *Luscinia megarhynchos*   Page 288
Palearctic migrant. Dense scrub and thickets. **Adult.** Upperparts brown, rich rufous broad-ended tail.

6. WHITE-CROWNED ROBIN-CHAT *Cossypha albicapilla*   Page 290
Widespread resident in riverine forest, gardens, thickets. **Adult.** Striking chestnut underparts and sides to long tail. Black back and sides of head; white crown flecked grey.

7. SNOWY-CROWNED ROBIN-CHAT *Cossypha niveicapilla*   Page 289
Local and uncommon; dense woodland and thickets. **Adult.** Smaller and shorter-tailed than White-crowned. Cap pure white; nape chestnut. Black cheeks contrast with dark grey upperparts.

8. RUFOUS SCRUB ROBIN *Cercotrichas galactotes*   Page 291
Uncommon, mainly north of the Gambia River. Scrub. **Adult.** Sandy-brown upperparts with bold white supercilium; tail ginger with white tips, often cocked. Underparts whitish.

9. BLACK SCRUB ROBIN *Cercotrichas podobe*   Page 291
Arid thorny scrub, northern Senegal. **Adult.** All-black scrub robin with white-tipped tail, often flicked and fanned; shows dark tawny primary patches in flight.

10. COMMON REDSTART *Phoenicurus phoenicurus*   Page 291
Common Palearctic migrant to wooded areas. **10a First-winter male.** Olive-brown above, buff with chestnut markings on underparts; orange-chestnut tail, frequently flicked. **10b Adult female.** Duller.

11. BLUETHROAT *Luscinia svecica*   Page 289
Infrequent Palearctic migrant. Scrub. **Non-breeding male** (white-spotted form). Brown above with white supercilium; darker tail, chestnut sides at base. Whitish underparts, with variable blue and orange gorget.

12. AFRICAN THRUSH *Turdus pelios*   Page 297
Very common in well wooded places throughout. **Adult.** Large, soft brown above, yellow bill. Pale buff below, faint breast markings, white throat with fine black streaks.

13. EUROPEAN ROCK THRUSH
Monticola saxatilis   Page 296
Uncommon Palearctic migrant on open ground. **13a Non-breeding male.** Grey and rufous with copious crescentic barring above and below. Tail orange-chestnut. **13b Female.** Mottled brown above, scaled below.

14. BLUE ROCK THRUSH *Monticola solitarius*   Page 296
Irregular Palearctic migrant, possibly resident northern Senegal. Rocky areas, buildings. **14a Non-breeding male.** Body dark blue with paler crescentic barring. **14b Female.** Browner, narrowly barred below.

PLATE 38: WARBLERS

1. GREAT REED WARBLER *Acrocephalus arundinaceus* — Page 300
Rarely seen Palearctic migrant in swampy areas. Massive size; bold extended supercilium, powerful bill.

2. AFRICAN REED WARBLER *Acrocephalus baeticatus* — Page 300
Poorly known localised resident; reedbeds, swamps. Slightly smaller and shorter-winged than Reed Warbler.

3. REED WARBLER *Acrocephalus scirpaceus* — Page 299
Widespread Palearctic migrant; reedbeds, swamps. Unstreaked olive above; whitish below with buff flanks.

4. GREATER SWAMP WARBLER *Acrocephalus rufescens* — Page 300
Locally common in reedbeds and swamps. Large and robust; short supercilium from eye to bill.

5. AFRICAN MOUSTACHED WARBLER *Melocichla mentalis* — Page 297
Resident in southern Senegal. Dense rank herbage. Huge size; shy; chestnut crown, dark malar stripe.

6. GRASSHOPPER WARBLER *Locustella naevia* — Page 298
Uncommon Palearctic migrant. Swamps, dense herbage. Secretive. Heavily streaked above; buffy below.

7. SAVI'S WARBLER *Locustella luscinioides* — Page 298
Regular Palearctic migrant to Senegal. Skulks in freshwater reedbeds. Unstreaked; rounded tail.

8. OLIVACEOUS WARBLER *Hippolais pallida* — Page 301
Common Palearctic migrant. Widespread. Grey-brown above, high-peaked crown, long broad-based bill.

9. MELODIOUS WARBLER *Hippolais polyglotta* — Page 301
Common Palearctic migrant. Widespread. Rounded head, pale lores and long, orange-based bill.

10. ORPHEAN WARBLER *Sylvia hortensis* — Page 317
Uncommon Palearctic migrant. **10a Male**. Grey with darker head, white eye. **10b Female**. Duller, yellow eye.

11. SARDINIAN WARBLER *Sylvia melanocephala* — Page 319
Rare Palearctic migrant. **Male**. Grey above, black hood, red eye, white-edged black tail.

12. BLACKCAP *Sylvia atricapilla* — Page 318
Common Palearctic migrant. Woodland. **12a Male**. Black cap, no white in tail. **12b Female**. Rufous cap.

13. GARDEN WARBLER *Sylvia borin* — Page 317
Common Palearctic migrant. Woodland, tamarisk belts. Plump, featureless, blue-grey legs.

14. COMMON WHITETHROAT *Sylvia communis* — Page 318
Common Palearctic migrant. Wooded scrub. White throat. Broad chestnut panel on closed wing.

15. SPECTACLED WARBLER *Sylvia conspicillata* — Page 319
Rare Palearctic migrant. Coastal scrub. **15a Non-breeding male**. Chestnut wings, greyish head and cheeks, white eye-ring, pale buff underparts. **15b Female**. Browner on head and cheeks.

16. SUBALPINE WARBLER *Sylvia cantillans* — Page 319
Common Palearctic migrant. Widespread, favours acacias and mangroves. **16a Non-breeding male**. Grey above, russet below, white moustachial. reddish eye-ring. **16b Female**. Paler, duller, reddish eye-ring.

17. WILLOW WARBLER *Phylloscopus trochilus* — Page 316
Frequent Palearctic migrant. Widespread. **First-winter**. Olive-green above, yellowish below; legs brown.

18. CHIFFCHAFF *Phylloscopus collybita* — Page 316
Common Palearctic migrant. Widespread. **Adult winter**. Olive above, buff below; legs dark. First-winter is yellower below.

19. WOOD WARBLER *Phylloscopus sibilatrix* — Page 316
Rare Palearctic migrant. Coastal scrub. Yellow supercilium and breast.

20. WESTERN BONELLI'S WARBLER *Phylloscopus bonelli* — Page 317
Regular Palearctic migrant. **Adult winter**. Greyish-olive and white, yellowish rump and fringes to tail, greenish wing panel on most.

11

PLATE 39: CISTICOLAS AND WARBLERS

1. RED-FACED CISTICOLA *Cisticola erythrops*   Page 302
Local in rank vegetation near freshwater. **1a Breeding**. Warm brown above, extensive burnt-orange bloom to face and underparts. **1b Non-breeding**. Orange face; creamy-apricot central underparts. Song of ringing notes.

2. SINGING CISTICOLA *Cisticola cantans*   Page 304
Common throughout in scrub and woodland. Chestnut crown and wings contrast with grey back. Underparts whitish, buff on flanks. Simple repetitive clinking call.

3. WHISTLING CISTICOLA *Cisticola lateralis*   Page 304
Localised but widespread in open woodlands, old fallow land. Uniform warm brown crown to mantle, chestnut tone on wings, whitish below. Warbled rising song of short duration, finishes with a flourish.

4. CROAKING CISTICOLA *Cisticola natalensis*   Page 306
Scarce resident, predominantly coastal. **4a Breeding**. Grey-brown above, streaked on crown, nape and mantle; chestnut in wings. **4b Non-breeding**. Browner, stronger buff supercilium. Song of explosive froglike notes.

5. WINDING CISTICOLA *Cisticola galactotes*   Page 306
Locally common in swamps. **5a Breeding**. Chestnut crown, dark-streaked grey back, chestnut in wings, whitish below. **5b. Non-breeding**. Black mantle streaks on warm brown back. Song a forced, rising nasal buzz.

6. PLAINTIVE CISTICOLA *Cisticola dorsti*   Page 305
Very local in open savanna woodland. Dumpy with rufous cap, plain pale brown upperparts, chestnut in wings, whitish below. Song a sustained quavering tremolo on a monotone.

7. SIFFLING CISTICOLA *Cisticola brachypterus*   Page 307
Widespread in open habitats with scattered trees. Small, generally plain, rufous-brown above, chestnut in wing, whitish below. Song a hurried repetitive wispy sequence of 3-4 notes from a high tree.

8. BLACK-BACKED CISTICOLA *Cisticola eximius*   Page 309
Localised in short grasslands, near water. **Non-breeding**. Small and short-tailed, strong crown streaks, gingery nape and flanks, heavy black mantle streaks. Thin calls in high display flight.

9. RUFOUS CISTICOLA *Cisticola rufus*   Page 307
Frequent in open dry woodlands. Unstreaked warm rufous-brown upperparts; underparts whitish-buff. Song a slow, repetitive, soft two-note whistle.

10. DESERT CISTICOLA *Cisticola aridulus*   Page 308
Local in dry grass and scrub; widespread in northern Senegal. Small with streaky upperparts; like washed-out, pale version of Zitting. Wingsnaps in display flight.

11. ZITTING CISTICOLA *Cisticola juncidis*   Page 308
Common and widespread resident in grasslands, often adjacent to swamps. Small, brown, heavily streaked crown and back. Song crisp, short buzzes coordinated with dipping low flight.

12. AQUATIC WARBLER *Acrocephalus paludicola*   Page 298
Uncommon winter migrant to riverine reedbeds and swamps in northern Senegal. **First-winter**. Prominent median crown stripe and supercilium. Streaking on upperparts extends to rump; fanned tail feathers spiky.

13. SEDGE WARBLER *Acrocephalus schoenobaenus*   Page 299
Uncommon migrant and winter visitor; swamps and lush vegetation. **First-winter**. Brown upperparts with dark streaking, prominent whitish supercilium. Plain rump.

PLATE 40: WARBLERS, TITS AND WHITE-EYE

1. RED-WINGED WARBLER *Heliolais erythroptera*   Page 310
Local resident, scarcer inland; dense bush. Loud song. **1a Adult breeding**. Grey head and back, dark chestnut wings, brown tail. Throat white, buff underparts. **1b Adult non-breeding**. Brown above, buff below.

2. TAWNY-FLANKED PRINIA *Prinia subflava*   Page 309
Widespread and common resident in grassy places. Swings tail from side to side. **2a Adult breeding**. Greyish upperparts, whitish supercilium, russet flanks. Whitish throat and breast. **2b Adult non-breeding**. Olive above, yellow-buff below.

3. GREEN-BACKED EREMOMELA *Eremomela pusilla*   Page 312
Widespread, common resident. Savanna woodland, thicket edges. Small parties; warbler-like, constant twittering. Pastel grey, green and yellow; throat white.

4. GREEN CROMBEC *Sylvietta virens*   Page 312
Restricted to a few coastal forest thickets. Tiny, plump, dull and tailless; olive above, greyer on head, grey-white supercilium, underparts yellowish (variable).

5. NORTHERN CROMBEC *Sylvietta brachyura*   Page 313
Widespread resident in open bush and woodland. Tiny, tailless. Grey back, buffy-orange below, grey-white supercilium.

6. GREY-BACKED CAMAROPTERA *Camaroptera brachyura*   Page 311
Common widespread resident. Upperparts and short cocked tail grey, short wings olive, underparts whitish-grey. Constant bleating call.

7. OLIVE-GREEN CAMAROPTERA *Camaroptera chloronota*   Page 311
Niokolo Koba; thick woodland. Dark olive upperparts and short cocked tail, chestnut cheeks. Underparts grey.

8. YELLOW-BREASTED APALIS *Apalis flavida*   Page 310
Locally common canopy dweller in a few coastal forests. Throat white, breast band yellow, belly white. Crown, cheeks and nape grey. Upperparts yellow-olive.

9. GREEN HYLIA *Hylia prasina*   Page 315
Locally common in a few coastal forests; shy in thick high cover. Upperparts olive, conspicuous pale broad arched supercilium. Underparts greyish.

10. YELLOW WHITE-EYE *Zosterops senegalensis*   Page 337
Frequent and widespread resident in wooded savanna. Gregarious. Tiny bright canary-yellow bird with white eye-ring. Yellow-green upperparts, bright yellow underparts.

11. YELLOW PENDULINE TIT *Anthoscopus parvulus*   Page 330
Uncommon resident; open wooded habitats; gleans over buds. Minute size. Yellow-olive upperparts, crown flecked black. Pale brown flight feathers and brown tail. Underparts yellow.

12. WHITE-SHOULDERED BLACK TIT *Parus leucomelas*   Page 329
Resident in dry savanna woodlands. Stout. Body and tail uniform black. White shoulder-bar on black wing. Striking pearl-white eye.

PLATE 41: FLYCATCHERS AND ALLIES

1. YELLOW-BELLIED HYLIOTA *Hyliota flavigaster* — Page 314
Uncommon resident. Forest edges, open woodland. **1a Male.** Short-tailed, glossy blue-black above, peachy below, bold white wing-bar. **1b Female.** Sooty-grey above, buffy below, bold white wing-bar.

2. AFRICAN BLUE FLYCATCHER *Elminia longicauda* — Page 325
Locally common in some *Rhizophora* mangroves and riverine forest. Droops wings. Upperparts caerulean-blue, nominal crest, black eye patch. Long wedged-shaped tail blue. Underparts whitish. Scratchy song.

3. PIED FLYCATCHER *Ficedula hypoleuca* — Page 323
Palearctic passage migrant. Favours acacias and open woodland. Flicks wings and tail. Silent. **Adult non-breeding.** Brownish above, whitish below; white-edged tail, white wing-bar.

4. PALE FLYCATCHER *Bradornis pallidus* — Page 320
Uncommon resident in dry savanna and woodland. Slender profile. **4a Adult.** Uniform pale brown above, whitish below. **4b Immature.** Brown above, speckled white; whitish below spotted brown.

5. SPOTTED FLYCATCHER *Muscicapa striata* — Page 321
Palearctic passage migrant. Mixed habitats with trees. Silent. Pale brown above with dark-streaked head; long, dark brown wings with pale fringes. Streaked, whitish underparts.

6. SWAMP FLYCATCHER *Muscicapa aquatica* — Page 322
Common along freshwater reaches and tributaries well up-river. Forages low over water. Bubbly song. Uniformly drab brown above, whitish below with distinct breast-band.

7. COMMON WATTLE-EYE *Platysteira cyanea* — Page 324
Common resident. Mangroves, thicker wooded areas. **7a Adult male.** Red wattle over eye, blue-black breast-band. **7b Adult female.** Chin white, throat dark chestnut. **7c Immature male** Greyer version of female.

8. BLACK-AND-WHITE FLYCATCHER *Bias musicus* — Page 323
Vagrant. Forest edges. Thickset. **8a Male.** Crested, yellow eye, iridescent black above, greyish below. **8b Female.** As male but upperparts and tail tawny-brown.

9. SENEGAL BATIS *Batis senegalensis* — Page 324
Localised resident in dry savanna. Small with prominent pale eye. **9a Male.** Black-and-white-striped head, grey back. Black breast band. **9b Female.** Black-and-orange-striped head, chestnut breast band.

10. NORTHERN BLACK FLYCATCHER *Melaenornis edolioides* — Page 320
Widespread resident; open woodland with damp margins. **10a Adult.** Uniformly black. Long square-ended narrow tail. Small head. **10b Juvenile.** Sooty, spotted and scaled with buff; often shorter tailed.

11. AFRICAN PARADISE FLYCATCHER *Terpsiphone viridis* — Page 326
Widespread, in sparser habitat than Red-bellied. **11a Male.** Crested head and underparts dark grey-black, back chestnut, white in wings; very long chestnut tail. **11b Female.** Shorter tail, no wing-bar.

12. RED-BELLIED PARADISE FLYCATCHER *Terpsiphone rufiventer* — Page 325
Locally common resident associated with forest. **12a Male.** Black hood and crest, chestnut body and long tail, wings chestnut and black. **12b Female.** Duller with shorter tail.

13. HYBRID RED-BELLIED X AFRICAN PARADISE FLYCATCHER — Page 326
Black hood and crest, underparts chestnut blotched black. Chestnut, black and white wings.

14. SHRIKE-FLYCATCHER *Megabyas flammulatus* — Page 323
Vagrant. Forest thickets. Hooked bill with bristles. **14a Male.** Uniform glossy blue-black above, white below, red eye **14b Female.** Chestnut upperparts, below white smudged grey.

15. WHITE-BROWED FOREST FLYCATCHER *Fraseria cinerascens* — Page 321
Casamance forest. Streams. Dark bluish-grey above, white supercilium. White below scalloped with dark grey.

16. LEAD-COLOURED FLYCATCHER *Myioparus plumbeus* — Page 322
Widespread but rarely seen. Regularly fans tail. Grey and white plumage. Longish tail black with white sides.

PLATE 42: SUNBIRDS

1. PYGMY SUNBIRD *Anthreptes platurus* Page 333
Widespread in dry savanna woodland, mostly inland. Tiny. **1a Male breeding.** Long, lax tail streamers (sometimes absent). Shiny plumage. Green hood, bright yellow belly. **1b Female.** No streamers, pale yellow below, olive above, short bill.

2. COLLARED SUNBIRD *Anthreptes collaris* Page 332
Common in coastal forest thickets. **2a Male.** Shiny green upperparts and throat, separated from bright yellow belly by narrow purple band. **2b Female.** Duller, entirely yellow below.

3. OLIVE SUNBIRD *Nectarinia olivacea* Page 333
Mostly Casamance, near water. Warbler-like. Sexes alike. Dull olive upperparts, buffish breast. Inconspicuous yellow pectoral patches.

4. VARIABLE SUNBIRD *Nectarinia venusta* Page 335
Locally common in coastal parts, less common inland. **4a Male breeding.** Yellow or whitish belly, shiny green hood, purple-black breast-band. Transitional plumages common. **4b Female.** Dull. Grey-brown above, pale yellow below, yellowish supercilium.

5. MOUSE-BROWN SUNBIRD *Anthreptes gabonicus* Page 331
Locally common in mangroves, sometimes in nearby woodlands. Short-billed, warbler-like. Sexes alike. Overall dull grey-brown. Dark streak through eye, bordered by white lines above and below.

6. SCARLET-CHESTED SUNBIRD *Nectarinia senegalensis* Page 334
Very common and widespread inland. Large. **6a Male.** Black body, shiny green face and scarlet breast. **6b Female.** Dull. Chocolate above; dark underparts mottled with yellow and buff.

7. OLIVE-BELLIED SUNBIRD *Nectarinia chloropygia* Page 335
Mostly Casamance; forest edges. **7a Male.** Shiny green hood and mantle, yellow and scarlet breast-band, olive belly. **7b Female.** Dull. Olive above, white throat, yellow belly with faint streaks. No supercilium.

8. WESTERN VIOLET-BACKED SUNBIRD *Anthreptes longuemarei* Page 332
Uncommon throughout; mixed woodland. Short, straight bill. **8a Male.** Shiny violet above, white below, wings black. **8b Female.** Greyish above, white below; white supercilium; shiny violet tail and primrose undertail-coverts.

9. SPLENDID SUNBIRD *Nectarinia coccinigaster* Page 336
Common at the coast; woodland and gardens. Large. **9a Male.** All dark. Shiny purple head and throat, green mantle, green wing-bar, crimson breast patch. **9b Female.** Pale olive above; pale yellow below, streaked with grey. **9c Immature male.** Purple throat patch.

10. BEAUTIFUL SUNBIRD *Nectarinia pulchella* Page 337
Very common throughout. **10a Male breeding.** Shiny emerald-green, yellow and red breast-band, long tail streamers (sometimes absent); green-shouldered transitional plumages frequent. **10b Female.** Pale olive above, yellowish below, narrow yellowish supercilium, pale outer tail. **10c Immature male.** Dark bib patch.

11. GREEN-HEADED SUNBIRD *Nectarinia verticalis* Page 334
Scarce; forest thickets and edges. Long bill. Large. **11a Male.** Shiny green head, blue breast, grey belly. **11b Female.** Grey throat and breast.

12. COPPER SUNBIRD *Nectarinia cuprea* Page 336
Common on coast, seasonal inland. Eclipse plumages frequent. **9a Male breeding.** All dark, shiny copper close-up. **9b Male non-breeding.** Black wings and tail contrast with olive back; uneven dark belly patch on yellowish underparts. **9c Female.** Shiny black bill, olive above, yellowish below, dark wings and tail.

PLATE 43: BUSH SHRIKES, HELMET SHRIKES AND SHRIKES

1. BLACK-CROWNED TCHAGRA *Tchagra senegala*　　　　　　　　Page 342
Widespread and common in dry scrubby bush. Bold striped head, chestnut wings, white tail-tips.

2. GREY-HEADED BUSH SHRIKE *Malaconotus blanchoti*　　　　　　Page 343
Widespread, uncommon, reclusive; woodland. Large size. **Adult.** Grey head, heavy hooked bill, green wings spotted yellow, underparts yellow. Haunting monotone whistle.

3. SULPHUR-BREASTED BUSH SHRIKE *Malaconotus sulfureopectus*　　Page 343
Frequent throughout. Lower levels in woodland and forest edges. **3a Adult.** Small grey head with black mask, yellow underparts suffused orange. Rhythmic, mellow-toned, three-note whistle. **3b Immature.** Ghosts Grey-headed but smaller proportions.

4. BRUBRU *Nilaus afer*　　　　　　　　　　　　　　　　　　Page 341
Common in dry wooded savanna. Elusive, very vocal. **Adult.** Small. Black, white and chestnut plumage. No breast bands.

5. NORTHERN PUFFBACK *Dryoscopus gambensis*　　　　　　　　Page 341
Common in mixed woodlands. **5a Male.** Black above, black wings fringed white, red eye, erectile powder-puff rump. **5b Female.** Pale grey above, apricot-buff below.

6. YELLOW-CROWNED GONOLEK *Laniarius barbarus*　　　　　　Page 342
Common and widespread. **Adult.** Black above, striking scarlet below, dull golden crown. Diverse vocalistions.

7. WHITE-CRESTED HELMET SHRIKE *Prionops plumatus*　　　　Page 344
Frequent to common throughout; mixed woodlands. Social. **Adult.** Black and white; crested white head, yellow wattle around eye.

8. SOUTHERN GREY SHRIKE *Lanius meridionalis*　　　　　　　Page 339
Uncommon to rare winter visitor; arid places, fallow land. **Adult.** *L. m. leucopygos.* Black mask, pale grey above, wings black and white, white below.

9. ISABELLINE SHRIKE *Lanius isabellinus*　　　　　　　　　　Page 339
Rare; coastal scrub, dry open habitats. **First-winter.** Very pale sandy above, browner on wings, pale rufous tail; creamy supercilium, smudgy cheek patch and white underparts, faintly scalloped.

10. WOODCHAT SHRIKE *Lanius senator*　　　　　　　　　　　Page 340
Common Palearctic migrant throughout. Coastal scrub, farmland. **10a Adult male.** Chestnut crown and nape, black and white body. **10b First-winter.** Cinnamon hood, sooty cheeks, blackish back and wings, pale wing-bar.

11. RED-BACKED SHRIKE *Lanius collurio*　　　　　　　　　　Page 339
Uncertain status in Senegambia. **11a Male.** Grey hood, black mask through eye, chestnut back, pinkish-buff below. **11b Female.** Greyish hood, brown mask and back, whitish underparts, barred on flanks.

12. YELLOW-BILLED SHRIKE *Corvinella corvina*　　　　　　　Page 340
Very common throughout. Woodlands, roadsides, telegraph wires. Noisy and gregarious. **Adult.** Mottled brown above, large dark eye-mask, waxy yellow bill; long mobile tail, cinnamon wing patch.

PLATE 44: STARLINGS

1. SPLENDID GLOSSY STARLING *Lamprotornis splendidus*   Page 349
Uncommon and localised in forest thickets, near tall trees or fruiting figs. **Adult**. Intense glossy greenish above, purplish below. Black eye-patch around striking white eye. Dark band across tail.

2. GREATER BLUE-EARED GLOSSY STARLING *Lamprotornis chalybaeus*   Page 348
Common resident; widespread in bush, farmland and gardens. **Adult**. Glossy blue-green head, blue upperparts, flanks royal blue. Folded wing-tips lie less than half way along blue-green tail. Eye yellow.

3. LESSER BLUE-EARED GLOSSY STARLING *Lamprotornis chloropterus*   Page 349
Resident in woodlands and farms; more common in north and inland. **3a Adult**. Blue-green head, narrow black eye-mask; bottle-green upperparts, magenta flanks. Folded wing-tips lie less than half way along blue-green tail. Eye yellow-orange. **3b Immature**. Body dull sooty-brown, wings iridescent green, eye grey.

4. PURPLE GLOSSY STARLING *Lamprotornis purpureus*   Page 347
Common resident throughout. **Adult**. Bulky, flat-crowned with long, strong bill. Head and body glossy purple, wings glossy blue-green. Folded wing-tips lie close to end of short purple tail. Eye yellow.

5. BRONZE-TAILED GLOSSY STARLING *Lamprotornis chalcurus*   Page 348
Widespread throughout, often present in small numbers with other starlings. **5a Adult**. Iridescent blue head and breast, greenish wings, purplish belly. Folded wing-tips lie half way along short purplish tail. Eye red-orange. **5b Immature**. Dull grey body, greenish on wings, eye grey.

6. LONG-TAILED GLOSSY STARLING *Lamprotornis caudatus*   Page 350
Widespread and common resident. Bush, farmland, gardens. **Adult**. Iridescent blue-green upperparts, blue-violet underparts; long, lax, graduated metallic tail; pale yellow eye appears white in black mask.

7. CHESTNUT-BELLIED STARLING *Lamprotornis pulcher*   Page 350
Uncommon localised resident, especially fields and near villages. **Adult**. Grey hood; blue-green iridescence on upperparts, breast and tail; chestnut belly. White panels in spread wings. Eye whitish.

8. VIOLET-BACKED STARLING *Cinnyricinclus leucogaster*   Page 351
Variable numbers year-round with wet season influx to all parts; mainly open woodland, anywhere with fruiting figs. **8a Adult male**. Upperparts and breast iridescent purple, belly white. **8b Adult female**. Upperparts brown, underparts white boldly marked with drop-like spots and streaks.

9. YELLOW-BILLED OXPECKER *Buphagus africanus*   Page 352
Widespread resident associated with large mammals, usually N'dama cattle in The Gambia. **Adult**. Olive-toned upperparts and breast, buff underparts and pale rump. Strong, waxy-looking, red-tipped yellow bill.

PLATE 45: SPARROWS, FINCHES, WEAVERS, WIDOWBIRDS AND BISHOPS

1. SUDAN GOLDEN SPARROW *Passer luteus*   Page 353
Erratic dry season visitor; arid country, waterholes. Mostly north of Gambia River. Flocks.**1a Male.** Soft golden-yellow body, chestnut mantle. **1b Female.** Pale sandy above, two pale wing-bars.

2. WHITE-RUMPED SEEDEATER *Serinus leucopygius*   Page 383
Frequent inland. **Adult.** Nondescript grey, drop-like breast streaks, white rump in flight, two wing bars, no head barring.

3. YELLOW-FRONTED CANARY *Serinus mozambicus*   Page 383
Widespread throughout, very common in dry woodlands. Flocks. **3a Male.** Yellow below, black malar stripe and eye-stripe. Yellow rump. **3b Female.** Weaker head markings.

4. HOUSE SPARROW *Passer domesticus*   Page 352
Common near urban settlements. **4a Male.** Grey cap, silvery cheeks, black bib, underparts greyish, upperparts warm mottled browns. **4b Female.** Nondescript, pale thick supercilium, dull pale streaky brown above, pale buff-grey below.

5. GREY-HEADED SPARROW *Passer griseus*   Page 353
Widespread, often abundant, in savanna woodland. **Adult.** Chestnut back and tail, head and body grey apart from white throat.

6. BUSH PETRONIA *Petronia dentata*   Page 354
Widespread, often abundant inland. **6a Breeding male.** Chestnut superciliary stripe, bill black. **6b Female.** Smaller than House Sparrow, more noticeable supercilium.

7. CHESTNUT-CROWNED SPARROW-WEAVER *Plocepasser superciliosus*   Page 355
Uncommon in dry savanna woodland. Pairs. **Adult.** Chestnut crown and cheeks, black malar stripe.

8. SPECKLE-FRONTED WEAVER *Sporopipes frontalis*   Page 354
Increasingly common north of the Gambia River. **Adult.** Blackish forehead, chestnut nape; cheeks and underparts whitish.

9. WHITE-BILLED BUFFALO-WEAVER *Bubalornis albirostris*   Page 355
Widespread, very common. Social. Large. **Adult breeding.** Largely black, whitish flecks on back and wings, swollen white bill and frontal shield in breeding plumage.

10. YELLOW-SHOULDERED WIDOWBIRD *Euplectes macrourus*   Page 365
Thinly distributed throughout. Wet grassland, rice fields, fallow land. **10a Breeding male.** Large, glossy black, crested nape, yellow mantle and shoulders. **10b Non-breeding male.** Streaked brown above, yellow shoulders. Longish rounded tail. **10c Female.** Similar to non-breeding male.

11. RED-COLLARED WIDOWBIRD *Euplectes ardens*   Page 366
Uncertain status. Paddies, grassland. **11a Breeding male.** All black with long tail broadening at tip. **11b Female.** Sparrow-like; streaky above, yellowish supercilium, plain buff underparts.

12. NORTHERN RED BISHOP *Euplectes franciscanus*   Page 365
Widespread, often abundant; cropland. **12a Breeding male.** All black head, brown wings and tail, black waistcoat, rest scarlet. **12b Non-breeding male.** Straw-coloured, streaky above, buffy-white supercilium, lightly streaked breast. **12c Female.** Similar but smaller.

13. BLACK-WINGED RED BISHOP *Euplectes hordeaceus*   Page 364
Widespread. **13a Breeding male.** Scarlet head and crown, black face, wings black. **13b Non-breeding male.** Darker than Northern Red Bishop; retains blackish wings. **13c Female.** Similar, but not as dark.

14. YELLOW-CROWNED BISHOP *Euplectes afer*   Page 364
Locally common; wetlands. **14a Breeding male.** Golden-yellow and black. **14b Female.** Broadest, most striking supercilium of the bishops; narrow dark line through eye.

PLATE 46: BUNTINGS AND WEAVERS

1. CINNAMON-BREASTED BUNTING *Emberiza tahapisi*   Page 384
Uncommon but more frequent inland; rocky scrub and laterite ridges. **Adult**. Body cinnamon, head black-and-white-striped, chestnut wing panel. Small blue-grey bib.

2. BROWN-RUMPED BUNTING *Emberiza affinis*   Page 385
Locally common in dry savanna woodlands. **Adult**. Black-and-white head pattern, brown back, bright yellow underparts.

3. ORTOLAN BUNTING *Emberiza hortulana*   Page 384
Rare Palearctic migrant. Savanna and open scrub. **Female**. Droopy pale yellow moustachial stripe on olive head, pale eye-ring, streaky brown body and white outer tail feathers. Faint streaking on crown. Pink bill.

4. HOUSE BUNTING *Emberiza striolata*   Page 384
Rare vagrant. Open scrub, near settlements. **Adult**. Grey head and breast finely streaked, body cinnamon-rufous, streaking very weak.

5. VITELLINE MASKED WEAVER *Ploceus velatus*   Page 358
Locally common. Dry savanna woodlands. **5a Breeding male**. Golden body, olive back, red eye, black mask narrowly includes forecrown, crown rusty. **5b Adult female**. Olive-grey above, buffish face, whiter below.

6. HEUGLIN'S MASKED WEAVER *Ploceus heuglini*   Page 358
Local throughout in farmland and woodland. **6a Breeding male**. Yellow body, yellow eye in black mask, yellow forecrown and crown. **6b Adult female**. Streaky olive upperparts, pale below, chunkier build than Vitelline Masked Weaver.

7. VILLAGE WEAVER *Ploceus cucullatus*   Page 360
Very abundant throughout. **7a Breeding male**. Black hood with broad chestnut wedge at nape, black and yellow above, yellow below. **7b Non-breeding male**. Olive crown, yellow throat, grey back, whitish underparts (variable). **7c. Breeding female**. As non-breeding male but more extensive yellow below.

8. LITTLE WEAVER *Ploceus luteolus*   Page 356
Local throughout; fallow land with small acacias. Smallest weaver, feeds like a warbler. **8a Breeding male**. Canary yellow with black mask to forecrown, olive mantle. **8b Female**. Lacks black mask, whitish belly.

9. COMPACT WEAVER *Ploceus superciliosus*   Page 361
Casamance; grassland and swamp. **9a Breeding male**. Dark eye in black face mask, chestnut forehead and yellow crown. **9b Breeding female**. Dark olive wedge on crown above broad yellow supercilium. Dark mask.

10. BLACK-NECKED WEAVER *Ploceus nigricollis*   Page 357
Patchily distributed in dense habitat throughout. **10a Adult male**. Yellow body with pale eye in narrow black eye mask, black bib, olive upperparts. **10b Adult female**. Olive crown, narrow eye mask, no bib.

11. YELLOW-BACKED WEAVER *Ploceus melanocephalus*   Page 360
Very abundant, mainly near freshwater. **11a Breeding male**. Clean-cut black hood, dark eye, green upperparts, yellow below. **11b Female**. Neatly streaked grey-brown upperparts, pale below with grey-white iris.

12. RED-HEADED WEAVER *Anaplectes rubriceps*   Page 362
Uncommon inland in dry woodlands. **12a Breeding male**. Head and breast red, face mask black, pink bill. Brown back, white below. **12b Female**. Brown head to mantle, white below, red in wings and bright pink bill. Non-breeding male similar.

13. GRAY'S MALIMBE *Malimbus nitens*   Page 361
Casamance; forest. **Adult male**. Black with scarlet breast patch. Slim bill, no white eye-ring. Female similar but less glossy.

PLATE 47: ESTRILDIDS

1. GREEN-WINGED PYTILIA *Pytilia melba*   Page 368
Uncommon, more frequent in the north; open dry woodlands. **1a Adult male**. Red face, bill and tail; grey nape, greenish back; yellow breast. **1b Adult female**. Pale grey head and breast.

2. RED-WINGED PYTILIA *Pytilia phoenicoptera*   Page 367
Widespread but uncommon; mainly dry savanna woodland. **Adult**. Grey body with red suffusion on wings and tail. Barred grey and white flanks and belly.

3. ORANGE-CHEEKED WAXBILL *Estrilda melpoda*   Page 373
Resident, locally common near damper grassy places in savanna. **Adult**. Grey head, neat bright orange cheeks, reddish bill. Brown back with red rump and black tail; whitish below.

4. LAVENDER WAXBILL *Estrilda caerulescens*   Page 373
Common resident; gardens, thicket edges, open woodlands. **Adult**. Lavender-grey with red rump, undertail-coverts and tail. Dark line through eye.

5. BLACK-RUMPED WAXBILL *Estrilda troglodytes*   Page 373
Widespread but local resident throughout. Savanna, thickets, grassland and farmland. **Adult**. Brown upperparts, black rump and tail. Red stripe through eye. Pale underparts lightly barred on flanks.

6. RED-CHEEKED CORDON-BLEU *Uraeginthus bengalus*   Page 372
Widespread and common resident; gardens, scrub, woodlands. **6a Adult male**. Brown back, pale blue breast, flanks and tail, yellowish belly. Red spot on cheek. **6b Adult female**. Lacks red cheek spot.

7. BAR-BREASTED FIREFINCH *Lagonosticta rufopicta*   Page 370
Uncommon local resident. Near freshwater with tangled vegetation. Sexes alike. **Adult**. Brown crown and upperparts; tail red; face and underparts wine-red, buff undertail.

8. BLACK-FACED FIREFINCH *Lagonosticta larvata*   Page 372
Rare, usually near freshwater. Woodland with clearings. **8a Adult male**. Grey head with clean black mask and grey bill, vinous-pink flush over body, black undertail. **8b Adult female**. Similar but no mask, less pink.

9. RED-BILLED FIREFINCH *Lagonosticta senegala*   Page 371
Abundant widespread resident. Gardens, farmland, woodland. **9a Adult male**. Scarlet body, wings brown, pink bill, yellow eye-ring. **9b Adult female**. Brown above, buff below, red spot in front of eye, pinkish bill.

10. BLACK-BELLIED FIREFINCH *Lagonosticta rara*   Page 371
Infrequent; south-east Senegal. Grassland savanna and thickets. **Adult male**. Red body, browner on wings. Belly and undertail black, lacks white flecking. Bill black with pinkish on base of lower mandible.

11. ZEBRA WAXBILL *Amandava subflava*   Page 374
Rare; northern and eastern areas. Swamps and rice paddies. Tiny. **Adult male**. Olive-brown above, red eyestripe (missing in female), golden belly, barred flanks.

12. AFRICAN SILVERBILL *Lonchura cantans*   Page 375
Uncommon local resident; usually drier habitats. Sexes alike. **Adult**. Pale buff-brown above, whitish below; dark wings and black, pointed tail. Silver-grey bill.

13. MAGPIE MANNIKIN *Lonchura fringilloides*   Page 376
Mostly south-east Senegal. Thickets and rice fields. Large. **Adult**. Black hood and tail, brown wings. Large grey bill. Below white with thick bars on flanks.

14. BLACK-AND-WHITE MANNIKIN *Lonchura bicolor*   Page 376
Recorded once, coastal Casamance; forest, edges and clearings. **Adult**. Black upperparts and breast; silver bill. White belly barred black on flanks.

15. BRONZE MANNIKIN *Lonchura cucullata*   Page 375
Abundant resident. Towns, villages, gardens, farmland, bush. **15a Adult**. Brown above, purple sheen to dark head. Iridescent green shoulder patch looks black. White below, barred brown on flanks. **15b Juvenile**. Plain brownish, buffer below; hint of developing barring on flanks; grey bill.

PLATE 48: WHYDAHS, INDIGOBIRDS, QUELEAS, ESTRILDIDS

1. EXCLAMATORY PARADISE WHYDAH *Vidua interjecta* Page 382
Locally common in dry savanna woodlands. **1a Breeding male**. Black above, chestnut collar, yellow belly; enormous tail. **1b Transitional male**. Brownish above, flecked black; striped crown; black wings. **1c Female**. Sparrow-like; pale crown stripe and white supercilium; reddish bill. **1d Immature**. Drab grey-brown, paler on belly.

2. PIN-TAILED WHYDAH *Vidua macroura* Page 381
Widespread; damp grassland, farmland. **2a Breeding male**. Black above, white collar, white below; long tail; coral-red bill. **2b Female**. Striped head pattern, usually with red bill.

3. WILSON'S INDIGOBIRD *Vidua wilsoni* Page 380
Uncommon; distribution should coincide with Bar-breasted Firefinch. **Breeding male**. Purplish black, whitish bill, light purplish legs, dark brown primaries.

4. VILLAGE INDIGOBIRD *Vidua chalybeata* Page 379
Common throughout, often around villages. **4a Breeding male**. Greenish-black, legs orange, bill white. **4b Female**. Streaky, crown stripes, pinkish legs. **4c Immature**. As female but plainer; weaker supercilium.

5. CUT-THROAT FINCH *Amadina fasciata* Page 377
Uncommon dry season visitor; farmland, scrub, savanna woodland. **5a Adult male**. Crimson throat gash, otherwise buff, finely barred or scaled; small chestnut patch on belly. **5b Adult female**. Lacks red throat.

6. QUAIL-FINCH *Ortygospiza atricollis* Page 374
Locally common resident; grassland near swamps. Small flocks. **6a Adult male**. Brown above, black face and red bill, barred flanks. **6b Adult female**. Duller, lacks black face.

7. RED-BILLED QUELEA *Quelea quelea* Page 363
Uncommon dry season visitor in south, breeds northern Senegal. Swamp, farmland, open grassland. **7a Breeding male**. Red bill, suffused pink around black face; body mottled pale buff. **7b Female**. Reddish bill; supercilium creates hint of spectacle around eye on greyish head. Non-breeding male similar.

8. RED-HEADED QUELEA *Quelea erythrops* Page 363
Uncommon; mainly wet season breeder. **8a Breeding male**. Scarlet hood, dark grey bill, streaky brown above, pale buff below. **8b Female**. Lacks red hood; yellow-stained supercilium.

9. CRIMSON SEED-CRACKER *Pyrenestes sanguineus* Page 368
Rare in freshwater swamp forests. **Adult male**. Scarlet body and tail, brown wings and belly. Massive, triangular gunmetal bill on large head. Broken white eye-ring.

10. WESTERN BLUEBILL *Spermophaga haematina* Page 369
Uncommon in coastal forest patches. Thickset. **10a Adult male**. Black with crimson throat, breast and flanks. Red-tipped conical blue bill. **10b Immature male**. Black with diffuse red below. **10c Adult female**. As male but red duller and white-spotted belly.

11. CHESTNUT-BREASTED NEGROFINCH *Nigrita bicolor* Page 367
Very rare in coastal forests. **Adult**. Very small; dark, two-toned pattern. Grey above, deep russet below.

12. DYBOWSKI'S TWINSPOT *Euschistospiza dybowskii* Page 370
Rare in south-east Senegal. Rocky outcrops in savanna. **12a Adult male**. Grey head, red back, black wings. Black belly finely spotted with white. **12b Adult female**. Flanks dark grey and more extensively spotted.

OSTRICH

Struthionidae

One species. Huge flightless bird with head and neck almost naked. Body entirely covered with soft plume-like feathers. Wings and tail similar. Thighs powerful and naked, feet with two toes.

OSTRICH
Struthio camelus   Not illustrated

Fr. Autruche

IDENTIFICATION 200-240cm. The world's largest bird. Sexes differ. Male black and white; female greyish-brown and white; immature resembles female. Unmistakable.

HABITS Favours open short-grass plain and semi-desert. Flightless and capable of fast sprints over long distances. Sociable, moving in groups when numbers sufficient.

VOICE Varied vocal reportoire includes a booming, roaring call and some hissing.

STATUS AND DISTRIBUTION There are no Gambian records. Known only from the Ferlo district of northern Sengal where previously common; has bred, now considered practically exterminated; some occasional singles recently reported from there (B. Tréca *in litt.*).

PETRELS AND SHEARWATERS

Procellariidae

Pelagic seabirds characteristically with two tube-shaped nostrils associated with specialist salt excretion mechanism on the bill. Wings are long and narrow, often held stiff and straight in flight. They glide low over the sea with occasional wing beats. Food is largely fish and squid, obtained by surface collection or plunge-diving. Information on this group of birds off Senegambia is currently very poor. The following species are known or could be expected throughout Senegambian waters; others might also occur.

FEA'S PETREL
Pterodroma feae   Plate 1

Fr. Pétrel gongon

Other names: Gon-gon, Cape Verde Petrel

IDENTIFICATION 35cm (wingspan 85cm). A small-medium gadfly petrel. Upperparts and sides of chest a sooty ashy-grey creating the appearance of a partial breast-band. Wings and rump are darker. Forehead mottled grey and white. Face white with a heavy dark smudge behind the eye. Underparts white, underwings dusky brown-grey. Bill black and stubby. Legs flesh-coloured. Flight pattern is rapid and low over the water with a zig-zagging action. Tail noticeably wedge-shaped. **Similar species** Size, flight action and pale coloration preclude confusion with other species within Senegambian waters.

STATUS AND DISTRIBUTION Unsubstantiated in Senegambian waters but known to disperse from breeding area on Cape Verde along entire West African seaboard to latitude of Sierra Leone. Formerly treated as a subspecies of Soft-plumaged Petrel *Pterodoma mollis* of southern oceans.

BULWER'S PETREL
Bulweria bulwerii   Plate 1

Fr. Pétrel de Bulwer

IDENTIFICATION 27cm (wingspan 65cm). Uniformly dark plumage with a paler band on the upperwing, difficult to see at long range when wings look uniformly dark. Wings long, slim and narrow. Tail is comparatively long and wedge-shaped. Slightly larger than storm-petrels with flight comparable to shearwaters. **Similar species** Sooty Shearwater is much larger with silvery-white underwing patches.

STATUS AND DISTRIBUTION Pelagic. A number of recent Senegalese sightings e.g. seven in October 1990 off Dakar. Breeds in the Canary Islands and on Cape Verde.

CORY'S SHEARWATER
Calonectris diomedea   Plate 1

Fr. Puffin cendré

IDENTIFICATION 50cm (wingspan 110cm). *C. d. borealis*. Largish, broad-winged and heavily built shearwater. Upperparts ashy-brown, sides of face and chest white and mottled with ashy-brown. Underparts white, underwings white with dark borders. Yellow bill with a darkish tip is an important characteristic; legs pink. Cape Verde Shearwater, *C.* (*d.*) *edwardsii*, often considered a full species, is smaller, slimmer, longer-tailed and has a slightly dark-capped appearance and darker, grey-brown upperparts. The thin grey bill looks dark-tipped at distance. Flight is slow with effortless glides, wings are held slightly drooped and projected forwards. Follows ships and trawling vessels. **Similar species** Lacks the marked cap and pale collar of Great Shearwater.

STATUS AND DISTRIBUTION Pelagic. Senegal records April-November with up to 450 in November 1990 and up to 240 moving north in April 1992 off Dakar (race not specified). In October 1996, up to 1,700 Cape Verde Shearwaters off Dakar, but only a single Cory's.

GREAT SHEARWATER
Puffinus gravis Plate 1

Fr. Puffin majeur

IDENTIFICATION 50cm (wingspan 105cm). Largish, heavily built shearwater, with a conspicuous dark cap and pale collar. Cheeks white, rest of upperparts ashy-brown contrasting with dark outer primaries and trailing edge, and white horseshoe at the base of the tail. Underparts white with a conspicuous dark smudge on the belly. Bill slim, long and dark. Distinctive underwing pattern of white wing edged in black with dark diagonal bar across inner section of wing, visible when seen well. Legs flesh-coloured. **Similar species** Dark-capped appearance, pale nape, more prominent white band at the base of the tail and a more rapid flight pattern separate Great from Cory's Shearwater.

STATUS AND DISTRIBUTION Pelagic; recorded northern Senegal in January/June, July and November. Trans-equatorial migrant breeding in South Atlantic.

SOOTY SHEARWATER
Puffinus griseus Plate 1

Fr. Puffin fuligineux

IDENTIFICATION 45cm (wingspan 100cm). Medium-sized but heavy-bodied shearwater with long narrow wings. All-dark, but paler under the chin. Contrasting silvery-white central portion of the underwing is diagnostic, but may only be visible

intermittently in flight. Slim bill long and blackish, legs bluish-grey. **Similar species** Other shearwaters have much paler underwings (when seen well). Bulwer's Petrel is smaller with wholly dark underwings.

STATUS AND DISTRIBUTION Pelagic; seen off Senegal on passage: 280 and c. 1000 off Dakar September and October 1990 and two offshore April 1992.

MANX SHEARWATER
Puffinus puffinus Plate 1

Fr. Puffin des Anglais

IDENTIFICATION 35cm (wingspan 80cm). Small-medium shearwater, strikingly black above and white below with differentiation cleanly defined. Bill slim, long and dark. Wings long and narrow. Flight swift and stiff-winged. **Similar species** Mediterranean Shearwater *Puffinus yelkouan* (Fr. Puffin yelkouan) recently reported for the first time off northern Senegal; it appears similar but sooty-brown, not black, and not sharply demarcated as in Manx. See also Little Shearwater.

STATUS AND DISTRIBUTION A bird swimming close to the shore off Fajara WD in October 1992 and a washed-up corpse in good condition 5km further south in late March 1995 were the first confirmed records. Considered common off Senegal during the Palearctic winter e.g. c. 40 August-September 1990.

LITTLE SHEARWATER
Puffinus assimilis Plate 1

Fr. Petit Puffin ou Puffin obscur

IDENTIFICATION 29cm (wingspan 60cm). *P. a. baroli*. Small. Upperparts blue-black, underparts white. Recalls a smaller Manx Shearwater with the face whiter and the wings much shorter. Looks chubby with a rounded head that may be 'jerked' back in flight recalling Common Sandpiper flight. Flies with characteristic bursts of rapid wing beats followed by short glides. **Similar species** Apart from considerable size difference the most useful feature differentiating from Manx is that white extends above the eye in Little (below the eye in Manx).

STATUS AND DISTRIBUTION Pelagic; uncommon off Senegal in Palearctic winter. In The Gambia, sightings in August-September 1990. Breeds in the Canaries.

STORM-PETRELS

Hydrobatidae

Diminutive pelagic birds characterised by strongly hooked bills and tube-shaped nostrils. Their flight mode is a diagnostic delicate fluttering above the waves, sometimes pattering directly onto the sea. They feed by picking krill, crustaceans, fish and animal fat directly from the surface. Usually all-dark with differing amounts of white on the wings and rump. Successful identification requires a combination of behaviour plus structural details such as tail shape and toe projection as well as plumage features. Five species listed two of which are known only from Senegalese waters; likely to be under-recorded and others may also occur.

WILSON'S STORM-PETREL
Oceanites oceanicus   Plate 1

Fr. Océanite de Wilson

IDENTIFICATION 18 cm (wingspan 40cm). Small. Plumage overall sooty-brown. Very conspicuous white rump extends down the sides of the underwing-coverts. Obvious greyish-white panel on the upperwing. There may be an indistinct pale grey stripe on the underwing. Tail square with rounded corners; legs are black and noticeably long. In flight in good conditions the toes can be seen extending well behind the tail and this an important identification feature. At very close range the yellow webbing between the toes can be seen. Habitually feeds by pattering with the wings held high. Wings rather rounded and paddle-shaped. This species is a regular attender of ships and trawlers. **Similar species** Intermediate in size between smaller British Storm-Petrel and larger Leach's Storm-Petrel. Does not exhibit conspicuous white underwing-bar of British Storm-Petrel. Wings held straight and erect when pattering, not angled as in British Storm-Petrel. Pure white rump lacks the central grey stripe seen in Leach's Storm-Petrel.

STATUS AND DISTRIBUTION May be expected in Senegambian waters April-September; hundreds off Dakar in April 1992; also recorded during these months off Banjul. Breeds in the southern Atlantic and migrates northwards.

WHITE-FACED STORM-PETREL
Pelagodroma marina   Plate 1

Fr. Océanite frégate

IDENTIFICATION 20 cm (wingspan 42cm). Small and unmistakable; the only storm-petrel in the region with a patterned face and white underparts. Plumage overall grey-brown, with a pale face, upperwing-bar and a grey rump. Tail square, legs project well behind the tip of the tail in flight. Distinctive flight involves the bird swinging pendulously whilst wind-hovering low over the ocean. When feeding it skips and dips low over the waves with legs loosely dangling. Not known to follow ships. **Similar species** May occur in small parties or rafts swimming on the sea when superficial resemblance to phalaropes could cause an identification pitfall. Dissimilarity of bill structure and shape separates them immediately.

STATUS AND DISTRIBUTION Highly pelagic with no records as yet from Gambian waters; included here as likely to occur. Race *P. m. eadsii* breeds on the Cape Verde islands, fledging May-July.

BRITISH STORM-PETREL
Hydrobates pelagicus   Plate 1

Fr. Pétrel tempête

Other names: European Storm-Petrel, Storm Petrel

IDENTIFICATION 17cm (wingspan 38cm). Smallest storm-petrel in our waters and darkest-plumaged. Overall sooty-black, more brownish on the underparts. Very obvious white band on the underwing and pale line on the upperwing sometimes visible. Square-cut white rump not as extensive as in Wilson's Storm-Petrel. Tail square and blunt-ended, black legs do not extend behind tail in flight. Flight is fluttering, rather hurried and bat-like. Does not glide and is a regular follower of ships. Will come close to the shore and there is an interesting record of birds collecting food directly from a beach which is used as a fish-landing site at Gunjur in southern Gambia.

STATUS AND DISTRIBUTION The commonest storm-petrel in Senegambian waters in recent years. Some years noted in large numbers from land, usually after adverse weather. Seen frequently across the estuary at Banjul and penetrates up to James Island. Expected in the dry season, especially February-May e.g. large numbers in March off Banjul and off Dakar in April. Breeds in Europe, migrating to West African waters in the northern winter.

MADEIRAN STORM-PETREL
Oceanodroma castro   **Plate 1**

Fr. Pétrel de Castro

IDENTIFICATION 20cm (wingspan 45cm). Small. Separation difficult from the other three species of storm-petrel which combine dark plumage and a white rump. Plumage overall sooty-black including underwing. White rump clearly defined and extends both onto the lower flanks and sides of the undertail-coverts but is not as extensive as in Wilson's Storm-Petrel. **Similar species** Lacks pale underwing-bar of British Storm-Petrel and feet do not project as in Wilson Storm-Petrel; slightly larger than both. Compared to Leach's Storm-Petrel rump patch plain, not divided by a central line, less contrasting on upperparts, and much less forked tail. Flight buoyant and steady with a series of quick beats followed by shearwater-type gliding. These field characteristics are subtle and meticulous observation is required. Not usually attracted to ships and trawlers.

STATUS AND DISTRIBUTION Highly pelagic and occurs offshore along the entire West African seaboard; seldom near to our coastline, but has been seen near Dakar in April. Breeds in the North and South Atlantic.

LEACH'S STORM-PETREL
Oceanodroma leucorhoa   **Plate 1**

Fr. Pétrel cul-blanc

IDENTIFICATION 21cm (wingspan 46cm). Largest of the white-rumped storm-petrels. Upperparts blackish-brown; underparts sooty-brown. Important field marks are visible pale bands across the upperwing in flight; wings long, pointed and noticeably broad at the base, and tail is prominently forked. White rump divided down its centre by a diffuse line of grey feathers which is not always visible in flight over the ocean. Flight pattern is both rapid and buoyant with frequent changes in direction recalling a miniature Black Tern in shape. Wings are held well back and low whilst feeding which is conducted in a pattering fashion. It does not generally associate with ships.

STATUS AND DISTRIBUTION Pelagic, favouring the region of the continental shelf. Seen off Senegal throughout the year, particularly in April with a Gambian record in September. Occurs along the whole length of the West African coast with main movements arriving off Africa in November and peaking in the Gulf of Guinea in January. Breeds in the North Atlantic.

GREBES

Podicipedidae

Three species. Grebes are highly accomplished divers, their legs placed well back on the body. Lobed toes facilitate swimming. Tail very short, flight swift with head held low. Note: Great Crested Grebe *Podiceps cristatus* (Fr. Grèbe huppé) is a vagrant to northern Senegal. A 1902 record of Great Crested Grebe in the upper region of the Gambia River is no longer acceptable.

LITTLE GREBE
Tachybaptus ruficollis   **Plate 1**

Fr. Grèbe castagneux

Other name: Dabchick

IDENTIFICATION 25cm. *T. r. capensis*. Sexes similar. Small and plump showing no apparent tail; sits high in the water. In breeding plumage mostly during the rainy season, dark brown with cheeks, neck and chin rich russet-brown, gape bright greenish-yellow. Non-breeding and immatures grey-brown, seen in all seasons. White patch in the wing is visible in flight. Eye dark red, stubby bill black, legs black. **Similar species** Black-necked Grebe *Podiceps nigricollis* (Fr. Grèbe à cou noir) is a rare visitor to northern Senegal, October-April; locally frequent to common. Differs from Little Grebe in larger size, greyer plumage, pale cheeks and distinctive head

shape with slightly upturned bill.

HABITS Favours tidal creeks, flooded rice fields, sewage outfalls and occasionally the main river, usually singly or small loose groups. Dives from the surface to feed or escape. Runs along water for several seconds before taking flight. Feeds principally on small fish, crustacea and aquatic insects.

VOICE Far-carrying whinnying trill; alarm note a loud *chik*, also a contact call *bee-heeb, bee-heeb.*

STATUS AND DISTRIBUTION Locally common in coastal WD (e.g. Bund Road, Kotu Sewage Farm) with seasonal and irruptive increases in numbers. Occasional sightings in freshwater of river or swamp in CRD and URD, most frequently in the dry season. Small numbers breed in Senegal; common to abundant winter visitor to northern Senegal. Known to be nomadic and fluctuating populations probably result from intra-African migrations.

BREEDING Nest, a floating platform of vegetable matter, loosely fixed, allows the nest to rise and fall with changing water levels. A few known nesting attempts in The Gambia have succumbed to either storm wash-outs or predation.

TROPICBIRDS

Phaethontidae

One species. Large, pelagic, tern-like seabird; graceful and elegant with rapid, strong wing beats. Wedge-shaped tail and two long central rectrices, longer than the body length; plumage mainly white. Forages by plunge-diving, often after hovering at a considerable height. Feeds mainly on fish and squid. Note: previously listed White-tailed Tropicbird *Phaethon lepturus* (Fr. Phaéton à queue blanche) for The Gambia now considered erroneous.

RED-BILLED TROPICBIRD
Phaethon aethereus   Plate 2

Fr. Phaéton à bec rouge ou Grand Phaéton

IDENTIFICATION 50cm (100cm with tail elongation). Wingspan 105cm. Sexes alike. Red bill, and very long white tail streamers diagnostic. Overall plumage white, sometimes flushed pinkish with irregular diagonal black bar across coverts; mantle and back barred black. Prominent black mask through eye extends to nape. Primaries black, white on innner webs. Tail white, wedge-shaped in all stages. Immature is heavily but delicately barred black on white, lacks streamers. **Similar species** Wedge-shaped tail separates tropicbirds from larger terns.

HABITS Oceanic, only likely to be seen from land during storms or after seasonal harmattan winds. Likely to be encountered singly. Characteristic flight with quick deep wingbeats, occasional gliding. Not usually attracted to ships.

VOICE Silent at sea.

STATUS AND DISTRIBUTION An inconclusive record over coastal mangroves, near Abuko in March 1996, after strong winds, and a recent confimed record off Tanji. Also seen off Dakar; a small population breeds on nearby Iles de la Madeleine.

BREEDING Nests colonially on bare ground, often amongst tree roots.

GANNETS AND BOOBIES

Sulidae

Two species. Large seabirds with long wings, streamlined cigar-shaped bodies and wedge-shaped tails. Bill stout, conical and sharply pointed. Dive spectacularly to catch fish prey.

NORTHERN GANNET
Sula bassana   Plate 2

Fr. Fou de Bassan

Other names: Gannet, *Morus bassanus*

IDENTIFICATION 87-100cm. Wingspan c. 170cm. Sexes similar; male larger than female. Adult with body and tail entirely white, head and nape washed yellow, primaries black. Bill pale blue, feet black. First-year birds, frequent in West African waters, dark brownish-grey, with small white spots on the upperparts, paler below with distinctive dark diffuse breast-band. A dense cluster of white spots at the base of the tail join to form a rump-crescent. Most individuals at the end of their first year have a white head. Full adult plumage is attained during a three-year period at sea. **Similar species** Cape Gannet *Sula capensis* (Fr. Fou du Cap) yet to be positively identified in Senegambian waters but a subadult was recently suspected off Dakar in August. Breeds in the South Atlantic and migrates trans-equatorially. Differs from Northern Gannet in having the tail, primaries and secondaries black (also visible on speckled blackish-brown immatures). At a distance, large tern species, especially Caspian Tern, may be confused with gannets.

HABITS Pelagic, occasionally seen from the shore, at coastal headlands, bays and in the mouth of the river. Most frequently recorded after strong winds which regularly occur in the early part of the year. Flight pattern is strong, with a noticeable tilting action in flight.

VOICE Silent at sea.

STATUS AND DISTRIBUTION In some years temporarily common off coastal WD after strong harmattan winds, particularly January-February; additional records November-March and occasionally in May and July. Sightings of full adults exceptional. Probably under-recorded.

BREEDING Breeds in huge colonies in North Atlantic; Scottish-ringed birds have been recovered in The Gambia.

BROWN BOOBY
Sula leucogaster   Plate 2

Fr. Fou brun

IDENTIFICATION 65-75cm. Wingspan c. 140cm. Uniformly dark chocolate brown upperparts, lacking any speckling or rump mark. Distinctive white 'waistcoat' and inner underwing on otherwise brown underparts. Immature dull brown above, dirty buff below. Sexes similar, except colour of

orbital skin around eye blue in male and yellowish with a dark frontal spot in female. Feet greenish. Immature facial skin greyish. **Similar species** Immature separable from young Northern Gannet by completely uniform plumage above, without speckling or pale rump mark. Subadult Northern Gannet always shows some white on upperparts.

HABITS Pelagic. Very rarely sighted in coastal bays from land, but worth looking for after strong dry-season winds. Glides with deliberation, rarely tilting, usually lower over the ocean than Northern Gannet. Brown Booby plunges more at an angle when fishing and will also pluck flying-fish as they leave the water. Follows ships and uses both masts and rigging as perching posts.

VOICE Silent at sea.

STATUS AND DISTRIBUTION Rare vagrant; five sight reports since 1990 from land at Sanyang, Tanji Bird Reserve and Fajara cliffs. In Senegal, several records, September and February–April.

BREEDING c. 1000 pairs breed on Cape Verde. Nests amidst rocks on bare ground.

CORMORANTS

Phalacrocoracidae

Two species; aquatic birds with dense and glossy plumage. Tail long and stiff; feet webbed and toes hooked. Bill long with upper mandible strongly hooked. Dive whilst swimming to catch their prey. Flight strong with neck extended. As a result of only partially effective waterproofing feathers tend to become waterlogged, impairing flight until properly dried out. This is achieved by holding wings outstretched, while perched upright in the sun, either on a stump, on a rock or on the shoreline.

GREAT CORMORANT
Phalacrocorax carbo   Plate 2

Fr. Grand Cormoran

Other names: White-breasted Cormorant; Cormorant

IDENTIFICATION 90cm. *P. c. lucidus.* Sexes similar. Large and blackish with pure white breast. Adult is blue-black with a greenish sheen except the chin to the upper belly which is clean white and appears as a large bib. Facial skin yellow-green, bill grey. In breeding dress there is a patch of white on the upper thigh and the crown is flecked with white. Immature duller above, greyer below. **Similar species** When feeding with body submerged, the clean white on the chin and throat and a much stouter neck separate it from smaller Long-tailed Cormorant.

HABITS Both solitary and gregarious. Perches erect

and conspicuously. Swims and dives very strongly for fish, crustaceans and occasional molluscs. Flight powerful and generally low over the water with head and neck outstretched. Perches on overhead powerlines and shipwrecks at the coast, and forages in narrow mangrove creeks.

VOICE Usually silent; a variety of growls, grunts and hisses at the nest.

STATUS AND DISTRIBUTION Seasonally common at the coast, peaking during the rains at the mouth of the main estuary; occasional birds along the river to CRD, becoming infrequent in URD. Records in every month. In Senegal found along the whole coast, in riverine and delta habitats.

BREEDING Unrecorded in The Gambia; known off Dakar, Djoudj and in the delta of Senegal River. Colonial; builds a large stick nest, which becomes covered in guano and is often used again year after year.

LONG-TAILED CORMORANT
Phalacrocorax africanus   Plate 2

Fr. Cormoran africain

Other names: Long tailed Shag; Reed Cormorant

IDENTIFICATION 60cm. Sexes similar. Typical cormorant shape, a third smaller than Great Cormorant. Breeding plumage blackish with a glossy, oily green sheen on the upperparts, short nuptial crest and pink swollen area of skin in front of the eye is flecked with fine white feathering. Tail is long; bare facial skin red to orange. Immatures and non-breeders are without the sheen above and are extensively dirty greyish-white on the foreneck, throat and underparts. Bill horn-yellow with some black markings. Eye ruby-red, legs and feet black. **Similar species** In water, avoid possible confusion with rarely seen African Finfoot, (see text for that species).

HABITS Inhabits all inland and coastal waters.

Swamps, paddies, mangrove creeks and sewage outfalls are all likely habitats. When not breeding it is found singly or in loose parties, sometimes in small flocks. Feeds in same style as its larger relative. Evening flights in good numbers to mixed species communal roosts are a prominent feature on freshwater stretches of the river and in coastal mangroves.

VOICE Generally silent. Vocal when going to roost and at the nest, where it gives a hissing cackle.

STATUS AND DISTRIBUTION Common to abundant in suitable habitat throughout. Considerable post-breeding dispersal occurs towards the coast early in the dry season, less often seen in URD later in the dry season. In Senegal widespread along all coastal and riverine delta habitats.

BREEDING Many birds carrying material near Basse in mid-September. Builds nest of sticks and reeds on islands in the river at varying heights, depending on site availability. Often in mixed waterbird colonies.

DARTERS

Anhingidae

One species. Generally similar to the cormorants in appearance and habits, with a long snake-like neck, held crooked at rest in characteristic serpentine pose. Dagger-like bill, straight and tapering; fine backward serrations on inner bill tip surfaces grip fish prey efficiently but also snare easily in nylon net and rope. Dives, submerging for long periods.

AFRICAN DARTER
Anhinga rufa   Plate 2

Fr. Anhinga d'Afrique ou Anhinga du Sénégal

Other names: Anhinga, Snake-Bird

IDENTIFICATION 80cm. Sexes differ. Male glossy black to brown with a chestnut head and neck, with a conspicuous fawn stripe running down the side of the face to the neck. Female and immature are pale buffish-brown. Chin and throat buff-white. Tail distinctively long, stiff and held fanned when resting. Wings and mantle are finely streaked white, prominently on elongated black scapulars. Iris golden, bill horn-brown, legs and feet brown. **Similar species** Shorter-necked African Finfoot could be confused with Darter fishing with body submerged.

HABITS Found in fresh and brackish waters, particularly bolons and mangrove creeks. Often perches on a bare branch or stump, dropping vertically into the water at the approach of danger. Neck held kinked or straight and tail flared fully open in flight; soars well. Skilled swimmer, fishes with body submerged and head and neck showing above water. Dries out after fishing with wings outstretched.

VOICE Usually silent; utters a harsh croak at the nest.

STATUS AND DISTRIBUTION Seasonally common to locally abundant throughout wetland Senegambia. Frequent in WD in early dry season with an increase in LRD and NBD (Bao-bolon) in the second half of the dry season; rare or absent in URD. Locally concentrated in CRD when breeding.

BREEDING Breeds during the rains, building a nest of sticks and reeds which is lined with finer plant material. Colonial, often with egrets, herons and cormorants. Incubates with eggs placed on top of its large webbed feet.

NOTE This species is sometimes considered conspecific with Oriental Darter *A. melanogaster.*

PELICANS

Pelecanidae

Two species. Huge, unmistakable, aquatic birds, stately on the water and in flight. Long straight bill with characteristic distensible throat pouch. Legs short and stocky with fully webbed feet. Awkward on land but graceful in the air and capable of impressive aerial manoeuvres. They are efficient gliders and soar effortlessly. Sexes broadly similar.

GREAT WHITE PELICAN
Pelecanus onocrotalus　　　Plate 2

Fr. Pélican blanc

Other name: White Pelican

IDENTIFICATION 150-185cm. Huge pelican, all-white body with yellowish tinges, flushed pale pink in breeding season, when a small crest develops on nape. Bill long, horn at base, yellow to flame towards tip, with substantial yellow pouch. Bare facial skin orange in male, yellow in smaller female. Legs orange. Immature buffish-brown, attaining adult plumage over a series of moults. In flight long broad wings show distinctive all-black flight feathers, forming a broad black tip and trailing edge, contrasting with white underwing-coverts. Immature in flight largely brown above with whitish rump; underside of wing brown with central white bar. **Similar species** Pink-backed Pelican is rather smaller, with smaller pouch, and greyer in plumage. Lacks strong black-and-white contrast on wings in flight.

HABITS Favours coastal and riverine areas, but often encountered soaring on midday thermals. Feeds communally, herding fish until whole fleet dips pouches below water in synchrony.

VOICE Noisy at the nest and when coming to roost; emits a range of bays and moos.

STATUS AND DISTRIBUTION Frequent from coastal WD to CRD, but most frequent in LRD and NBD especially in the wet season; no modern records in URD. Mainly encountered in groups of 20-50 adults with a record count of 250. Widely distributed in suitable habitat in Senegal.

BREEDING Not recorded in The Gambia. Major colonies in Djoudj, northern Senegal and at Kalissaye, southern Senegal number several thousand pairs. Nest of sticks, sparsely lined. Normally only one chick survives.

PINK-BACKED PELICAN
Pelecanus rufescens　　　Plate 2

Fr. Pélican gris

Other name: Grey Pelican

IDENTIFICATION 130-160cm. Huge pelican, but detectably smaller than Great White. Greyish-white overall, with bristling crest on nape. Bill grey with pinkish pouch, legs pinkish. Eye brown with bare grey skin around eye. In flight shows far less contrast between pale whitish-grey wings and blackish flight feathers than does Great White. Pink back rarely seen; most strongly developed in nuptial plumage (end of dry season, early rains) when bare parts also brighten, including narrow diagonal yellow and black striation on throat pouch, grey crest more prominent and body whiter. Alternative name of Grey Pelican is far more apt. Immature grey-brown, with greyish-pink feet, grey bill with reduced greenish pouch. **Similar species** See Great White Pelican; structural difference in bare skin area around eye useful quick separator.

HABITS Favours any aquatic habitat, fresh or saline, including the sea, where conditions favour fishing. Often perches on wrecks off Banjul coast. Feeds cooperatively as Great White. Soars high and frequently, often with raptors.

VOICE Silent when not breeding. Bill clapping, hisses and croaks at the nest.

STATUS AND DISTRIBUTION Common to abundant resident found in large numbers along the coast throughout the year e.g. Bund Road. Locally abundant in LRD when breeding e.g. Kwinella. Becomes progressively less common inland towards URD. Widely distributed in suitable habitat in Senegal.

BREEDING Colonial. Breeders return to nest sites early in the rainy season in LRD. Well established village baobab plantations are regular sites for platforms of sticks and branches which may become substantial with time. Colonies are also sited on low mangrove-covered islands in the estuary. Young fledge at c. 3 months when many disperse to the coast.

FRIGATEBIRDS

Fregatidae

One species. Oceanic, piratical seabird with long angular wings and deeply forked tail. May be more regular offshore than currently indicated, as it breeds on Cape Verde.

MAGNIFICENT FRIGATEBIRD
Fregata magnificens   Plate 2

Fr. Frégate superbe

IDENTIFICATION 90-115cm (wingspan 240cm). Sexes differ. Very large, with dark plumage and unmistakeable rakish silhouette. Long deeply forked tail in both sexes. Long narrow wings held half extended when soaring. Characteristically opens and closes tail in flight. Long slender hooked bill; legs and feet tiny. Male glossy black above, duller below; red throat patch is only inflated when displaying on breeding territory, and not usually visible. Female dull and blackish-brown with a large whitish patch on the breast. Immature pale brown above, white below, and takes four to six years to reach adult phase. **Similar species** Head and bill shape precludes problems with any fork-tailed raptor species when high over the ocean.

HABITS Pelagic, occasional in coastal waters. The most aerial of seabirds, gliding and soaring with incredible manoeuvrability. The long hooked bill is used to pick off flying fish and squid from the ocean surface, but frigatebirds are also highly piratical on various gull and tern species.

VOICE Silent at sea.

STATUS AND DISTRIBUTION Vagrant. Two records from coastal WD: off Banjul March 1965 and off Cape Point near Bakau in October 1980.

BREEDING Nests colonially, building a platform of twigs, usually in the top of low mangrove. Breeding season prolonged, on restricted islets on Cape Verde where it is declining.

HERONS, EGRETS AND BITTERNS

Ardeidae

Eighteen species. Long-legged, long necked wading birds, mostly of aquatic habitats. All have dagger-shaped bills, sometimes long. Identification problems within the group principally involve the white egrets, and between immatures of some smaller herons and bitterns.

EURASIAN BITTERN
Botaurus stellaris   Plate 2

Fr. Butor étoilé

Other names: Common Bittern, Great Bittern, Bittern

IDENTIFICATION 75cm. Largest of Senegambian bitterns, golden-brown overall, streaked and striped with black and darker brown. Heavy and powerful greenish-yellow bill. Legs comparatively short and greenish. **Similar species** Possibly immature Black-crowned Night Heron and White-backed Night Heron, but both of these are considerably smaller with varying brownish-green plumage spotted with white. Much darker and larger than Squacco Heron which also lacks black crown of Eurasian Bittern. See also White-crested Tiger Heron.

HABITS Adopts famed 'bittern-posture' when disturbed, holding the bill vertically erect and the body frozen. Superbly camouflaged. Flight infrequent, with neck retracted but flushes with neck extended and legs dangling. In flight appears owl-like with rounded downcurved wings. Most likely to occur in mangroves or marshes.

VOICE Usually silent outside the breeding season.

STATUS AND DISTRIBUTION Rare. Two modern records from coastal mangroves, in February 1974 and April 1977, and an inland record in CRD, December 1996. Several recent records from northern Senegal, January-March.

LITTLE BITTERN
Ixobrychus minutus   Plate 2

Fr. Blongios nain

IDENTIFICATION 36cm. Sexes differ. Diminutive bittern with large creamy patches contrasting with otherwise dark wings. Male of the resident African race *I. m. payesii* has the crown, back and the flight feathers blackish-green, with contrasting rufous neck. Female has a brown back with pale feather edges. Both sexes are buff below. Immature more boldly streaked above and below, with pale wing panels heavily mottled brown and buff. Migratory Palearctic race *I. m. minutus* has wing patches and neck pinkish-buff. In both races eye yellow-orange, bill, legs and feet greenish. **Similar species** Pale wing panel, although obscure in immatures, is always detectable with care and distinguishes from other similarly coloured herons and bitterns. All except Dwarf Bittern are separable on size.

HABITS Favours swamps, reedbeds, marshes and overgrown waterside margins. Secretive, often only glimpsed in flight. Recognisable by its tiny size; typically flushes from rank vegetation with wings and neck outstretched and legs dangling untidily. Freezes and fixes in unconventional positions as well as upright when alarmed. An agile climber amongst waterside vegetation.

VOICE A rapid *gak-gak-gak-gak* when disturbed. Advertising call is a *guhrrrr-guhrrrr*.

STATUS AND DISTRIBUTION Uncommon in WD, recorded at the coast January-May. Frequent in suitable habitat in CRD during the wet season, June to November, becoming locally common in some of the more extensive reedbeds, notably in July. Information on races involved not available. Widespread in coastal Senegal north and south of The Gambia.

BREEDING Coastal WD record in The Gambia in September 1962. A platform of reedstems placed low over water in reeds or tangled vegetation.

DWARF BITTERN
Ixobrychus sturmii   Plate 2

Fr. Blongios de Stürm

Other name: Rail Heron, Rail Bittern

IDENTIFICATION 30cm. Sexes similar. Diminutive. Unusual dark slaty-grey above, pale buff below with throat and breast marked with well spaced longitudinal black stripes and a bolder central stripe from below the chin. Stubby short crest more apparent when neck stretched; when pale chestnut patch on the angle of the wing is exposed from beneath the neck feathers. Facial skin bluish-grey; bill, legs and feet yellowish-green. Immature paler, with a rufous flush on underparts and pale-barred back. **Similar species** Immature Striated Heron distinguished by broader black throat stripes, shorter bill, continuous dark colouring from cap through hind neck to back, fine transverse barring, not white spotting, on upperparts. Adult Striated Heron has pale grey

throat with single narrow white and chestnut streak and black cap attenuating to a point on nape. Little Bittern has obvious colour differences and large creamy patch in the wing.

HABITS Favours freshwater streams, flooded land and ponds, usually with nearby tall creeper-covered vegetation. Largely crepuscular and nocturnal. Readily perches in large trees after being flushed. Round-winged in laboured slow flight; adopts 'bittern-posture' when alarmed.

STATUS AND DISTRIBUTION Rare at the coast with all recent records occurring in late rains and early dry season. Frequent in URD and CRD during the wet season with all records July-November. In Senegal recorded February-December in the east with wet season records from both north and south.

VOICE A deep trisyllabic, whoot-whoot-whoot or *hoo-hoo-hoo*; deep guttural croak when flushed.

BREEDING Not proven, but suspected in URD. In northern Senegal nest with eggs in September. Scant platform of twigs placed low over the water.

WHITE-CRESTED TIGER HERON
Tigriornis leucolophus   Plate 2

Fr. Butore à crête blanche

Other name: White-crested Tiger Bittern

IDENTIFICATION 65-80cm. Sexes differ. A largish dark sandy-buff bird very heavily and transversely barred black above and below, darker above than below. Barring on the female is narrower, thus giving overall darker appearance. Some birds show cinnamon tones. Large white nape patch partially obscured below elongated black crown feathers. Two recent sightings in The Gambia mention two or three long thin white plumes growing out from the nape, hanging laxly around the foreneck when perched and loosely down in flight. Blackish-brown bill long, slender, sharp and yellowish-green on the lower mandible. Immature broader-barred appearing paler and sandier in colour. Eye yellow, facial skin greenish; legs brownish, toes with some yellow. Variation in size and plumage in this species requires further study. **Similar species** Paler Eurasian Bittern lacks white crest and is striped not barred.

HABITS Known only from a single tidal, dense *Rhizophora* mangrove tributary. In other parts of its West African range it is localised on dense forested freshwater streams. Secretive, generally poorly known. Gambian sightings have been diurnal and of solitary individuals. Elsewhere it has been seen to raise and lower the crest between perching and taking flight and calls have been reported through the year.

VOICE A single or double nocturnal booming *whoo whoo*, described as hollow-sounding or as a low deep moaning. See also Pel's Fishing Owl.

STATUS AND DISTRIBUTION Prior to 1996, only three observations from NBD, in November 1982, November 1991 and February 1993, all from the same locality. This is an extension of its previously published range. Sightings from coastal Casamance in southern Senegal include a breeding record. Range restricted to West Africa.

BREEDING Unsuccessful breeding in NBD; downy chick on 14 November 1996, recovered dead at point of fledging on 23 December. Has bred Casamance in November 1980. Nest consists of a loose twig platform.

WHITE-BACKED NIGHT HERON
Gorsachius leuconotus   Plate 3

Fr. Bihoreau à dos blanc

IDENTIFICATION 50-55cm. Sexes differ slightly when breeding. Head indigo-black with a short nape crest, neck distinctively cinnamon-hued rufous-brown. Wings and back dark chocolate, tinged indigo. Silky-white patch of central back usually concealed when the wings are closed, but visible when the bird is climbing through the undergrowth or in flight. Underparts faded patchy pink-brown. Large eye, rich chestnut when breeding, surrounded by a conspicuous patch of bare ghostly white skin, sometimes greenish. Bill black; longish gangly legs greenish-yellow, fading to grey-green. Plumage, bill and legs appear brighter and cleaner just prior to breeding when female has paler lower belly to undertail-coverts. Immature retains spotted brownish plumage for only a few weeks post-fledging (on leaving nest sometimes shows distinct gingery facial skin, soon becoming greyish-white). **Similar species** Immature can be confused with immature Black-crowned but White-backed is overall darker brown and less heavily white-spotted.

HABITS Favours densely vegetated freshwater ponds and courses enlarged by the rains and tall *Rhizophora* mangrove-fringed creeks along the main river. Occasional records from low *Avicennia* coastal mangroves particularly during post-breeding dispersal. In display white back plumes are fanned over closed wings. Sits at rest with head and neck withdrawn into shoulders. Roosts deep inside creeper-clad vegetation, remaining motionless. If disturbed, climbs silently and warily up into the canopy of the

tree cover before flying. Diurnal sightings occur mostly during breeding. Otherwise the most nocturnal of all Africa's herons and is especially secretive. Waits at water's edge for prey or may stab downwards whilst perched.

VOICE Most frequent call when breeding is a soft throaty purring *ow* (beware harsher Striated Heron); also a number of grating noises, croaks and growls.

STATUS AND DISTRIBUTION Uncommon, becoming very locally common when breeding from August-October at a few well-watched sites in WD and NBD. Records from all months and all divisions. Senegalese records both north and south of The Gambia near the coast.

BREEDING Most records during the mid-late rains; half-finished nest mid-August, eggs mid-September in WD, most fledging records in late September or early October, with an exceptional LRD record of feathered young in May. Nest of sticks variable in size and depth, over water.

BLACK-CROWNED NIGHT HERON
Nycticorax nycticorax   Plate 3

Fr. Héron bihoreau

Other name: Night Heron

IDENTIFICATION 58-65cm. Sexes similar. A stocky grey and white heron with relatively short legs. Crown and back are a glossy inky-black; wings silver-grey. Forehead white and a white line over the eye giving a capped appearance. Bill relatively heavy, slightly decurved, black legs and feet greenish-yellow. White nape plumes reach 25cm in breeding season. Distinctive large red eye. Immature grey-brown, with large whitish spots on back and wings, underparts streaked. **Similar species** Immature White-backed Night-Heron is darker and less spotted; Eurasian Bittern is golden-brown and larger.

HABITS Favours fresh and brackish watercourses with adjacent dense tree cover for roosting, during the day, usually quite high-up, perched with head and neck tucked into the shoulders and preening frequently. Gregarious, particularly immatures. Generally active at dusk, flying from the roost on large rounded owl-like wings, with harsh croaks. Sometimes forages by day.

VOICE Harsh loud croaks *whock-whock-whock-whock*.

STATUS AND DISTRIBUTION Common to abundant throughout particularly in WD and URD; numbers comparatively stable in former but with considerable increase in numbers in latter in early dry season. Population peaks December to March, swollen by presence of Palearctic migrants. Absent from arid central Senegal.

BREEDING Breeds during the mid-late rains in small colonies, often on islands in the river or beside freshwater. A frail nest of sticks, low over water in a tree or reedbed.

SQUACCO HERON
Ardeola ralloides   **Plate 3**

Fr. Héron crabier

Other name: Common Squacco Heron

IDENTIFICATION 46cm. Sexes similar with seasonal plumage changes. A smallish, sandy dark-streaked and hunched-up egret. Breeding birds develop drooping darker crest and cinnamon back plumes, with bright cobalt-blue bill with black tip. Erectile crest plumes long and bushy; when flared with plumes of the breast and dorsal patch, they create an outlandish loose 'feather duster' look. Dramatic change in appearance when bird takes flight revealing all white wings contrasting with dark brown saddle. Non-breeding birds and immatures are darker brown, and shorter-crested; bill, legs and feet greenish-yellow. **Similar species** In flight, resemblance to Cattle Egret resolved by dark brown back and body streaking on Squacco. Considerably larger than Little Bittern and much smaller and paler than Eurasian Bittern.

HABITS Favours freshwater fringes, flooded fields, ponds and mangrove swamps. Generally solitary, but congregates in groups when feeding conditions are favourable. Stands motionless at the water's edge or in tall grassy vegetation where it is difficult to detect; often only head and neck are visible.

VOICE Soft vibrant, sometimes grating, croaks.

STATUS AND DISTRIBUTION Common to abundant in wetlands throughout with little seasonal variation. Palearctic winter migrants augment the resident population during the dry season. Present throughout Senegal, along the coast and along rivers.

BREEDING Several hundred pairs breed in CRD in September. Colonial. Nest low, over water, constructed of twigs and sticks and quickly built.

CATTLE EGRET
Bubulcus ibis   **Plate 3**

Fr. Héron garde-boeufs

Other name: Buff-Backed Heron

IDENTIFICATION 53cm. Sexes similar with seasonal plumage changes. A smallish white egret with a comparatively short yellow bill. In breeding plumage crown and nape, plus plumes on chest and back, rich tawny-brown. Bill, facial skin and legs briefly bright red before returning quickly to a normal dark greenish colour. Non-breeding birds are plain white, often dirty; some retain signs of breeding plumage as tawny patches on head and breast. **Similar species** Lacks height of always pure white Little Egret and has shorter legs, with yellow not black bill. Similarly separated from the white phase of Western Reef Heron. Intermediate Egret is larger with a stouter bill and is immaculate white. In flight resembles Squacco Heron but this shows striking colour contrast between wings and body.

HABITS Ubiquitous in grassland, along watercourses, cultivated gardens and open woodland. Attracted to bush fires and congregates at rubbish-tips in enormous numbers, even occasionally seen in closed forest. Generally feeds on dry, even arid, ground. Highly gregarious and in the absence of plains game it attends herds of cattle, goats and sheep, snatching the insects they disturb.

VOICE Usually silent. Harsh croaks and chatterings at the nest and to a lesser extent at roost.

STATUS AND DISTRIBUTION Abundant to very abundant in all divisions; observed more frequently November-March. Widespread throughout Senegal.

BREEDING Breeds colonially in wet season, often in mixed heronries. Nest sparse construction of twigs and stems.

STRIATED HERON
Butorides striatus   **Plate 3**

Fr. Héron à dos vert

Other name: Green-backed Heron

IDENTIFICATION 40cm. *B. s atricapillus*. Sexes similar. Small dark and compact heron, and very short-tailed in flight. Upperparts blackish-green; underparts soft grey-green. Throat whitish with a central stripe mottled with chestnut running down front of neck. Ear-coverts whitish. Wings green speckled pale yellow. A dark green crest is raised like a spiky brush when alarmed or in display. Bill blackish, lower mandible paler than the upper, legs greenish-yellow. Immature brown overall with lines of spots across the wings, upperparts heavily spotted white and underparts heavily streaked with white and brown. Looks noticeably stumpy-tailed in flight, legs trailing. **Similar species** Immature Black-crowned Night Heron is larger, browner and lacks a dark cap. See smaller and much rarer Dwarf Bittern.

HABITS Widespread favouring the river and its adjoining creeks, freshwater ponds, paddyfields and rarely in rock pools on the sea shore. Significant arrivals of breeding visitors herald the rains with considerable north-south movement over estuary at Banjul. Solitary when not breeding. Typically seen crouched low in the shallows or on an overhanging branch. An accomplished climber in tangled vegetation. Skulking but not shy.

VOICE A forceful *tchack-tchack*.

STATUS AND DISTRIBUTION Common throughout in the wet season, temporarily locally abundant. Numbers much reduced in the dry season, especially inland. Widespread in suitable habitat in Senegal.

BREEDING Begins mid-rains and peaks in October. Nest a fragile platform of thin twigs, over water or in a large tree some way from water, even in urban areas. Nest prone to destruction in heavy rains and replacements commonplace, usually more substantial and built higher. Several pairs may be loosely congregated along the same watercourse.

BLACK EGRET
Egretta ardesiaca   Plate 3

Fr. Héron noir ou Aigrette ardoiseé

Other names: Black Heron; Umbrella Heron

IDENTIFICATION 66cm. Sexes alike. A medium-sized heron, entirely black with a velvety sheen in bright sunlight. Crest at the back of the head blows freely in the wind looking characteristically unkempt. Bill black, slim and sharp, eye bright yellow. Legs dark, feet and toes yellow-orange. Immature duller, lacks plumes. **Similar species** Black Egret shows no white allowing ready separation from bluer Western Reef Heron which always has a white throat.

HABITS Favours shallow waters, fresh and saline. Gregarious; feeding groups collect in temporary swamps and rice fields after the rains where prey becomes trapped in the subsiding pools. Sophisticated and characteristic foraging strategy involves the bird spreading its wings forwards to create a

canopy or umbrella, groups appearing to do this in synchrony. This takes place in all light conditions.

VOICE An unobtrusive *kak-kak-kak.*

STATUS AND DISTRIBUTION Common, occasionally locally abundant especially at the coast and CRD. Unpredictably converges in large numbers at suitable sites in any season. In Senegal mainly in the west, both north and south of The Gambia.

BREEDING During the rains in CRD. Builds stick platform nest over water.

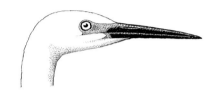

Little Egret. Bill sharp, straight and dagger-like. Colour typically shiny black.

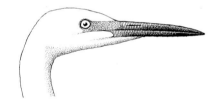

Western Reef Heron (white morph). Bill heavier with slightly curved culmen. Colour typically dark slate and matt.

WESTERN REEF HERON
Egretta gularis   Plate 3

Fr. Aigrette des récifs

Other names: African Reef Heron, Reef Heron

IDENTIFICATION 56cm. Sexes alike. A medium-sized heron with a considerable range of plumage colour. Much commoner dark form always shows a white chin and throat, and ranges from charcoal-blue to a pale smoky-grey. There is often a patch of white on the leading edge of the wing, and variable white on the coverts. Less common white form may show variable irregular grey feathering. Breeding birds have two nape plumes. Bill grey-black, stout, and slightly arched at the culmen. Legs black, feet yellow regularly extending up the legs. **Similar species** Black Egret lacks white on throat, shows

125

shaggy crest in flight. Little Egret has finer, straighter bill; note bill colour.

HABITS Largely restricted to saline and brackish waters, usually near mangroves at the coast and estuary. Follows other water birds in a crouched posture, darting at prey that larger birds ignore.

VOICE Harsh crow-like *kaawww*.

STATUS AND DISTRIBUTION Common to very abundant in coastal WD, common in LRD becoming gradually less frequent further east. Common along the Senegal coast.

BREEDING Colonies form mid-late rains on islands in lower section of the river. Nest a platform of twigs, usually low in mangrove.

NOTE This species is sometimes considered conspecific with Little Egret.

LITTLE EGRET
Egretta garzetta　　　　　Plate 3

Fr. Aigrette garzette

IDENTIFICATION 56cm. Sexes similar. Slim, medium-sized egret with immaculate white plumage. Bill generally jet black, sharp, straight and dagger-like, not subtly convex on the culmen as in a typical Western Reef Heron. In breeding plumage develops two or three plumes on the nape, and many protracted lanceolate plumes on the neck and recurved ones on the back. Legs black, feet distinctively yellow. Orbital skin and lores grey-green, orange when breeding. **Similar species** White phase Western Reef Heron generally separable, but some individuals very difficult; see Western Reef Heron. A dark phase of Little Egret is known, which may or may not have a white throat.

HABITS All aquatic habitats, including the shoreline in shallow surf. Appears more skilful and manoeuvrable than Western Reef Heron and other egrets when hunting and has an agile ability to jump, twist and turn to catch prey items that are attempting to escape. This can assist separation from white phase Western Reef Heron.

VOICE When disturbed and at the nest, utters *karks* and *krarks*.

STATUS AND DISTRIBUTION Common year-round in WD. Recorded in all divisions with resident breeding population supplemented by winter visitors from Europe. Widespread in suitable habitat in Senegal, largely absent from central parts.

BREEDING During the rains in CRD. Nest a small platform of sticks at varying heights.

INTERMEDIATE EGRET
Egretta intermedia　　　　　Plate 3

Fr. Aigrette intermédiaire

Other name: Yellow-billed Egret

IDENTIFICATION 66cm. *E. i. brachyrhyncha*. Sexes alike. Medium-sized with entirely white plumage at all times. Bill yellow, when breeding red with an orange tip. In breeding plumage the filamentous plumes on the back extend beyond the tail. Lower neck and breast are also heavily plumed. Upper legs yellowish, remainder brown-black including the toes. **Similar species** Taller and longer-necked Great White Egret has extension of facial skin below eye; smaller Cattle Egret has buff-brown in plumage. Little Egret and Western Reef Heron have dark bills.

HABITS Favours open water at the coast and along the river system, adjacent flooded fields, mangrove creeks and bolons. Often solitary.

VOICE Generally silent; peculiar buzzing call in display.

STATUS AND DISTRIBUTION Common to occasionally locally abundant in wetlands with little seasonal variation. Present in all suitable habitat in Senegal.

BREEDING Nests colonially in mixed heronry in CRD in November; nest built of sticks lined with reeds and stems.

GREAT WHITE EGRET
Egretta alba　　　　　Plate 3

Fr. Grande Aigrette

Other name: Great Egret

IDENTIFICATION 90cm. *E. a. melanorhynchos*. Sexes similar. Large, entirely white egret with long neck held in a characteristic S-shape. Bill yellow, long and strong; changing to black in breeding season when orbital skin turquoise. In full breeding plumage heavily plumed on the back, plumes sometimes retained when not breeding. Legs black. **Similar species** Extended bare facial skin behind eye separates from Intermediate Egret. Nearly twice the size of Little Egret and white phase Western Reef Heron.

HABITS Favours open water at the coast and along the river mangroves and adjacent marshes and flooded fields. Solitary but numerous individuals may be dispersed over an open stretch of water when foraging. Adopts a stiff erect posture, leaning slightly forwards with the neck held in alignment to the body. Remains motionless in this position for prolonged periods. Will wade into deep water up to its underside to fish. Individuals are occasionally found on pasture hunting for insects.

VOICE Deep, hoarse croaks. Softer low pitched calls during courtship and nesting.

STATUS AND DISTRIBUTION Common to abundant throughout in suitable habitat. Similarly widespread in Senegal. Slightly larger Palearctic nominate *E. a. alba* may occur.

BREEDING Often in mixed heronries in tall trees, near to freshwater, sometimes in towns. A substantial stick nest lined with grasses, upto 1m in diameter.

PURPLE HERON
Ardea purpurea   Plate 3

Fr. Héron pourpré

IDENTIFICATION 80cm. Sexes similar. Slim, dark, elegant heron with a frail-looking neck. Adult has black crown with short crest, rufous neck, throat white with three prominent black neck and facial streaks. Upperparts and wings soft brownish-grey. Bill slender, pale browner or yellow; legs and feet yellowish-brown. More frequently seen immature browner; neck streakings only faintly defined and upperparts conspicuously scaled. In flight shows a noticeable downwards U-shaped bend in the retracted neck. **Similar species** Goliath Heron is much larger.

HABITS Favours mangrove creeks and densely vegetated freshwater courses, only occasionally on open stretches of water. Shy and solitary. When alarmed adopts bittern-posture with head and neck held bolt upright affording excellent camouflage. May stand with body hidden in the vegetation, only the head and neck peering out.

VOICE Harsh *kwark-kwark* when flushed.

STATUS AND DISTRIBUTION Frequent to common, recorded in all months in all divisions, possibly less numerous in the late dry season. Widespread throughout Senegal apart from arid zones. African population is probably supplemented by Palearctic migrants.

BREEDING Not proven in The Gambia. Recorded from two sites in northern Senegal including Djoudj, July-August and December. Solitary or in mixed heronries. Nest is a platform of twigs, sticks and reeds low over water.

GREY HERON
Ardea cinerea   Plate 3

Fr. Héron cendré

IDENTIFICATION 92cm. Sexes alike. A large grey and white heron. Grey upperparts, conspicuous black 'epaulettes' on wings. Prominent black supercilium extending into a crest. Forehead and front of neck white with heavy black flecking, flushed orange-pink in breeding season. Bill large and yellow. Legs yellow-brown. Immature has crown grey, not white, and the black shoulder patches, crest and body plumes are absent. In approaching flight there are two conspicuous white pips visible on the forewing. African race *monicae* of Grey Heron is noticeably paler than the Palearctic race *cinerea* of which British and Polish-ringed birds have been recovered in The Gambia. **Similar species** See Black-headed Heron.

HABITS Ubiquitous in aquatic habitats. Often seen along the shoreline and occasionally on dry agricultural land. Capable of deep wading and will both swim and occasionally dive. May rest on the ground in a sitting position with wings flexed open to the sun. Generally wary.

VOICE A harsh *kroak* or *fraank* when taking flight.

STATUS AND DISTRIBUTION Common to abundant in all divisions year-round. Most numerous in the early dry season, with large numbers in WD. In Senegal absent only from central arid areas.

BREEDING Not recorded in The Gambia. Breeds Senegal (June-July) and Mauritania. Colonial. Substantial nest of branches and twigs, on ground or in vegetation.

BLACK-HEADED HERON
Ardea melanocephala   Plate 3

Fr. Héron mélanocéphale

IDENTIFICATION 90cm. Sexes similar. Large, lightly built dark heron. Neck longer and more willowy than Grey Heron. Head black from the crown to the nape and down the length of the hind neck. Pure white on the front of the neck restricted to the throat. Short drooping black crest. Bill grey-brown with paler lower mandible. Whitish facial skin. Legs and feet black. Immature duller, with greyish-brown head and neck. **Similar species** Confusion most likely to occur between immature Black-headed and immature Grey Heron. Look for continuation of clean demarcation of dark hindneck along full length of otherwise pale neck of a young Black-headed. In flight, the underwing of Black-headed is a well-defined contrasting dark grey and white, unlike the uniform grey underwing of Grey Heron.

HABITS Often in drier habitats, favouring pasture land with scattered acacias. The only large heron in the region regularly encountered away from water. Often on the wing at first light. Congregates at freshwater with nearby trees for roosting and nesting. Far more approachable than Grey Heron.

VOICE A range of raucous croaks and growls.

STATUS AND DISTRIBUTION Common to seasonally locally abundant in WD and URD especially November-March. Numbers reduced throughout during the rains. Seasonally common in Senegal.

BREEDING Nesting in URD in August and WD in November. Colonial, nests high in forest trees and village kapoks, building a substantial platform of sticks. Spectacular courtship display.

GOLIATH HERON
Ardea goliath   Plate 3

Fr. Héron goliath

IDENTIFICATION 148cm. Sexes similar. Huge, dark heavy-billed heron with chestnut head, hindneck and underparts. Wings slate-grey appearing immense in flight. Foreneck whitish, streaked black, short crest. Large and formidable blackish bill. Legs and feet black. Immature much browner with less defined foreneck markings. **Similar species** Confusion can occur with Purple Heron, particularly if only glimpsed deep inside vegetation. Look for the head proportions and pattern. Despite similar plumage, Goliath is much larger and more robust.

HABITS Favours coastal mangroves, creeks and, inland swampy margins along the river. Less frequent on well-wooded freshwater streams. Shy and solitary when not breeding. Prefers the narrower creeks to open areas but is occasionally seen standing well out in open stretches of water. Slow and laboured flight on bowed wings.

VOICE A baying bark similar to that of Western Baboon *Papio papio*.

STATUS AND DISTRIBUTION Locally frequent to common in WD and LRD, most records January-April. Fewer records from further E along the river. In Senegal mainly in coastal areas close to Gambian borders, also in the southeast and north.

BREEDING No recent information. Normally solitary and nest usually a low platform of sticks.

HAMERKOP

Scopidae

One species. Curious African monotypic endemic whose evolutionary affinities remain vague. Widespread throughout sub-Saharan regions in suitable habitat, absent from arid zones. Builds huge haystack-like nests. A source of much legend and folklore.

HAMERKOP
Scopus umbretta   Plate 2

Fr. Ombrette

Other names: Hammerhead, Anvilhead

IDENTIFICATION 56cm. Sexes alike. Unmistakable, sturdy, uniformly dull brown bird with the stance of a large long-legged duck. Substantial slightly hooked bill and a large shaggy crest create an anvil- or pickaxe-headed appearance. Crest raised fully erect when alarmed or excited. Wings long and rounded extending well beyond the tail at rest. Legs and feet black. **Similar species** Soars raptor-like but long bill and legs preclude confusion. Flight buoyant and owl-like with flaps and glides; rounded wings, head shape and short tail produce a characteristic silhouette.

HABITS Frequents all types of aquatic habitats especially with nearby well established vegetation, including rice paddies, ponds, mangrove creeks and occasionally the sea shore. Usually in pairs or small groups. Forages by wading in the shallows, snatching at small animal prey, and using its feet to stir mud and vegetation to displace prey. Much time is spent on nest construction, for which the Hamerkop is famous. Vocal displays with bowing and wing-spreading ceremonies precede frequent copulations.

VOICE Rowdy. Harsh guttural *sikweesikwee-kwee-kwee* with increasing tempo. Cackles during display.

STATUS AND DISTRIBUTION Common and widespread in suitable habitats more frequent in inland Divisions. Widespread in Senegal.

BREEDING Nest huge and shaped like a haystack,

made of sticks, leaves and mud. Downwards-facing 60cm access hole. Whole structure is usually placed high in the fork of a tree and is used year after year. Empty nests may be used by other species e.g. Barn Owl.

STORKS

Ciconiidae

Eight species. Medium to huge birds with long necks and bare legs. Bill normally heavy and straight. Fly and soar with neck fully extended and not withdrawn as do the herons. They frequent a diversity of habitats both aquatic and terrestrial, feeding on a wide range of invertebrate and vertebrate life. Generally silent, they are well-known for their bill-clattering displays at the nest.

YELLOW-BILLED STORK
Mycteria ibis   Plate 4

Fr. Tantale africain

Other name: Wood Ibis

IDENTIFICATION 100cm. Sexes similar. Large pinkish grey-white stork with black flight feathers and tail. Conspicuous bare red face extending beyond the eye; bill long, yellow-orange and noticeably down-curved. Legs pink. Pink flush of the plumage is intensified when breeding. Immature resembles adult but is brownish and duller. **Similar species** Red face, yellow bill and black tail distinguish it from White Stork.

HABITS Favours the river system and the estuary, including tidal inlets and creeks. Feeds in loose assemblages (sometimes singly) and is frequently seen with African Spoonbill and Pink-backed Pelican. Forages by wading slowly in the shallows, sometimes cooperatively, scything with slightly opened mandibles through the water. Also hunts by standing motionless using this open-billed strategy whilst waiting for passing prey. Partial wing stretching casting shade over foraging space noted. At the coast uses wrecks as perching or roosting sites in association with other large waterbirds. Occasionally soars.

VOICE Silent. Bill clapping display at the nest.

STATUS AND DISTRIBUTION Frequent to occasionally common on the coast of WD and NBD in all seasons. Locally common in LRD and CRD. Extensively distributed across southern Senegal and in the delta of the Senegal River in the north.

BREEDING Colonial; breeds LRD and NBD during the rains in only a few colonies; also breeds in Senegal. Nest is a platform of sticks and twigs lined with grasses and reeds and regularly built in tall village trees.

AFRICAN OPENBILL STORK
Anastomus lamelligerus   Plate 4

Fr. Bec-ouvert africain

Other names: Open-billed Stork, Openbill

IDENTIFICATION 95cm. Sexes similar. Medium-large. Brownish-black overall with a greenish sheen. Bill large with diagnostic 'nutcracker' gap in distal half of mandibles; brownish-horn, paler towards the base. Orbital skin bluish. Immature develops the open bill with maturity and is duller. **Similar species** Wholly dark plumage and unique bill precludes confusion with any other species.

HABITS Favours water courses with nearby tall trees. Singly or small parties. Bill is adapted for holding, grasping and extracting molluscs, not crushing.

VOICE Loud croaks and honks. Bill-clatters in display.

STATUS AND DISTRIBUTION Not recorded in The Gambia. In Senegal, modern records from both the north as far as Senegal River in January and south in coastal Casamance in February. Infrequent throughout West Africa.

BREEDING Breeds mainly south of the equator, nearest Sierra Leone and Nigeria. Comparatively small nest of sticks and aquatic vegetation.

BLACK STORK
Ciconia nigra   Plate 4

Fr. Cigogne noire

IDENTIFICATION 98cm. Sexes alike. Large stork, predominantly black with a white breast. In good light upperparts show iridescent sheens. Legs, feet and bill bright red. Eye with red orbital skin. Immature duller lacking iridescence, legs yellowish-green. **Similar species** Smaller Abdim's Stork has a greenish bill and a blue face; when soaring high, identification is difficult; white on the underwing in Abdim's is more extensive and Black Stork has an entirely black back and rump (white in Abdim's).

HABITS Favours fringes of marshes and river estuaries. Extremely rare. If encountered, it is likely to be soaring. Normally solitary.

VOICE Silent; bill-clatters at the nest.

STATUS AND DISTRIBUTION A single bird over Fatoto, URD in October 1996 is first confirmed sighting since 1923. Recorded most years in northern Senegal.

ABDIM'S STORK
Ciconia abdimii   Plate 4

Fr. Cigogne d'Abdim

Other name: White-bellied Stork

IDENTIFICATION 81cm. Sexes similar. Medium-sized black and white stork with a colourful face. Upperparts, head, neck and breast black with an iridescent sheen of greens and violets. Lower back, rump, belly and underwing-coverts white. Bill greenish, tipped reddish; prominent blue facial skin extends around eye. Legs pink, redder at the joints and shortish. Immature duller. **Similar species** See Black Stork. When soaring, the legs of Abdim's extend less beyond the tail, and neck is shorter.

HABITS Favours grassland, feeding on large insects and small rodents, also wetlands for fish and crustaceans. Follows insect plagues. Highly gregarious elsewhere in its range.

VOICE Silent; bill-clatters at nest.

STATUS AND DISTRIBUTION One modern report in July 1990 in CRD, of a single bird; only four other records since 1959. Most Senegalese records are along eastern border. Trans-equatorial migrant.

BREEDING Breeds N of the equator prior to and during the rains. Colonial nester, building on village huts and woodland trees in Senegal from April-June, next nearest Sierra Leone. A platform of sticks stabilized with soil and mud.

WOOLLY-NECKED STORK
Ciconia episcopus   Plate 4

Fr. Cigogne épiscopale

Other names: White-necked Stork, Bishop Stork

IDENTIFICATION 85cm. *C. e. microscelis.* Sexes similar. Medium-large, the only black stork with a white neck. Neck feathering appears woolly at close range. Forehead dark and sooty with a white band at the base. Belly white and distinctively long undertail-coverts white; because these are longer than the tail they cause the tail to look distinctly black and white when the bird soars. Legs and feet black, bill black with dark red tone. Immature duller lacking white band at the base of the bill.

HABITS Favours Gambia River and adjacent mangrove systems, flooded fields, rice paddies and ponds. Attracted to bush fires, eating large quantities of locusts and grasshoppers; also to termite emergences. Occasionally congregates in numbers on thermals, soaring with large raptors.

VOICE Normally silent; bill-clatters at the nest.

STATUS AND DISTRIBUTION Seasonally and locally common in LRD and NBD during and just after the rains, July to early December, maximum c.80 birds in October 1996 near Kiang West. Occasional singles in WD, rare elsewhere. In Senegal, mainly to the south in coastal Casamance.

BREEDING Builds nest of sticks, small branches and grasses in a leafy riverine forest tree, but although this habitat is available, breeding is yet to be confirmed in Senegambia. Nearest known breeding Sierra Leone.

WHITE STORK
Ciconia ciconia   Plate 4

Fr. Cigogne blanche

IDENTIFICATION 102cm. Large. Sexes similar. All white except for black flight feathers. Bright red straight bill and bright red legs. Immature has dull red legs and bill, dark brown wings and looks greyer. **Similar species** Yellow-billed Stork has a down-curved yellow bill and bare red face, and black, not white, tail.

HABITS Favours dry cultivated land and bush fires, where it is attracted by the number of fleeing insects and small vertebrates. The few sightings in The Gambia are generally of soaring birds.

VOICE Silent. Bill-clatters at the nest.

STATUS AND DISTRIBUTION Vagrant from the Palearctic with five records 1970 to 1998, all in the dry season. Records of good numbers in northern Senegal with occasional birds further south nearer Dakar and east of The Gambia.

SADDLE-BILLED STORK
Ephippiorhynchus senegalensis **Plate 4**

Fr. Jabiru d'Afrique ou Jabiru du Sénégal

Other name: African Jabiru

IDENTIFICATION 150cm. Female smaller than male. A huge, elegant black and white stork; enormous black-banded red bill slightly upturned with a yellow frontal shield (the 'saddle') at the base. Two small fleshy red wattles at the base of the bill. Head and neck, wings and back are iridescent black; rest of body white. In flight white primaries very prominent. Eye brown in male, golden yellow in female. Legs and feet black with pink knee joints. Immature duller sooty-grey overall. **Similar species** Unique pattern of bill and body plumage is unmistakable.

HABITS The estuary, floodplains and swampland are likely sites, less frequent on cultivated land. Stately and majestic in all its movements, usually solitary in The Gambia. Feeds similarly to the larger herons by careful stalking with swift stabs at fish, amphibians and crabs. Soars gracefully, with head, neck and legs held slightly lower than body.

VOICE Reputed to be virtually mute.

STATUS AND DISTRIBUTION Rare; a few records 1965-88 from WD, LRD (near Tendaba) and CRD with more recent records from the Bund Road in WD. These include repeated sightings of an immature late December 1990 and an adult in July 1991. In Senegal, most likely in south and east after the rains, infrequent in the north.

BREEDING Breeds in Senegal. A large nest of sticks, branches, reeds and turf at the top of a tall tree, deep enough to conceal the incubating bird.

MARABOU STORK
Leptoptilos crumeniferus **Plate 4**

Fr. Marabout d'Afrique

Other name: Marabou

IDENTIFICATION 150cm. Sexes similar. Huge. A grotesque and unmistakable bird with massive tapered bill. Upperparts dark slate, underparts white. Head and neck naked, fleshy-pink with sparse tufts of woolly down and hair. A prominent pink gular-sac hangs down the foreneck, with partial ruff of white feathers at the base of the neck. Legs and feet black often covered with white powder. Black wings contrast strikingly with white body feathers in soaring flight. Immature duller with smaller bill, crown more profusely covered in woolly tufts. **Similar species** None in Africa.

HABITS Prefers semi-arid areas, but individuals could occur anywhere including the coast. Usually gregarious and often in association with vultures with which they share carrion. High in the pecking order of scavenger communities. Also ingests animal faeces and follows locust swarms. Frequently rests in a hunched posture with lower part of leg flat to the ground, a characteristic of some other storks. Particularly at the nest under the open sun they splatter their legs with white liquid excreta which acts as a reflector, reducing heat absorbtion. Unlike other storks, soars with the neck retracted. Principally a bird of open country where, vulture-like, it can locate carrion whilst on the wing.

VOICE Silent. Bill-clatters whilst breeding.

STATUS AND DISTRIBUTION Frequent to common in CRD, peaking in early dry season at regular nesting colonies, otherwise numbers seasonally stable. Uncommon elsewhere with immatures at the coast in some years towards late dry-season. Nowhere common in Senegal, with records in southern Casamance, the north and east.

BREEDING Colonial. A platform of sticks often high in a kapok tree. Sites returned to annually in CRD, October-April; colony location periodically shifted a short distance. Has bred in northern Senegal.

IBISES AND SPOONBILLS
Threskiornithidae

Two species of spoonbills and three ibises. Bill long and decurved in the ibises, straight and spatulate in the spoonbills. Diet includes fish, crustaceans, amphibians, insects and carrion. Forage in aquatic shallows or on grassland. Fly with neck outstretched, legs trailing. Most nest colonially, often in mixed heronries. In Senegal, the critically endangered Northern Bald Ibis or Waldrapp *Geronticus eremita* (Fr. Ibis chauve) has been seen in Niokolo Koba in March 1985 and at Matam in December 1988–January 1989.

GLOSSY IBIS
Plegadis falcinellus   **Plate 5**

Fr. Ibis falcinelle

IDENTIFICATION 56cm. Sexes similar. Smallish, recalling a dark curlew in shape and size. Overall dark brownish to dull bronze-brown, head and neck streaked whitish, underparts speckled greyish. Glossy bronzy-green sheen of much brighter nuptial phase (rarely seen in The Gambia) is retained to varying degrees on the back and wings and most likely on birds seen in late dry season. White crescent-shaped patch at the base of the bill. Legs, feet and bill dark green, eye brown. Immature, duller and sootier, sometimes with white markings on head and face. **Similar species** Hadada Ibis is larger and more heavily built.

HABITS Freshwater margins and adjacent wet grassland. Favours baobabs for perching. Graceful both in flight and on the ground. Feeds by probing in mud or on flooded land for a range of aquatic and terrestrial invertebrates. Occurs singly or in flocks, maximum count c. 70 in The Gambia.

VOICE Usually silent, flight call a croaking *graa-kaa*.

STATUS AND DISTRIBUTION Uncommon, with all recent records from November-June in WD and CRD; a June record in WD relates to an immature. In Senegal mainly in the far north with 400 reported in December; scarce in the south.

BREEDING No Gambian records. Breeds in Mali and a small population is suspected in northern Senegal. Builds a nest of sticks, twigs and reeds.

HADADA IBIS
Bostrychia hagedash   **Plate 5**

Fr. Ibis hagedash

Other name: Hadada

IDENTIFICATION 76cm. Sexes similar. *B. h. brevirostris*. A bulky dark ibis with a white crescentic cheek stripe. Head, neck and underparts grey-brown. At close quarters shows a metallic bronzy-green iridescence on wings and upperparts. Distinct white malar stripe. Legs black, eye whitish. Bill black, and reddish on the basal half of the upper mandible. Thickset and uniformly dark in flight with the head and bill held down, feet scarcely projecting beyond tail. Immature is duller with a shorter bill. **Similar species** Much bulkier than smaller Glossy Ibis which lacks white malar stripe and has longer trailing legs in flight.

HABITS Normally associated with water, found in marshes, rice paddies and riverine forest. Feeds on molluscs, aquatic worms and diving-beetles by either probing or picking up food from the surface. Perches high, often in pairs, sometimes in small groups. Extremely noisy at dawn and dusk, particularly if disturbed.

VOICE Loud and raucous *ha-ha, ha-ha-ha* or *ha-DA-ha*, hence its common name.

STATUS AND DISTRIBUTION Frequent to common in CRD with no obvious seasonal population changes; only occasional records in other divisions. Frequent across southern Senegal and up along the eastern border to the north.

BREEDING No Senegambian information. Nest a platform of sticks lined with grass placed on a leafy bough usually over water. Usually solitary.

SACRED IBIS
Threskiornis aethiopicus   **Plate 5**

Fr. Ibis sacré

IDENTIFICATION 68cm. Sexes similar. Distinctive, white-bodied with naked black head and neck. Black trailing edge bordering the immaculate white of the wings is prominent in flight. Adults have rump covered in long dark filamentous plumes creating the distinctive bulky silhouette. Decurved bill, legs and feet black, iris brown, ruby-red whilst breeding. Immature dirty white; head and neck partially feathered; smaller, less decurved bill. **Similar species** Unmistakable.

HABITS Tidal mudflats, mangrove fringes and freshwater margins, rubbish tips, occasionally wanders onto cultivated land and open bush. Probes for marine invertebrates in soft mud at low tides. Catches insects by snatching when foraging in a terrestrial habitat. Will also scavenge, especially on rubbish tips. Roosts colonially, frequently in the crown of a baobab.

VOICE Generally silent, occasional croaking calls.

STATUS AND DISTRIBUTION Locally common at river estuary in coastal WD throughout the year. Former records of large flocks in CRD and URD in the dry season not recently reported. In Senegal nowhere common, but widespread on all major water-courses.

BREEDING Modern record of c. 20 pairs in CRD in November-January. Breeds in northern Senegal in September-November. Nest a lined platform of sticks. Colonial and in mixed heronries.

EUROPEAN SPOONBILL
Platalea leucorodia   Plate 4

Fr. Spatule blanche

Other names: White Spoonbill, Eurasian Spoonbill

IDENTIFICATION 88cm. Sexes similar. A white spoonbill lacking red facial skin. Non-breeding plumage all-white. Ridged bill black and barred horn-grey from base to middle, yellowish towards the tip. Forecrown feathered with a thin black line on lores to eye. Facial skin yellow-orange, eye red, legs and feet black. An incomplete crest and partial yellowish collar become complete in breeding plumage. Immature resembles non-breeding adult but with primaries tipped black, bill dirty yellowish-pink, and legs and feet dull flesh or horn-yellow. Two races: the more yellowish *P. l. leucorodia* breeds Europe and *P. l. balsaci* breeds Mauritania. **Similar species** African Spoonbill has striking red facial skin.

HABITS Similar to the much more frequent African Spoonbill. Dutch-ringed birds observed in coastal Guinea-Bissau may pass through The Gambia. Dispersal of the Mauritanian population along the West African coast is not understood and requires further research.

VOICE Usually silent; bill-clattering when breeding.

STATUS AND DISTRIBUTION Rare passage migrant with few recent records, all October-March, involving both immatures and adults. Most occur at river estuary in WD (Bund Road); single birds in LRD and URD. Recorded all along the northern coast of Senegal, Djoudj being an important site, and inland along the Casamance River.

BREEDING Mauritanian population of c. 1000 pairs nests directly on ground late April-July. Builds nest of sticks, reeds and grass.

AFRICAN SPOONBILL
Platalea alba   Plate 4

Fr. Spatule d'Afrique

IDENTIFICATION 91cm. Sexes similar. Large white water bird with bare red face and long blue-grey spatulate bill. Plumage whitish to dirty-grey. Crest on nape cream-coloured and occasionally a yellowish tinge extends to chin and throat. Legs and feet pinkish-red, eye pale blue-grey. Immature has more extensive feathering on the forehead, black underwing-coverts and wing tips, with a horn-yellow bill and blackish legs. **Similar species** European Spoonbill lacks red face.

HABITS Favours open water both fresh and saline, including flooded fields. Usually remains close to cover of mangroves or tall andropogon grass. Often in small groups, occasionally singly. Feeds in co-operative working parties wading slowly through pools and gullies. Hunts also like Yellow-billed Stork by standing motionless with its partially opened mandibles submerged, waiting for passing prey. Feeds on fish, crustaceans and a variety of other aquatic invertebrates. Shy. Rests for long periods during the day standing with bill, head and neck tucked well into the back feathers.

VOICE Generally silent. Croaks at the nest where it also bill-clatters.

STATUS AND DISTRIBUTION Locally common at the river mouth and at a few coastal watercourse sites in WD, where non-breeders are seen throughout the year. Largest numbers most likely to occur in LRD and CRD in the dry season. Sporadically distributed south and east of The Gambia in Senegal, and at the delta of the Senegal River.

BREEDING Gambia: suspected once in November, and in northern Senegal in November-December. A nest of sticks and aquatic vegetation, usually over water. Elsewhere colonial, often in mixed heronries.

FLAMINGOS
Phoenicopteridae

Two species. Unmistakable large wading birds with elongated neck and legs. Feet are webbed which assists swimming. Fly with neck and legs outstretched. Gregarious but usually encountered in small numbers in The Gambia with greater concentrations in northern Senegal.

GREATER FLAMINGO
Phoenicopterus ruber   Plate 4

Fr. Flamant rose ou Grand Flamant

IDENTIFICATION 160cm. *P. r. roseus*. Sexes similar,

female shorter-legged than male. Tall, with sinuous neck, long coral-pink legs, curious pink and black angular bill. Plumage generally soft pink or whitish-pink, sometimes very white. Upper- and underwing-coverts flame-red contrasting with black primaries in flight and when wings spread. Eye yellow.

Immature plumage greyish-brown, bill grey tipped black. **Similar species** Lesser Flamingo is appreciably smaller but note size variation between males and females in a flock of Greaters. Bill colour is the best distinction.

HABITS Feeds on open stretches of coastal waters particularly on the estuary and along the tidal muddy fringes of the river. Sometimes seen flying over the sea near the beach. Feeds by wading with the bill held upside down, partially or totally submerged in water. When swimming occasionally up-ends in the manner of some waterfowl. Filters minute aquatic organisms through lamellae along the edge of the bill. Nuptial flashing wing-salute display often performed by groups on non-breeding grounds. Rests with neck folded and one leg tucked up into the body feathers.

VOICE A goose-like honking and nasal grunting.

STATUS AND DISTRIBUTION Locally common to occasionally abundant in coastal WD and NBD with records in every month. Fewer records east along the river up to LRD. Flocks exceeding 60 are unusual and often consist of mixed ages. Population in northern Senegal fluctuates seasonally with 6,000-19,000 counted in recent years, inluding some Palearctic migrants.

BREEDING Breeds in dense colonies with 200 pairs nesting in February in the Saloum delta just north of Gambian border. Nest is a cone of mud up to 45cm tall.

LESSER FLAMINGO
Phoenicopterus minor   Plate 4

Fr. Flamant nain ou Petit Flamant

IDENTIFICATION 100cm. Sexes similar. Much pinker and considerably smaller than Greater Flamingo. Bill coral-red and much darker than Greater, appearing almost blackish from a distance. Upper- and underwing-coverts a blotchy deep red not uniform bright scarlet. Immature grey to brown. **Similar species** Mixed flocks of Greaters and Lessers are uncommon in The Gambia and separation of individuals on size alone is complicated by substantial height differences between the sexes of each species. The most reliable field characteristic is the much darker and uniformly coloured bill of Lesser.

HABITS Similar to Greater Flamingo with which it sometimes associates. Feeds on muddy banks and puddles at low tide. Prefers to feed by skimming the surface whilst wading.

VOICE A soft honking.

STATUS AND DISTRIBUTION Local in The Gambia with records in all seasons from riverside locations in LRD. There are records throughout coastal Senegal both north and south of The Gambia with flocks ranging from 100 to 8,500 in the north in recent years.

BREEDING Nearest known breeding colony is in SW Mauritania where nesting recorded July. Nest a low mud cone with a concave top.

DUCKS AND GEESE

Anatidae

Twenty one species, including seven in northern Senegal only. Ranging from small to very large, they are highly adapted for an aquatic lifestyle. Variations in bill structure, and leg and neck lengths enable exploitation of a diverse range of foraging niches. Shared characteristics include dense waterproof plumage, flattened bill, shortened tail, webbed feet and swift flight with neck extended. Clutch size is usually large. If seen, the colour, shape and position of the wing panel (speculum) is often an important means of flight identification. In most Palearctic species, males moult into an 'eclipse' female-type plumage after breeding. This is comparatively short-lived, but migrants reaching Africa may be still in partial eclipse. Courtship and display may occur in wintering areas. Despite the presence of extensive freshwater habitats in The Gambia this group is relatively poorly represented, for reasons not well understood. Immense rafts of Palearctic waterfowl on the Gambia River reported in the dry season earlier this century are now only rarely encountered. Both Mallard and Common Pochard are only occasionally recorded in Senegambia: they are included in the text but not illustrated. Common Shelduck *Tadorna tadorna* (Fr. Tadorne de Belon), Eurasian Wigeon *Anas penelope* (Fr. Canard siffleur), American Wigeon *Anas americana* (Fr. Canard d'Amérique), Gadwall *Anas strepera* (Fr. Canard chipeau), Blue-winged Teal *Anas discors* (Fr. Sarcelles à ailes bleues) and Marbled Duck *Marmaronetta angustirostris* (Fr. Sarcelle marbrée) are vagrants to Djoudj National Park in northern Senegal and none is illustrated or discussed here.

FULVOUS WHISTLING DUCK
Dendrocygna bicolor   Plate 5

Fr. Dendrocygne fauve

Other names: Fulvous Tree Duck, Fulvous Duck

IDENTIFICATION 46cm. Sexes similar. A medium-sized perching duck with disproportionately long legs and distinctive upright stance; never any white about the face. Overall golden-brown with a darker brown back, and conspicuous brown line down the rear of the neck. Relatively broad creamy stripes on dark flanks are diagnostic. In flight the black rump is characteristic as is the well-defined white V of the uppertail-coverts. Slate-grey legs and feet extend beyond the tail in flight. Immature similar but upperparts paler, also lighter below with faint streaks. **Similar species** White-faced Whistling Duck is a similar shape but darker with a white face (duller in immature).

HABITS Prefers still waters with surface vegetation. Dives, dabbles and upends when feeding on the seeding heads of aquatic plants. Gregarious, sometimes in huge flocks, though in The Gambia most likely to be found in small numbers among much larger groups of other waterfowl. Occasionally irruptive.

VOICE Characteristic disyllabic flight whistle *tsuee*.

STATUS AND DISTRIBUTION Uncommon in CRD, rare elsewhere. Possibly under-recorded but historically never common. Records November-April, most in February. Found in northern Senegal on the delta of the Senegal River and in the east on the Gambia River in Niokolo Koba.

BREEDING No Gambian records; breeds erratically in northern Senegal with records in dry and wet season. Ground nest built from soft grasses and reeds, often concealed by a dome.

WHITE-FACED WHISTLING DUCK
Dendrocygna viduata   Plate 5

Fr. Dendrocygne veuf

Other names: White-faced Tree Duck, White-faced Duck

IDENTIFICATION 48cm. Sexes similar. A medium-sized perching duck with long legs and neck and upright stance. Overall dark chocolate brown with conspicuous neatly demarcated white face and dark flanks finely barred white. Legs and feet blue-grey, extending beyond the tail in flight. Note white of the face may become discoloured when feeding. Immature duller with a buff-brown face. **Similar**

species Fulvous Whistling Duck is paler, golden-brown with creamy flank stripes and lacking a pale face.

HABITS Prefers freshwater parts of the river or rain-flooded habitats. Upright stance in tight groups, giving alert, nervous appearance when out of the water. Particularly active at night. Forages by dabbling, upending and on occasions diving. Wary and quick to take off when approached. An aerial display involving several birds diving and plunging on each other has been observed in September when nesting occurs.

VOICE Vocal. Trisyllabic high-pitched whistle *wishiwishiwishi* given by both sexes.

STATUS AND DISTRIBUTION Locally abundant in all seasons in coastal WD where population has increased in recent years. Seasonally abundant in other divisions especially CRD. The commonest duck in The Gambia. Numbers increase throughout the rains. Abundant in parts of Senegal.

BREEDING Basic nest-scrape on the ground, normally away from water in grass or under low thorny vegetation, with little or no down. Creches of ducklings from amalgamated clutches are usually seen during the rains.

WHITE-BACKED DUCK
Thalassornis leuconotus   Plate 5

Fr. Canard à dos blanc

IDENTIFICATION 40cm. Sexes similar. Small-medium grebe-like duck. Overall mottled rufous-brown, darker above, paler below. Most distinctive fieldmark when swimming or at rest on the water is a large white crescent between eye and bill; this is detectable from a considerable distance, contrasting with the blackish speckled face. White patch on the back only obvious in flight. Appears slightly hump-backed on the water. Tail shortish, stiff and tapered, greyish legs extend beyond tip in flight; feet are peculiarly large. Bill lead-grey speckled yellowish-green. **Similar species** Little Grebe is smaller, plainer and dumpier with a short bill.

HABITS Prefers freshwater, to be expected in similar areas to African Pygmy Goose. Food obtained by diving; forages for waterlily seeds and aquatic invertebrates. Highly aquatic and only leaves the water when forced to do so. Flight swift and strong. Usually in pairs or small family groups.

VOICE A disyllabic musical whistle *kurweekurwee*.

STATUS AND DISTRIBUTION Not recorded in The Gambia. Occasional isolated sightings from along the coast and Djoudj in northern Senegal.

Suitable habitat is widespread in poorly surveyed regions in The Gambia.

BREEDING End of wet season record in northern Senegal. Nest a rather grebe-like floating mass of tangled stems and roots.

EGYPTIAN GOOSE
Alopochen aegyptiacus   **Plate 5**

Fr. Oie d'Egypte

IDENTIFICATION 60cm. Sexes similar. Overall plumage light chestnut-brown to greyish-brown. Conspicuous field marks are irregular large chestnut patch on the belly and oval dark chestnut eye-patches. In flight extensive white in the wing contrasts with black flight feathers. When swimming the rear-end appears higher than the shoulders. No seasonal plumage variation. Bill is pink with black tip, legs pinkish. Immature duller with chestnut belly and eye-patches lacking or vague. **Similar species** Unique and unmistakable.

HABITS Most sightings are from the freshwater parts of the Gambia River. Forages mainly by grazing in fields.

VOICE Varied wheezes and honks in breeding season.

STATUS AND DISTRIBUTION Rare. Most likely in CRD and URD. Considered common in the early part of the century. More frequent in northern Senegal with a few thousand birds in the Senegal River delta.

BREEDING Recorded in northern Senegal from the end of the rains to February. Nest in a variety of locations, including tree holes or under a rock on the ground, a simple depression lined with down.

SPUR-WINGED GOOSE
Plectropterus gambensis   **Plate 5**

Fr. Oie-armée de Gambie

Other name: Gambian Goose

IDENTIFICATION 75-95cm. Male larger than female. A powerful glossy black and white goose with a metallic sheen of green and violet on the dark upperparts; underparts white. Facial area lacks feathers, bare skin purplish-pink. A conspicuous knob on the forehead is variable in size. Wings broad with a large white shoulder patch; pronounced spur on the carpal joint of the wing visible in laboured flight. Long legs pink, bill pink. Immature smaller,

duller and brownish. **Similar species** The largest endemic waterfowl species in Africa. Knob-billed Duck is considerably smaller and shorter-necked with a dark bill.

HABITS Frequents the main river, freshwater marshes and occasionally tidal creeks. Wary. Congregates in considerable numbers in ricefields where it may cause considerable crop damage; also feeds on fish. Skeins often fly to and from their feeding grounds on the river at dawn and dusk.

VOICE Generally silent, but occasionally a high-pitched whistle.

STATUS AND DISTRIBUTION Generally uncommon on the coast but common to abundant in other divisions during the rains and into the early dry season. Numbers lower February to June. Widespread in suitable habitat throughout Senegal.

BREEDING Most records towards the end of the rains, October-December. Opportunistic in its choice of nest site, ranging from a high tree hole to a ground scrape protectively placed under low thorns, often some distance from water.

KNOB-BILLED DUCK
Sarkidiornis melanotos   **Plate 5**

Fr. Canard à bosse

Other names: Knob-billed Goose, Comb Duck

IDENTIFICATION 50-65cm. Sexes similar in plumage but male larger than female. Large and goose-like, dark iridescent greens and blackish above, dirty white and dark-flecked below. Head and neck white, heavily speckled black. Breast white, flanks grey. Only the male possesses a comb, the large fleshy grey-black protuberance at the base of the bill. Legs and strongly-clawed feet grey-brown. Immature has practically the entire head black and lacks comb. **Similar species** Spur-winged Goose is considerably larger and longer-necked with a pink face and bill.

HABITS Essentially a freshwater duck favouring swampy ricefields and riparian habitats in The Gambia. When swimming it looks hump-backed, with the tail held up well out of the water.

VOICE Flight call a high-pitched wheeze.

STATUS AND DISTRIBUTION Non-breeding dry season visitor to The Gambia, where numbers are much reduced compared with earlier this century. Uncommon at the coast, frequent to locally common on freshwater in CRD and URD, recorded every month with January-May peak. Widely distributed in appropriate habitat throughout Senegal.

BREEDING Not proven in The Gambia, despite large congregations of immature birds in CRD in March and May. Nesting suspected in Senegal. Prefers a treehole; clutch large, up to 20 eggs.

AFRICAN PYGMY GOOSE
Nettapus auritus Plate 5

Fr. Anserelle naine ou Sarcelle à oreillons

Other names: Pygmy Goose, Cotton Teal

IDENTIFICATION 32cm. Sexes differ. A small, attractive teal-sized duck with bright distinctive facial markings in the male. Male dark green above, rusty tawny-brown below fading to whitish on the belly. White face and throat, with pale jade ear-coverts edged black in breeding plumage. Female duller and smudgy greyish-white on the face. Bill short and stubby, bright yellow in male, duller in female, which also lacks the black nail at the tip. Legs and feet grey. In flight, black wings with prominent white speculum usefully distinguish it from all other diminutive waterfowl in the region. Immature closely resembles female. **Similar species** None.

HABITS Favours seasonal lakes, ponds covered with waterlilies, and flooded rice fields. Ripened waterlily seeds constitute the mainstay of its diet. Sometimes takes insects from the surface and fish fry are caught by diving. Confiding, but spends much time motionless amongst packed and dense vegetation when difficult to see. Usually in small parties, which separate into pairs to breed. Flight swift and low. Rarely seen out of the water, preferring to perch on a partially submerged log to preen.

VOICE A high-pitched soft twittering whistle, *choochoopee -wee*; subdued quacking in courtship.

STATUS AND DISTRIBUTION Uncommon wet season visitor in WD and URD, July-October, locally frequent in CRD where it is present in all seasons with peak numbers in the rains. In Senegal most common along the Senegal River basin; otherwise along the coastal strip and the eastern interior.

BREEDING Immatures recorded in The Gambia (CRD) December-January. A cavity nester utilising treeholes, termite mounds and even the thatch of the nest of Hamerkop. Prefers locations over water.

COMMON TEAL
Anas crecca Plate 5

Fr. Sarcelle d'hiver

Other name: Teal

IDENTIFICATION 35cm. Sexes differ. Smallest migratory duck to the region. In breeding male, chestnut head and well demarcated metallic green patch around the eye are diagnostic. Both eclipse male and female speckled brown on the breast with a dark bill, showing a wedge of buff below the tail. Separated from other small brown dabbling ducks by well pronounced pale eye stripe, pale chin and dark crown. Black and green speculum prominent in flight and a useful aid to identification. **Similar species** Female Garganey is greyer with a more strongly-masked face pattern.

HABITS Coastal sewage outfalls and freshwater regions of the main river are likely locations. Often solitary, sometimes in small groups. Both surface feeds and upends. Flight is typically swift and agile, often quite high, descending to water at a steep angle.

VOICE A disyllabic melodious whistle *kidickkidick* and a high pitched *quack*.

STATUS AND DISTRIBUTION Palearctic breeder, migrant to Senegambia. Uncommon on the coast, and infrequent sightings inland anytime during the dry season. Senegal River delta accommodates a few thousand birds around January and there are records from southern Casamance.

MALLARD
Anas platyrhynchos Not illustrated

Fr. Canard colvert

IDENTIFICATION 58cm. Sexes differ. Heavy looking surface-feeding duck. Male in full breeding plumage has glossy green head, greenish-yellow bill, a narrow white collar, and purplish-brown breast. Underparts pale grey, tail white with distinctive curled-up black central feathers. Female is mottled brown with brownish bill. Both sexes have broad purple speculum bordered with white bars which are very conspicuous in flight and an important identification feature. Legs orange. Male in eclipse resembles a dark female but with brighter speculum, darker crown and ruddier breast. **Similar species** Female Mallard from female Northern Pintail by heavier bill and head, thicker less elegant neck, white border on both sides of purple speculum and a shorter tail. Mallard sits noticeably lower in water.

STATUS AND DISTRIBUTION Palearctic breeder with no proven modern records in The Gambia. Birds in northern Senegal in recent years are possibly reared in captivity and not of wild origin.

NORTHERN PINTAIL
Anas acuta   Plate 5

Fr. Canard pilet

Other name: Pintail

IDENTIFICATION 60cm. Sexes differ. A largish, slender duck with a long slender neck. Male occasionally seen in breeding plumage: distinctive facial and neck markings; head is chocolate brown with a white stripe on the side of the neck that extends onto the breast. Upperparts and sides pale grey. Female cinnamon-brown with darker speckling, elegantly slim. Elongated tail shorter outside breeding season, but the tail still tapers to a point and together with sharply pointed wings gives the bird a distinctively streamlined flight jizz. In eclipse, overall plumage is grey-brown; bill dull pale grey and a useful fieldmark. Speculum green with white trailing edge. **Similar species** Female can be distinguished from female Mallard by slimmer, more elegant shape (especially neck, plainer face and pointed tail; from female Northern Shoveler by smaller, grey bill.

HABITS Favours freshwater regions of the main river, occasionally elsewhere in brackish conditions. May be alone, more usually in small flocks, both of single species and mixed. Feeds with the head submerged, also upends and dabbles. Noticeably wary.

VOICE Mostly silent in Africa, vocal on breeding grounds.

STATUS AND DISTRIBUTION Palearctic breeder, migrant to Senegambia. Records from all river divisions and coastal WD from December to February, but not in large numbers. Tens of thousands in northern Senegal in same period, with records also from Casamance.

GARGANEY
Anas querquedula   Plate 5

Fr. Sarcelle d'été

IDENTIFICATION 38cm. Small. Sexes differ. Male readily identified at rest by a conspicuous white stripe on dark brown head extending from above the eye down the nape. Important diagnostic flight features are the dull green speculum bordered with white, long black and white scapulars and pale blue forewing, all generally retained on birds seen in The Gambia outside its Palearctic breeding season. Female and eclipse male overall speckled brown. **Similar species** Female Garganey needs care in separation from female Common Teal; note distinct dark eyestripe, paler in Common Teal, also pale

patch at the base of a longer greyish-green bill on Garganey visible at close range.

HABITS Typically the freshwater regions of main rivers, occasionally elsewhere in brackish conditions. Normally feeds whilst swimming and with the head submerged but sometimes grazes on open land. Agile in flight moving in tight flocks. Courtship activity occurs amongst the floating rafts of ducks whilst they are wintering in The Gambia. Often joined by smaller numbers of less common waterfowl, so large flocks are worth surveying.

VOICE Some vocalisation during courtship but generally silent in Africa.

STATUS AND DISTRIBUTION Palearctic breeder, migrant to Senegambia. Locally and erratically common to very abundant in CRD and URD and further east along the Gambia River into Senegal during the dry season up to mid March. Annual records of singletons and small groups in WD. Commonest European duck overwintering on the Gambia River. 100,000+ birds in northern Senegal in 1970s, now possibly declining.

NORTHERN SHOVELER
Anas clypeata   Plate 5

Fr. Canard souchet

Other name: European Shoveler

IDENTIFICATION 50cm. Sexes differ. Characteristic disproportionately large spatulate bill; a thickset duck with a shortish neck, swims with its front low in the water. Most males seen in The Gambia are in partial eclipse: sides reddish to rufous, breast white and head greenish-black with white smudging around the face. On the wing, a large blue shoulder patch with a green speculum bordered with white is diagnostic in both sexes. Female and elipse male speckled brown. **Similar species** No other duck has the unique spatulate bill-shape of Northern Shoveler.

HABITS Normally singly or in very small parties on freshwater. Usual feeding technique is to sweep the bill from side to side filtering small aquatic organisms whilst swimming with the head and neck lowered and outstretched; will also filter soft mud.

VOICE Generally silent in Africa.

STATUS AND DISTRIBUTION Palearctic breeder, migrant to Senegambia. Annually frequent in WD; infrequent along the river from LRD to URD. Singles and groups numbering up to ten, found north and south of the Gambia River. Seasonally far more numerous in northern Senegal, with population stronghold of several thousand in Senegal River delta.

COMMON POCHARD
Aythya ferina   **Not illustrated**

Fr. Fuligule milouin

Other name: Northern Pochard

IDENTIFICATION 45cm. Sexes differ. Breeding male unmistakable, with uniform dark chestnut head and neck contrasting with black breast and pale grey body. High crown and long sloping profile to head common to both sexes, and an important fieldmark. Bill black. Female has brown head and foreparts and is distinguished from female Tufted Duck by head shape, pale cheek patch, and broad grey not white wingbar. Eclipse male resembles female but is greyer above. Accomplished and energetic divers.

STATUS AND DISTRIBUTION Palearctic breeder, rare in The Gambia. Recorded from both northern and southern Senegal.

FERRUGINOUS DUCK
Aythya nyroca   **Plate 5**

Fr. Fuligule nyroca

Other name: White-eyed Pochard

IDENTIFICATION 38cm. Sexes differ. Breeding male dark compact chestnut-brown diving duck; in eclipse moults to a greyish-brown, similar to the female. A flat forehead with a peaked crown and distinctive white vent are prominent fieldmarks common to both sexes. Eye dark in the female but always conspicuously white in the male. **Similar species** Tufted Duck lacks the white vent. In flight white wingbar is broader and brighter than Tufted Duck

HABITS Elsewhere in West Africa prefers vegetated

freshwater and flooded fields

VOICE Not heard in The Gambia; elsewhere a croaking flight call noted.

STATUS AND DISTRIBUTION Palearctic breeder, migrant to Senegambia. Rare, most recent record of two immature birds shot in coastal WD in 1983. Possibly under-recorded. Small numbers in Djoudj, northern Senegal.

TUFTED DUCK
Aythya fuligula   **Plate 5**

Fr. Fuligule morillon

IDENTIFICATION 42cm. Sexes differ. A small, compact, round-headed, crested diving duck. Breeding male glossy black on the upperparts with an immaculate white rectangle on the sides; crest droops loosely over the back of the head. Female and elipse male dark brown with a shorter crest; female may show a white band of varying size at the base of black-tipped grey bill. Both sexes have a conspicuous radiant yellow eye in all seasons. White wingbars visible in flight, but are not as conspicuous as in Ferruginous Duck. **Similar species** Male unmistakable. Female lacks white vent of Ferruginous Duck and often shows white at base of bill.

HABITS Favours freshwater habitats. Active when feeding, tuck-diving with great regularity, coming to the surface only briefly before resubmerging. Flight is usually high and at speed.

VOICE Not recorded in The Gambia.

STATUS AND DISTRIBUTION Palearctic breeder, migrant to Senegambia. Rare, a few recent records of 1-3 birds in WD (Tanji Bird Reserve and Kotu Sewage Farm) and CRD. Flocks of less than 50 birds in northern Senegal, and also recorded from southern Senegal.

KITES, VULTURES, HARRIERS, HAWKS, BUZZARDS, EAGLES AND ALLIES

Accipitridae

A diverse family of birds of prey, characterised by hooked bills, sharp talons and broad, rounded wings. Often solitary (except vultures). Females generally larger than males and some species are sexually dimorphic. Identification sometimes difficult (especially when perched) and some species are notoriously variable. Immature plumages can also be confusing. The family can be conveniently subdivided into several groups (with the exception of a few unique species).

Kites (*Milvus, Elanus, Chelictinia*)

Four species. Graceful fliers, often gregarious. *Milvus* are long-winged with long, forked tails, noted for their scavenging habits. *Elanus* and *Chelictinia* are more graceful in flight, frequently hovering.

Vultures (*Gypohierax, Neophron, Necrosyrtes, Gyps, Torgos, Trigonoceps*)

Eight species including one with an atypical semi-vegetarian habit; otherwise a group of medium to huge carrion feeders. Mostly associated with semi-arid open areas and the margins of human settlement. Generally characterised by dull plumage, bare faces and usually bare necks, and powerful hooked bills. Soar and glide at high altitudes, becoming highly gregarious once carrion located. There is a well defined pecking order at these gatherings of scavengers. Sexes alike in plumage with females usually slightly larger than males. Reproductive maturity mostly reached at not less than six or seven years in the larger species. Generally silent, but can be raucous when sparring for position at a carcass.

Snake Eagles (*Circaetus*)

Three species. Large-headed and thick-necked with staring yellow eyes and bare tarsi. As their name suggests, snakes form a significant part of their diet.

Harriers (*Circus*)

Three species, all non-breeding dry-season Palearctic migrants. Medium-sized raptors, sexes differ in all species. Long-winged, long-tailed and long-legged with characteristic flight, quartering low over open-ground, and dropping suddenly onto prey. Flight buoyant with slow wingbeats, gliding on stiff wings held in a shallow V. Communal roosting may occur. Identification problems are mostly associated with the females and immatures of the two 'ring-tail' species.

Sparrowhawks and Goshawks (*Micronisus, Melierax, Accipiter*)

Eight species. These are the true hawks, ranging from small to large in size. Characteristic features include comparatively short rounded wings, smallish sharp bills and long unfeathered legs. Chanting goshawk (*Melierax*) is longer-winged than true *Accipiter* species, perching prominently on trees or posts. *Micronisus* resembles *Melierax* but is more like *Accipiter* in behaviour. Prey is normally birds, pursued with great agility. Females can be up to a third larger in size than males, so species may overlap considerably in size which can contribute to identification problems, particularly when immatures are involved.

Buzzards (*Buteo*)

Three species, all migrants; two European and one African. Medium-large, thickset raptors soaring on broad, rounded, wings and short fanned tails; often with wings held in ashallow V. Rather sluggish often perching prominently. Sexes alike. Notoriously variable at all ages, ranging from all-pale to all-dark, which may make identification difficult.

Eagles (*Aquila, Hieraaetus, Lophaetus, Stephanoaetus, Polemaetus, Haliaeetus*)

Aquila eagles (two species) are large, dark raptors with long wings, shortish tails and powerful bills. Hawk Eagles *Hieraaetus* (four species) are medium-large raptors of woodland and savanna. They have long, broad wings, barred tails and feathered tarsi. The remaining four genera are each represented by a single species.

The following species do not conveniently fall into the above groups: Cuckoo-Hawk (*Aviceda*), Honey Buzzard (*Pernis*), Bat Hawk (*Macheiramphus*), Bateleur (*Terathopius*), Harrier-Hawk (*Polyboroides*), Grasshopper Buzzard (*Butastur*) and Lizard Buzzard (*Kaupifalco*).

AFRICAN CUCKOO-HAWK
Aviceda cuculoides　　　Plates 8 & 12

Fr. Faucon-coucou

Other names: West African Cuckoo-Falcon, Cuckoo-Falcon, Cuckoo-Hawk

IDENTIFICATION 40cm. Sexes similar. A medium-sized grey hawk recalling an oversized African Cuckoo. Upperparts grey-brown. Wings, head and well demarcated upper-breast all grey. Belly white with thick brown barring. Small pigeon-like head is slightly crested, with a rufous patch on the nape detectable at close quarters. Tail dark above, tipped white, with three black bars. Wings long, nearly reaching tip of tail when perched. Eye bright yellow and prominent like Honey Buzzard. Immature dark brown above with buff edgings and small irregular pale patches; below whitish with variable large brown streaks and blotches. Slight crest already detectable.

FLIGHT CHARACTERISTICS Wings conspicuously long and pointed, tail longish. Languid direct flight, kite-like with deep beats. Underwing-coverts solid chestnut-burgundy in local nominate form, with lightly barred pale underwing. **Similar species** Long wings and slight crest separate African Cuckoo-Hawk from *Accipiter* species.

HABITS A solitary and secretive bird, most records associated with creeper-clad watercourses. Moves around in tangled undergrowth rather like a coucal. Automated head movements when surveying for food rather like a foraging bee-eater. Largely insectivorous sometimes making aerial sallies in the mode of a fly-catcher; normally swoops down to the ground.

VOICE Characteristic penetrating repetitive whistle *peeee-oo* and a *choo-titti-two*.

STATUS AND DISTRIBUTION Rarely recorded, all recent records from WD and LRD, March-August; older records also November-January. Five sightings in Senegal, December, January and June.

BREEDING Immature, April 1994 in western Foni District, WD and March 1997 in Banjul. Nearest proven breeding Sierra Leone. Nest a frail platform of leafy twigs well hidden in thick cover.

EUROPEAN HONEY BUZZARD
Pernis apivorus   Plate 8

Fr. Bondrée apivore

Other names: Honey Buzzard, Western Honey Buzzard

IDENTIFICATION 52-60cm. Sexes separable. Male has slate-grey head, female usually brown-headed but may show grey around the eyes. Considerable plumage variation; jizz all-important aid to identification. Pigeon-like protruding head looks disproportionately small; tail long and narrow, tip rounded. Upperparts usually dark brown, underparts white or buffish streaked and mottled brown. Conspicuous bright yellow eye. Immature plumage is paler, eye brown.

FLIGHT CHARACTERISTICS Wingspan 135-150cm. Tail always looks long. Slow deep, soft and elastic wingbeats, soars with wings held straight out or drooping. Broad underwing pale with conspicuous black edges, usually prominently barred, always shows dark carpal patch. Undertail usually with two dark bars near base and single dark band near tip. **Similar species** Booted Eagle has a rounder, larger head and powerful bill; in both phases it shows contrasting dark wing feathers with neat pale wedge created by three innermost primaries, while Honey Buzzard shows irregular white patches on the primaries. From true buzzards by yellow, not brown eye, small head, long tail and overall slimmer appearance; soars on flat wings, not raised like true

buzzards. Immature brown-phase Harrier-Hawk is larger and bare-faced.

HABITS Sits relaxed, wing-tips reaching tip of tail, moves sluggishly. Usually solitary. Could be encountered anywhere but particularly in the vicinity of immense arboreal bees nests, or on damp ground walking and digging for grubs, when confusion with immature Harrier-Hawk is most likely. Generally insectivorous but also catches small birds, amphibians and lizards.

VOICE Probably silent in Africa. Musical flight call on European breeding grounds.

STATUS AND DISTRIBUTION Rare Palearctic migrant, occasional recent records in WD November-January. A few older and more easterly records November-April; once in June. Infrequent in eastern Senegal, scattered records from coast and eastern interior, in most months. Possibly underrecorded in Senegambia.

BAT HAWK
Macheiramphus alcinus   Plates 8 & 12

Fr. Faucon des chauves-souris

Other name: Bat-eating Buzzard

IDENTIFICATION 45cm. Sexes similar, female larger. Recalls a large dark falcon with slim head and insignificant drooping crest. Upperparts blackish-brown overall with prominent white streak above and below prominent yellow eye. Varying amount of white on the throat and breast, with a dark mesial stripe. Rest of underparts rufous-brown. Tail dark, indistinctly barred. Long wings nearly reach tail-tip when at rest. Legs bluish-white, long middle toe assists capture of swift-flying prey. Immature more blackish above, with more white on breast.

FLIGHT CHARACTERISTICS Wings long, narrow and sharply pointed but especially broad at the base. Hunting flight is powerful and rapid, otherwise languid and kite-like. Underwing-coverts characteristically spotted brown and white. **Similar species** Long pointed wings possibly confusable with falcon species but dark underwings distinctive.

HABITS Hunts over open areas, favouring cliffs, quarries, river banks, bolons, dried-up wells and buildings. Known bat roosts well worth watching for this species. Reclusive; spends the day hidden at roost. Emerges at dusk hunting until dark. Rarely feeds at dawn. Prey consists largely of bats with a preference for those small enough to be captured and swallowed whole whilst still in flight. Ignores large fruit-bats.

VOICE Generally silent; occasional high-pitched kite-like whistles and falcon-like *kek-kek-kek.*

STATUS AND DISTRIBUTION Rare. Several older records, from NBD and coastal WD. Recently recorded from NBD, WD, LRD and CRD. Similarly infrequent in Senegal.

BREEDING Nearest proven breeding Sierra Leone. Nest of fresh branches lined with vegetation.

BLACK-SHOULDERED KITE
Elanus caeruleus   Plate 8

Fr. Elanion blanc

Other name: Black-winged Kite

IDENTIFICATION 33cm. Smallish raptor. Sexes alike. Upperparts pale dove-grey with a conspicuous black patch on the outer edge of the upperwing. Underparts whitish, throat and cheeks white. Appears thickset at rest, with large owl-like head and distinctive dark mask through eye. Tail short. Immature looks scaly, washed in chestnut with wings and coverts brown-grey tipped rufous; black shoulder flecked white and two buffy partial breast bands.

FLIGHT CHARACTERISTICS Wingspan 76cm. Long pointed wings; underside white contrasting with black primaries. Black median and lesser coverts on upperwing constitute the black shoulders of the closed wing. Habitual hovering hunting strategy draws immediate attention at long range. Wings angled for gliding and held noticeably forwards and raised when soaring. **Similar species** Male Montagu's and Pallid Harriers have much longer tails and totally different flight. Dark Chanting Goshawk also has long tail and dark wingtips but on rounded wings.

HABITS Prefers open grassland and lightly wooded habitats. Nomadic; erratic population dynamics dictated by a variety of ecological factors. Utilises telegraph wires and poles as hunting watch-posts. Often active at dusk. Graceful in its movements.

VOICE A variety of high pitched whistles and low grating screams.

STATUS AND DISTRIBUTION Seasonally common year-round, with greatest numbers in all divisions in the early dry season, December-February, particularly in URD. Considerably fewer in heavy rains away from WD. Irruptive; in some years also very common in Senegal especially in far north.

BREEDING Can capitalise swiftly on sudden food abundance. Frequent observations of copulations and stick material stealing from buffao-weaver nests

in The Gambia. Two juveniles with adults late January 1996 near Tanji Bird Reserve, WD. Breeding recorded in northern Senegal February-April and October, in the coastal south in January. Nest is a frail structure lined with grasses in a leafy tree.

AFRICAN SWALLOW-TAILED KITE
Chelictinia riocourii   Plate 8

Fr. Elanion naucler

Other names: Scissor-tailed Kite, Swallow-tailed Kite

IDENTIFICATION 30cm. Sexes similar. Unmistakable, small, tern-like insectivorous kite. Upperparts, tail and wings silver-grey, smudged sooty-grey on back. Face white with small black patch behind bright red eye. Delicately structured, diminutive when perched, looks slim with tail extending far beyond folded wing tips. Immature body feathers edged rufous-brown, and shorter tailed.

FLIGHT CHARACTERISTICS Wingspan 78cm. Exceptionally graceful. Virtually pure white below with a small black oval patch at the carpal joint, absent in some birds. Long and deeply forked tail. Spends most daylight hours on the wing. **Similar species** Tail shape unique; black carpal patches and pale wing-tips further distinguish from Black-shouldered Kite.

HABITS A bird of semi-desert and large expanses of dry fallow land. Nomadic. Attracted to bush fires for insects, and hunts harvested land. Catches some prey on the wing but usually lands to feed. Hovers and glides. Sometimes gregarious, but usually singles or small groups in Senegambia.

VOICE Mostly silent; mews and whistles at the nest.

STATUS AND DISTRIBUTION Uncommon; all records December to June, most clustering in the late dry season to about first rainfall. High proportion of modern records from URD with occasional sightings in WD and NBD e.g. Bao-bolon. Probably under-recorded along the north bank of the river. Widespread but sporadic in same months in Senegal, especially in the north and east, recorded also in the south.

BREEDING No local evidence, nearest Mauritania, Mali and Nigeria.

BLACK KITE
Milvus migrans   **Plates 8 & 12**

Fr. Milan noir

Other name: Yellow-billed Kite

IDENTIFICATION 55cm. Sexes similar. Two subspecies occur *M. m. parasitus*, the yellow-billed African race, and *M. m. migrans*, the black-billed European race. General plumage dark chocolate brown suffused rufous, with black streaking. Primaries black, secondaries browner. Slightly forked brown tail palest towards the tip, with indistinct barring. Underparts paler than upperparts, usually with dark streaking, visible at close quarters. Short unfeathered legs and weak-looking feet. European race is slightly larger than yellow-billed form with a paler head and generally darker above and below. Immatures of both races show only shallow notch in the tail.

FLIGHT CHARACTERISTICS Wingspan 160-180cm. Long and narrowish wings typically held angled back from the carpals, and rounded at the tips. Long tail slightly forked; when fully spread looks triangular with sharp corners. Flight buoyant with wings slightly arched, tail often flexed. Hangs effortlessly on open wings. Greyish patch near the carpal on underwing also shows pale upperwing-covert panel. **Similar species** Red Kite has much deeper forked, cinnamon-orange tail and is more conspicuously and contrastingly marked with a more elastic wing action. Grasshopper Buzzard has white eye, lacks fork in tail, and shows extensive rufous upperwing in flight. Dark-phase Booted Eagle when soaring shows rounded tail with whitish uppertail-coverts. In distant views female and immature European Marsh Harrier is most effectively separated on flight characteristics, holding wings in a shallow V, and gliding extensively.

HABITS Extremely gregarious. Eats almost anything from refuse and carrion to emergent termites hawked on the wing. Twists and turns to pick items off the surface of open water and the ground. Often seen after first light quartering main roads for night-kills. Appears to be more a bird of the bush than of urban sprawl in The Gambia, and less daring in its efforts to scavenge than elsewhere. Gathers with other scavengers at fish-landing sites and abbatoirs, and is particularly attracted to bush fires.

VOICE Tremulous gull-like mewing trill *killl-errr, killll-errrr.*

STATUS AND DISTRIBUTION Common to seasonally abundant throughout Senegambia; distinct population peak in the early dry months in WD where the greatest numbers occur. A second peak in sightings usually occurs around July after which numbers decline steeply until late October-November. Early November record of c. 2,500 birds on passage high over the coast (Banjul) and northward migratory passage in NBD (Bao-bolon) in early March. Information regarding the races involved not yet available.

BREEDING A few records, in second half of dry season. Nest of sticks and rubbish built high in a tree.

RED KITE
Milvus milvus   **Plate 8**

Fr. Milan royal

IDENTIFICATION 62cm. Sexes similar. A slender, angular reddish-chestnut raptor with a noticeably pale head and characteristic long deeply-forked tail. Crown and nape whitish-rufous with faint black streaking. Neck and upper back brown-black, uppertail strikingly orange-chestnut with black striations on the shafts. Underparts pale rufous with bold darker streaking. Immature paler, head more rufous, underparts buff with reduced streaking.

FLIGHT CHARACTERISTICS Wingspan 180cm. Agile. Long and rather narrow wings are often held back from the carpal joints, sometimes slightly arched with tips upswept. Tail long and deeply forked when closed, and notched even when fully fanned. Large whitish patch across inner primaries and at the base of outer primaries. Conspicuous pale panel across upperwing-coverts. **Similar species** Much brighter coloured at all ages than Black Kite, and flight both more elegant and agile on more elastic wings. Forked tail distinguishes it from any buzzard or eagle, and from male Eurasian Marsh Harrier.

HABITS Practically unknown in West Africa. Gambian sightings all of single birds. Generally solitary.

VOICE Probably silent in sub-Saharan Africa. In Europe less vocal than Black Kite, with musical mewing call.

STATUS AND DISTRIBUTION Rare vagrant from Palearctic. Four contemporary observations: first in December 1987, on two dates, 10km apart in WD near Abuko. Other records in February 1996 at the same location, with soaring Hooded Vultures, and mid-May 1996 5km north of Tanji Bird Reserve. These are the only West African records. Also recently recorded in both East and South Africa. Origin of these birds uncertain: European birds often sedentary, but there is also a small breeding population in north-west Africa.

AFRICAN FISH EAGLE
Haliaeetus vocifer   **Plates 7 & 11**

Fr. Aigle pêcheur

Other names: Fish Eagle, West African River Eagle

IDENTIFICATION 63-73cm. Sexes similar. Large, unmistakeable wetland eagle with white head and tail contrasting with warm chestnut belly and upperwing-coverts. Rest of wings and back black; short square tail white. Immature and subadult appear rather vulturine when perched, drab dark brown overall with white streaks on the head and throat. Tail characteristically white with dark sub-terminal band.

FLIGHT CHARACTERISTICS Wingspan 190cm. Majestic; soars on flat wings. Broad, rounded underwing is chestnut contrasting with black flight-feathers and immaculate white upper breast and tail. Immature underparts whitish heavily blotched and streaked blackish-brown. Underwing leading edge white rest brown-black with white patch on primaries. **Similar species** See Palm-nut Vulture; confusion is most likely in immature plumages.

HABITS An eagle of river systems, coastal lagoons and wide mangrove creeks; ranges away from water when soaring, and may be encountered overhead elsewhere. Diet principally fish, supplemented with other animals including birds which may be killed by drowning. Adults usually occur in pairs, immatures are mostly alone and are more likely to scavenge. Usually perches prominently. The sight and sound of this bird close to the meandering Gambia River at daybreak is evocative, encapsulating birdwatching in Africa.

VOICE Very vocal, a celebrated loud far-carrying, ringing *WE AH WHOW-KYow-kow-kow* given at first light and whilst soaring. Often calls in duet.

STATUS AND DISTRIBUTION Frequent to common resident along the river from LRD to URD; less frequent from LRD to the coast. In Senegal, population stronghold in north-west regions of the Senegal River, but found throughout in all wetlands.

BREEDING Massive riverside nest annually re-occupied, typically from late rains to around January.

PALM-NUT VULTURE
Gypohierax angolensis   **Plates 6 & 11**

Fr. Vautour palmiste ou Palmiste d'Angola

Other name: Vulturine Fish Eagle

IDENTIFICATION 60cm. Sexes alike, female slightly larger. A white and black eagle-like vulture of uncertain affinities. Face and head feathered except for some bare orange-red skin around yellow eye and bill. White head, neck, back, upperwing-coverts and underside. White primaries tipped black, secondaries and intermediate scapulars black. Short, rounded tail black tipped white. Immature tawny-brown, darkest on the back, bare skin dull yellow, progresses to maturity through a series of plumages over five years.

FLIGHT CHARACTERISTICS Wingspan 140cm. Bulky. Entire underparts white, shortish rounded tail black, tipped white. Rounded white wings with black-tipped primaries, all-black secondaries; characterisitc shallow wristy wingbeats. Primaries semi-translucent against a bright sky. **Similar species** Could be confused with African Fish Eagle but entirely white underparts and black tipped white tail of the Palm-nut are obvious differences. Fish Eagle has upperwing-coverts chestnut not white. Intermediate immature Fish Eagle at distance can cause problem, but note different head and bill shape. Brown immature Palm-nut distinguished from large brown eagles by shorter tail, bare facial skin, heavy head and bill, short tail, heavy wing beats and thick legs.

HABITS Favours coastal woodlands and forest, and mangrove belts along the river. Feeds on the fleshy pulp from the fruits of the Oil Palm *Elais guineensis* in season; thus considered a partial vegetarian. Other dietary mainstay is dead fish which it has been seen to pick off open water with its feet, plus offal and carrion. Can sustain flapping flight, although cumbersome, for long periods and thus not dependent on thermals; as a result is frequently active in the early mornings. Not normally gregarious.

VOICE Mostly silent; modest growls and high-pitched squeals associated with copulation and threat behaviour.

STATUS AND DISTRIBUTION Common to occasionally locally abundant year-round at the coast and throughout WD. Common in all other divisions decreasing further east. Widespread in southern and south-eastern Senegal and in the north, infrequent elsewhere.

BREEDING Substantial branch and stick nest lined with fresh greenery and fortified with clods of turf. Recorded in The Gambia collecting nesting material, flying at dead wood and clutching with both feet, breaking off the bough to carry it away in flight.

EGYPTIAN VULTURE
Neophron percnopterus   **Plates 6 & 11**

Fr. Vautour percnoptère

IDENTIFICATION 58-68cm. Adult mostly creamy-white with yellow facial skin, usually with some buff stains on the body, and contrasting jet black wing feathers. Small untidily crested head and feathered hindneck creates a dishevelled appearance. Tail white and wedge-shaped. Smallish bill horn-brown, tipped black, legs and feet yellow, eye red. Immature dark brown with buff head, becoming white with age. Intermediate plumages can be confusing.

FLIGHT CHARACTERISTICS Wingspan 164cm. Adult: body and tail distinctively cream-white, contrasting with blackish primaries and secondaries. Diagnostic silhouette of long wings which are relatively narrow at the tips, and wedge-shaped tail. **Similar species** On the ground, immature could be confused with immature Hooded Vulture but immature Egyptian has a slimmer bill. Bare facial skin of immature Egyptian is greenish whereas Hooded is red-pink. Face, head and neck appear larger than Hooded which also lacks the long brown feathering on the nape of the neck. From below in flight Hooded has six primary 'fingers' on wing, Egyptian five.

HABITS Favours sub-desert with nearby mountains; habitat not available in The Gambia. Feeds on carrion and fish offal; visits rubbish tips. Also exploits emergent termites and will dig for insect larvae amidst animal dung. Cannot compete with larger vultures at carrion and opportunistically gleans dropped scraps. Renowned for its tool-using technique of breaking open the eggs of ground-nesting birds with stones. Not dependent on thermals to get airborne after dawn.

VOICE Rarely heard. High-pitched hisses when excited.

STATUS AND DISTRIBUTION Rare visitor. All records from WD (near Brufut and Sanyang) and LRD (Kiang West) December-June; modern records all subadult birds. Coastal records in the extreme north and south in Senegal, and also in the southeast where breeding is a possibility.

BREEDING Builds a small nest of sticks which is lined, on a mountain ledge or in a cave. A female close to laying south-east Senegal in March but no nest found there.

HOODED VULTURE
Necrosyrtes monachus   **Plate 6 & 11**

Fr. Percnoptère brun

IDENTIFICATION 70cm. Adult dark brown with a small head and weakish narrow bill, often scruffy looking. Bare pink face skin varies in colour and intensity; dirty-white down from nape down neck (the hood) is scarcely prominent. Immature has paler pink head with denser blackish-brown down. Wing and tail feathers blackish. Thighs and edges of the crop patch white, more pronounced in immatures. Bill black-brown, legs and feet blue-grey.

FLIGHT CHARACTERISTICS Wingspan 175cm. The commonest soaring raptor over much of The Gambia. Broad-winged, tail short and square with rounded edges. Forewing paler than contrasting darker primaries and secondaries. Whitish fringes to crop patch often detectable. **Similar species** Small head and narrow bill distinguishes from larger vultures e.g. immature Palm-nut Vulture. See Egyptian Vulture (which has feathered head) for separation from immature Hooded.

HABITS Ubiquitous, even found in forest thickets. Often on the wing soon after first light and comes to roost well before dark. Omnivorous, food ranging from oil-palm fruit, to insects, excrement and any scraps it can scrounge from a carcass. Gorges on winged termites when they emerge. Abundant in proximity to urban human settlements and fishing villages.

VOICE An excitable high-pitched squeak but mostly silent. Far-carrying squeaky whistle at the nest, like a harrier-hawk.

STATUS AND DISTRIBUTION Common to abundant throughout especially in WD and URD. Abundant throughout Senegal.

BREEDING Most records in dry season. Not as communal as other vultures; fragile nest constructed of sticks lined with a variety of oddments. Crown of a tall *Borassus* palm is a favourite site.

WHITE-BACKED VULTURE
Gyps africanus   **Plates 6 & 11**

Fr. Gyps africain

Other names: African White-backed Vulture

IDENTIFICATION 90cm. Large typical vulture. Adult dark brown with faint paler streaks, old birds becoming increasingly paler, not scaled as in Rüppell's Griffon Vulture. Markedly white back and rump not visible at rest, but conspicuous at take-off and in flight. Sides of chest white, flight and tail feathers black. Head and neck smaller and shorter than Rüppell's, black sparsely covered in whitish down. Poorly developed white ruff. Bill black, eye dark brown, legs and feet black. Crop patch chocolate brown. Immature darker, more streaked, lacking conspicuous white rump and back. Head and neck feathered. Darker and smaller than immature Rüppell's.

FLIGHT CHARACTERISTICS Wingspan 210cm. White leading edge of underwing-coverts contrasts strikingly with dark flight feathers. Body dark brown and streaked. White back clearly visible from above. Immature is more mottled and distinction between white and black underwing is insignificant. **Similar species** Smaller than adult Rüppell's Griffon Vulture which is copiously spotted.

HABITS Normally over open dry savanna with scattered trees, but often in proximity to villages inland. Powerful scavenger of soft tissues and intestines: like Rüppell's inserts head and neck into open body cavities to forage. Waits for thermals to develop to get airborne.

VOICE Hisses and squeals are sometimes loud in feeding disputes. Otherwise silent.

STATUS AND DISTRIBUTION Common to locally abundant throughout. Highest densities in CRD in the late dry months, when it is also most likely to occur in WD. Numbers fall during rains. Widespread throughout Senegal.

BREEDING Mostly latter half of the dry season. Small tree-nest of sticks, sometimes lined, splattered with excreta. Colony nesting recorded in Senegal; loose communities gather in The Gambia along upper reaches of the river in April.

RÜPPELL'S GRIFFON VULTURE
Gyps rueppellii Plates 6 & 11

Fr. Vautour de Rüppell

Other names: Rüppell's Vulture

IDENTIFICATION 98cm. Adult dark grey-brown, strikingly scaled whitish over both upper- and underparts. This characteristic dense cover of broad whitish crescentric spots most pronounced on the wing coverts and underparts in older individuals. Primaries, secondaries and tail black. Head and long neck covered in a dirty grey down. Modest white ruff. Eye yellow, large bill reddish-horn, feet grey. Crop patch dark brown or black. Immature lacks extensive scaling and thus appears darker, with streaked underparts.

FLIGHT CHARACTERISTICS Wingspan 240cm.

Underwing-coverts black-brown, with obvious white frontal bar behind leading edge of inner wing and two complete rows of white spots across secondary and primary coverts. Immature underwing pattern less well-defined but with off-white frontal bar and at least three rows of pale bands on coverts. On banking, scaly spotting of the back and upperwing-coverts is clearly visible. **Similar species** Adult easily separable from White-backed Vulture by spotted plumage and its longer neck. Immature White-backed is smaller, darker and shorter-billed.

HABITS A particularly aerial vulture soaring up to 2000m and capable of foraging up to 150km daily from nest or roost. Dependent on thermals to get airborne. Roosts and breeds communally. Frequently the most numerous vulture seen at a corpse, adapted to eat soft flesh and viscera. The long neck enables it to probe deep into body cavities to withdraw intestines. Gorged individuals rest on the ground or sit along the boughs of nearby trees.

VOICE Cackles and hisses whilst scavenging and at the nest, otherwise silent.

STATUS AND DISTRIBUTION Frequent in WD, common elsewhere with greatest numbers in CRD in the dry season; overall most numerous December-January. Numbers reduced in the wet season in years of heavy rain. Widespread through most of Senegal, less so in the south.

BREEDING Breeds in dry-season. Nests in loose colonies in tall trees, particularly in the crowns of *Borassus* palms along the river. Nest a sparse construction of sticks, meagre lining.

EUROPEAN GRIFFON VULTURE
Gyps fulvus Plates 6 & 11

Fr. Vautour fauve

Other name: Griffon Vulture

IDENTIFICATION 100cm. Overall body colour of adult soft ginger-rufous to khaki-brown, contrasting with much darker wings; head and neck have a bluish tinge; recalls a very large White-backed Vulture. Spiky rufous feathers around the neck contrast with a creamy-white ruff. Bill horn-yellow, cere grey, eyes yellow-brown. Legs and feet dark grey-brown with the anterior part of the upper tarsus covered in white down. Immature is darker with marked pale streaking on the underparts. Adult plumage is attained after seven years.

FLIGHT CHARACTERISTICS Wingspan 270cm. Huge; with relatively small head and neck, and shortish tail. Soars with the wings held in a slight V. Uniform khaki underwing-coverts and underbody contrast with blackish flight feathers and tail. Immature much sandier with whitish streaking, contrasting even more with the dark underwing. **Similar species** Paler and more sandy than smaller, darker brown White-backed. Lacks dark scaly appearance of Rüppells. On the ground, European Griffon appears to tower above other vultures.

HABITS Not well known in Africa. Up to eight seen together when feeding in mixed groups. Approachable when feeding, flying off to sit in a nearby tree when disturbed. Threat displays and quarrels regular at food.

VOICE Hisses, low cackles and screeches whilst squabbling at a communal scavenge.

STATUS AND DISTRIBUTION First recorded in 1991. Uncommon Palearctic migrant with all records from CRD and URD from late October to early May; highest frequency January-March in CRD where eight birds together in May. Single bird in WD, February 1997. In Senegal reported since 1970s; a bird ringed in Spain was recovered in Dakar in 1977.

LAPPET-FACED VULTURE
Torgos tracheliotus   Plates 6 & 11

Fr. Vautour oricou

Other name: Nubian Vulture

IDENTIFICATION 105cm. Massively proportioned vulture with huge bill on bare head. Overall brown, darker on the back. Feathering of the lower breast noticeably lanceolate, with a ruff of shortish paler brown feathers below the neck. Naked head variable purplish-pink with ridges of skin (the lappets) on the sides of the face and neck. Immense bill greenish-brown, paler towards the tip. Legs greyish. Crop patch is very dark brown. Immature similar, usually with paler blue-grey head, sometimes with whitish down.

FLIGHT CHARACTERISTICS Wingspan 262cm. Huge silhouette. All-dark wings with jagged trailing edge and distinctive narrow white band along the leading edge of the underwing. Pronounced white thighs and streaked belly. Shortish broad tail. **Similar species** On the ground the massive size with huge bulbous tipped bill and naked angular head readily separate it from other large vultures.

HABITS Usually in small numbers, often alone. Capable of breaking into carcasses earlier than other vultures using deliberate sideways blows of its huge bill; will also ingest sinew, bones and hide too tough for other species. Typically waits at the periphery of a mixed feeding group, periodically

charging feeding birds with characteristic lumbering gait to pirate their food. Its size and dominance generally ensures success. Capable of killing sick and wounded animals by the impact of a lunging attack. Dependant upon thermal currents to get airborne. Spends long periods standing after feeding.

VOICE Generally silent.

STATUS AND DISTRIBUTION Locally frequent at CRD-URD boundary, fewer records in LRD; one in WD. Especially attracted to dry season concentrations of migratory cattle herds in CRD. Modern records evenly distributed November-June. Widespread in Senegal.

BREEDING No Gambian information but recorded December in northern Senegal. Nest a massive platform of branches, carpeted with skin and hair collected from carcasses, often in the top of a thorny tree.

WHITE-HEADED VULTURE
Trigonoceps occipitalis   Plates 6 & 11

Fr. Vautour huppé ou Vantour à tête blanche

IDENTIFICATION 85cm. Thickset and medium-large, an unmistakable blackish and white vulture with characteristic angular head covered in white down. Bill red, tipped black, extensive blue cere. Lower belly and thighs white, contrasting sharply with the black band of the chest. Upperparts black, grey-white secondaries clearly visible in closed wing. Legs flesh, crop patch white. Immature browner with greyish-white plumage becoming whiter in a succession of moults. Head downy and brown, but angular shape and bill remain diagnostic.

FLIGHT CHARACTERISTICS Wingspan 210cm. Contrasting black breast and white belly, and black tail distinctive. Clear thin white line along the middle of the underwing is diagnostic in immatures that have not acquired the clean white lower belly and inner flight feathers. **Similar species** Adult's dark underwing-coverts and white secondaries unique amongst vultures, as is thin white line on underwing of immature.

HABITS Usually solitary, occasionally in pairs. Usually depends on scraps if larger vultures are at a carcass. In The Gambia most likely to be encountered scavenging on the corpses of smaller domestic animals, or whilst soaring.

VOICE Occasional hissing.

STATUS AND DISTRIBUTION Uncommon to frequent in LRD and CRD, with slight peak April-July. Rare or absent elsewhere. Widespread but uncommon in Senegal, largely unrecorded from central parts.

BREEDING Recorded in The Gambia November-December and May-June, in Senegal in January. Generally infrequent across West Africa. Nest a large platform of sticks lined with grasses and animal hair, placed high, often in a baobab or silk-cotton tree.

SHORT-TOED EAGLE
Circaetus gallicus　　　Plates 8 & 12

Fr. Circàete Jean-le-Blanc

Other name: European Snake Eagle

IDENTIFICATION 65cm. *C. g. gallicus*. Sexes similar. Medium-large, upright eagle. Heavy-bodied, with large rounded owl-like head on thick neck, grey-brown back, creamy-white front with darker upper breast and head giving hooded effect. Flanks often with brown barring. Uppertail grey-brown with four darker bars. Distinctively long bare scaly legs blue-grey. Large staring eyes are orange-yellow with frowning appearance at close range. Immature browner overall with bonnet-shaped head whitish from nape upwards, underparts dirtier white smudgily blotched and indistinctly barred.

FLIGHT CHARACTERISTICS Wingspan 180cm. No dark patch at the carpal joint where wing is broadest. Strikingly pale body contrasting with brown hood in typical plumage. Underwing also pale, lightly barred with darker grey primary tips. Narrowish dark-banded tail square-ended with sharp corners. **Similar species** Dark hooded head and white underparts recall miniaturised Martial Eagle but Martial is strongly spotted, with feathered legs and dark undersides to huge wings in flight. Similarly pale but small-headed Osprey shows angled wings and dark carpal patches. Pale *Buteo* species also show prominent dark carpal patches. African migrant or nomadic subspecies Beaudouin's Harrier-Eagle *C. g. beaudouini* (Fr. Circaète de Beaudouin) may occur; shows multiple continuous barring across the whole underparts below a dark hooded head. Also has darker upperparts and darker, better defined bars on tail.

HABITS A reptile-hunting specialist, favouring arid areas with scattered trees (note Beaudouin's Harrier-Eagle prefers damper habitats). Circles, quarters and occasionally runs for prey on the ground. Looks out from a prominent perch, or hovers clumsily with strenuous wingbeats, often with legs dangling. Usually solitary.

VOICE Rarely vocal; soft mewing call on breeding grounds.

STATUS AND DISTRIBUTION Palearctic-breeding *C. c. gallicus* frequent to common in the dry months in CRD and URD with peak in November-December. Less frequent towards the coast. Recorded also May-July but no information from heavy rains in August-September. Individuals characteristic of African race *C. c. beaudouini* recorded in January in URD, and February and July in CRD. In Senegal, *gallicus* recorded most months, with *beaudouini* recorded in north during the rains and to the south of The Gambia in both dry and wet periods.

BROWN SNAKE EAGLE
Circaetus cinereus　　　Plates 8 & 12

Fr. Circaète brun

Other name: Brown Harrier-Eagle

IDENTIFICATION 68cm. Sexes similar. Medium-large, uniformly dark brown eagle with large owl-like head and neck, prominent staring yellow eyes and relatively long bare legs. Crest of lax feathers accentuates head size. A black patch over and through the eye produces a frowning appearance. Bare pale grey legs appear characteristically stiff when perched. Immature has irregular white fringes to brown feathering (including the underwing), contributing to a paler mottled appearance. Eyes paler and duller.

FLIGHT CHARACTERISTICS Wingspan 200cm. Underwing is striking in flight: generally brown underparts and underwing-coverts contrast with pale grey-white unbarred flight feathers, which may look translucent against a bright sky. Shows brown tail with three regularly-spaced narrow grey bars and narrow whitish tip. **Similar species** Perched all-brown juvenile Bateleur can be confused on head shape. *Aquila* eagles have feathered legs.

HABITS Favours dry wooded country with tall trees. Sluggish but shy; typically stands prominently silhouetted on the crown of a large tree or bare baobab. When disturbed moves on a few trees. Usually solitary, occasionally in pairs, sometimes joining other raptors soaring. Also hovers clumsily. Feeds on snakes and lizards, carrying them by the head in flight; also eats other small animals.

VOICE Vocal in flight, particularly so in display and when carrying a snake. Far-carrying throaty *khok-khok khok-khok-khok* often terminating in a lengthy *kwee-oo*.

STATUS AND DISTRIBUTION Infrequent at the coast, becoming frequent to common in LRD and CRD, less so in URD. Sightings peak December to

148

April, present but considerably reduced in rains. Widespread throughout Senegal, but largely absent from northern coast.

BREEDING Birds seen carrying nest material in rocky hilly area of CRD in February, display in LRD November, immature CRD February. In northern Senegal nesting recorded September and November. Nest usually small.

WESTERN BANDED SNAKE EAGLE
Circaetus cinerascens   **Plates 8 & 12**

Fr. Circaète cendré

Other names: Smaller Banded Snake Eagle, Smaller Banded Harrier-Eagle

IDENTIFICATION 58cm. Sexes similar. Medium-large, portly eagle with rounded owl-like head and neck. Grey overall, with faint paler barring most evident on belly. Tail shows characteristic broad white bar above a broad black subterminal band and narrow dirty white tip. Shoulder of closed wing narrowly white. Conspicuous dark-tipped orange-yellow bill and cere. Eye large, whitish, legs yellow. Immature paler overall, more streaked, almost white on breast and belly.

FLIGHT CHARACTERISTICS Wingspan 150cm; Grey above and below, front-heavy in appearance, white-barred black tail distinctive; underwing barred. **Similar species** Short-toed Eagle lacks broad white tail band and is paler. Much slimmer African Harrier-Hawk can pose problems because of its white tail bar and similarly drab plumage, but easily separated by weaker head, unbarred underwings with black trailing edge, fluttery, lighter flight and mewing call.

HABITS Usually solitary, favouring damper habitats with plentiful trees or along the banks of the freshwater parts of the river. Feeds on snakes and amphibians, may sit with a snake dangling like a bootlace from the bill. Normally perches prominently in the top of a solitary tree. Pairs display with one perched and other flying high above calling.

VOICE Not particularly vocal. Repeated disyllabic honking *aw-aw* with upwards inflection noted in display. Loud flight call *kok-kok-kok-kok-ko-ho* fading out.

STATUS AND DISTRIBUTION Seasonally frequent at a few regular sites in WD, most likely during and after the rains but could occur anytime. Frequent in other divisions and recorded in every month, particularly in CRD where it is probably resident. In Senegal recorded just over both borders at the coast and to the east in Niokolo Koba.

BREEDING Not yet recorded but likely. Displays seen in May and July; juvenile with adult and alone at Tanji Bird Reserve in WD in a number of years.

BATELEUR
Terathopius ecaudatus   **Plates 8 & 12**

Fr. Aigle bateleur

Other name: Bateleur Eagle

IDENTIFICATION 60-75cm. Sexes differ. A colourful, strikingly short-tailed eagle. Adult back black, mantle and tail chestnut. Wing-coverts grey. Bare face, legs and feet tomato-red. Wingtips extend well beyond tail when perched. Nape crested, looks flared and hooded in wind. Sexes distinguishable perched; secondaries pale grey in female, black in male. Immature brown dappled white, with rather longer tail than adult. Facial skin pale greenish-blue. Maturity reached after seven to eight years.

FLIGHT CHARACTERISTICS Wingspan 175cm. Unmistakable in flight. Appearing almost tailless with feet extending beyond tail in adult. Upswept wingtips and rocking flight reflect the French name meaning tightrope-walker. Underwing largely white in female, broad black trailing edge in male. Cruises with head held downwards, surveying its hunting territory, often at low level.

HABITS Favours open grassland, thorn scrub and light woodland, usually with low human populations. Usually solitary or in small groups, sometimes with other raptors. Flight circling or direct, rarely flaps after take-off. Pigeons and sandgrouse likely to be preferred food in The Gambia. Takes carrion, also attends termite swarms at the beginning of the rains, and may pirate other raptors.

VOICE Generally silent, occasional barks and screams recorded.

STATUS AND DISTRIBUTION Uncommon overall, recorded year-round in all other divisions. Most easily seen in LRD where a few birds regularly encountered, with peak sightings July-September. In Senegal, mostly in east, but diminishing in numbers.

BREEDING Breeding not proven but plentiful records of juveniles with adults. In northern Senegal nest with egg in November. Bulky nest lined with fresh leaves.

AFRICAN HARRIER-HAWK
Polyboroides typus   **Plates 6 & 11**

Fr. Gymnogene ou Gymnogene d'Afrique

Other names: Harrier-Hawk, Gymnogene

IDENTIFICATION 60-66cm. *P. t. pectoralis*. Sexes similar. A largish grey, light-bodied hawk, broad-winged with longish tail, noticeably small head with a bare unfeathered face and partial nape-crest makes the head look pinched and vulturine. Upper-parts, head and breast plain grey. Belly finely barred black and white. Characteristic large black spots on wing-coverts. Tail black with a distinctive single broad white band across middle and a narrow white tip. Bare facial skin, an excellent field character, varies from bright yellowish-green to orange-red. Distinctively long yellow legs. Immature brown above with some faint whitish flecking; wings darker, paler below, ginger on flanks. Facial skin greyish-blue, cere and gape yellow. Reaches maturity at about three years, passing through transitional dappled brown and grey plumages.

FLIGHT CHARACTERISTICS Broad wings, under-side finely barred with a broad black trailing edge, fringed with a narrow white line. Undertail-coverts similarly barred black and white. Conspicuous broad white band across middle of grey tail. Imma-ture underwing-coverts rufous-brown dappled with darker brown; brown tail with five narrow darker bands. Flight slow and buoyant, even fluttering. **Similar species** Adult is unmistakable. Brown immatures can be confused with Honey Buzzard and smaller eagles.

HABITS Favours wooded areas with plentiful palm trees. Omnivorous, ingests large quantities of oil palm fruits in season. Varied hunting tactics are characteristic including soaring and gliding, observation from a perch, and walking and digging in soft earth. Above all famed for double-jointed inter-tarsals which allows its long legs and feet to be inserted at extraordinary angles into otherwise impenetrable crevices. Climbs well, using wings as well as feet. Raids the nests of hole-nesting wood-hoopoes, barbets and parakeets for nestlings.

VOICE A far carrying characteristic quivering *mew sueee-sueee-suueeee* rising in volume.

STATUS AND DISTRIBUTION Common in all divisions year-round, occasionally congregates in high densities. Widespread in Senegal, but largely absent from drier central areas.

BREEDING Usually in latter part of the dry season. Nest in a large tree, often a palm.

PALLID HARRIER
Circus macrourus   Plate 9

Fr. Busard pâle

IDENTIFICATION 40-46cm (wingspan 99-117cm). Male characteristically light on the wing. Overall very pale grey, almost whitish above, with wedge of black primaries, leading primary characteristically pale. Sides of face, throat, and underparts includ-ing tail and wing whitish. Eye, cere and feet pale yellow. Female 'ring-tailed' with a well demarcated facial pattern: whitish collar behind a dark crescent on the ear-coverts which extends to base of the bill, coupled with a well-defined black eye-stripe and white patch round eye. Creamy nape patch may be apparent creating a capped look. Underparts grey-buff, streaked chestnut. Back and underwing-coverts dark grey-brown. Narrow white band on uppertail-coverts and usually five darker bands on grey-brown tail. Eye, cere and feet pale yellow. Immature similar to female with even more pro-nounced head pattern and collar. Upperparts brown, edged rufous, especially on lesser wing-coverts. Underparts rich rufous, uniform or streaked dark chestnut. **Similar species** Absence of black tips to secondary-coverts and lack of any chestnut under-wing markings on male Pallid differentiate it from darker and larger Montagu's. Separation of adult female and immature is difficult and complex: nar-row blackish secondary bands on underwing of Pallid are more widely spaced on Montagu's. Montagu's face pattern is paler and less well de-fined, and Pallid has a more pronounced whitish collar. A record of Hen Harrier *C. cyaneus* (Fr. Busard Saint-Martin) from northern Senegal in February 1982 is considered doubtful.

HABITS Favours grassland and open woodland.

VOICE Generally silent in Africa.

STATUS AND DISTRIBUTION Seasonally uncom-mon throughout. Records October-May; most modern records from URD. Generally common in Senegal, but largely absent from central arid areas.

MONTAGU'S HARRIER
Circus pygargus   Plate 9

Fr. Busard cendré

IDENTIFICATION 41-46cm (wingspan 97-115cm). Male has the upperparts, head and breast blue-grey, belly pale, flanks lightly but distinctively streaked chestnut. Wing-tips extensively black. Characteris-tic conspicuous black bar on secondaries distinct in flight from above. Underside of secondaries lightly barred. Eye, cere, legs and feet yellow. Female streaked brown overall, paler below with small white 'ring-tail' rump patch. Dark markings on the head limited to a crescent of tawny feathers over the ear-coverts which are not edged with a pale

150

collar. Eye-stripe thin and small. Immature similar to female; white feathering around a dark brown eye with dark ear-covert crescent, lacking pale collar. **Similar species** See Pallid Harrier.

HABITS Favours arid open grasslands.

VOICE Generally silent in Africa.

STATUS AND DISTRIBUTION Seasonal and uncommon from WD to CRD becoming locally common in URD in the early dry season. Widespread in Senegal, especially Senegal River delta.

EURASIAN MARSH HARRIER
Circus aeruginosus   Plate 9

Fr. Busard des roseaux

Other names: European Marsh Harrier, Marsh Harrier

IDENTIFICATION 48-55cm (wingspan 115-13cm). Male has an overall dark chocolate brown back and wing coverts with contrasting blue-grey tail and flight feathers, primaries black-tipped. Head, nape and breast streaked yellowish-buff. Underparts richer brown. Underwing-coverts rufous. Eye, cere, legs and feet yellow, bill dark. Female is dark, chocolate brown overall with characteristic cream crown and face, and leading edge to wing. Immature similar to female but darker. **Similar species** Black Kite is slightly fork-tailed and has a uniform brown head. A colourfully-marked male Marsh could recall vagrant Red Kite (but lacks forked tail).

HABITS Favours marshes and swamps but could occur anywhere.

VOICE Generally silent in Africa.

STATUS AND DISTRIBUTION Widespread and seasonally common. Numbers peak in early dry season, but fluctuate annually. Locally abundant October-March at communal roosts in URD. Occasional non-breeders seen during the rains. A wing-tagged bird sighted in LRD in January 1994 was of Spanish origin. Widespread in Senegal but largely absent from central arid areas.

GABAR GOSHAWK
Micronisus gabar   Plates 9 & 13

Fr. Autour gabar

IDENTIFICATION 28-36cm. Recalls scaled-down Dark Chanting Goshawk but much more *Accipiter*-like in posture. Dark dove-grey above with densely and finely barred belly, bars thicker on flanks. Breast, throat and chin uniform grey. Larger female

more heavily marked. Prominent white rump; uppertail dark slate-grey with four dark bands. Dark-shadowed 'eyebrow effect' around red-brown eye. Bill black; cere, legs and feet red-orange. Immature as adult, but brown replaces grey, with throat and breast streaked brown, tail greyish with narrow darker bars. Well-defined pale supercilium, legs and eye yellowish to dull orange. Melanistic morph is regularly reported; all-black with paler barred tail and white flashes in primaries.

FLIGHT CHARACTERISTICS Flying away, white rump (present also on immature) is conspicuous. **Similar species** Adult Shikra has plain grey rump and underparts barred in rufous-pink. Rare Ovambo Sparrowhawk has barred throat and breast and white central marks on tail shafts.

HABITS Favours lightly wooded habitat, often but not always near watercourses. Shy; sightings often brief of birds dashing in chase of prey. Subject to sudden and temporary local increases in numbers which suggests some regional movements. Hunts for small birds whilst on the wing and is also a nest robber. Also eats lizards.

VOICE High-pitched repeated whistle *kik-kik-kik-kik* in display.

STATUS AND DISTRIBUTION Frequent throughout, becoming occasionally common east of WD; recorded in all divisions in all seasons. Overall most regular in latter half of dry season particularly in CRD and URD. Erratic increases in numbers not yet explained. In Senegal widespread apart from coastal districts.

BREEDING Not proven in The Gambia. Young bird in October, LRD.

DARK CHANTING GOSHAWK
Melierax metabates   Plates 9 & 13

Fr. Autour-chanteur sombre ou Autour-chanteur

IDENTIFICATION 45-55cm. Upperparts, throat upper breast and wing-coverts pale-mid grey, occasionally with white freckling. White belly and thighs copiously and finely barred grey. Darker charcoal-grey facial feathering around the dark brown eye creates a masked effect. Relatively long, slender rounded tail edged and tipped white. Cere, feet and very long legs bright red-orange. Immature dull brown-grey, throat and chin vertically streaked darker brown, lacking face mask; broad pale supercilium. Underparts and thighs buffish barred brown.

FLIGHT CHARACTERISTICS Wingspan 140cm. Black primary tips. Wings broad relative to

Gabar Goshawk

African Goshawk

Shikra

Western Little Sparrowhawk

Ovambo Sparrowhawk

Great Sparrowhawk

Tail patterns of Senegambian sparrowhawks and goshawks.

comparatively slender body. Rump grey and characteristically finely barred. Central feathers of tail black tipped white, outer ones broadly tipped and banded black and white. In immature, pale brown wings contrasts with darker tips, rump pale. **Similar species** Harrier-Hawk has a bare face, large black spots on the wing-coverts and is much broader-winged. Smaller Lizard Buzzard has a vertical black stripe on a white throat. Immature Dark Chanting could be confused with ring-tail harriers on overall plumage colour, but has the rump pale, not white, and lacks owl-like facial pattern.

HABITS Favours arid open land with scattered trees. Perches conspicuously on bare limbs with characteristic horizontal body posture. Not shy of passing vehicles.

VOICE Vocal when breeding – a musical piping cry *wheee-whew-whew-whew*, silent otherwise.

STATUS AND DISTRIBUTION Frequent to common resident throughout in open woodland; significant population stronghold in LRD. Widespread throughout Senegal.

BREEDING Builds high in the leafy canopy of a tall tree. Immatures appear regularly in February-March.

AFRICAN GOSHAWK
Accipiter tachiro   **Plates 9 & 13**

Fr. Autour tachiro

Other name: West African Goshawk, Red-chested Goshawk

IDENTIFICATION 36-40cm. *A. t. macroscelides.* Sexes differ. A medium-sized, well-marked forest hawk about the size of Lizard Buzzard. Powerful legs and feet, with characteristically huge hind talons. Variable forest light conditions can affect appearance of plumage tones. Adult male has head dove-grey merging into darker slate-grey back, with a charcoal tail with three variably-sized bright spots down its central feathers. Chin white, rest of underparts whitish with bold broad rufous bars (sometimes appearing pinkish) becoming uniform ginger on thighs and flanks. Bill relatively large, eye orange, legs yellow-orange. Female appreciably larger, browner and duller. Immature sleeker looking than adult (beware large immature female Shikra). Dark liver-brown above (larger female) or blackish (smaller male); underparts white or cream with variable large drop-like brown blotches and streaks, barred on flanks, and mesial stripe.

FLIGHT CHARACTERISTICS Bright central tail spots diagnostic if glimpsed in restricted forest conditions. Soaring above the canopy, tail long and longish wings chequered white and grey from below. **Similar species** Adult male from smaller adult Shikra by richer broader barring below, darker back and prominent white tail spots. Any small immature African Goshawk has bolder markings than immature Shikra. Black-backed immature male African Goshawk from larger adult Great Sparrowhawk by barred and blotchy underparts (latter is all-white below). Ovampo Sparrowhawk (less likely in closed forest) has brown-black barring, not rufous, and shows white flecking on central shafts of uppertail, not well-defined spots.

HABITS Favours forest-thicket, faithful to specific sites and perches in mid-levels, remaining stationary for long periods. Regular bathers, preferring a puddle under overhanging foliage. Soaring flight displays over WD forests July to September have been observed.

VOICE A well-spaced *kwik, kwik* call given also in circling display. A soft falcon-like *kek kek kek kek* or *kew-kew kew kew* recorded from recently fledged birds.

STATUS AND DISTRIBUTION Locally frequent breeding resident in a few coastal forest sites, occasional elsewhere. No recent records more than 50kms from the coast in The Gambia. In Senegal present in coastal forest in Casamance; also recorded east of The Gambia on the upper reaches of the river.

BREEDING Known at three WD sites (e.g. Abuko Nature Reserve). First recorded in early July 1992, nest in the crown of an oil-palm. Nest built of small and medium twigs embellished with fresh green locust bean leaves. Three recently fledged young in Abuko, September-November 1995.

NOTE The form in Senegambia is sometimes considered to be part of a separate species, *A. toussenelii*, or even as a separate species in its own right, *A. macroscelides*.

SHIKRA
Accipiter badius   Plates 9 & 13

Fr. Epervier shikra

Other name: Little Banded Goshawk

IDENTIFICATION 26-30cm. *A. b. sphenurus*. Female larger and rather darker than male, some males appearing very small. A small blue-grey hawk with a white breast profusely and finely barred pinkish-rufous which can appear uniform orange even at close range. Upperparts and wings grey (never black), mantle sometimes suffused smoke-grey; primaries black. Cheeks pale grey. Pale supercilium apparent on most birds. Throat and thighs whitish. Tail above is plain grey, with grey rump, not white. Eye yellow. Bill relatively small. Immature has upperparts grey-brown edged buff; underparts pale and blotched and streaked rich brown. Throat whitish with a broken dark central line. Shows two white nearly triangular-shaped spots on nape.

FLIGHT CHARACTERISTICS Flap and glide flight pattern, sometimes dove-like, rapidly accelerating when chasing prey. Soars frequently, wings looking distinctly small and rounded. Underwing-coverts white barred russet, tips edged blackish. Undertail pale and barred. **Similar species** Absence of white rump eliminates any confusion with Gabar Goshawk.

HABITS A fearless little hawk with a dashing nature. Catholic in habitat choice, found anywhere with well-spaced trees, less frequently within forest. Normally solitary, but immatures may gather to hunt dragonflies towards the end of the rains. Regular bather, attracted to garden ponds. Unobtrusive at rest, spending long periods perched in the foliage of a large tree; may use telegraph wires as a watch post. Views are often brief with the bird appearing dramatically at speed, then sailing up on open wings circling briefly before disappearing.

VOICE Vocal: rapid loud, high-pitched *ki-ki-ki-ki* or *weep weep weep*.

STATUS AND DISTRIBUTION Frequent to common year-round resident throughout, marginally more common in WD. In Senegal, widespread in all seasons south of 15°.

BREEDING Sightings of immatures most frequent towards the end of rains. Suspected breeding WD, late October 1995. Nest a frail platform of twigs high in a tree lined with leaves.

WESTERN LITTLE SPARROWHAWK
Accipiter erythropus   Plates 9 & 13

Fr. Epervier de Hartlaub

Other name: Red-thighed Sparrowhawk

IDENTIFICATION 24-26cm. Sexes differ. A very small but striking black and pinkish-white forest hawk. Male especially diminutive with upperparts and closed wings slate-black, contrasting white throat creating capped effect. Underparts grey, with prominent bloom of pinkish-chestnut on flanks and thighs. Uppertail grey-black with cross bars of white spots, absent from central feathers. Slightly larger female is brown-black above with heavier chestnut barring below. Eyes and legs orange in male, yellow in female. Immature brown above with rufous edgings; underside buffish erratically barred and spotted chestnut-black.

FLIGHT CHARACTERISTICS Wingspan c. 48cm. Underwing-coverts grey with multiple fine chestnut barring. White spots on uppertail create broken bars, distinctive when spread. Narrow white rump plus three rows of large white tail spots, though appearance of latter may vary significantly with moult and feather position. Flaps and glides with a fluttering Laughing Dove-like action. **Similar species** Dark-backed immature African Goshawk is much larger, more powerful, and more boldly marked.

HABITS A sedentary forest hawk, rarely observed and little known. Secretive, darting for cover if disturbed. Occasionally works forest edge at first light. Threatened by habitat destruction.

VOICE A shrill *kew-kew-kew*; more information sought.

STATUS AND DISTRIBUTION Rarely seen. Reliably recorded from only two forest areas in WD; most records are in the dry season. Suitable forest habitat also exists NBD and riverine forest in CRD. Isolated records throughout Senegal, including immediately north and south of The Gambia at the coast in both dry and wet seasons.

BREEDING No recent information. Previous record in May in The Gambia. Likely nesting site would be high in a liana-clad forest tree.

NOTE This species is sometimes considered conspecific with African Little Sparrowhawk *A. minullus* (Fr. Epervier minule).

OVAMBO SPARROWHAWK
Accipiter ovampensis   **Plates 9 & 13**

Fr. Epervier de l'Ovampo

Other name: Ovampo Sparrowhawk

IDENTIFICATION 33-40cm. Lightly-built smallish hawk, larger than Shikra and nearer Eurasian Sparrowhawk in size, appearing relatively small-headed. Male pale phase has upperparts slate-grey, paler on the head with white flecks on the nape. Throat and chin grey, rest of underparts and thighs white with dense narrow grey-brown barring. Tail brown above with distinctive small white central shaft-spots, whitish below with three broad dark bars. Female browner, larger and more heavily barred, with black-streaked crown and cheeks, and

bold white supercilium. Eye red. Cere, legs and feet unusual dull orange-yellow. Immature brownish with rufous scalloping. Streaky about head, prominent pale supercilium contrasts with dark ear-covert patch, nape coppery with white mottling, breast pale rufous-buff. Striking dark phase is sooty-black overall with greyer crown and supercilium. All show characteristic white tail-shaft triangles.

FLIGHT CHARACTERISTICS More agile and graceful than other hawks, tail-shaft spangling best seen when flaring tail to land; underwing-coverts white barred grey. **Similar species** Gabar Goshawk also has melanistic phase, but is smaller, has red legs, unbarred throat and chin and lacks white flecks in uppertail. Smaller Shikra has rufous barring not grey on underparts and an unmarked uppertail. Female European Sparrowhawk has similar supercilium but is heavily barred dark brown below. Much larger Dark Chanting Goshawk has breast, throat and chin uniform grey, not barred.

HABITS Favours well-wooded areas with adjacent open bush, and riparian woodland with well established creeper-clad trees which attract flocks of drinking passerines. Considered secretive, movements in West Africa not yet understood.

VOICE Described as a high-pitched whistle *weeeet-weeet.*

STATUS AND DISTRIBUTION First recorded 1991, all records from WD and probably also NBD. Uncommon but with an increasing number of records, June to December. Two recent records from southern Senegal, June and July. Previous nearest records from Ghana and Togo.

BREEDING No West African records, but juvenile with dark phase adult in WD in June 1991 and a single juvenile November 1993 at Tanji Bird Reserve WD. In South Africa nests high in a tree.

EURASIAN SPARROWHAWK
Accipiter nisus   **Not illustrated**

Fr. Epervier d'Europe

Other names: European Sparrowhawk, Northern Sparrowhawk

IDENTIFICATION 28-38cm. Sexes differ in size and colour. Male slate-blue above and white, finely barred rufous below; throat white. White patch on nape often concealed, short white supercilium. Tail grey, with four to six darker bars. Sturdier, larger female has darker greyish upperparts, underparts whitish with dense brown barring. White supercilium. Cere greenish-yellow, eyes bright yellow, legs and feet yellow. Immature dark brown above, dark barring on underparts more irregular and ragged.

FLIGHT CHARACTERISTICS Wingspan 55-70cm. Agile in flight, showing short rounded wings and long square-ended tail. Male underparts appear uniform orange at a distance. **Similar species** Gabar Goshawk has a white rump; Shikra is smaller and much paler with unbarred tail when seen flying away. African Goshawk is heavier-headed with distinctive tail pattern conspicuous in flight.

HABITS Favours woodland of all types and grassland with scattered trees.

VOICE Usually silent in sub-saharan Africa, alarm call a rapid *kyi kyi kyi.*

STATUS AND DISTRIBUTION Vagrant; two recent records both in February in URD. Not recorded in Senegal; passage migrant in Mauritania.

GREAT SPARROWHAWK
Accipiter melanoleucus   Plates 9 & 13

Fr. Autour noir

Other names: Black Sparrowhawk, Black Goshawk

IDENTIFICATION 46-58cm. *A. m. temminckii.* A massive, harrier-sized black and white hawk, the largest *Accipiter* in the region. Upperparts and head dark slate-black. White nape patch generally concealed, but visible well if neck moved e.g. tearing prey. Underparts white with some irregular heavy dark flecking. Flanks, thighs and sides of breast black with white blotches. Relatively long tail paler than mantle with four darker bands. Eye dark red, bill black, cere pale yellow and legs and feet yellow. Immature dark brown above edged rufous, and below buff, with dark streaks and elongate light teardrop markings.

FLIGHT CHARACTERISTICS In typical fleeting view seems all-black above; white nape patch and chin may be glimpsed. Underwing white barred black, black axillaries contrasting with white underparts. **Similar species** Adult or immature female Great Sparrowhawk is much the same size as an adult or immature male African Hawk Eagle. Bare yellow legs in all stages of Great Sparrowhawk precludes confusion as hawk eagles always have feathered legs and very different behaviour and habitat preference.

HABITS A forest hawk, favouring dense cover. If seen, remains perched turning and twisting its head to inspect the observer. Most sightings are near water in the dry season where prey congregates to drink. Habits in The Gambia are little known and based on a handful of sightings.

VOICE Generally silent, but *kek-kek-kek* and a *koooo* when breeding.

STATUS AND DISTRIBUTION Rare; modern records from March-May in or near coastal forest-thicket in WD. Also recent record in LRD in November. All Senegalese records are from southern coastal Casamance, with sightings in June and December–March.

BREEDING No records. Single modern record of juvenile mid-April in WD. Nests in tall forest trees elsewhere in West Africa, during late rains.

GRASSHOPPER BUZZARD
Butastur rufipennis   Plates 8 & 12

Fr. Busard des sauterelles

IDENTIFICATION 45-50cm. Sexes similar. Medium-sized, long winged and long-tailed hawk. Upperparts ashy-brown, darker around the face and head. Underparts paler rufous-chestnut with grey-brown streaks. Eye rich golden-yellow with orange orbital ring (a useful field-mark). Immature pale rufous below, upperparts darker brown, subtly scalloped paler; head foxy-brown, darker streaked, with obvious dark moustachial stripes.

FLIGHT CHARACTERISTICS Distinctive and diagnostic rufous-ginger patch on dark-bordered upperwing. Stoops like a falcon, soars on flat wings. **Similar species** See rare Fox Kestrel. Immature recalls a large female Common Kestrel.

HABITS Favours mixed dry savanna woodlands. Usually occurs in loose parties, with up to 100 birds in attendance at bush fires, foraging amidst swarms of fleeing winged insects. Very approachable, a regular visitor to receding waterholes. Nomadic to an ill-defined degree.

VOICE Generally silent. Young birds in Kiang West uttered a soft *kek, kek, kek.*

STATUS AND DISTRIBUTION Seasonally common, sometimes locally abundant in LRD and CRD April to July, after which numbers fall dramatically until November to January. Infrequent in WD. Widespread throughout Senegal.

BREEDING Nest with three freshly-fledged juveniles 30 July 1996 in LRD (Kiang West) is first Gambian record. Aerial display, possibly associated with breeding in WD (Yundum) during mid-rains.

LIZARD BUZZARD
Kaupifalco monogrammicus Plates 9 & 13

Fr. Buse unibande

IDENTIFICATION 36cm. Sexes alike. A small-medium relatively short-tailed, thickset hawk. Obvious central black vertical line on a white throat is diagnostic, offering instant separation from all similar raptors. Upperparts slate-grey, darker towards lower back. Head and breast uniform pale grey, rest of underparts white finely barred dark brown. Blackish tail with white tip and single bold white band. Legs red-orange. Immature similar to adult but with buff edges to body feathering giving a scaly appearance.

FLIGHT CHARACTERISTICS Soars regularly on short, slightly pointed wings, looking heavy-headed. White bar across tail striking. Underwing pale with darker tips. Demarcation of grey upper breast and barred belly often clearly visible, as is streak. **Similar species** Gabar Goshawk is smaller and has a white rump; tail has four black bars and is longer, but note that immature Lizard Buzzard may also show vague tail barring.

HABITS Favours habitats with a variety of well-established broad-leaved trees and is particularly fond of perching in the crown of an oil-palm where it sits unobtrusively for long periods. Frequently sits on telegraph wires. Vocal both at rest and in flight, especially while soaring and in display. Diet consists largely of grasshoppers, skinks and lizards caught on the ground or picked from the trunk of a tree. Sedentary and territorial. Behaviourial and ecological features align Lizard Buzzard closer to accipiters than to buzzards.

VOICE A repeated descending musical whistle *kli-o, klu-klu-klu-klu* that often attracts attention to its presence before it is seen. Very vocal in flight.

STATUS AND DISTRIBUTION Common throughout and year-round, diminishing in numbers eastwards. Common and widespread in Senegal south of 14°.

BREEDING In the second half of the dry season in a leafy tree or in the crown of a palm. Small stick nest lined with leaves.

COMMON BUZZARD
Buteo buteo   Plates 7 & 12

Fr. Buse variable

Other names: Steppe Buzzard, Buzzard

IDENTIFICATION 56cm. Sexes similar. Generally uniform warm brown above with some rufous tones on the tail; below brown unevenly barred white or conversely mainly white unevenly barred and blotched brown; a diffuse band across the middle chest may be evident. Tail usually with multiple darker bars. Immature similar to adult.

FLIGHT CHARACTERISTICS Wingspan 110-128cm. Pale underwing barred brown with characteristic dark carpal patch and dark band along trailing edge. Short tail often fanned when soaring. **Similar species** Booted Eagle has feathered lower legs, and wings are not markedly rounded as the buzzards; see that species. Buzzards are half the size of Brown Snake Eagle which has a bonnet-shaped head and less rounded wings lacking carpal patch. Brown immature Harrier-Hawk is larger with a distinctively slim head, and not barrel-chested like Common Buzzard.

HABITS Most records are of birds in flight over wooded habitats. Chooses prominent perches, may hover inexpertly. May be seen regularly over a small area for a few weeks before moving on.

VOICE Generally silent in Africa; mewing *kee-oo* when breeding.

STATUS AND DISTRIBUTION Rare; recorded in November and December in WD and December and January in LRD. Subspecies involved not determined in The Gambia. Rustier plumaged *B. b. vulpinus* 'Steppe Buzzard' is migratory mainly to East and South Africa from eastern Europe, while browner *B. b. buteo*, though largely sedentary in Europe does migrate to north-west Africa, supplementing a small breeding population there. Both subspecies are reported from northern Senegal November-February, and *buteo* subspecies has been reported as far south as Liberia.

LONG-LEGGED BUZZARD
Buteo rufinus   Page 157

Fr. Buse féroce

IDENTIFICATION 65cm; sexes similar. Larger and heavier than Common Buzzard and similarly variable in plumage. Generally pale rufous with tail uniform pale rufous, but ranging from nearly white to all dark (tail cinnamon-brown with a broad subterminal blackish bar). Immature closely resembles adult but with distinctly barred tail. Bright yellow legs when perched are a good field character.

FLIGHT CHARACTERISTICS Wingspan 126-155cm; longish broad wings; dark band along underside trailing edge on adult is partial or lacking on immature. Otherwise underside pale except for dark wingtips, thighs, and sometimes black carpal patches. Head, throat and breast often very pale. Tail uniform pale cinnamon. **Similar species** The barred tail of immature Long-legged makes separation from a pale rusty 'Steppe Buzzard' doubly difficult; immature Long-legged is likely to be paler and more longer-winged. Creamy immature Snake Eagle has bulkier head and is considerably larger.

Long-legged Buzzard

HABITS Favours drier open areas, including fallow rice-fields. Perches on poles, also may stand on bare ground for long periods like a harrier.

VOICE Generally silent in Africa, high pitched mewing when breeding.

STATUS AND DISTRIBUTION Vagrant to The Gambia. One LRD in December 1973, one URD in January 1983; neither identified to subspecies. Slightly better known in northern Senegal, mostly October-March with both Palearctic migrant *B. r. rufinus* and smaller, paler Saharan resident *B. r. cirtensis* recorded.

RED-NECKED BUZZARD
Buteo auguralis   Plates 7 & 12

Fr. Buse à queue rousse

Other names: African Red-tailed Buzzard

IDENTIFICATION c. 56cm; sexes similar but male smaller. Prominent rusty or chestnut-rufous wash over the dark brown plumage of the back; crown, nape and sides of face washed dark chestnut. Throat and breast dark brown variably marked with white, rest of underparts white with bold brown blotches on breast, flanks and thighs varying in extent. Characteristic uniform chestnut tail with darker subterminal band and whitish tip. Immature darker brown above showing chestnut traces, tail brown and barred. Underparts white, unevenly spotted brown, lacking dark breast.

FLIGHT CHARACTERISTICS Wingspan c. 120cm; dark head and chest of adult contrasts with white below; underwing whitish with dark carpal patches. Reddish-chestnut tail distinctive. **Similar species** Common Buzzard lacks rusty-chestnut plumage and dark head, contrasting with brown-spotted white breast. Contrasting dark head and neatly demarcated breast and white underparts could cause confusion with snake eagles, but Red-

necked Buzzard is always much smaller.

HABITS Favours open areas with well grown trees, but prefers wood and forest fringes in its breeding range elsewhere in West Africa. Usually solitary in Senegambia. Catches most of its insect and small animal prey on the ground.

VOICE A piercing, plaintive mewing *peeee-ah*.

STATUS AND DISTRIBUTION Rare, with a few records distributed over all seasons and scattered through WD, LRD and URD. In eastern Senegal most frequent during the rains in Niokolo Koba but also seen there in the dry season; also recorded in the dry season in the north with an immature in August in the extreme north-west.

TAWNY EAGLE
Aquila rapax   Plates 7 & 11

Fr. Aigle ravisseur

IDENTIFICATION 65-75cm. Sexes similiar. *A. r. belisarius.* A medium-large eagle ranging in colour from dark to light brown with a tendency for the dark phase to predominate in our area. Best identified using a combination of plumage, structural and behavioural characteristics. Strong dark brown bill with wide yellow gape extending beneath the eye. Head powerful, profile changeable from flat to rounded but with no crest (beware wind-blown lax nape feathers). Legs entirely feathered and heavily trousered. Tail brown and faintly but detectably barred. Immature broadly similar to adult.

FLIGHT CHARACTERISTICS Wingspan 160-240cm. Wings broad and rounded; rounded tail fanned when soaring. Lack of prominent contrasting plumage features is in itself characteristic. In optimum conditions seven well-spaced 'fingers' are discernible, often darker than the underwing-coverts. Upperwing-coverts often pale and mottled. Important is the head shape which looks sturdy; characteristically waved meaningfully side to side and downwards when scanning for food on the wing. **Similar species** In flight, other brown but always weaker-headed raptors to eliminate include smaller vultures (especially immature Palm-nut Vulture) and immature Harrier-Hawk as relevant text. The occasional presence of the larger migratory Palearctic Steppe Eagle *Aquila nipalensis* (Fr. Aigle des steppes) in Senegambia warrants further investigation. When perched the wider gape of this species can be seen to extend well beyond the underside of the eye.

HABITS Ranges over a diversity of habitats, but

most favour lightly wooded country. Most often seen in flight. Piratical harassment of other raptor and stork species for food is another useful pointer to its identification. Comes to carrion, gorges on termite emergences, and will grub in soft soils for terrestrial invertebrates.

VOICE An occasionally barked *kah* or *kwohk*. Does not whistle or scream which may assist in separating from a high-soaring immature Harrier-Hawk or *Buteo* buzzard species.

STATUS AND DISTRIBUTION Frequent throughout and year-round, but most likely to be seen in early dry season. Widespread in Senegal where a recent reduction in numbers noted.

BREEDING Records December-February. Substantial nest which is lined with fresh leaves and grass, typically on top of an acacia or baobab.

WAHLBERG'S EAGLE
Aquila wahlbergi   Plates 7 & 11

Fr. Aigle de Wahlberg

IDENTIFICATION 55-60cm. Sexes similar. A medium slim-looking eagle of variable plumage. Usually appears dark brown (darker in general than larger, often dishevelled Tawny Eagle); only very occasional occurrences of pale morphs in our area. Typically paler on wing-coverts and sides of face and darkest on square-cornered tail, which may be indistinctly barred. When perched a small but conspicuous and diagnostic crest is evident (absent on Tawny Eagle). Baggy feathered legs appear longer, and face smaller than Tawny. Immature is more streaked but diversity of plumages makes ageing difficult.

FLIGHT CHARACTERISTICS Wingspan 130-160cm. Silhouette and flight pattern diagnostic and more useful than plumage colour. Wing to tail ratio and slim body produce cross-shaped effect when soaring; wings are held very flat. Narrow parallel-edged square-ended tail (rarely flared when banking) is key pointer. **Similar species** Shape of dark unflared tail eliminates smaller Booted Eagle (confusion most likely between pale phases). Juvenile all-brown Harrier-Hawk has a weak-looking head and an often-repeated mewing call. See also Tawny Eagle.

HABITS Ranges over a diversity of habitats in Senegambia tending towards wetlands and forested habitats. Often seen soaring when readily distinguished from Tawny Eagle on flight silhouette.

VOICE Generally silent, occasionally whistles a drawn out rising *kleeee-ay, kleeee-yay* and a rapid series

of *kwip-kwip-kwip-kwip* in greeting displays. No barking calls.

STATUS AND DISTRIBUTION Frequent throughout and year-round, possibly less numerous in wet season. Sparsely distributed in Senegal, largely absent from north-east.

BREEDING A nest with two eaglets in CRD in late December 1975; copulation recorded late June and a soaring aerial display in September, both WD.

AFRICAN HAWK EAGLE
Hieraaetus spilogaster   Plates 7 & 12

Fr. Aigle fascié

IDENTIFICATION 55-65cm. Sexes similar. A medium-large powerful black and white eagle with no crest. Upperparts blackish (female browner) with sparse white flecking on mantle and rump. Underparts white with narrowish black streaks (more profuse on female). Well feathered legs and thighs white and largely unmarked in both sexes. Immature overall tawny-brown above and orange-rufous below.

FLIGHT CHARACTERISTICS Wingspan 130-160cm. Often soars very high but distinctively black and white. Rounded underwing largely white with narrow dark trailing edge, dark upperwing shows white patch (translucent against bright clear sky and mirrored below) at the base of the primaries, a feature shared by Long-crested Eagle. Relatively long tail dark grey-brown narrowly barred with broad black terminal band. Immature has underbody noticeably orange-rufous with faint darker streaks; underwing-coverts similarly rufous well demarcated from rest of barred wing by a narrow dark band. Tail shows barring but lacks broad terminal black band. **Similar species** Small male could be confused with Ayres's Hawk Eagle, which is slightly crested. African Hawk Eagle has previously been considered a subspecies of Bonelli's Eagle *Hieraaetus fasciatus* (Fr. Aigle de Bonelli), a Palearctic breeder which has been reported in northern Senegal (November-April) since 1990.

HABITS Ranges widely over mixed habitats, including open mature woodland with tall trees. Usually in pairs, typically perches in tall leafy trees. Capable of tackling, sometimes jointly, both avian and mammalian prey of heavy and sizeable dimensions. May hunt like a giant hawk with a combined surprise and dash technique.

VOICE Not particularly vocal; occasional musical *klu-klu-klu-kluee* and an explosive *kwee-oo* fading away

STATUS AND DISTRIBUTION Frequent in all

divisions, but less numerous during the latter part of the rains. Sightings peak in late dry season, April-May, especially in LRD and CRD. Widespread south of 15° in Senegal.

BREEDING Eyrie in WD in 1970s regularly occupied February-April; recently-fledged immature in May in CRD. Recorded in eastern Senegal in December.

BOOTED EAGLE
Hieraaetus pennatus   Plates 7 & 11

Fr. Aigle botté

IDENTIFICATION 45-53cm. A small Palearctic eagle, sexes similar, occurring in two colour phases plus a rarer intermediate; pale form appears most frequent in The Gambia. In all phases upperparts dark brown. Pale phase birds buff-white below with some streaks, with black flight feathers. Dark phase brown below with paler tail. At close range head shows a dark hooded effect extending below eye. At rest buffy wing-coverts show as a broad pale bar on folded wing on most birds of all phases. Legs feathered, feet yellow. Immature hardly separable from adult in the field.

FLIGHT CHARACTERISTICS Wingspan 110-130cm. Appears long-tailed and slender-winged, primaries may be deeply separated or pointed when gathered. In both phases pale area on the outermost part of the innermost three primaries detectable from above and below; white shoulder markings on the front of each wing are good field marks. Upperside of wing looks variegated, with broad buffy band recalling Black Kite. Tail shows poorly-defined barring, darker subterminal band, and tipped white. **Similar species** Dark phase Booted best separated from flying Marsh Harrier (note female has creamy leading edge) on flight technique: Booted flaps a few times quickly then glides on flat wings. Rounded tail of Booted distinguishes it from the common, usually gregarious, Black Kite which has a long, slightly notched tail. Small-headed Honey Buzzard and true *Buteo* buzzards do not show inner primary patch on underwing when soaring. Pale-phase Booted differs from larger pale phase Wahlberg's Eagle in flight silhouette, and Wahlberg's lacks pale broad wing-bar. Extremely rare Ayres's Hawk-Eagle also shows white 'head-lamps' feature and could be confused with pale phase well-streaked Booted.

HABITS Normally seen soaring and cruising on thermals, most sightings in the afternoons over open grassland and cultivated fields with scattered tall trees. Cruising direct flight is often low, allowing

close scrutinity for identification features. Several separate localities producing dry season records in successive years suggests likely establishment of a wintering territory and site-faithfulness. Observed with Black Kites on migration in March in NBD.

VOICE Probably silent in The Gambia although one of the most vocal of the European eagles with a variety of melodious whistles and cackles.

STATUS AND DISTRIBUTION Frequent to seasonally common Palearctic migrant with records in all divisions from July-March. Most likely November-March in CRD and WD; a few wet season records July-August in URD. A number of records throughout Senegal, mostly from north. Most records of pale phase birds but identification problems may reduce dark phase records.

AYRES'S HAWK EAGLE
Hieraaetus ayresii   Plates 7 & 12

Fr. Aigle d'Ayres

Synonym: *Hieraaetus dubius*

IDENTIFICATION 47-53cm. Sexes similar. Medium-sized, stocky hawk-eagle slightly but distinctly crested. Upperparts dark uniform chocolate brown, characteristically crested head may appear black at a distance. White shoulder patch on closed wing. Underparts white, heavily spotted and streaked including the belly, flanks and thighs, particularly female. Uppertail grey with broad dark tip and three or four thinner dark bars. Considerable individual variability occurs in density of markings on underparts, and also in head colour which may be very pale. Immature back and wings darker grey-brown with an overall scaly effect; underparts buff with darker streaking less evident on flanks.

FLIGHT CHARACTERISTICS Wingspan 110-130cm. Adult has relatively short wings, heavily barred below, with white 'head-lamps' on leading edge; upperwing uniformly dark. Shortish ashy-brown undertail with three or four dark bands and black tip; crest may not be detectable in flight. Immature (unlikely in Senegambia) has strongly barred underwing. **Similar species** Adult separable from commoner adult African Hawk Eagle by smaller size, crest and white 'lamps' on leading edge. African Hawk Eagle also has white panels in the outer area of the longer upperwing, while Ayres's is far more heavily blotched on underparts. Commoner and notoriously variable Booted Eagle shares white 'lamps' but lacks crest and heavy barring on the underwing.

HABITS All confirmed Gambian sightings are of adult birds in flight, once over dry savanna wood-

159

land, twice over seasonally flooded watercourse with scattered high trees and liana-clad understorey. Shows remarkable aerial skills, stooping at avian prey and small mammals with wings closed, falcon-like. Soars high, perches unobtrusively in the canopy where it is able to move at speed through the branches in pursuit of prey.

VOICE The few Gambian records have been silent. Elsewhere *hueeep hueeep* during display.

STATUS AND DISTRIBUTION Very rare. First recorded in The Gambia in March 1991 at Yundum in WD; also at Kampant in April 1991 and March 1994 and Bansang in CRD in November 1997. Uncertain status in Senegal.

LONG-CRESTED EAGLE
Lophaetus occipitalis   Plates 7 & 12

Fr. Aigle huppard ou Aigle huppé

Other name: Long-crested Hawk Eagle

IDENTIFICATION 53-58cm. Sexes similar. Medium-sized dark brown (black at a distance) eagle. Long floppy crest, sometimes held erect at all ages, identifies it at all ages. Legs and feet greyish. Immature similar with buff edgings to body feathers creating scaly-fringed effect.

FLIGHT CHARACTERISTICS Relatively easy to identify in the air. Large white 'window' or patch on upperside of wing, underwing also shows pale primary patch, with dusk, barred secondaries, tail whitish with broad dark bars. Often soars. **Similar species** Long crest is unique. In flight, Brown Snake Eagle is larger with entirely pale, unbarred flight feathers. Dark morph Honey Buzzard may show a similar pattern but note differences in shape.

HABITS Favours almost any area with tall trees. Perches prominently, long crest blowing in the wind.

VOICE Very vocal, particularly in flight. A prolonged screamed *keeee-eh* and a higher pitched *kik-kik-kik-kee-eh*. Also a single *kow* repeated every few flaps in direct flight.

STATUS AND DISTRIBUTION Present year-round, but most frequent throughout all divisions November-April; most sightings in URD in the early dry season. In Senegal widespread in eastern and southern Senegal, with a few records in the north.

BREEDING A small tree platform of sticks, lined with green leaves, in late dry season, usually April.

CROWNED EAGLE
Stephanoaetus coronatus   Plates 7 & 11

Fr. Aigle couronné

Other name: Crowned Hawk Eagle

IDENTIFICATION 80-90cm. Sexes similar. Huge and powerful eagle, dark overall with a crested head. Adult upperparts and upperwing-coverts blue-black. Sides of face, throat and neck paler rufous-brown. Underside, thighs and legs whitish-buff heavily and richly barred blackish. Tail long, upperside blue-black with two broad grey bars. Immature has head, nape and the whole of underparts white with black spotting on thighs and legs. Rufous tinge on upper breast; feathering of upperwing-coverts and back grey, edged paler creating scaly effect. Tail heavily barred.

FLIGHT CHARACTERISTICS Wingspan 210cm. Broad-winged and long-tailed. Adult underwing rufous on coverts, rest of wing white and conspicuously heavily barred brown-black. Undertail greyish, barred and tipped black. Immature underwing light rufous on coverts boldly barred black on white on rest of wing. **Similar species** Confusion could occur between immature Crowned and immature Martial Eagle, but perched Martial is broader-shouldered, showing primaries reaching tip of tail, while wings of Crowned only reach the rump and reveal a conspicuously barred tail. Nape of Crowned white, Martial flushed grey. Underside of soaring immature Crowned shows rufous forewing with black and white barring on flight feathers. Martial shows contrasting white forewing and body with uniform pale grey flight feathers.

HABITS Forest eagle sometimes associated with river valleys and gallery trees. Difficult to find as it prefers to perch high in the canopy of a large tree where it is remarkably agile despite its size. Soars high and then readily visible. Kills monkeys and similarly-sized mammals, dismembering them and storing parts in tree-tops to return to eat over a period of days.

VOICE Very vocal, a musical *kewee-kewee-kweee*.

STATUS AND DISTRIBUTION First seen in The Gambia late December 1991 in CRD; no other information. In southern Casamance in Senegal, there are a number of recent records including a pair displaying in June over coastal forest; also recorded in Niokolo-Koba in March 1992.

MARTIAL EAGLE
Polemaetus bellicosus   Plate 7 & 11

Fr. Aigle martial

IDENTIFICATION 78-83cm. Sexes similar. Huge powerful eagle of commanding appearance. Head, throat and upper breast dark brown. White belly sharply demarcated, faintly spotted as are the thighs. Larger female more heavily marked. Crown is noticeably flattened in silhouette with a short crest. When perched appears conspicuously broad-shouldered, body tapering into a shortish looking tail. Legs feathered, eye yellow. Formidable black bill, cere and feet greenish-white, legs feathered. Immature appears scaly in a combination of greys, darker on the back, whiter on the head and underparts. Crest more prominent than adult.

FLIGHT CHARACTERISTICS Wingspan 190-260cm. Immense, broad and long wings. Adult underwing-coverts mottled brown and white. Tail above dark barred grey-brown, below barred grey. Characteristic clean-cut bib effect of brown upper breast contrasting with white belly. **Similar species** Immature when soaring can recall an aged thus very pale White-backed Vulture, but mottled underwing and barred tail distinguish it. See Crowned Eagle text for separation of immatures. In flight at a distance Martial could at first sight be confused with smaller Short-toed Eagle, which has whiter underwing.

HABITS Generally sedentary but could occur anywhere in Senegambia, particularly roaming immatures. Normally associated with vast open stretches with occasional big trees. Spends a good proportion of the day on the wing soaring and is able to hover. Perches prominently on a favoured bare limb, typically with the whole head and body leaning deliberately forwards in a horizontal position with legs placed well apart. Partial to guineafowl and is capable of killing prey up to the size of a small antelope.

VOICE Generally silent; vocal near the nest, *kwi-kwi-kwi-kluee-kluee*.

STATUS AND DISTRIBUTION Rare resident recorded from all divisions, year-round in LRD, most likely to occur in the other divisions in the early dry season. Widespread throughout Senegal but sadly declining.

BREEDING Builds nest high in a solitary tree. In The Gambia a well-watched nest fledged a single chick in early April. A mid-December laying date recorded in northern Senegal.

OSPREY

Pandionidae

Monotypic and cosmopolitan. Considerable West African non-breeding population of Palearctic migrants including immatures that may linger year-round. Exclusively a fish-eater, respected in local fishing communities. Soles of feet adapted to grip fish.

OSPREY
Pandion haliaetus   Plates 6 & 11

Fr. Balbuzzard pêcheur

IDENTIFICATION 60cm. Sexes alike. Small-headed, medium-sized brown and white fish-eating raptor. Upperparts, wings and tail brown. Forehead brown with white streaks, white crown with a dark blackish-brown eye stripe above a white throat. Underparts white with a smudged brown breast band, darker in female. Immature generally paler with upperparts fringed buff creating a scaly effect; stronger breast-band.

FLIGHT CHARACTERISTICS Wingspan 150cm. Underbody and underwing-coverts strikingly white; breast shows streaky band. Dark band across the centre of the wing, pronounced blackish carpal patches. Long, angled wings are frequently bowed.

Head looks smallish and narrow. **Similar species** Similarly-coloured Short-toed Eagle has less tapered shape and larger head; lacks angular wing shape in flight. Beware large immature gulls.

HABITS Frequents a wide range of aquatic environments. Commonly seen sitting on the beach or on a post in the water. Hunting technique is to soar and circle, adopting a flapping hover before plunging in with legs and feet outstretched resulting in a substantial splash. Shakes powerfully on rising from the water. Aligns prey head-first to direction of flight and hauls it to a favoured post, often well away from water. Non breeders stay on throughout the rains in reasonable numbers.

VOICE Relatively silent in Africa. Occasional alarm call *kew-kew-kew-kew*.

STATUS AND DISTRIBUTION Seasonally common Palearctic visitor throughout, most numerous in the dry months. Commonest along the WD coast line from October to April but numbers fluctuate. Becomes less frequent from LRD towards URD along the river. Tagged and ringed birds from northern Britain observed and recovered in The Gambia. Widespread in Senegal along all major watercourses.

SECRETARY BIRD

Sagittariidae

One species. Name once thought to have been derived from the appearance of the bird's crest which resembles a bunch of quill pens of an old-fashioned secretary, but more likely to be an adulteration of the Arabic for a hunter-bird 'saqr-et-tair'. Unique, long-legged bird supremely adapted for patrolling the open savanna. Usually terrestrial but soars well.

SECRETARY BIRD
Sagittarius serpentarius

Fr. Serpentaire

IDENTIFICATION 150cm. Very large; unmistakable, long-legged and long-tailed stately looking largely terrestrial bird with a marching gait. Legs feathered black to knee-joint; bare legs grey-pink. Loose and fluttering erectile crest and long extended tail rectrices. Face bare, skin bright orange, yellow in immature. Strongly hooked bill blue-grey. Short toes slightly webbed with powerful claws.

FLIGHT CHARACTERISTICS Broad-winged. Grey diamond-shaped tail tipped black with extended projecting central feathers. Grey body and under-wing-coverts contrast with black primaries, secondaries and pink legs. Seldom takes to the air but flies well. Soars like a vulture.

HABITS Essentially a bird of open savanna grassland with occasional flat-topped trees for roosting and nesting. Stalks prey including insects, small birds, rodents, snakes and lizards. Dispatches quarry with deliberate kicks and stamps of the foot. Makes short running dashes when hunting, kicking and wing-flapping.

VOICE Mostly silent.

STATUS AND DISTRIBUTION No recent records in The Gambia; older information from WD, CRD and URD. Rare to uncommon in eastern Senegal (PNNK) and a record in coastal Casamance, January-May.

BREEDING Nest is a platform of sticks crown of an acacia.

FALCONS

Falconidae

A group of sleek hunters with long, pointed, often scimitar-shaped wings giving anchor-like flight profile. Swift in flight and capable of exceptional speeds. Most species catch their prey on the wing, executing fast stoops to strike in mid-air. Smallish to medium in size, females often larger than males. Breed on bare cliff ledges, natural cavities or in the disused nests of other birds; do not build their own nests. Thirteen species, including Saker *Falco cherrug* (Fr. Faucon sacré) and Merlin *Falco columbarius* (Fr. Faucon émerillon) which are vagrants to northern Senegal, not described here.

LESSER KESTREL
Falco naumanni **Plates 10 & 13**

Fr. Faucon crécerellete

IDENTIFICATION 29-32cm. Sexes differ. Smallish, slim and long-tailed. Both sexes show whitish claws, diagnostic but visible only at close range. Male has pale blue-grey crown; upperparts rufous-chestnut, diagnostically unspotted. Greater wing-coverts blue-grey creating a band between dark brown flight feathers and reddish-brown upperwing. Underparts cleaner white than Common Kestrel. Lacks defined moustachial streak.

FLIGHT CHARACTERISTICS Wingspan 60-74cm. Male has characteristic whitish underwings contrasting with a creamy-buff body; shows black wing-tips and unspotted back from above. Female Lesser can with experience be distinguished from Common Kestrel by paler underparts and reduced barring on flight feathers, finely spotted underwing-coverts, and less pronounced moustachial streak. Central tail feathers are frequently slightly elongated, particularly in the male giving a distinctive tail shape. **Similar species** Common Kestrel, which is not usually gregarious, has black claws, and male has spotted back.

HABITS Gregarious, attracted to grassland fires with flocks of up to 30 birds reported. Takes insect prey on the wing and swoops dangerously close to the flames. Hovers less but is capable of 'holding-still' for longer than Common Kestrel.

VOICE Silent in Africa except at communal roosts where squabbling occurs.

STATUS AND DISTRIBUTION Frequent Palearctic migrant with all recent records December-March, from all divisions except LRD. Probably under-recorded. Many hundreds on passage through NBD early March 1994 during seasonally unusual light rain was the first significant proof of migratory passage through The Gambia. In Senegal, most frequently seen along the coast with records also in the east.

COMMON KESTREL
Falco tinnunculus **Plates 10 & 13**

Fr. Faucon crécerelle

Other names: Rock Kestrel, Kestrel

IDENTIFICATION 32-35cm. Sexes differ. Small, slim falcon with longish rounded tail and head. Male has spotted chestnut upperparts and greater wing-coverts. Underparts warm buff with scattered

blackish spots and flecks. Head, rump and upper-tail grey with subterminal broad black band and white tip. Female and immature have underparts pale sandy-buff streaked blackish. Rusty-brown upperparts barred dark blackish-brown, tail with multiple blackish bars. Slight but obvious moustachial streak.

FLIGHT CHARACTERISTICS Wingspan 68-82cm. Characteristic foraging strategy, hovering with wings outstretched and tail fanned. Male from below is buff, lightly spotted with black on the breast and flanks. Underwing-coverts buff spotted black. Flight feathers whitish barred grey, wing-tips darker. Female from below more heavily streaked than male, underwing-coverts appear spotted. **Similar species** See Lesser Kestrel: claw colour, male back colour and hovering habits are best separation features.

HABITS Usually solitary. Hunts for terrestrial prey, scrutinising the ground whilst hovering. Favours open country with scattered trees, agricultural land, coastal scrub and saltmarsh. Often seen resting on telegraph wires.

VOICE Shrill, piercing *kee-kee-kee.*

STATUS AND DISTRIBUTION Frequent Palearctic migrant, seasonally common in WD and LRD. Recorded in all divisions November-April with pronounced peak December-January; only records outside these months are in June. Common and widespread throughout Senegal.

FOX KESTREL
Falco alopex **Plate 10**

Fr. Crecerélle renard

IDENTIFICATION Size 38cm. Sexes alike. A long-winged, long-tailed falcon. Entire body above and below a dusky foxy-red, with faint darker streaking on breast and upperparts except rump. Black primaries conspicuous on perched bird. Tail incompletely barred.Orbital skin pale blue. Legs, feet and toes yellowish. Claws black. Immature has more pronounced streaking on wing-coverts and scapulars and more boldly barred tail.

FLIGHT CHARACTERISTICS Wingspan 95cm. Very slim and long-tailed. Tail graduated, rarely fanned, more than half the length of the entire bird. Underwing characteristically whitish with black wing-tips. Innerwing barred black. Profuse narrow barring on tail. **Similar species** No other falcon is so rufous. See Grasshopper Buzzard.

HABITS Favours semi-arid savanna and woodland. Could be expected at bush and grass fires. Intra-

African movements not well understood, but probably nomadic in West Africa, associated with rains, fires and insect plagues.

VOICE Vocal when breeding, high-pitched *kree-kree kree.*

STATUS AND DISTRIBUTION Rare vagrant, three records; one each from WD and URD, both October, in 1986 and 1988, and May 1957 in NBD. A few modern records east of The Gambia in Senegal in January, March and April.

GREY KESTREL
Falco ardosiaceus　　　　**Plates 10 & 13**

Fr. Faucon ardoisé

IDENTIFICATION 32-36cm. An unmistakable, thickset, tapered and largish-headed falcon. Entirely slate-grey with some fine vermiculations detectable in good light, and conspicuous yellow spectacles of bare skin around brown eye. Immature darker grey with sparse but clear black streaking on breast and belly, and broader, whiter spectacles.

FLIGHT CHARACTERISTICS Wingspan c. 90cm. Uniform grey, primaries noticeably blackish, appearing faintly barred. Flies low and fast over open ground, sweeping upwards to a perching post. Does not hover, rarely soars. **Similar species** No other falcon is entirely grey.

HABITS Found in various habitats but favours woodland with some tall trees. Often perches on telegraph wires. General behaviour is falcon-like with dashing stoops at prey, totally different from other kestrels. Often perches on the ground. Generally sedentary.

VOICE Infrequently heard piercing *keek-keek-keek.*

STATUS AND DISTRIBUTION Common resident throughout and year-round, less frequent in LRD and NBD. In Senegal infrequent N of 15°; common elsewhere.

BREEDING An April record at the coast, another prospecting potential hole in an oil palm in LRD in July. May also utilise old nest of Hamerkop.

RED-NECKED FALCON
Falco chicquera　　　　**Plates 10 & 13**

Fr. Faucon à cou roux

Other names: Red-necked Kestrel, Red-necked Merlin

IDENTIFICATION 30-36cm. *F. c. ruficollis.* Sexes

similar. A smallish, thickset falcon, grey overall with a distinctive rich chestnut cap and nape. Buffish belly finely barred black distinguishes it from small male Lanner. Moustaches and supercilium black contrasting with white throat and face. Tail bluegrey tipped white, with black barring and a broad dark subterminal band. Immature grey-brown above, crown and nape brown not chestnut. Barring on underparts untidy, sometimes streaked.

FLIGHT CHARACTERISTICS Wingspan c. 85cm. Underwing-coverts white barred black, rest of underwing dark grey-brown. Thighs finely barred. Dashing flight, with rapid beats of long pointed wings. Often hunts below canopy along woodland edges and occasionally soars. **Similar species** Adult Lanner is larger, darker, lacking barring on underparts and with less rufous on crown and nape.

HABITS Associated frequently with rhun palms *Borassus aethiopum,* also with riverine forest and scattered acacia woodland. Perches well-concealed in the crown of a palm tree. Hunts birds, bats and occasionally large insects, sometimes cooperatively. Appears to be sedentary. Particularly active at dawn and dusk.

VOICE Infrequently heard screaming *kek-kek-kek-kek kek.*

STATUS AND DISTRIBUTION Frequent, resident year-round; recorded in all divisions. Population stronghold in WD, much less common in LRD, becoming more regular inland to URD. Widespread in Senegal to the north and south but largely absent across the north central part between 14° and 15°.

BREEDING Lays eggs on debris accumulated in the base of rhun palm frond, or in nest. Re-uses same site annually in latter part of dry season.

RED-FOOTED FALCON
Falco vespertinus　　　　**Plate 13**

Fr. Faucon Kobez

Other name: Western Red-footed Falcon

IDENTIFICATION AND HABITS 30cm (wingspan 60-70cm). Sexes differ. Adult male dark grey with chestnut thighs and undertail-coverts, facial skin and legs bright red. Female and immature resemble African and Eurasian Hobby but female has a rufous crown and nape, and red legs. Immature has a pale forehead and white collar is more extensive. Similar in shape to hobby but wings shorter. Tail recalling kestrel is tapered and rounded but slightly shorter. Graceful in flight, gliding on outstretched wings. Hovers, but less often and with deeper wingbeats than Kestrel.

STATUS AND DISTRIBUTION Palearctic migrant, en route to wintering grounds in southern Africa. Generally gregarious. No Gambian records, but likely. Several recent passage records in coastal Senegal and Guinea-Bissau border in September, December, January and March. Main flight path is well east of The Gambia.

EURASIAN HOBBY
Falco subbuteo   Plates 10 & 13

Fr. Faucon hobereau

Other names: European Hobby, Hobby

IDENTIFICATION 30-36cm. Sexes similar. A smallish dark, slim and elegant falcon with sickle-shaped wings. Upperparts dark slate-blackish to peregrine-blue. Underparts buff prominently flecked blackish, undertail and 'trousers' russet. Black moustachial stripes contrast with a prominent whitish face and throat. There is also another less uniform streak behind the ear-coverts, a definitive fieldmark. Immature dark brown above, narrowly scalloped; below more heavily streaked on pale buff, lacking russet on thighs and lower belly.

FLIGHT CHARACTERISTICS Wingspan 70-85cm. Underparts appear heavily streaked, with russet 'trousers' (absent in young birds). Face is much whiter than African Hobby and moustaches are thus more pronounced; may look almost collared due to pale nape. Immature is very dark brown above. Swift-like flight silhouette characteristic. Similar species African Hobby is smaller and darker.

HABITS Ecologically and behaviourally similar to African Hobby. Seen chasing swallows and martins on passage and foraging amidst swarms of flying insects. Feeds on latter in classic hobby fashion, passing prey from feet to bill.

VOICE Silent in Africa.

STATUS AND DISTRIBUTION Uncommon; possibly under-recorded. All recent records from coastal WD in December-April; expected in association with passage hirundines. Rare in Senegal.

AFRICAN HOBBY
Falco cuvieri   Plates 10 & 13

Fr. Hobereau africain ou Faucon du Cuvier

Other name: African Hobby Falcon

IDENTIFICATION 28-30cm. Sexes similar. A dark, smallish, sleek and very streamlined falcon. Upper-

parts dark slaty-blue to charcoal. Underparts a characteristic warm orange to brick-red with fine black streaking which sometimes appears as dots. Head pattern typically falcon-like with black moustaches and buff-whitish face. Immature similar to adult but feathering of upperparts edged rufous, underparts more heavily streaked.

FLIGHT CHARACTERISTICS Wingspan 68-80cm. Dark rufous underparts; streaking difficult to detect at speed. Underwing-coverts buff with black streaking. Undertail paler grey than uppertail and tipped pale rufous. Powerful and rapid flight with sudden changes of direction when hunting; extremely manoeuvrable. Resembles a large swift when glimpsed at high speed. Identification made easier when bird 'hangs still' in mid-air whilst feeding. Similar species See paler, slightly larger Eurasian Hobby.

HABITS Most active at dawn and dusk. Usually solitary, occasionally seen in pairs. Uses the crown of a favoured tall tree as a look-out, and propels itself into flight with great acceleration and power. Feeds largely on insects caught aerially, which are dismembered and eaten whilst remaining on the wing. Also hunts swallows, martins and swifts and if successful goes to pluck and eat quarry on a regular plucking post. Observations seem associated with insect emergence, especially dragonflies and termites, and with migration of swallows and martins. Probably nomadic throughout its Afro-tropical range.

VOICE High pitched ki-ki-kee or kik-kik-kik-kik.

STATUS AND DISTRIBUTION Frequent, with year-round records in all divisions. Generally most frequent in early dry season in WD. Not known further north than 15° in Senegal.

BREEDING Not recorded, but suspected. Would probably exploit disused nest of Pied Crow in correlation with the rains.

LANNER FALCON
Falco biarmicus   Plates 10 & 13

Fr. Faucon lanier

Other name: Lanner

IDENTIFICATION Female 50cm, male 42cm, otherwise sexes similar. F. b. abyssinicus. Adult upperparts blue-grey, underparts pinkish-buff with some sparse brown spotting on the flanks and breast. Characteristic rufous crown and nape diagnostic, present at all ages; moustachial stripe and sides of crown blackish. Immature has upperparts brown, edged buff; underparts buff and heavily streaked (not barred) brownish-black. Rufous crown and nape paler than adult, long narrowish moustachial streak prominent on pale cheeks.

FLIGHT CHARACTERISTICS Wingspan 95-115cm. Longer and narrower wings than Peregrine with blunter less pointed wing-tips, tail somewhat longer. Underside of wings pale, looking unbarred from a distance; closer views show fine barring on underwing-coverts. Tail above brownish-grey, tipped whitish, with 10-12 dark grey bands, terminal bar broader. Tail below, paler grey and banded darker. Soars to considerable heights. Immature shows heavy dark streaking on underparts and contrasting underwing; forewing streaked dark with a paler broad trailing edge. **Similar species** More blue-grey-backed Peregrine is bulkier in stature, shorter-tailed and more dynamic in flight. Peregrine (usually) lacks rufous crown, and has more pronounced moustachial streaks (see also Barbary Falcon).

HABITS Ranges over several habitat types, dry savanna woodland, coastal sand-bars and mudflats being favoured haunts. In The Gambia shows a particular preference for birds as prey, particularly, pigeons and doves, francolins and waders. Typically takes prey in flight, but also on the ground.

VOICE A rapidly delivered, high-pitched *kii-kii-kii-kii.*

STATUS AND DISTRIBUTION Uncommon to frequent, with records in all divisions in most months; most frequent in WD and LRD in early dry season months. In Senegal, most frequent along the northen border with records also south and east of The Gambia. Migrant North African *F. b. erlangeri* occurs in northern Senegal.

BREEDING Unrecorded but likely in Senegambia. Juveniles in WD and CRD in January and April. Nest on a cliff ledge or in the disused nest of another bird.

PEREGRINE FALCON
Falco peregrinus   **Plates 10 & 13**

Fr. Faucon pèlerin

Other name: Peregrine

IDENTIFICATION Female 46cm; male 38cm, otherwise sexes similar. A thickset heavy-shouldered falcon. Upperparts slate fading to a paler blue-grey on the rump. Underparts creamy-buff, spotted and densely barred on the belly and undertail, more heavily in female. Crown and nape black; prominent broad black moustachial streaks contrast with white cheek patches. At rest, wing-tips extend beyond tip of tail; wings look darker than back. Immature dark brown above with slight rufous scaly

edgings. Underparts buff, heavily streaked blackish. Facial characteristics as adult. There are a number of subspecies to consider adding to the already considerable size range: the nominate Palearctic *F. p. peregrinus* and the Mediterranean *F. p. calidus* in which individual females may be particularly large, and the smaller African *F. p. minor* where the male can be markedly small. *F. p. brookei* of southern Europe closely resembles Barbary Falcon (see below).

FLIGHT CHARACTERISTICS Wingspan 80-115cm. Broad-based wings appear distinctly triangular with sharp tips. Soars on stiff wings with tips slightly bowed. Tail medium in length, shorter and more tapered than Lanner Falcon. Underwing and tail greyish-white with dark grey barring. Body looks heavy and cigar-shaped. Head and facial markings remain diagnostic. **Similar species** See paler, desert-dwelling Barbary Falcon.

HABITS Sightings are usually of solitary birds and are generally in the vicinity of wader haunts, but pigeons, doves and francolins are also taken. The acceleration of the diving stoop when the bird is hunting is legendary. Quarry is killed on impact. Near the coast uses the crowns of ancient baobabs as perching posts.

VOICE Screeching, raucous *kek-kek-kek-kek* and a gruff *archk-archk.*

STATUS AND DISTRIBUTION Uncommon to frequent, records in all divisions with a peak in November-February. Isolated July records, but largely absent May-October. Less frequent than Lanner. Records from Senegal are coastal in both north and south, with one record in extreme south-east.

BREEDING Not recorded in Senegambia. Normally nests on a cliff ledge.

BARBARY FALCON
Falco pelegrinoides   **Plates 10 & 13**

Fr. Faucon de Barbarie

IDENTIFICATION 34-40cm (wingspan c. 90cm, male smaller). Smaller than Peregrine and has a pale grey cap with a trace of rufous in the centre; nape and supercilium also show some rufous. Moustachial streak less well-defined. Underparts buff with dark flecks and partial barring only in the area of the flanks. Plain buff 'trousers' and barred underwing in flight; tail darkens towards the tip, subterminal bar noticeably broader than any race of Peregrine. Overall appears chunky and compact with quick wingbeats. Immature is pale brown above with fine dark breast streaks.

STATUS AND DISTRIBUTION A few recent
records of falcons showing these features from LRD
to URD, December-March; vagrant to Senegal.
Barbary is easily confused with adult Lanner which
has crown and nape totally rufous and lacks the
broad subterminal bar in the tail.

NOTE Barbary Falcon was formerly treated as a
race of Peregrine Falcon.

QUAILS, PARTRIDGES, FRANCOLINS AND GUINEAFOWL

Phasianidae

Seven species ranging from small to large, including the guineafowl which are sometimes treated as a
separate family, Numididae. Subfamily Gallinae of quails, partridges and francolins has two *Cortunix* quails,
one in northern Senegal only, and Stone Partridge in the monotypic genus *Ptilopachus*. Francolins (genus
Francolinus) are represented by three species, one abundant, two rare and local.

COMMON QUAIL
Coturnix coturnix　　　Plate 14

Fr. Caille des blés

IDENTIFICATION 18cm. Sexes differ. Short-
winged, with a globular oval body. Overall sandy-
straw coloured, strongly streaked with whitish-buff
and black above, noticeably paler below, with light
and dark streaks on flanks. Well-marked small neat
head and throat, markings less pronounced in the
female. Darkish brown crown with creamy stripe
down centre and a long creamy stripe above the
eye. Male has blackish stripes down chin and throat
and across the neck. Female has buffer throat and
chest appears lightly spotted with black. Immature
resembles female but flanks more spotted and
barred than streaked. **Similar species** See Harlequin Quail. Common Quails can be distinguished
from smaller and longer-necked Little Button-quail
by the latter's contrasting pale wing panel and dark
flight feathers in flight. Note that francolin chicks
are both very quail-like and able to fly when still
small.

HABITS Favours grassy pasture and drying-out seasonal swampland. Extremely furtive nature. Most
sightings are of flushed birds when flight is likely
to be fast and low. May briefly flutter above the grass
before dropping down out of sight.

VOICE Generally silent in our area but may shriek
when flushed.

STATUS AND DISTRIBUTION Palearctic migrant, contemporary data scanty; the few records
are all November to February, from URD with infrequent sightings in WD and LRD; considered not
uncommon earlier this century. Frequent in considerable numbers in north-west Senegal in some
years, September to mid-March, with records from
the south also. Italian-ringed bird recovered in
Senegal.

HARLEQUIN QUAIL
Coturnix delegorguei　　　Not illustrated

Fr. Caille arlequin

IDENTIFICATION 15cm. Sexes differ. Male has
dark chestnut underparts, black belly and black and
white throat markings; female ghosts male and is
separable from paler female Common Quail when
seen well by a necklace of throat spots. In flight
separation difficult but Harlequin darker.

STATUS AND DISTRIBUTION Recorded from N
and SE Senegal, November-April, with one possible
occurrence in The Gambia. This intra-African migrant with nomadic tendencies could occur again.

STONE PARTRIDGE
Ptilopachus petrosus　　　Plate 14

Fr. Poule de roche

Other name: Stone Bantam

IDENTIFICATION 25cm. A small dark brown
bantam-like bird with a prominent cocked tail and
creamy-white belly patch, palest in female; often
seen scuttling away. Upperparts and wings overall
grey-brown heavily flecked with buff-white when
well. Crown feathers, often like a comb, are nearly
black as is the tail. Legs dark red with no spurs, eye
brown surrounded by bare red skin. Bill red at the
base, horn-yellow towards the tip. Immature similar to adult but extensively barred. **Similar species**
Inadequate views of Stone Partridge may lead to
confusion with the rare forest-dwelling Ahanta
Francolin (see that species).

HABITS Found in a diversity of habitats, not restricted to rocky environments in The Gambia.
Good cover in a well wooded area seems to be a

167

prerequisite and it is common near remnant forest edges. Lives in coveys typically of five to twelve birds. Shy, best views secured at dawn. Not often seen flying unless threatened, preferring to walk, with tail held erect, often on forest trails. Roosts off the ground and is a nimble climber through the branches of a tree. Looks pigeon-like or sandgrouse-like in shape when squatting down to feed. Calls in groups from a raised vantage point, members of covey characteristically shuffling shoulder to shoulder upward on a fallen log or termite mound as they do so. Also calls from the ground; especially vocal in the first hour of daylight, less so towards dusk.

VOICE A loud but pleasant outburst of *weet-weet-weet-weet* gathering in pace and volume as several birds join the chorus, more whistling in quality than the three francolin species in Senegambia.

STATUS AND DISTRIBUTION Common resident throughout with greatest numbers in WD and URD. Widespread in Senegal but absent from dense forested areas in Casamance.

BREEDING Well-hidden scrape lined with leaves, rarely found. In WD half-grown chicks in late December and tiny recently hatched chicks in mid-February.

WHITE-THROATED FRANCOLIN
Francolinus albogularis   Plate 14

Fr. Francolin à gorge blanche

IDENTIFICATION 23cm. Small francolin with obvious white chin and throat, sometimes bordered with black dots. Short, rounded chestnut wings (brightest on male) conspicuous in flight, which is fast and low. Upperparts rufous-chestnut, noticeably rusty-orange on the hindneck, paler buff on the mantle with broad white streaks and variable black markings on the back. Outer tail feathers rufous, others grey, barred buff. Underparts buff with some darker streaking on the breast and flanks, more barred on the female. Dark line through lores and eye, and broad cream-white supercilium. Bill black with yellow base, legs orange-yellow. Immature similar but with more barring than female. **Similar species** Beware immatures of others in this group.

HABITS A rather quail-like francolin of dry savanna woodlands, associated particularly with the laterite plateau in the far east of the country. If disturbed, crouches, on close approach bursting into the air with whirring wings and extended neck for short, low, flight before dropping back into the vegetation. Usually in pairs or small coveys, probably as family parties.

VOICE Distinctive *che-cheer-che, che-cheer-che*, also a high pitched trumpeting *ter-ink-inkity-ink* and a *kili-kili-kili*. Typically most vocal at both dawn and dusk.

STATUS AND DISTRIBUTION Uncommon and local in URD with recent records, January, September and December and once LRD in November. No information from NBD; absent from WD. Considered more extensively distributed earlier this century. Extends to 15° in Senegal. Local and uncommon throughout West Africa.

BREEDING A leaf-lined scrape under cover. Pair with six chicks early December in URD; also eastern Senegal at the end of the rains.

AHANTA FRANCOLIN
Francolinus ahantensis   Plate 14

Fr. Francolin d'Ahanta

IDENTIFICATION 33cm. A forest francolin, dark brown above, milky coffee-brown below with white U-shaped feather margins. Legs and bill pale orange (not bright red as sometimes stated). Weak supercilium, back of neck dark with some white streaking, rest of upperparts finely pale-spotted, faintly barred on wings. Immature has greyer underparts, streaked white with black arrow markings on the upperparts. **Similar species** Double-spurred Francolin will also enter forest. Ahanta differs in coffee-brown underparts with narrow white streaks, not creamy with heavy black teardrop spots; leg colour pale orange not greenish. Immature Ahanta is darker than Double-spurred. Stone Partridge is smaller and uniformly darker and has very red legs, not dull orange.

HABITS Strictly associated with forest thicket within which it is faithful to certain restricted sites. Usually in pairs or family parties. Always elusive, seldom uses open trails; best seen by looking through tangled undergrowth in direction of call. Calls most vigorously at dawn and in the first hours of the day, when neighbouring groups may interact; calls sporadically at other times.

VOICE Produces a loud burst of sudden hard clattering notes associated with territorial display; not usually persistent, often in relatively short and isolated bursts; calls in all seasons. Also a quieter, throaty and rythmic *chup chup chip chip chup chup prrrr*.

STATUS AND DISTRIBUTION Rare year-round resident. Restricted to small areas of appropriate forest habitat in WD including Bijolo Forest Park (J. Hammick pers. comm.), within which it is recorded in all seasons and may be locally frequent. Gravely threatened by habitat loss. In Senegal, restricted to coastal forest in Casamance. This species is a West African endemic.

BREEDING Adult with three or four half-grown young late June 1992 in southern Gambia was the first breeding record for The Gambia. In southern Senegal a female on the point of laying in January, and chicks seen in September. Nest scrape probably in dense ground cover.

DOUBLE-SPURRED FRANCOLIN
Francolinus bicalcaratus   Plate 14

Fr. Francolin à double éperon

Other names: Bush-fowl, Two-spurred Francolin

IDENTIFICATION 35cm. Earthy-brown with prominent white stripe over the eye. Female slightly smaller than male. Male has two large spurs often of different lengths on relatively long legs. Head markings important in forest habitats to separate from rare Ahanta Francolin; forehead black, crown brown shading to sandy-rufous at the nape, separated from white supercilium by a thin black band; black stripe passes through eye. Chin and throat whitish. Underparts buffy with bold teardrop chestnut and black spots on the breast feathering. Tail predominately grey, mottled brown. Bill and legs olive-green. Immature has yellower legs with a single or no spur, and is more blotched. Note that the Senegambian population is considered to be paler than birds from wetter areas in West Africa. **Similar species** See Ahanta Francolin.

HABITS Ubiquitous, most numerous near cultivated arable land and open fallow ground. Found singly, in pairs or coveys but rarely exceeding twelve birds. Enters forest edges as the dry season progresses, may penetrate deeper to locate water. Ardent dust-bather. Wary, making an urgent run or flight dash across open spaces if disturbed in order to seek protective cover.

VOICE Strong and forceful *ke-rak, ke-rak, ke-rak* from the male. Other calls are a strident *qua-air, qua-air*. Often calls from stump, rock or mound, sometimes also in flight.

STATUS AND DISTRIBUTION Common to occasionally abundant in all divisions with no significant seasonal changes; numbers greatest in WD. Less common in heavily forested areas in southern Senegal, otherwise widespread in all districts.

BREEDING A scrape lined with a few leaves in low shady cover. Extended season from late rains throughout the dry season.

HELMETED GUINEAFOWL
Numida meleagris   Plate 14

Fr. Pintade commune

Other name: Grey-breasted Helmeted Guineafowl

IDENTIFICATION 56cm. *N. m. galeata*. Sexes alike. Large, with bulky body and distinctively rounded back, the only Senegambian guineafowl. All black plumage profusely spotted with white dots. Wings short and rounded. Primaries and secondaries dark grey barred with white. Plain grey chest and upper mantle creates grey collar. Small head and foreneck naked, bare skin brown-black and blue-white. Red wattles and black hair-like bristles on hind neck. Characteristic red horny casque maroon at the tip, basally yellowish. Short tail grey and spotted. Strong arched bill red-brown, lower mandible grey. Immature browner with white streaking on neck feathers, and more modest casque. Note that guineafowl are often domesticated and thus to be expected in the vicinity of rural and urban dwellings. Farmed and captive-bred birds often show irregular patches of white in the plumage.

HABITS Favours dry wooded grassland with areas of stony laterite, often close to mangrove-fringed bolons in which it roosts. Plumage pattern affords outstanding cryptic camouflage on burnt ground. Highly gregarious but coveys exceeding ten birds in The Gambia now unusual. Largely terrestrial, wary habits reflect hunting pressures; numbers are declining. Feeding activity and waterhole drinking peaks at both ends of the day. Runs for cover in a noisy disorderly dash. A major prey item of Martial Eagle and its presence may influence Martial's range. Precocial young able to fly at an early age.

VOICE A cacophonous nasal *kek, kek, kek, krrrrrr* or *chrr rrr* when disturbed or when coming to roost.

STATUS AND DISTRIBUTION Locally frequent, sometimes common, in LRD and NBD, less frequent in CRD. Considered very common earlier this century when found in large flocks; now in need of some local conservation measures. Widespread in Senegal, commoner in areas where hunting is controlled, e.g. abundant in Niokolo Koba.

BREEDING A lined and well hidden scrape. Lays eggs towards end of rains with half-grown young seen during first half of October in LRD.

BUTTON-QUAILS AND QUAIL-PLOVERS

Turnicidae

Two species. Very small quail-like birds of grassland and open woodland with grassy margins. Difficult to find and watch, and most records involve brief sightings of disturbed birds flying away. Wings are short, rounded and bowed. Plumages are cryptically patterned, mostly in browns, buffs and yellows and the hind toe is lacking.

LITTLE BUTTON-QUAIL
Turnix sylvatica   Plate 14

Fr. Turnix d'Afrique ou Turnix d'Anadalousie

Other name: Kurrichane Button-quail

IDENTIFICATION 15cm, female slightly larger than male. *T. s. lepurana*. Very small, quail-like, with spotted sides to breast. Lacks striking markings on plain face. Upperparts sandy-brown with darker streaking, underparts buffy-orange, belly more tawny. Black heart-shaped spots along the flanks are important field mark. In flight, rounded wings exhibit creamy upperwing-coverts (not white) which contrast with largely buff-grey wing. Short tail barred. Bill blue-grey, darker towards the tip, eye pale yellow with pale blue eye-ring, legs pinkish-white. Immature more spotted, especially across the breast. **Similar species** Larger Palearctic Common Quail (dry season) has well marked face markings. Quail-plover is smaller with characteristic black and white wings.

HABITS Normally encountered singly or in pairs. Several loosely grouped pairs may occur erratically in short grass, also found in elephant grass *Andropogon*. Reluctant to fly, flight low on rigid wings, legs dangling. If disturbed in the open, after standing alert may crouch and rely on cryptic coloration for concealment. Walks jerkily. The female courts the male 'pumping-up' her body and lowering her head into her chest to deliver her curious enticement call.

VOICE Female utters a deep, resonant, extended advertising *oo-oo* or *hoom-hoom-hoom*; male answers with a falcon-like *kek-kek-kek-kek*.

STATUS AND DISTRIBUTION Infrequent and local, recorded in all seasons but most frequently November-February in LRD, occasional local increases in relative frequency in some years, e.g. 1988. Sparsely distributed in northern Senegal, with sightings from near the coast in southern Senegal.

BREEDING Adult with five recently-hatched chicks in LRD (near Jali, KWNP) in late November. In northern Senegal recorded October to February. Shallow grass-lined scape protected by a tussock. Possible polyandry requires further study.

QUAIL-PLOVER
Ortyxelos meiffrenii   Plate 14

Fr. Turnix de Meiffren ou Turnix à ailes blanches

Other name: Lark Quail

IDENTIFICATION 14cm. Sexes similar. Very small, resembles a diminutive courser in posture, butterfly-like jerky flight recalls a bush-lark. Characteristic broad cream-white superciliary stripe, chin and throat also white. Upperparts pale rufous-brown with darker markings. Breast golden-buff, white and dark brown feather tips creating a spotted effect. In flight, upperwing pattern is diagnostic: all-white shoulders, white trailing edge and patch in outer primaries contrast with black flight feathers which may show some buff. Striking wing pattern not visible on standing bird. Bill yellowish to pale green, eye brown, long legs (good field character) flesh to cream-yellow. Immature significantly paler above showing more extensive white fringing. **Similar species** Flight jizz and wing pattern, coupled with dumpy miniature courser-like appearance should exclude other quail and button-quail species.

HABITS Favours short dry grassy habitat. Extended views are unlikely. Extremely difficult to find on landing, but a stationary observer has the best chance of a second view. The two contemporary Gambian sightings have been in sedge grass near saltpan and mangrove interface, and amidst stubble in a recently harvested millet field. Secretive nature and unwillingness to fly may lead to being under-recorded.

VOICE No calls or wing-whirring noted on flushing. A very low soft whistle described when going about its business undisturbed; as yet not tape recorded.

STATUS AND DISTRIBUTION Rare; two modern records each involving single birds: coastal WD (Cape Point) in December 1986 and LRD (near Keneba) in November 1989. Inconsistent in north-western Senegal but recorded every month.

BREEDING In northern coastal Senegal breeds October to March. Nest scrape lined with leaves and rimmed with small pebbles.

RAILS AND ALLIES

Rallidae

Twelve species, of which two are restricted to northern Senegal. A group of small to fairly large birds predominantly associated with freshwater and swampy grassland; the single flufftail also likes thick forest undergrowth. Secretive and semi-nocturnal; may be under-recorded. Ground-dwelling birds with short rounded wings and long legs. Weak looking fliers, but great travellers notwithstanding. Common characteristics include habitual jerking of a short cocked tail, and a period of flightlessness as flight-feathers are moulted simultaneously.

WHITE-SPOTTED FLUFFTAIL
Sarothrura pulchra   **Plate 15**

Fr. Râle perlé

Other names: White-spotted Pygmy Rail, White-spotted Pygmy Crake

IDENTIFICATION 15cm. Sexes differ. Miniature rail (sparrow-sized) of moist forest floor, chestnut foreparts and tail; black body cleanly spotted white (male) or narrowly barred ginger (female). Spots of male smaller on browner lower belly and thighs. Tail chestnut, weakly barred black on female. Bill blackish, legs and long toes dark grey. Immature is duller version of adult, sexes distinguishable at an early age. **Similar species** Sexes respectively unique. Buff-spotted Flufftail *Sarothura elegans* (nearest records Guinea) male similar but spots smaller, denser, buff-toned; female spotted and barred brown lacks any chestnut; single note tuning-fork call; could be borne in mind in forest of southern Senegal.

HABITS Forest-adapted and sedentary, the only flufftail in Senegambia. A bird of seasonally wet forest floor, known from only two forest islands in The Gambia where a few pairs persist. Rarely seen outside the rains when not calling or parading well staked-out territorial boundaries. At its finest when entering a shaft of light that has penetrated the canopy. Terrestrial, moves by creeping and picking its way through the undergrowth without disturbing the leaves; very rarely flies. Famously difficult to see and may be heard at a site for a whole wet season and never seen. Elsewhere in West Africa reported to have adapted to habitat degradation.

VOICE In tone and resonance the repetitive, ventriloquial *poo-poo poo-poo poo* of the male sounds like a cross between the call of a wood-dove and a tinker-barbet. A higher pitched and faster *wuwuwuwuwuw* maintained for up to 7 seconds amidst a bout of *poo poo poo* calls sounds excitably rushed. Calls throughout the day, sometimes for long periods, most vociferous July-late October. Also calls during the dry season.

STATUS AND DISTRIBUTION Very restricted, coastal forest patches of WD only, e.g. Abuko. Most records mid-July to early-December when territorially vocal. Modest Gambian population is the most northerly of the species' range in West Africa. In Senegal only known from coastal Casamance forest.

BREEDING A male watched collecting and carrying well rotted vegetation in August. Builds a small, usually wet-looking domed mound camouflaged with leaves, and lined with dry material.

AFRICAN CRAKE
Crex egregia   **Plate 15**

Fr. Râle des prés

IDENTIFICATION 22cm. Sexes alike. Upperparts olive-brown mottled darker; pale supercilium, dark eye-stripe and whitish chin are good field marks. Throat and upper breast grey, belly and flanks boldly barred white and black. Short bill purplish-red at base, rest grey; legs greyish. Looks quite dark in flight. **Similar species** Size alone separates from smaller Baillon's Crake which has white streaks on upperparts and wings. Smaller Little Crake has bill yellow with red spot at base, legs greenish, and much less barring. Similar African Water Rail is bigger with longer bill and narrower white flank barring. African Crake is considered to be the African counterpart of the Corn Crake *Crex crex* (Fr. Râle des genêts) an erratic Palearctic migrant to West Africa, with conspicuous chestnut wings, as yet unrecorded in Senegambia.

HABITS Swampy, metre-high grassland with small hidden pools is a favoured habitat.

VOICE Varied calls, a high pitched whistling eight-note sequence repeated at short intervals. In South Africa a nocturnal *ke, ke, ke, ke* when on the move and *kerek-kerek-kerek* when breeding.

STATUS AND DISTRIBUTION Very locally frequent in freshwater swamps, with most records from CRD and URD in the rains; a few records from WD towards end of dry season. Very few records in Senegal.

BREEDING No evidence, but adults present during rains when reproduction could be expected. Nest is a domed cup of dry grass in a tussock.

AFRICAN WATER RAIL
Rallus caerulescens   Plate 15

Fr. Râle bleuâtre

Other name: Kaffir Rail

IDENTIFICATION 35cm. Sexes similar but male slightly larger. Only Senegambian water rail. Upperparts brown, head and underparts dark slate-grey with flanks heavily barred white and grey-brown; undertail-coverts white. Characteristic strikingly long slightly decurved bill, bright red when breeding, duller otherwise. Red legs and feet dangle conspicuously in flight. Immature similar but duller, whitish on chin and throat. **Similar species** Palearctic European Water Rail *Rallus aquaticus* (Fr. Râle d'eau) is shorter-billed, and not recorded in sub-Saharan Africa.

HABITS Shy. The first Gambian record was atypically confident, using well-worn tracks and ignoring nearby humans. Quarrelsome, typically betraying its presence deep inside extensive reedbeds by its squabbling noises. Swims well, runs head down, and walks with a high striding gait.

VOICE A familiar rail in southern Africa. Call is often first indication of its presence, described as a prominent sequence of a dozen or so loud piping *kreeea* notes expressed in short cascading phrases, each phrase falling in pitch and dropping in volume. Others nearby may join in. Various other squeals, grunts and snores.

STATUS AND DISTRIBUTION First recorded in The Gambia May 1994 in CRD; a huge north-west extension to its known range. No records in Senegal. Vagrant in West Africa, recorded only occasionally.

BREEDING In southern Africa an untidy bowl of dried grass with a flattened 'doorstep' entrance and exit, low over water in a sedge or tussock.

LITTLE CRAKE
Porzana parva   Plate 15

Fr. Marouette poussin

IDENTIFICATION 18cm. Sexes differ. Small, dumpy crake, with distinctively pointed wings and tail. Male brown with black streaking above, mantle with some white flecking, sometimes striking on scapulars and tertials. Underparts grey, barred paler on flanks. Female brown above with white markings, rich buff below with faint barring on flanks and paler throat. Important identification features are short horn-yellow bill, darker at tip, with red spot at base, and green legs.

HABITS Normally secretive. In the open inclined to walk slow and jerkily with head held low and tail cocked; if disturbed departs quickly with combined dash, run and flight into cover. Could be expected on the muddy fringes of reedbeds, in tussocky grass-land flooded in the rains, or the grassy interface between saltpans and woodland. Good swimmer and able diver.

VOICE In Europe a harsh and high pitched trilling *kick-kik-krrr*, loud *kwuk* or *kweck*.

STATUS AND DISTRIBUTION Palearctic migrant. First Gambian record near Sapu in CRD, December 1998. Recorded from northern Senegal December-April.

BAILLON'S CRAKE
Porzana pusilla   Plate 15

Fr. Marouette de Baillon

Other name: Lesser Spotted Crake

IDENTIFICATION 17cm. Sexes similar. Small, short-billed, typically furtive. Warm olive-brown above with inconspicuous small white spots or irregular streaks bordered with black, densest on mantle. Bluish-grey below with distinctive bold black and white barring on flanks. Immature paler and duller. Bill greenish-yellow with dark tip, no red at the base; short legs yellowish-pink, conspicuous deep red eye shows well against dark grey cheeks. Forewings appear rich chestnut in flight. **Similar species** See Little Crake, and considerably larger (and more probable) African Crake.

HABITS Favours densely vegetated swamps, reedbeds and edges of lakes, always difficult to see.

VOICE In Europe a shrill descending jarring trill, also a husky churr and a frog-like bubble.

STATUS AND DISTRIBUTION Palearctic vagrant to northern Senegal where mist-netted in November and January. No known records south of The Gambia, but has an extensive global distribution.

SPOTTED CRAKE
Porzana porzana   Plate 15

Fr. Marouette ponctuée

IDENTIFICATION 22cm. Smallish and rotund from the side, but head-on views show a body compressed laterally to slip easily through vegetation. Upperparts olive-brown heavily streaked black with some white flecking, face grey with dark mask through lores and conspicuous dark grey superciliary stripe. Underparts grey with diagnostic white spots on sides, breast and also on wings. Flanks barred brown and white when seen well, not as clear-cut as in other crakes. Short, stout and stubby yellow bill with orange-red base useful field character. In flight, the white leading edge of the wing is also characterisitc. **Similar species** See smaller Little and Baillon's Crakes (both vagrants). African Crake has red and grey bill, and greyish legs.

HABITS Much as other crakes, to be expected in grassy swamps and reedbeds; widespread over much of The Gambia. Movements in West Africa little understood.

VOICE In Europe a far-carrying *quip-quip-quip* likened to the dripping of water into a half full vessel.

STATUS AND DISTRIBUTION No recent records in The Gambia, other than a vocal identification in October 1988 in WD. Very well known from northern Senegal River delta; also southern Senegal, and further south in West Africa.

BLACK CRAKE
Amaurornis flavirostris   Plate 15

Fr. Marouette noire ou Râle à bec jaune

IDENTIFICATION 21cm. Sexes alike. Unmistakable, uniform inky-black plumage with bright yellow-green bill and fleshy coral-red legs. Tail constantly flicked and jerked. Eye and surrounding skin red. Immature dull olive-brown, bill greenish-black, legs brownish-black. Downy chick jet black, bill pale pink. **Similar species** None.

HABITS Frequents a wide array of freshwater habitats, rarely in mangroves. Not shy, tolerates human disturbance. Sedentary, only moving in drought conditions.

VOICE Loud piercing croaks and clucks, often in duet, increasing in volume; often delivered from dense waterside vegetation. Also a *pruk* and *chipp* when alarmed.

STATUS AND DISTRIBUTION Locally common resident in freshwater habitats throughout; most

frequent in WD and least common in LRD. Found across southern Casamance in Senegal but infrequent in the north to delta of Senegal River.

BREEDING Immatures seen throughout the year, highest incidence in the dry months. Downy chicks in January. Nest a tidy bowl of grasses, usually well above water.

PURPLE SWAMPHEN
Porphyrio porphyrio   Plate 15

Fr. Poule sultane

Other names: Purple Gallinule, Purple Reedhen

IDENTIFICATION 45cm. *P. p. madagascariensis.* Sexes alike. Unmistakable and large, adult overall purple-blue with a purple sheen, back greenish. Noticeably strong-looking neck and head with massive bright waxy-red bill and frontal shield. Undertail-coverts white, tail constantly flicked. Legs and very long toes red. Immature duller and brownish, legs red-brown, attains adult plumage via a transitional moult with patches of buffish white underparts. **Similar species** Possibly much smaller Allen's Gallinule if view poor or distant. Note that American Purple Gallinule *Porphyrio martinica* (Fr. Taève pourprée) 33cm, is intermediate in size between Allen's and Purple, with a red-and-yellow-tipped bill with blue-white shield, and diagnostic yellow legs. Vagrant to southern Africa and Liberia and a renowned wanderer.

HABITS Favours marshy areas. Bold and approachable it will stand and continue to forage in very exposed situations if feeding conditions are suitable. Probable nomadism dictated by fluctuation in regional water levels. Although overlap occurs, frequents a more diverse variety of aquatic habitats than Allen's, occasionally occurring in brackish mangrove waters. Will catch, drown and swallow small chicks of aquatic birds and also steals eggs to augment largely vegetarian diet. A good climber, encountered ascending swamp vegetation to sunbathe in the early hours. Threatened by wetland drainage.

VOICE A variety of cackles, grunts and snorts, also a rippling bubble.

STATUS AND DISTRIBUTION Uncommon overall, but locally frequent in CRD late dry season to mid rains, April-August. A few annually in WD towards the end of the dry season and occasionally into the rains. Senegalese stronghold in northern coastal quarter, May-June.

BREEDING Nest a substantial structure of reeds and sedges well above water, normally with a canopy. In The Gambia only evidence is one bird carrying possible materials in July.

ALLEN'S GALLINULE
Porphyrio alleni   **Plate 15**

Fr. Talève d'Allen ou Poule sultane d'Allen

Other name: Lesser Gallinule

IDENTIFICATION 25cm. Sexes differ in breeding season. A much smaller, Moorhen-sized, darker version of Purple Gallinule. Adult with dark blue head, neck and underparts, white undertail-coverts; wings and back olive-green. Bill red, legs and feet bright red. In breeding plumage, the normally dull brown eye becomes dark red and the frontal shield bright blue in male, lime green in female; bill remains red. Immature sandy-brown, darker above, edged buff, wings and back boldly streaked with darker brown. Legs and feet dull yellow or flesh-coloured, bill horn with a trace of pink at the base and white near tip. **Similar species** Adult is much smaller than Purple Swamphen. Immatures could be confused with larger *Porzana* species and Lesser Moorhen, best seperable on head and bill shape and colour.

HABITS Prefers dense marshy vegetation. Limited data in The Gambia suggest most likely to be seen in the first two hours of daylight, after which it seems to retire into thick cover. Generally but not inevitably secretive. Regarded as an erratic nomad throughout its range.

VOICE A series of seven or eight clicks rapidly delivered *ki-ki-ki-ki-ki-ki-ki* or *duk-duk-duk-duk-duk-duk-duk*; can sound frog-like.

STATUS AND DISTRIBUTION Rare, with a few records of single immature birds in wetlands in WD in the latter half of the dry season, and several together in half grown ricefields in CRD during early rains. Known in coastal Casamance and disjunctly in northern Senegal.

BREEDING Nest of grasses and sedges in the form of a domed bowl low over water, suspected in some years in The Gambia but not proven.

COMMON MOORHEN
Gallinula chloropus   **Plate 15**

Fr. Poule d'eau commune

Other names: Moorhen, Common Gallinule

IDENTIFICATION *G. c. meridionalis.* 33cm. Sexes similar. A medium-sized crake, with rounded red frontal shield on short red, yellow tipped bill. Conspicuous short and cocked black and white undertail repeatedly flicked when swimming or walking. Upperparts and wings dark olive-brown, separated by distinctive white lateral line from grey-black underparts. Head and neck darker sooty-black. Eye red, gangly legs and unlobed toes yellow-green, red 'garter' above tarsal joint. Immature brown with whitish chin and foreneck, breast buffy-grey, frontal shield poorly developed, lateral flank line creamy. **Similar species** See Lesser Moorhen and Eurasian Coot.

HABITS Frequents freshwater margins and flooded rice paddies, in loose aggregations when not breeding. Swims with much head-nodding. Generally intolerant of disturbance and make a brisk dash for cover.

VOICE An assortment of loud, abrupt nasal *krrrnk* or *pruuk* notes, but rarely vocal in The Gambia.

STATUS AND DISTRIBUTION Seasonally common in WD, particularly close to southern Senegal; locally common in CRD and URD with high proportion of records in January-March. Immatures have occurred in January in well-spaced locations, but there are no records of downy chicks. Widespread in Senegal except arid central and eastern regions.

BREEDING Nest hidden in thick vegetation over water, but breeding not proven in The Gambia.

LESSER MOORHEN
Gallinula angulata   **Plate 15**

Fr. Gallinule africaine

IDENTIFICATION 23cm. Sexes alike. Appreciably smaller than Common Moorhen, greyish-black above and below, shows white lateral flank line and white undertail-coverts. Legs greenish-yellow or orange-brown, eye red. When breeding, bill bright yellow with pointed frontal shield and ridge of bill red; dull yellow in non-breeding plumage. Immature greyish-brown above; pale buffy-grey below with creamy-white undertail-coverts; dull brownish-yellow bill and dull yellowish-green legs. Chick sooty-black, upper part of bill chestnut with an ivory-white tip. **Similar species** Only two-thirds the size and greyer than more frequent Common Moorhen, and comparable in size to Black Crake. Lesser Moorhen best identified on size, bill colour pattern, and by absence of red garters. See also immature Allen's Gallinule.

HABITS An African endemic, typically furtive. Usually in ones or twos in seasonal low grassy swamps. In other parts of Africa its presence is ephemeral and it will settle promptly to breed if conditions are favourable, implying constant nomadism. Little known in Senegambia.

VOICE Five or six hollow *do-do-do-do-do* notes repeated with pauses and sounding like a pump.

STATUS AND DISTRIBUTION Rare to uncommon with a few modern records in CRD and URD mostly at the advent of the rains. Found across southern Casamance to just east of The Gambia early in the year and in the north during the rains.

BREEDING Favours temporary flooded and growing grassland; immature bird seen late October in CRD. Basket-shaped nest with blades pulled over to create a dome, low over water and well hidden.

EURASIAN COOT
Fulica atra   Plate 15

Fr. Foulque macroule

Other name: Coot

IDENTIFICATION 38cm. Largish, distinctively rounded shape, sits high on the water. Plumage entirely velvet sooty-black; very short tail and diagnostic white short bill and forehead shield

prominent at considerable distance. Out of the water stands upright on grey-green legs, with lobed grey toes. Flies low and heavily with much splashing, legs trailing, revealing striking white trailing edge to wing. Immature greyer, paler below, shield poorly developed. **Similar species** Black Crake is much smaller, lacks white on the face and rarely swims. Common Moorhen has red frontal shield, conspicuous white undertail constantly flicked, and white lateral flank line. Note the nearest records for disjunctly distributed Red-knobbed or Crested Coot *Fulica cristata* (Fr. Foulque à crête) are Morocco and S Africa; Red-knobbed is separable in non-breeding plumage by absence of white in the open wing and by pattern of black feathering at base of bill.

HABITS Favours extensive open stretches of freshwater with adjacent reedbeds, most likely in the dry season. Dives frequently, sometimes with a forward off-the-surface leap.

VOICE In Europe an explosive *kook* or *kawp*.

STATUS AND DISTRIBUTION Vagrant; first recorded in The Gambia in February 1994, a singleton in CRD near Kaur, but extensive suitable under-watched habitat exists. Currently increasing throughout West Africa with considerable numbers in wetlands in northern Senegal where up to 350 in January in some years.

CRANES

Gruidae

One species. Superficially stork-like, but cranes have bulky back and rump feathers and much shorter, stouter toes. A group of elegant and graceful birds with ritualised courtship dancing, accompanied by trumpeting vocalisations. Numbers are declining, as earlier writers refer to flocks of hundreds in CRD. In much need of protection from the drainage of wetlands.

BLACK CROWNED CRANE
Balearica pavonina   Plate 4

Fr. Grue couronnée

Other names: Northern Crowned Crane, Crowned Crane

IDENTIFICATION 115cm. Sexes similar. Huge and unmistakeable. At rest appears entirely black with large pale patch in the wing. Head adorned with striking golden spiky crest. Conspicuous pink and white cheek patches contrast with velvety black crown. Two rosy-red throat wattles. Long neck, breast and belly covered in slaty-grey lanceolate feathers. White and straw coloured wing patch is prominent in flight. Shortish bill; legs and feet black. Immature duller, more rufous above, sandier

below. **Similar species** None.

HABITS Favours wet fields and swampy land with nearby tall trees, e.g. baobabs for communal roosting. Slow, stately walk with the neck slightly crooked and head gently nodding are typical field characteristics. In flight the head, neck and legs are held outstretched. Forage gregariously. The species has important local cultural significance.

VOICE A loud trumpeting honk. Calls in flight and is especially vocal in nuptial display; also a soft purring call used in greeting and in communication with the chicks.

STATUS AND DISTRIBUTION Very locally frequent to common in WD and CRD at a few regular and well known sites; occasional records in other divisions. Recorded in all seasons. In Senegal widespread in suitable habitat, largely absent from central arid zone.

BREEDING Solitary. Substantial nest built of vegetation on the ground usually in a marsh.

FINFOOTS

Heliornithidae

One species. Grebe-like in appearance but with affinities to the rails. An elusive water bird, and one of the most sought-after birds on The Gambian list. Disproportionately small head; long, broad and stiff tail. Legs set well back, toes fleshy and lobed.

AFRICAN FINFOOT
Podica senegalensis   Plate 15

Fr. Grébifoulque du Sénégal

Other names: Peter's Finfoot, Finfoot

IDENTIFICATION 55cm. Sexes differ. Breeding male has contrasting bicoloured foreneck and hindneck, separated by a thin grey-white line from the eye down side of neck. Crown, forehead, nape and hindneck blackish with a green gloss. Chin, throat and foreneck blue-grey with whitish flecking on the throat. Underparts dark brown, barred and streaked whitish, paling to uniform creamy-white on belly. Upperparts and wings dark brown, spotted white, greenish tinge on mantle. Tail black, tipped buff. Bill coral-red, eye red-brown. Feet and legs red-orange. Non-breeding male has chin, throat, foreneck and lores with white replacing grey; hindneck remains dark. Female in all seasons has forehead to hindneck dark chocolate brown, lores whitish; chin, throat and foreneck whitish, broad greyish-brown stripe down side of neck bordered by two narrow white stripes. Overall barred and spotted white. Bill red-brown, upper mandible darker. Iris pale and pupil red. Immature similar to female but paler and more uniform brown, spotting much reduced, neck stripe ill-defined. Note that if possible all records should attempt to sex and age the individual concerned.

Similar species Most likely confusion is between semi-submerged dark-billed, brown immature Finfoot and snake-necked African Darter or oily-sheened, hooked-billed Long-tailed Cormorant; see text and illustrations for these species.

HABITS Favours the aquatic fringes of riverine forest; mangrove habitats and woodland streams. Furtive, but some well-watched individuals become approachable. Swims with neck-pumping motion. Swims fast and direct towards cover if disturbed, with head and slightly kinked neck held almost flat on the water. Sometimes seen well out on wide stretches of the river.

VOICE A sharp *skwak*, a duck-like *kwark* and a fluty *pay pay*; elsewhere a roaring boom has been described; more information needed.

STATUS AND DISTRIBUTION Locally and seasonally uncommon to frequent with records from all divisions. Mostly seen during peak rains, least likely in late dry season. Becomes temporarily sedentary on flood courses until the water level begins to recede in late November; dispersal then ensues. Recorded from north-east and southern Senegal.

BREEDING Female with two downy chicks on 21 October 1995 in WD, fully grown on 3 December is first confirmed Gambian breeding record; although well-watched no male was ever observed. Well-grown young in Senegal in November. Nest placed low over water in tangled roots and creepers.

BUSTARDS

Otididae

Five species. Small to very large, long-necked, long-legged birds of open, often arid, habitats. Only one species seen with any regularity in The Gambia during recent times. Chronic persecution from hunting, habitat loss and possibly climatic change have combined to cause these reductions. Conservation measures are urgently required. Some species show marked sexual dimorphism, coupled with elaborate courtship displays. Eggs (usually 2) laid directly onto the ground. Gait characteristically stately. In flight powerful with slow beats on a substantial wingspan, with neck outstretched. Omnivorous. The considerable identification problems are compounded by differences in size between the sexes, and plumage dimorphism.

DENHAM'S BUSTARD

Neotis denhami   Not illustrated

Fr. Outarde de Denham

Other names: Stanley Bustard, Jackson's Bustard

IDENTIFICATION 76cm. Sexes differ. Very large, dark-looking with thickset neck and distinctively patterned head lacking a crest. Head with prominent facial stripes, and black crown with broad white-grey central stripe. Chin white, neck neatly and contrastingly bicoloured, anterior half grey (barred buff to dark brown in smaller female), posterior half distinctively unmarked rufous-brown. Black closed wing has conspicuous broad white wing panels and bars, in flight shows extensive white on black wings. Upperparts brown with fawn vermiculations; underparts white; tail barred black and white. Bill grey, legs pale yellow. **Similar species** See Arabian Bustard.

HABITS Favours open habitats; dried out marshes and recently burnt grassland with emergent growth are the most likely sites. South-north movements are rain related and it is most likely to be solitary on its Senegambian non-breeding quarters.

VOICE Normally silent; booms when displaying; also barks *kha-kha* or *kho-kho*.

STATUS AND DISTRIBUTION Considered not uncommon earlier this century in The Gambia. There are less than ten records for the whole of the country since 1979, widely scattered from June to December. In eastern and northern Senegal considered a non-breeding visitor mid-July to mid-November.

BREEDING Breeds July-October in Mali; reproduction opportunistic.

ARABIAN BUSTARD

Ardeotis arabs   Not illustrated

Fr. Outarde arabe

Other name: Sudan Bustard

IDENTIFICATION Upto 90cm. *A. a. stieberi*. Huge (female considerably smaller and greyer), overall looks pale sandy-brown with white underparts and powerful neck and head with straggling black crest. **Similar species** From Denham's Bustard by noticeable crest, lack of prominent black eye-stripe; by white finely barred blackish neck of Arabian, not dark brown and grey; by closed wing of Arabian appearing sandy-brown with some white, not black with white, spots on wing-coverts.

HABITS A bustard of wide open dry areas including semi-desert with dry river beds. Could possibly occur in Senegambia with Denham's with which it mixes on migration further east in Chad.

VOICE Normally silent. Croaks and barks recorded in display and when alarmed.

STATUS AND DISTRIBUTION Not recorded in The Gambia since 1928. Now rare in Senegal when up to the 1960s it was considered common in parts of the north, in July-November.

BREEDING Old northern Senegal record of a female ready to lay in July, chicks in May-June and September and a half grown bird in March.

SAVILE'S BUSTARD

Eupodotis savilei   Plate 16

Fr. Outarde houppette

Other name: Savile's Pygmy Bustard

IDENTIFICATION 51cm. Sexes differ. A small, short-necked bustard more like a thick-knee in size, recalling a miniature Black-bellied Bustard. Rusty crest on nape of male is prominent only in display. Face greyish-white washed buff, chin and throat white, latter with a black line, neck grey. Underparts from base of neck to belly glossy black with contrasting all-white collar patch, upperparts sandy-rufous with buff and blackish arrowhead markings. In flight, shows pale band across the whole wing contrasting with black primaries. Female has neck to chest more uniformly buff speckled blackish, white breast and black belly. Immature resembles female. **Similar species** Much the smallest bustard in the region. Male Black-bellied Bustard is longer-necked and longer-legged and has white stripe down side of neck. Female White-bellied and Black-bellied have no black on the belly.

HABITS Favours dry savanna woodland and light scrub, often near expanses of tall *Andropogon* grasses. Difficult to locate even when calling within a few metres, and creeps away secretively. Calling in large tracts of tall maize fields in September. Considered sedentary where better known.

VOICE Notably ventriloquial. Advertisement call from male consists of a long whistled note succeeded by a series of short clear *peep-peep-peep* whistles; female replies with a tremulous *whi-whi-whi*. Also an accelerating series of short whistles, and a series of frog-like notes, both delivered in the same rythmn, and a harsh chattering alarm call.

STATUS AND DISTRIBUTION First recorded in The Gambia in October 1994: a brief sight record of a single bird south of the river in LRD on a tar road in East Kiang (N. Borrow pers. comm.). Several calling just W of N'jau on the north bank in CRD in mid-September 1996 were taped, with two birds briefly seen walking and in flight. Widespread north of 14° in Senegal; records also from extreme south-east and at the coast over the Gambian border.

BREEDING Nest with 2 eggs and a record of a female ready to lay in September in northern Senegal.

NOTE This species was formerly considered conspecific with Buff-crested Bustard (Red-crested Korhaan) *E. ruficrista*.

WHITE-BELLIED BUSTARD
Eupodotis senegalensis   **Plate 16**

Fr. Outarde du Sénégal

Other names: Senegal Bustard, White-bellied Korhann

IDENTIFICATION 59cm. Sexes differ. A medium-sized stocky bustard. Male distinguished by combination of conspicuous blue neck and white belly. Forehead and crown black; face, chin and upper throat white; inverted V-shaped black stripes on chin. Upperparts finely vermiculated orange-buff and black, looking uniform biscuit-coloured at a distance. Tail with two thin dark bars. In flight shows a white panel on dark primaries; underwing whitish-grey. Darker tips to the secondaries and dark primaries detectable on closed wing. Female has sooty-brown crown, nape speckled tawny and black; the whole throat white (black V-mark rudimentary or absent), neck biscuit-coloured. **Similar species** Larger, long-legged female Black-bellied shows finely vermiculated upperparts. In flight Black-bellied underwing is mainly black; White-bellied is mainly white.

HABITS Favours damper regions in Senegal; comparable habitat in The Gambia seemingly unoccupied. When considered common it was regarded as 'untame'. Especially vocal at dawn and dusk in areas of reasonable population density. Activity most likely in the early morning or evening, and like other bustards rests-up during the hottest periods of the day.

VOICE A frog-like croak *aaa* succeeded by a roller-like *kakaawara*.

STATUS AND DISTRIBUTION No recent records in The Gambia; historically (up to early 1960s) considered the commonest bustard. Recorded along the whole southern and northern Senegalese borders with scattered records between; largely absent from east. Declining throughout our area and in immediate need of protection.

BREEDING In northern Senegal breeds July-October; old Gambian record in February.

BLACK-BELLIED BUSTARD
Eupodotis melanogaster   **Plate 16**

Fr. Outarde à ventre noire

Other name: Black-bellied Korhaan

IDENTIFICATION 61cm. A medium-sized bustard and the only one regularly seen in The Gambia. Sexes differ. Very long diagnostically thin and frail snake-like neck with bulbous head and noticeably long thin dull grey-yellow legs. Male has a blackish silvery-grey throat that merges into a broadish black stripe down the centre of foreneck, in turn joining the all-jet black underparts; neck stripe bordered whitish from ear-coverts to upper breast. Thin black stripe from behind the eye to nape where it is vestigally crested black. Head and hindneck finely barred with buff and black; mantle, scapulars and inner secondaries tawny-buff with broad black centres to the feathers; lower back and rump finely vermiculated with black and fawn, looking mottled. Greater wing-coverts white (conspicuous at a distance and useful field mark); flight feathers black and white, including underwing-coverts and axillaries. Tail with four dark bars, outer feathers plain dark brown. Conspicuous white markings on the closed wing distinguish it at a distance. Striking in flight with white patches in the primaries, secondaries and upperwing-coverts; underwing black to white on the coverts. Female has crown, hindneck and upperparts same but differs considerably by having no black on the belly and lacks black stripe on foreneck; neck and underparts are fawn-whitish to brown-buff. In flight shows a combination of white underparts and a black underwing. Eye yellow or pale brown, bill brown to yellow, darker above. Immature resembles female and has more buff fringes to wing feathers. **Similar species** For separation of female Black-bellied from female White-Bellied Bustard see that species. For separation of male Black-bellied from either sex of smaller Savile's Bustard, see also that species.

HABITS Favours wooded grassland often with nearby laterite bluffs but could turn-up in dry marshes and open cultivated fields. Camouflage conferred by the mottled plumage and its habit of freezing when approached makes it difficult to spot. In the dry months it is often discovered only by accident and then at very close range and alone. Frequency of encounters increases dramatically in the rains after the commencement of the male's nuptial flight display; a series of slow stiff wing-beats are followed by a long slow glide, with the wings held in a deep V above the back and the head in an arched-upwards position. Annually reappears at certain sites at the very onset of the rains after possible displacement caused by bush fires.

VOICE In terrestrial display, a softly whistled note succeeded by a popping bark *'m'pokh m'pokh'* otherwise mostly silent.

STATUS AND DISTRIBUTION Modern records in all months from all divisions except large stretches of underwatched NBD; no longer common in WD. Most frequent in LRD and URD where sightings peak in the former during the early rains when aerial displays are seen in most years; sightings much reduced throughout February to mid-June. Widespread throughout Senegal up to 15°; then rare or absent.

JACANAS

Jacanidae

One species. An unmistakable bird of lily-ponds, rice fields, and water-hyacinth or water-lettuce covered freshwater, appearing occasionally on brackish water. Rail-like in many ways, with immensely long toes well suited to walking across floating vegetation.

AFRICAN JACANA
Actophilornis africanus Plate 15

Fr. Jacana poitrine dorée

Other names: Lily-trotter, Jacana

IDENTIFICATION 30cm, female larger than male. Body rich chestnut, wings with glossy sheen in bright sunshine, hindneck black contrasting with conspicuously white foreneck that darkens to golden-yellow on breast. Vivid pale blue frontal head-shield, brightest on breeding male. Short tail. Long legs and extremely long toes olive-brown, often muddy and greyish. Immature with shield poorly developed or absent; depending on age has black stripe through eye and black crown. **Similar species** Lesser Jacana *Microparra capensis* not reported closer than Mali, is much smaller, usually with browner crown and yellowish bill.

HABITS Stalks, creeps and runs on floating surface vegetation of lakes and pools, occasionally forages

on open farmland. Flocks congregate when rain pools subside. Birds of agricultural wetlands become confiding, otherwise rather shy. Famed for carrying its progeny under the closed wing when they are still unable to fly, with legs of chicks dangling down. Frequently squabbles, raising wings in a threat posture to a potential aggressor. Flight normally low and for short distances on rounded wings, with legs hanging or held straight out behind.

VOICE Very vocal at all times, a screeched *kyowrrr* or a shorter, loud and grating *kreep-kreep-kreep* delivered with rapidity.

STATUS AND DISTRIBUTION Locally common to abundant breeding resident in suitable habitat throughout, with main numbers in WD, CRD and URD; numbers fall in the late dry season. Widespread in appropriate habitat in Senegal.

BREEDING Extended season. Nest of floating vegetation. Monogamy/polyandry influenced by population density. Suffers extensive predation, particularly from aquatic reptiles.

PAINTED-SNIPE

Rostratulidae

One species. Superficially resembles the Palearctic snipes but heavier in appearance and with a shorter slightly decurved bill. Habits rather rail-like and often hard to detect when crouching, due to superb camouflage. Nearest relatives considered to be the jacanas.

GREATER PAINTED-SNIPE
Rostratula benghalensis Plate 15

Fr. Rhynchée peinte

Other names: African Painted Snipe, Painted Snipe

IDENTIFICATION Sexes differ, male (24cm) is smaller and much less richly coloured than female. Both have prominent crown, face and neck markings. Shared features are a broad white 'harness-band' in front of dark wing, dark crown and buff coronal stripe, and striking buff V on the back.

179

Female (26cm), brightest when breeding, with throat and neck dark reddish-chestnut, blackish-brown breast-band contrasting with white underparts, and conspicuous white eye-patch. Back bronze-brown, metallic greenish on the shoulders, and intricately barred black. Male paler and duller with chestnut replaced by brownish-grey, wings differ dramatically being spotted golden-buff; eye-patch buff. Legs pale greenish-grey to dull slaty-blue. Bill slightly downcurved at tip, purplish-brown to horn. Immature resembles male but breast-band less well demarcated. **Similar species** Female unique. Male superficially similar to snipes *Gallinago* spp., but bill is slightly decurved; white breast-band obvious in all plumages.

HABITS Mostly active in the first two hours of light after which it retires until the cooler part of the evening, large eyes facilitate crepuscular activity. Continues to feed unperturbed for a few hours after first light in rice paddies, fields of lilies and muddy pools adjacent to reedbeds, and occasionally in saline habitats. Probes for mud-living invertebrates and also picks up germinated seeds. Sometimes feeds from the surface with a scything action and is a competent swimmer. May also forage on dry ground. Diurnal sightings usually of a few birds crouched just inside cover at the waters edge, which on being sighted normally cower and retreat from view. Easier to find when vegetation and water levels have receded. Broad and rounded wings in flight; short weak flights with legs dangling create a rail-like appearance.

VOICE Rarely heard. A *kek* when flushed, courting notes resonantly bullfrog-like.

STATUS AND DISTRIBUTION Locally common, sometimes abundant, at well-watched sites throughout. Recorded in all months but sightings reduced in the wet season when aquatic vegetation dense. Probably frequents most freshwater habitats with suitable emergent vegetation. Movements must occur from flooded areas that seasonally dry up, resulting in local concentrations on remaining wet sites. Widespread in all suitable habitat throughout Senegal.

BREEDING Proof of nesting in August 1988, also three unfledged young seen October 1990 in WD. Polyandrous, female dominating display. Female courts two or more males, laying in several nests, but incubation and parental care by male alone.

OYSTERCATCHERS

Haematopodidae

Two species, one a vagrant. Heavily built, cosmopolitan waders with disproportionately short legs. Sexes alike. Plumages usually black or black and white. Long, straight and strong orange bills for probing and manipulating shellfish: bivalve molluscs are an important dietary component.

EURASIAN OYSTERCATCHER
Haematopus ostralegus   Plate 19

Fr. Huîtrier-pie

Other names: Oystercatcher, European Oyster-catcher

IDENTIFICATION 43cm. Upperparts, head and breast black. Underparts white. Broad white wingbar striking in flight. Lower back, rump and uppertail-coverts white. Broad black terminal band across most of white tail. Bill strong, straight, blunt and orange-red. Eye and eye-ring red, legs pink. In non-breeding plumage duller, with white neck collar patches. **Similar species** None.

HABITS Favours sandy beaches and lagoons, retreating to mangrove creeks at high tide. Intertidal movements in compact groups low over the water. Food consists mainly of bivalves, especially favouring clams, but also marine and terrestrial worms.

VOICE Noisy. Flight-call *ke beek, ke beek*. Also loud *pick-pick-pick*.

STATUS AND DISTRIBUTION A Palearctic breeder and long-distance migrant to Senegambia. Numbers peak in October-April along the whole coast with smaller numbers recorded throughout the year. Numbers have increased since 1985.

AFRICAN BLACK OYSTERCATCHER
Haematopus moquini   Not illustrated

Fr. Huitrier de Moquin

IDENTIFICATION 44cm. Whole plumage black, glossy when breeding; eye red, eye-ring scarlet, long bill red tipped orange, legs and feet purplish-pink. Noticeably long in the wing from body to carpal joint. Immature duller and browner, eye-ring and eye brown. Orange bill tipped brownish, legs and feet greyish-pink. **Similar species** Eurasian Oystercatcher is always black and white.

HABITS Associated with beaches and rocky shore-lines.

STATUS AND DISTRIBUTION A sedentary South African species with nearest breeders in Angola. No Gambian records, but two records in Senegal in February 1970 and December 1975. Melanism in Eurasian Oystercatchers is unknown and the Canarian Black Oystercatcher *H. meadewaldoi* has not been reported since 1913 and is almost certainly extinct.

STILTS AND AVOCETS

Recurvirostridae

Two species. Fairly large graceful black and white waders with relatively slim bodies, long legs and short tails. Surface feeders, each with a distinctive slender bill-shape. Toes with some webbing. Gregarious at all times, intermixing on tidal mudflat habitats. Numbers of both species have increased recently.

BLACK-WINGED STILT
Himantopus himantopus　　Plate 19

Fr. Echasse blanche

IDENTIFICATION 38cm. Uniform black wings and mantle contrast with pure white body, tail grey-ish-white. Long thin neck and diagnostic extremely long pink legs. Straight, black needle-like bill. Eyes red. Head markings in breeding male variable from wholly white to black extending from eye to nape and hindneck. Immatures and winter adults have dusky markings on head and neck. In flight shows all-black, relatively short triangular wings contrasting with white body; legs project 18cm beyond tail-tip. **Similar species** Unmistakable.

HABITS Frequents a wide range of aquatic habitats, from muddy tidal lagoons along the coast (most abundant), along the river, and in rice-paddies, drainage ditches, roadside pools, sewage outlets and freshwater swamps. Wades in deep water picking food from the surface.

VOICE Noisy if disturbed: alarm and flight call *kik*.

STATUS AND DISTRIBUTION Both long and short-haul migrant. Common to locally abundant throughout, occurring singly or in small or large flocks with a maximum count of 5,000 in WD in January. Widespread in suitable habitat throughout Senegal where present all year. No Gambian breeding record but has attempted to breed in northern Senegal. Nest a scrape on the ground or on shallow floating vegetation.

PIED AVOCET
Recurvirostra avosetta　　Plate 19

Fr. Avocette à tête noire

Other names: Avocet, Eurasian Avocet

IDENTIFICATION 44cm. A white and black wader with a characteristic thin, upturned black bill and lead-blue legs. Looks black-capped with the dark crown extending to just below the eye and down to the nape. Side of mantle black, upperwing pied with extensive, central white area on secondaries and coverts. In flight looks pure white, boldly marked above with black 'straps' on the mantle and coverts; from below snowy white with black outer primaries. In non-breeding birds and immatures black areas are sooty-grey. Appears deep-bodied when feeding, and in flight accentuated body length due to extended neck and legs make wings look shortish. **Similar species** Unmistakable.

HABITS Prefers saline lagoons along the coast and open stretches of tidal mud in LRD. Wades in deep water and regularly swims. Feeds with characteristic sideways sweeps of the bill. May 'up-end' like a duck. Usually occurs in small, close groups but may number hundreds. Active early and late in the day but feeding primarily determined according to the tide; sometimes at night.

VOICE Noisy. A clear liquid flight call *kluit-kluit*.

STATUS AND DISTRIBUTION Palearctic breeder, migrant to Senegambia. Frequent and briefly locally abundant on the coast and along the river in LRD in October-March, with a maximum count of 400 in WD (Bund Road). Uncommon inland and rare in URD. In northern Senegal up to 5,000 in March; migratory also in southern Senegal and recorded on the Gambia River to the east. No breeding records for West Africa.

181

THICK-KNEES AND STONE-CURLEWS

Burhinidae

Four species. Large waders, taller and more heavily built than *Vanellus* plovers, having stout bills, longer necks and sturdier long legs. The yellow eyes are large, adapted for crepuscular and nocturnal habits, and their eerie calls are especially intense on moonlit nights. All are cryptically patterned in browns, greys, black and buff. When disturbed, thick-knees tend to crouch, before creeping away slowly and deliberately. The sparse nest scrape is often placed close to a fallen branch or small boulder. Wing panels and facial markings are important identification features.

STONE CURLEW
Burhinus oedicnemus　　　Plate 16

Fr. Oedicnème criard

Other name: Eurasian Thick-knee

IDENTIFICATION 42cm. A sandy, brown-streaked thick-knee with large, yellow eyes and large-jointed yellow legs. On the closed wing, the Stone Curlew shows a narrow white upper band bordered black with a broad pale grey panel below. **Similar species** See wing markings described for Water Thick-knee, Spotted Thick-knee and Senegal Thick-knee.

HABITS Favours semi-arid or stony deserts, but may frequent aquatic habitats at night.

VOICE Variations on a raucous Curlew-like *cur-lee* heard mostly at night, but unlikely to be heard in winter quarters.

STATUS AND DISTRIBUTION Palearctic migrant to northern Senegal and although unrecorded in The Gambia could well occur. A British-ringed bird has been recovered further south in Sierra Leone.

SENEGAL THICK-KNEE
Burhinus senegalensis　　　Plate 16

Fr. Oedicnème du Sénégal

IDENTIFICATION 38cm. The most abundant thick-knee in our area. Upperparts sandy to tawny-brown, finely streaked darker; underparts buffish-white, breast streaked dark brown. Folded wing shows a broad, pale greyish panel bordered above by a narrow black band. The yellow-based, black-tipped bill is proportionately the longest and heaviest of all the West African thick-knees. **Similar species** see wing panel markings described for Stone Curlew, Water Thick-knee and Spotted Thick-knee.

HABITS Prefers moving water, river banks, shorelines with mangroves and along bolongs where it may congregate in large flocks. Seen on dry bush tracks and roads at night. Largely crepuscular and

nocturnal, seeking shade by day, often under isolated clumps of mangroves. When approached stands alert, runs quickly, makes a short flight and runs for cover. Often seen in mixed flocks of other water birds, especially with Wattled Plover.

VOICE A haunting, wailing cry that rises and falls, often with several birds calling together, *pi, pi pi-pi-pi-pi-Pii-Pii-Pii-pii pi pi.*

STATUS AND DISTRIBUTION Common in all suitable habitat throughout The Gambia. Numbers fluctuate. Widespread in Senegal.

BREEDING Breeding season appears extended; nests with eggs in January in NBD and early July in LRD; downy chicks in late November. Regularly nests amongst leaf litter under shrubs on sand dunes or near water.

WATER THICK-KNEE
Burhinus vermiculatus　　　Plate 16

Fr. Oedicnème vermiculé

Other names: Water Dikkop

IDENTIFICATION 43cm. A typical thick-knee, greyish-sepia overall and looking rather dark. Wing panel consists of a narrow upper white bar separated from the lower grey panel by dark streaks and not a solid black bar. Also the upperparts are vermiculated, not streaked, and the legs and base of the bill are yellowish green. Toes extend beyond tail in flight. **Similar species** Wing panel pattern separates it from other species.

HABITS Always near water on rivers, mangrove fringes and estuaries. Reputed to mingle with crocodiles.

VOICE More metallic and hollow sounding than Senegal Thick-knee, a *ti-ti-tee-teee-tooo* falling and fading at the end.

STATUS AND DISTRIBUTION No Gambian records. Vagrant to southern Senegal.

SPOTTED THICK-KNEE
Burhinus capensis　　　　　Plate 16

Fr. Oedicnème tachard

Other name: Spotted Dikkop

IDENTIFICATION 43cm. Easily distinguished from other thick-knees by uniformly spotted black on buff appearance of upperparts (streaked in other species), and by the absence of any contrasting wing-bars or a panel. As with other thick-knees white wing-spots are conspicuous in flight. A useful aid to identification at night is its leggier, taller appearance which differentiates it from Senegal Thick-knee which also comes to the roadsides regularly at night.

HABITS Most sightings are on laterite roads and tracks well after dark. Favours dry, stony bush with expanses of dry grass. During the day remains well hidden and unlikely to be seen. Normally in pairs or small groups.

VOICE More abrupt than very similar Senegal Thick-knee, *whih-whih-whih-whih* typically dwindling at the end; also a rasping note.

STATUS AND DISTRIBUTION Frequent in LRD (stronghold is Kiang West) and URD; rare elsewhere. Probably more numerous than observations suggest due to nocturnal habits. Widespread in northern and eastern Senegal. No breeding records for The Gambia though almost certainly does so; in northern Senegal breeds April-July.

COURSERS AND PRATINCOLES
Glareolidae

Five species of small to medium atypical waders in two subfamilies. Coursers Cursoriinae: four species including the water-associated Eygptian Plover. True coursers are birds of dry habitats, cryptically coloured and fast sprinters with long legs. Shared features include a neat posture, shortish downcurved bill (straighter in Eygptian Plover), the absence of a hind toe, and a general reluctance to fly unless forced. Pratincoles Glareolinae (one species) are especially aerial, hunting insects on the wing, and highly gregarious. Long pointed wings, swallow-like forked tail, short legs and graceful flight are reminiscent of smaller terns.

EGYPTIAN PLOVER
Pluvianus aegyptius　　　　　Plate 16

Fr. Pluvian d'Egypte

Other names: Crocodile Plover, Crocodile Bird

IDENTIFICATION 21cm. Sexes alike. Unmistakable stunningly coloured plover-like bird. Upperparts blue-grey. Crown, mask, mantle, upper back and breast-band glossy greenish-black, latter separating peachy-buff belly from brilliant white chin and throat. Supercilium white, meeting in a short crest on nape to form a tidy V. Short, straight, relatively stout and pointed bill black, short legs blue-grey. Flies jerkily very low over water, when broad-based triangular grey wing reveals impressive area of white and black diagonal bar. Short grey tail tipped white. Immature duller but plumage pattern similar. **Similar species** None.

HABITS Confiding. Frequents the fringes of a diverse range of aquatic habitats, including river banks, concrete and wooden jetties, bridges, ferry terminals and landing sites, drainage ditches and culverts, wide open boggy flats and roadside tracks adjoining reedbeds and lakes of *Nymphaea* lilies. Solitary or more usually in small groups, occasionally in flocks. The habit of picking the teeth of gaping crocodiles (hence the alternative name 'Crocodile Bird') seems to be a myth initially perpetrated by the writers Herodotus and Pliny c. 500BC.

VOICE Flight call *chee-chee-chee* and *creek-creek*; when alarmed a loud and scolding *chersk*.

STATUS AND DISTRIBUTION Locally frequent to common in URD and CRD from June to February with a population peak September to December. Occasional records from all other divisions in corresponding months with a modern coastal record in September. Numbers fluctuate with seasonal flooding off the Gambia River. Recorded along the whole northern and eastern borders of Senegal then east along the Gambia River, occasional records south.

BREEDING Unrecorded in The Gambia. Records in eastern Senegal from February with eggs up to June. Eggs prone to heavy predation. Nest a scrape on a sandy shingle spit, several of which are ritually dug before the eggs are laid. The clutch is then covered partially with sand and the incubating bird sits over the nest, regularly bearing water in soaked body feathers to moisten the eggs. The eggs are completely covered if unattended. On hatching, the chicks are attended to in a similar manner.

CREAM-COLOURED COURSER
Cursorius cursor   **Plate 16**

Fr. Courvite isabelle

IDENTIFICATION 24cm. Tall, graceful, uniform sandy-buff courser, white on the vent, with creamy-white legs. Diagnostic blue-grey crown with contrasting long white supercilium and black eye-stripe, both of which meet at the nape. In flight upperwing shows white trailing edge to secondaries, primaries contrastingly black, underwing characteristically uniformly blackish. Immature has head pattern less well defined, looks overall lightly speckled or barred. **Similar species** Lack of black belly patch excludes smaller Temminck's Courser; pied head markings, red legs, and breast band distinguishes darker, taller Bronze-winged Courser.

HABITS Favours stony, sparsely vegetated sub-desert. In flight appears large; erractic track is useful field mark. Appears to stand on tip toes when alarmed, emphasising its height, interspersed with bouts of crouching and spurts of running. In The Gambia usually singly or in couples on bare fields.

VOICE A frog-like *quark* and piercing repeated *kuit-kuit* flight call.

STATUS AND DISTRIBUTION Rare in The Gambia. Four modern records between 1966-97, all December-February, from coastal salt flats adjacent to mangroves or on farmed land in WD. Considered both resident and migratory in northern Senegal in appropriate habitat.

BREEDING In extreme northern Senegal recorded March-April and December. No nest scrape.

TEMMINCK'S COURSER
Cursorius temminckii   **Plate 16**

Fr. Courvite de Temminck

IDENTIFICATION 20cm. Small and diurnally active, diagnostic black patch from belly to vent and chestnut cap, bordered from eye to nape by an arched black eyestripe and white supercilium, forming a neat V at the nape. Upperparts brown, neck and breast rufous-chestnut, black belly contrasting with white flanks. In flight, underwing black, and black flight feathers on upperwing contrast with brown back and wing-coverts. Hooked bill black-brown, long legs whitish-grey. Immature speckled and spangled buff over entire upperparts, black patch smaller and duller, crown finely streaked blackish. **Similar species** Cream-coloured Courser is paler and taller with no black belly.

HABITS Burnt grassland, well-grazed pasture and bare harvested agricultural ground in open country are favourite haunts, typified by a bird community including wheatears, Whinchat and woodchats and pipits. Site-faithful; regularly appears immediately conditions are right. If flushed flies quite high and attempts a boomerang-like return. Normal party size is four to eight, sometimes more. Runs and bobs; in sexual display runs up to a potential mate, adopts an erect posture then flashes the dark belly patch. In extreme heat typically cowers under the shade of a tussock or very low sapling.

VOICE Rarely vocal, sometimes a sharp penetrating *err-arr err-err* in flight.

STATUS AND DISTRIBUTION Locally common in all divisions, mostly December to June, rarer in south and south-east, probably moving north into Senegal during the rains.

BREEDING Season extended, both well grown young and downy chicks in early Febuary and eggs in late April and June. Recorded in northern Senegal practically every month. No nest scrape. Often lays eggs near cow or donkey dung in The Gambia.

BRONZE-WINGED COURSER
Rhinoptilus chalcopterus   **Plate 16**

Fr. Courvite à ailes violettes ou Courvitte à ailes bronzées

Other name: Violet-tipped Courser

IDENTIFICATION 30cm. Largest courser with prominent but soft pied facial markings and large bulbous eye. Dark crescentic band on a three-tone breast is diagnostic: broadish black breast-band separates pale brown breast from cream-white belly. Upperparts plain earth-brown, forehead, broad superciliary stripe and mark behind the eye all creamy-white, contrasting with black-brown ear-coverts and cheeks. When alert shows white throat with broad buffy neck band below. Heavyish-looking bill black, with dull red base. Legs purplish-red. In flight shows white band on rump and white tip of uppertail, white bar across upperwing, and white underwing. Primaries strikingly tipped metallic violet. Immature similarly patterned, but face markings and breast-band less prominent, body fringed rusty-buff, wingtips greenish. **Similar species** In poor light conditions and when Bronze-winged stands with its head and neck hunched up, be aware that this courser shares its habitat with all three common *Vanellus* plovers, particularly Black-headed Plover. Note similarity of call to thick-knees. All these birds come to bush-tracks and roads at night.

HABITS Favours lightly wooded areas, near to tracts of bush and at the edge of forest thicket. Crepuscular and nocturnal, diurnal sightings rare. During the day typically rests motionless beneath a low sapling; when disturbed makes short flight then runs. Most encounters are of birds coming to sandy tracks or murrom roads just after dusk or in the night, usually standing very close to the edge of the track whereas *Vanellus* plovers are more likely to walk all over the road. If care is taken will allow close approach.

VOICE Calls nocturnally or when flushed by day, *ji-ku-it* or *groraag* with an upward inflection; recalls a thick-knee.

STATUS AND DISTRIBUTION Uncommon to locally frequent in WD, LRD and CRD. Probably resident, with records in all seasons, peaking November to January. Probably under-recorded. Widespread in Senegal (excepting western coastal) both north and south of The Gambia, wet season visitor in the north.

BREEDING No scrape. Breeding not yet proved in Senegambia but likely; all group sightings should be scanned for immatures.

COLLARED PRATINCOLE
Glareola pratincola **Plate 16**

Fr. Glaréole à collier

Other names: Red-winged Pratincole, Common Pratincole

IDENTIFICATION 26cm. Brown and tern-like. Throat and breast orange-brown surrounded by black. Upperparts dull earth-brown, underparts whitish washed buff. Bill short, black with bright red base. In flight long wings show diagnostic red-brown underwing-coverts (in good light conditions), deeply forked tail and white-rump. Immature speckled black and buff, black bib surround is absent or

partial. Two geographical races in Senegambia are considered indistinguishable in the field, but possibly separable by season; Palearctic *G. p. pratincola* more numerous in dry season, and African *G. p. fuelleborni* in the wet season. **Similar species** Black-winged Pratincole *Glareola nordmanni* (Fr. Glaréole à ailes noires), not illustrated, has black underwing-coverts, suspected earlier this century but there are no recent reports from Senegambia. Its migration routes are far to the east in Africa; nearest vagrants in Togo and southern Cameroon.

HABITS Highly gregarious but sometimes in small numbers, occasionally singly. Favours open ground most often with damp or dried muddy pans preferably with easy access to open expanses of water. Wide gape facilitates the collection of insects whilst on the wing; large flocks are of importance in controlling insect pests, particularly locusts. Movements of regional African race possibly influenced by termite emergences which are in turn rainfall-related. Able runners despite short legs, will stand on tip-toe to investigate their surroundings like a courser. It is likely that huge numbers pass swiftly through The Gambia in under-observed locations. Spanish-ringed bird recovered in Senegal and a bird ringed in Senegal controlled in Morocco.

VOICE Flocks engage in loud and sharp group twittering calls *pirri pirree pirrip pip*; individuals shriek a *tit-irr-it* when agitated.

STATUS AND DISTRIBUTION Records in every month except October; peaking late November to March in CRD and URD. 1,000+ birds on passage in URD in mid-March and 2,000 in CRD in February are of likely Palearctic origin. Rainy season flock of 300 in mid-July in LRD made up largely of immatures of presumed African origin.

BREEDING Breeds colonially in north and east Senegal May-July. In The Gambia, pair protecting immature in URD in August, and older suspected breeding record in CRD in June. Eggs 2-4, usually laid directly onto the ground.

PLOVERS
Charadriidae

Seventeen species of small to medium shorebirds, characterised by compact shape, short bills and longish legs. Not all species are confined to wetland habitats with some being restricted to grasslands and fields. Sexes generally similar. The family is represented in Senegambia by three genera: *Charadrius* (eight species, five of which are Palearctic migrants) are the typical plovers, usually small and short-necked, and often with breast-bands. *Pluvialis* (three species, all migrants, two rare) have long pointed wings, spangled upperparts, and black bellies in breeding plumage. The lapwings *Vanellus* (six species, all but two resident) are larger with broad, rounded wings, a relaxed buoyant flight and strikingly patterned plumage. Always noisy and especially vocal at night.

LITTLE RINGED PLOVER
Charadrius dubius   Plate 17

Fr. Petit Gravelot

IDENTIFICATION 15cm. Typical small plover with a complete white collar. Striking facial pattern differs from Ringed Plover in having smaller white frontal patch, thin white line between brown of crown and black forecrown band, a conspicuous yellow eye-ring (dull and inconspicuous in Ringed Plover), narrower breast band and a black bill. Legs are muddy ochre. In non-breeding plumage tones muted as in Ringed Plover. Wings relatively long, showing no wing-bar in flight. Immature scaly but less so than Ringed Plover. **Similar species** Ringed Plover is plumper, larger, relatively shorter-legged and shows a conspicuous wing-bar in flight. The flight calls of the two are diagnostic.

HABITS Favours drier sandy ground and is partial to sports fields, although also frequents wet, muddy places. Usually in small loose groups.

VOICE Distinctive piping *tee-u* or *pee-oo* with downward inflection, lacking the liquid quality of Ringed Plover.

STATUS AND DISTRIBUTION Long-distance Palaearctic migrant to Senegambia. Frequent and locally common in WD and URD October-March, with most observations in December-January; uncommon in all other divisions. Widespread throughout Senegal but largely absent from arid central area.

RINGED PLOVER
Charadrius hiaticula   Plate 17

Fr. Grand Gravelot

Other names: Common Ringed Plover, Greater Ringed Plover

IDENTIFICATION 18cm. A small dumpy plover, in breeding plumage sandy-brown above, white below, with a striking facial pattern and bold black breast-band. Non-breeding birds (the norm in Senegambia) are duller with muted markings. Whitish forehead separated from the brown crown by a darker band; another dark brown band extends from the base of the bill broadening through and below the eye and extending to the back of the crown, enclosing a whitish supercilium above and behind the eye. Blackish-brown breast-band weakest centrally, and not always complete, extends to the lower nape. Complete white collar. The stubby bill is orange with a black tip in breeding plumage but brownish in winter with some yellow on the base of the lower mandible. Legs orange in the breeding season, drab ochreous in winter. Immature duller with scaly upperparts and browner, frequently incomplete, breast-band. Relatively long-winged with a conspicuous white wing-bar obvious in fast, low flight. The dark-tipped tail is edged white. **Similar species** See Little Ringed Plover.

HABITS Favours shoreline and open sandy or muddy estuaries, sometimes singly or in small groups, occasionally in hundreds. Also occurs inland on playing fields.

VOICE A fluty *tee-lee* or *tooeet* with upward inflection.

STATUS AND DISTRIBUTION Long-distance Palaearctic migrant. Seasonally common to abundant in coastal WD and NBD, recorded in all months but peaking October-March. Fewer records in other divisions and no recent sightings in URD. In Senegal occurs coastally, along the Senegal River and in the south east.

KITTLITZ'S PLOVER
Charadrius pecuarius   Plate 17

Fr. Pluvier de Kittlitz

Other name: Kittlitz's Sand Plover

IDENTIFICATION 13cm. Tiny fast-running plover, dark above and sandy below, lacking a breast-band, with a tallish leggy appearance for its small body size. In breeding plumage has white forecrown separated from brown crown by a black band. A dark loral stripe and eyestripe continues back to the nape to form a collar as does the adjacent contrasting white supercilium. Upperparts dark brown with paler fringes; whitish underparts strongly washed warm yellowish-buff. In immatures and non-breeding birds the plumage becomes faded and the head markings may be much duller. However the body retains its rufous-brown tones, but is more dappled on the back. Bill and legs dark. In flight shows dark leading edge to wings, a narrow white wing-bar and white sides to the dark tail. The toes extend well beyond the tail-tip which is unusual in this group of waders. **Similar species** Tawny appearance, more dappled back, long legs and characteristic supercilium should differentiate from Kentish Plover and White-fronted Plover.

HABITS Not shy. Varied habitats, often distant from water, including playing fields. May be seen on intertidal mud and flooded salt pans. Occurs singly or in small parties.

VOICE *pipip* or *kittip* when flushed, piping *tip-peep tip-peep* in flight. Not very vocal.

STATUS AND DISTRIBUTION Seasonally frequent in coastal WD and NBD, October-April. Uncommon in all other divisions. In Senegal occurs along the whole coast and in the Senegal River delta.

BREEDING Breeds northern Senegal, February-July; southern Senegal in March.

FORBES'S PLOVER
Charadrius forbesi   Plate 17

Fr. Pluvier de Forbes

Other name: Forbes's Banded Plover

IDENTIFICATION 20cm. Smallish, long-tailed plover, identified by its well-demarcated double dark brown breast-bands, separated by white, and by the white stripe extending as a continuous band from the back of the red eye-ring around the nape. Crown dark with no white on forehead. Upperparts dark uniform brown with greyish sides to face and throat; underparts white. Shortish slim red bill tipped black, legs yellowish-brown or flesh. In non-breeding plumage paler with less well-defined markings. Immature has buffy fringes on upperparts. Similar species Double breast-bands and broad supercilium behind eye are unique.

HABITS Prefers drier terrain with short grass, fallow and burnt ground; could occur on bare laterite outcrops. Bobs when anxious and zig-zags on taking flight.

VOICE A repeated and plaintive *pee-ooh*, and *pleuw-pleuw* in aerial display.

STATUS AND DISTRIBUTION Rare throughout Senegambia. A record in CRD in November and an unconfirmed recent record in coastal WD in January. Recorded in eastern Senegal along the Gambia River and in coastal Casamance in February.

KENTISH PLOVER
Charadrius alexandrinus   Plate 17

Fr. Gravelot à collier interrompu

Other name: Snowy Plover

IDENTIFICATION 16 cm. Small, pale, dumpy plover with a slim black bill, usually with dark grey or blackish legs. Upperparts grey-brown, underparts white, wing-tips extending to end of tail at rest. Breeding male forehead white, with narrow white supercilium extending above black line from bill to collar; black patch in front of rufous crown; blackish breast patches. Female duller, lacking

black on crown, partial breast-bars brown. Immature resembles female but paler with scaly feather edges. In flight shows narrow white wing-bar, tail centre darker towards tip, white tail edges. Similar species Can be confused with variably-plumaged White-fronted Plover, but Kentish is less active. White-fronted is heavier-billed and short-winged, generally paler in appearance and usually less clearly marked. In Senegambia separation of these two closely-related birds is further complicated by the lack of knowledge of Kentish Plover breeding and migration strategies. Ringed and Little Ringed Plovers have darker sandy backs, and in flight show broad wing-bar (Ringed) or no wing-bar (Little Ringed). Both have complete breast bands.

HABITS Favours sandy areas along the coast, sometimes inland. Usually singles or small loosely-associated groups. In The Gambia not seen to pursue prey with same vigour as White-fronted but does chase. Feeds with other waders of similar size by picking, sometimes after a short run.

VOICE An emphatic but quiet *fwhit* or disyllabic *whit-whit* given in flight or on ground.

STATUS AND DISTRIBUTION Palearctic migrant and proven breeder in Senegal; origins of Gambian birds uncertain. Frequent to locally common along the whole coast, recorded in all seasons with peak in early dry season, much reduced May-September. Occasionally singles in LRD and CRD. Along entire coast of Senegal and in delta of Senegal River.

BREEDING In northern Senegal, chicks in May and July; in Casamance, eggs in March. No Gambian information.

WHITE-FRONTED PLOVER
Charadrius marginatus   Plate 17

Fr. Gravelot à front blanc

Other name: White-fronted Sand Plover

IDENTIFICATION 16 cm. A very pale plover, sprinting far and fast when hunting. Shows conspicuous white patch on bulky, broad-skulled head. More robust black bill than Kentish. Upperparts uniform dull rusty-brown; complete hindneck collar buffy-white most visible when neck stretched, underparts all-white or creamy-white, with partial buffy chest bands sometimes completely absent. Wing-tips do not reach tip of tail on standing bird. Shows narrow white wing-bar in flight, tail dark along full length of centre, outer feathers white. Legs reddish-buff (adult) to greenish-olive (immature). Breeding male forehead broad and white, with black-brown bar before rufous-brown cap, dark eye-patches across lores to bill; female's head-bar browner. Immature lacks black on head, dark eye and loral patches present, underparts creamier, upperparts showing pale scalloped feather edges. **Similar species** Kentish Plover is thinner-billed, longer-winged, and not so fast running, generally with darker legs. Kittlitz's Plover is buffer-breasted, relatively long-legged (protrude in flight) and has different head and wing markings. Immature Ringed Plover is always more robust in build, darker-backed with darker breast patches.

HABITS Usually seen in ones or twos occupying a wide-open sandy area, normally distant from groups of migrant waders. Could occur in higher numbers. Not normally on the tideline, favouring tidal flats and raised spits with sparse vegetation. Most active and prominent at low water on recently exposed grey sand where it forages, often amidst fresh fiddler-crab casts. Alert-looking, crouching to scan for sandflies and small crustaceans which it pursues with a characteristically fast, low, swerving scuttle. May suddenly stop and freeze. Also feeds in short jerky movements.

VOICE Not vocal; a variety of soft notes *twit* or *tiewi* in greeting, chatters when excited. More sound-recording work required to compare with Kentish.

STATUS AND DISTRIBUTION Local and resident breeder along whole coast at sites where beach and estuarine habitats occur e.g. Tanji Bird Reserve, Old Cape Roads, Kartong in WD, and Jinnack in NBD. Could also occur on saltpans inland e.g. Bao-bolon in NBD and Soma in LRD. Recorded in all months. In Senegal recorded from coastal districts just north and south of The Gambia.

BREEDING Breeding records in May, June and November. Buries eggs in sand when away from nest; carries water to chicks in belly feathers.

GREATER SAND PLOVER
Charadrius leschenaultii   **Plate 17**

Fr. Pluvier de désert ou Pluvier de Leschenault

Other names: Great Sand Plover, Sand Plover

IDENTIFICATION 22cm. Considerably larger and leggier than Ringed Plover, characterised by a relatively long, strong bulbous-tipped black bill. Grey-brown above, with smudgy patches at sides of breast, which sometimes join to form a complete band. Lacks white collar, small white forehead patch and whitish supercilium. White wing-bar and white sides to dark rump. Legs olive-yellow to olive-grey. Unlikely to be seen in breeding plumage in West Africa (chestnut breast-band, black facial mask with white frontal patch).

HABITS A bird of the tideline and muddy mangrove fringes.

VOICE A short *treep* in flight.

STATUS AND DISTRIBUTION No confirmed records in The Gambia, but quite likely to occur, with possible records in January 1991 and two birds in January 1996 (Peter Colston pers. comm.) both from coastal WD. Vagrant to northern Senegal in November 1986. Has occurred as vagrant in several other countries in West Africa.

EURASIAN DOTTEREL
Charadrius morinellus   **Not illustrated**

Fr. Pluvier guignard

Other name: Dotterel

IDENTIFICATION 21cm. Confiding, plump plover. Long pale supercilium joins on the nape to form a distinctive V, offsetting the dark cap. In breeding plumage back and wings brown, with golden-buff feather margins, neck grey, crown blackish, with white throat patch and supercilium. Chestnut breast bordered above by narrowly dark-edged white band, below by broader black patch. Undertail white. Looks dark in flight, lacking wing-bar or rump markings, but with white-tipped tail and white shaft to outermost primary. Female brighter than male. Non-breeding birds brown above, with pale chestnut feather margins, darker crown and striking buff supercilium. Underparts grey-buff, pale beneath tail, with a distinctive narrow pale breast band. **Similar species** Distinguished from Golden Plover by smaller size, pale legs, lack of wing-bar, and black tail in flight. Temminck's Courser is not dissimilar, but has a curved bill and no pale breast-band.

HABITS Only Gambian record from tidal flats, but more likely to be expected on open dry habitat inland.

VOICE On breeding grounds a trilled *wit-a-wee*.

STATUS AND DISTRIBUTION Palearctic vagrant to The Gambia; a single record in March in coastal WD. Not recorded in Senegal but has occurred in Mauritania.

AMERICAN GOLDEN PLOVER
Pluvialis dominica   **Not illustrated**

Fr. Pluvier doré américain

IDENTIFICATION 26cm. Smaller and slimmer

than European Golden Plover, with longer legs and long primaries extending beyond the tail-tip at rest. In non-breeding plumage distinguished by greyer plumage and more distinct head pattern. At all times, greyish axillaries and underwing-coverts are diagnostic.

STATUS AND DISTRIBUTION Vagrant from North America. Has been recorded from coastal WD in January 1984 and November 1997; also once from northern Senegal in May 1979.

EUROPEAN GOLDEN PLOVER
Pluvialis apricaria   Not illustrated

Fr. Pluvier doré

Other name: Greater Golden Plover, Golden Plover

IDENTIFICATION 28cm. A plump wader with a smallish head and short bill. Non-breeding plumage copiously spangled yellowish-brown above; buff on breast with faint streaking, shading to white on belly. Whitish supercilium and chin. In flight shows pale wing-bar, dark rump and white axillaries. In breeding plumage, upperparts spangled bright gold, underparts and face black bordered by broad white band extending from forehead, down side of neck and along flanks. **Similar species** See American Golden Plover.

HABITS Coastal mudflats, tidal mangrove streams and drier habitats.

VOICE A distinctive plaintive *tlui* or *too-lee* flight call.

STATUS AND DISTRIBUTION Rare long distance migrant from the Palearctic to coastal Senegambia, with scattered records October to May.

GREY PLOVER
Pluvialis squatarola   Plate 17

Fr. Pluvier argenté

Other name: Black-bellied Plover

IDENTIFICATION 30cm. A large, bulky plover, with grey-brown spangled upperparts in non-breeding plumage, with breast and upper belly white, streaked grey-brown. Pale supercilium accentuated by sooty mark on ear-coverts. Large dark eye, and short stout black bill. In flight, white wing-bar and rump are conspicuous and black axillaries diagnostic. Unmistakable in breeding plumage, with white-bordered jet black underparts contrasting with silver-spangled back and wings. Immature similar

to winter adult. **Similar species** Golden Plover is smaller, browner and spangled yellowish-brown above, lacks the white rump and has white axillaries. American Golden Plover can be greyish but is smaller still, slimmer and relatively longer-legged, with a longer neck and smaller head, and grey axillaries.

HABITS Coastal, especially along the shoreline and on tidal mud. Walks slowly while feeding.

VOICE A clear far-carrying tri-syllabic *tluui* or *tee-oo-ee*, usually in flight.

STATUS AND DISTRIBUTION Long-distance Palearctic migrant. Seasonally common to abundant along the whole coast but common elsewhere. Numbers greatest September-March but some present throughout the year. In Senegal also restricted to the coast.

WATTLED PLOVER
Vanellus senegallus   Plate 17

Fr. Vanneau du Sénégal

Other names: Senegal Wattled Plover, African Wattled Lapwing

IDENTIFICATION 32cm. A tall robust brownish plover with long yellow legs and prominent white forehead spot. Conspicuous red-based yellow hanging wattles at base of bill. Upperparts brown, throat and chin black, neck and sides of face pale brown, streaked and flecked black-brown. Breast brown, paling to buff on belly. Bill yellow with black tip, eye whitish. In flight white patch on inner wing contrasts with black primaries and grey-brown lesser coverts. Immature has insignificant wattles, reduced white on forehead, and ill-defined black throat. **Similar species** Under poor or distant viewing conditions could be confused with Senegal Thick-knee with which it often associates. At night when size difficult to judge, confusion is also possible with coursers.

HABITS Favours a variety of fresh and saline aquatic habitats, particularly mangrove fringes with adjacent salt pans, deserted coastal beaches, dunes, and the tideline. Also frequents grassy cultivated land close to water. In the dry season mixes freely with Palaearctic waders at coastal lagoons and on inland freshwater habitats. Usually in pairs or small parties of up to ten birds, occasionally in larger flocks.

VOICE Alarm call repeated *ke-weep*; other calls based on a rising, shrill and screaming *peep-peep-peep*.

STATUS AND DISTRIBUTION Common to locally

189

abundant breeding resident throughout The Gambia. Population greatest in WD and URD, decreasing in the latter towards the end of the dry season. Absent from arid areas of Senegal but widespread in suitable habitat.

BREEDING Extended season, but generally before or early in the rains.

WHITE-CROWNED PLOVER
Vanellus albiceps   Plate 17

Fr. Vanneau à tête blanche

Other names: White-headed Plover, White-headed Lapwing

IDENTIFICATION 30cm. A boldly-marked river bird with long yellow beard-like pointed wattles. Crown patch, chin and underparts snow white. Sides of neck, head and sides of upper breast unmarked ash-grey; back earth-brown. Closed wing predominantly black with conspicuous white on coverts. In flight shows vast expanse of white on black wing, becoming conspicuously pied. Bill yellow, tipped black; eye yellow and legs yellowgreen. Prominent carpal spurs. Immature has buffy fringes, smaller wattles and is less strikingly white on the head. **Similar species** Facial features may initially recall Wattled Plover but White-crowned has pure white underparts and much larger wattles; in flight the extensive white in the wing is diagnostic.

HABITS Associated with large, rocky riverine systems sharing a number of ecological and behavioural characteristics with the Egyptian Plover with which it shares this habitat in Senegal. These similarities include wetting feathers to transport water to the chicks, flaring and presenting the expansive white area on the wings to deter mammalian intruders, and movement downstream during the rains. Remarkably aggressive when breeding, even the Spur-winged Plover being subordinate to it.

VOICE Vocal. A loud, fast and repeated *peep-peep-peep*.

STATUS AND DISTRIBUTION Rare but regular wet season visitor to river banks and marshes in CRD and URD from the Senegalese reaches of the Gambia River. Up to 90 have been recorded. Occasionally seen at other times. Occasional in northern Senegal but more frequent in east on the upper reaches of Gambia River.

BREEDING On shingle beaches, islands and spits. Nest with eggs in February, in eastern Senegal.

BLACK-HEADED PLOVER
Vanellus tectus   Plate 17

Fr. Vanneau à coiffe noire

Other name: Black-headed Lapwing

IDENTIFICATION 28cm. Sandy, earth-brown upperparts. Head distinctive, black but with white forehead, chin, and broad post-ocular stripe to back of neck. Smallish red wattles in front of eye and short, horizontal pointed black crest at back of head. Prominent staring golden-yellow eye. Black median line extends down front of neck to breast onto all-white underparts. Legs red, bill red, tipped black. Flight slow and usually low, showing white wing-bar contrasting with the broad black trailing edge, earth-brown back and secondary-coverts. Immature resembles adult but with buffy fringes on dark parts, shorter crest and smaller wattles. **Similar species** When seen in headlights at night could be confused with much rarer Bronze-winged Courser.

HABITS A plover of arid ground, not frequenting aquatic margins, although the dry sandy or short tussocky grass that it favours is often near to water or the coast. Particularly active at night, often standing in small groups in the middle of roads or tracks. During the day usually shelters in the shade.

VOICE Not as noisy as other plovers. A *kwairr* and a sharper *kiarr* when attacking intruders, and, when flushed, a high-pitched *kir*.

STATUS AND DISTRIBUTION Common breeding resident throughout, most frequent in URD, peaking in the early dry season. Numbers fall in the rains when some may move north out of The Gambia. Widespread throughout Senegal.

BREEDING Generally at the approach of the rains, in May-June.

SPUR-WINGED PLOVER
Vanellus spinosus   Plate 17

Fr. Vanneau éperonné

Other name: Spur-winged Lapwing

IDENTIFICATION 28cm. Familiar plover, distinctively plumaged in black, white and grey-brown. Black cap separated by sharply defined large white cheek patches and collar. Short, drooping crest formed by elongated, smooth, tapering black nape feathers. A black stripe extends from below the bill, broadening into a glossy black lower chest and onto the flanks; lower belly and undertail-coverts white. Upperparts greyish-brown with distinctive lanceolate feathering overlying closed wing. Bill and legs black, eye deep crimson, wattles absent. Immature has buffy fringes to duller black plumage. **Similar species** Black-headed Plover has a crest, small red wattles, yellow eye and a different pattern of black and white.

HABITS Widespread. Favours aquatic habitats, rarely far from the water's edge. Frequently seen in groups on the shoreline, or by busy coastal roads, walking indifferently out of the way of passing traffic. Sizeable flocks are not uncommon especially just after the rains. Aggressively defends its territory against all comers, even birds the size of a Great White Egret.

VOICE Alert and noisy: a screeched four-note *did-he-DO-it* delivered when adopting on-guard, head back, upright stance with wings arched in threat posture. Alarm call is a loud penetrating metallic *tick-tick-tick*; also a repeated *kik*.

STATUS AND DISTRIBUTION Common to abundant breeding resident throughout all divisions, showing local movements with flocks being most frequent in the early dry season when it becomes more numerous especially in WD and URD. Largely absent in the drier central parts of Senegal but otherwise widespread in all suitable habitats.

BREEDING Extended season; with nest losses and chick predation high, replacement is routine.

SENEGAL PLOVER
Vanellus lugubris   Plate 17

Fr. Vanneau demi-deuil

Other name: Lesser Black-winged Plover

IDENTIFICATION 22cm. Small and plain, lacking wattles and crest. Grey-brown head and neck, paler on lores, with white forehead and chin. Back brown. Grey-brown breast separated from white belly by narrow darker band. Eye pale yellow, bill black, reddish at base, legs brown. In flight upper and underwing show completely white secondaries, contrasting with purplish-black primaries. Tail white with incomplete narrow black subterminal bar, the outer feathers being white. Immature is fringed and

mottled buff particularly on the upperparts; forehead less white.

HABITS Prefers dry fallow cropped land and recently-burnt grass, and dry wooded savanna near water. Likely to be seen in the presence of cattle as it normally associates with plains game in East and South Africa. Confiding and approachable; runs in short, quick bursts. Nomadic.

VOICE Piping *thi-wit* and di-syllabic *teeyoo, tee-yoo*; calls at night.

STATUS AND DISTRIBUTION First Gambian record at Tanji Bird Reserve, December 1997–March 1998; known from a few records in northern and southern Senegal near the coast (close to The Gambian borders) where it has bred.

BREEDING Sole record from Saloum in northern Senegal in February 1980.

NORTHERN LAPWING
Vanellus vanellus   Plate 17

Fr. Vanneau huppé

Other names: Lapwing, Green Plover

IDENTIFICATION 29cm. Iridescent green-and-black-backed plover with a long wispy upswept crest, white belly and black breast-band. Undertail-coverts cinnamon. Legs brownish pink. Non-breeding birds have back and wing coverts scalloped buff, throat white, breast-band mottled, and a shorter crest. In flight shows broad, rounded black and white wings, black breast contrasting with white belly and cinnamon undertail. Tail white with broad black distal band. **Similar species** Black-headed and Spur-winged Plovers have pale brown or grey-brown backs, the latter having a black breast and upper belly.

VOICE Flight call *cheew-ep* or *pee-witt*.

STATUS AND DISTRIBUTION Rare vagrant from North Africa or Palearctic. Four modern records, all in coastal WD. In northern Senegal rare but regular on the delta of Senegal River around January, which corresponds with the few Gambian records. Could occur anywhere.

SANDPIPERS, SNIPES, GODWITS, CURLEWS, PHALAROPES AND ALLIES
Scolopacidae

Twenty nine species. A diverse group of shorebirds, typically with long legs and slender bills. Sexes are generally similar but most have distinct breeding and non-breeding plumages as well as identifiable juvenile plumages. All species in Senegambia are migrants, breeding at high latitudes in the northern hemisphere and wintering or passing through our region. Small numbers of some species remain throughout the year. Flight patterns and calls are important fieldmarks. The family can be subdivided into following groups.

Smaller sandpipers (*Calidris*)

Eight species, including the two smaller stints. Sexes are similar with seasonal variation; some species are difficult to identify.

Snipes (*Lymnocryptes, Gallinago*)

Three species. Plump, cryptically-patterned waders with long, straight bills and shortish legs. Most active at dawn and dusk, in freshwater habitats, and flushing only when closely approached.

Godwits (*Limosa*)

Two species. Tall elegant waders, with long necks, long legs and long bills. Bills are straight or slightly upturned, seasonal plumages differ and females are larger than males.

Curlews (*Numenius*)

Two species. Large, long-legged waders with long, strongly decurved bills. Mottled brown plumage does not show seasonal variation.

Larger sandpipers (*Tringa, Xenus, Actitis*)

Nine species. The seven species of shanks *Tringa* are medium-sized, elegant waders with long legs and long bills. Sexes generally similar in plumage but show some seasonal variation, which is dramatic in some species. *Xenus* and *Actits*, both represented by a single species, are similar but shorter-legged.

Phalaropes (*Phalaropus*)

Two species. Highly aquatic, graceful small waders with lobed toes, often confiding. Typically sit high on the water, spinning around picking planktonic food with delicate, rapid jabs. Females are larger and, in breeding plumage, brighter than males.

Three unique species in monotypic genera cannot be conveniently grouped into the above categories: Buff-breasted Sandpiper (*Tryngites*), Ruff (*Philomachus*) and Ruddy Turnstone (*Arenaria*). In addition, the monotypic Broad-billed Sandpiper (*Limicola falcinellus*) may occur. A vagrant dowitcher *Limnodromus* sp. from North America was recorded in The Gambia in December 1978 in WD, but was not specifically identified.

RED KNOT
Calidris canutus   Plate 19

Fr. Bécasseau maubèche

Other names: Knot; Lesser Knot

IDENTIFICATION 25cm. A largish dumpy wader with a stout, straight or very slightly decurved, medium-short, blackish bill. Rather featureless in non-breeding plumage: grey above with some white fringing; underparts whitish, suffused grey on breast and flanks. Obscure whitish supercilium and cheeks peppered darker grey. Immature has a browner grey back and appears scaly, and a buffish wash on finely streaked breast. In breeding plumage, feathering on upperparts shows black-brown centres edged rufous-buff; underparts distinctive coppery-red. Intermediate plumages showing varying amounts of rufous, seen in April-May and early August. In flight shows narrow but bold white wing-bar and blackish primaries. Whitish rump barred grey, looks pale in the field. **Similar species** Curlew Sandpiper shares the rich red on head, breast and belly in breeding plumage but is smaller, slimmer and has a longer decurved bill. In all plumages it shows a clear white rump. Distinguished from even

smaller Dunlin in non-breeding plumage by the relatively shorter thicker bill, greenish legs and lack of dark centre to rump and tail in flight.

HABITS Favours estuaries and tidal inlets with nearby sandbars. Occurs along the Senegambian coast, generally encountered singly or in small parties, never in the dense flocks seen in other parts of its winter range. Feeds slowly and methodically, probing exposed mud.

VOICE Mostly silent in Africa; an occasional grunting *knutt*.

STATUS AND DISTRIBUTION Palearctic breeder; long-distance migrant to Senegambia. Frequent along coastal WD. Recorded along the whole northern and southern Senegalese coast and in small numbers in the delta of the Senegal River.

SANDERLING
Calidris alba   Plate 19

Fr. Bécasseau sanderling

IDENTIFICATION 20cm. A small, highly active, whitish-grey wader with a blackish shoulder patch and short, straight, black bill. In non-breeding plumage upperparts pale grey with white markings, apart from the dark shoulder patch. Has whitish supercilium and dark smudge through eye. Feet and legs black; no hind toe. Immature has darker back with blackish and white chequered spotting. Breeding plumage rarely seen in Senegambia: head, back and breast rufous with darker mottling. In flight shows prominent broad, white wing-bar and blackish leading edge to wing and white-sided, darker-centred rump. **Similar species** Stout bill, pale plumage and larger size differentiate it from stints. Dunlin is darker and browner, with a longer slightly decurved bill and a more hunched feeding posture. Non-breeding Grey Phalarope is another possible confusion species, but should be separable on habits and habitat.

HABITS A bird of the tideline, actively running and following the water's edge when feeding. Tame, flying a short distance when disturbed. Congregates in very large numbers on northbound migration.

VOICE A *twick-twick* or *kip-kip* in flight but not vocal in Africa.

STATUS AND DISTRIBUTION Palearctic breeder; long-distance migrant to Senegambia. Common to abundant on passage along the whole coast October-February with records in all months, though scarce May-September. Very occasional records east to CRD.

LITTLE STINT
Calidris minuta   Plate 19

Fr. Bécasseau minute

IDENTIFICATION 13.5cm. A tiny, compact wader with medium-length black or greyish-black legs. In Africa, usually in non-breeding plumage, with a mousey-grey back and pale grey-brown wing-coverts, fringed paler and whitish on inner coverts. Underparts white with faint grey shading on sides of breast sometimes forming an ill-defined pectoral band. Whitish supercilium. Relatively long primaries, the closed wing extending just beyond the tip of the tail. Bill short, stout but fine-tipped. Immature has pale grey-brown nape, scalloped brown upperparts with whitish fringes on mantle creating a clear V mark on the back; warm buff breast shading to white belly. Occasionally seen in breeding plumage when upperparts mottled chestnut-rufous; underparts and throat white separated by an indistinct reddish and brown speckled band on the upper breast. In flight shows narrow white wing-bar, and characteristically centre of rump dark; outer tail feathers grey. **Similar species** Paler-legged

Temminck's Stint has a dull, comparatively featureless head, dark breast, dark tail centre with white sides and a distinctive call. The short straight bill eliminates the larger Dunlin. At a distance Sanderling is a possible confusion species, but habitat, overall paleness, bold wing-bar and smudgy eye-patch prevent confusion.

HABITS Favours open flooded marshy areas, both freshwater and saline, and estuarine mud. Picks food from the surface of the mud with a brisk, jerky jabbing action; probes occasionally. Normally gregarious, flocks being very fast and manoevrable on the wing.

VOICE A short repeated *chik* or twittering *see-seet* in flight.

STATUS AND DISTRIBUTION Palearctic breeder; long-distance migrant to Senegambia. Common, sometimes abundant, in all divisions particularly in WD, NBD and URD. Recorded throughout the year with an October-February peak. Distributed throughout Senegal both on the coast and inland.

TEMMINCK'S STINT
Calidris temminckii   Plate 19

Fr. Bécasseau de Temminck

IDENTIFICATION 14cm. In Africa usually in non-breeding plumage: a small stint with a featureless grey-brown head concolorous with breast and upperparts, distinctive by its dullness. Rest of underparts whitish. Looks short-legged with tapering body shape due to its rather long tail which extends beyond the wing tips. (cf. Little Stint). When clean, legs yellow or yellowish, clearly paler than Little Stint. When flushed may 'tower' like Snipe. In flight the long, dark centred, white-sided tail is diagnostic. Feeds with quick stabs and a plodding gait, sometimes teetering and recalling a small Common Sandpiper. **Similar species** See Little Stint.

HABITS Likely to be seen singly or very small numbers along the fringes of open saltpans or freshwater marshes inland.

VOICE A high-pitched ringing *treee* frequently repeated; trilling *tirriririr* on taking flight.

STATUS AND DISTRIBUTION Palearctic breeder; long-distance migrant to Senegambia. Rare to uncommon in coastal WD and LRD with all records October-May, but probably under-recorded. Frequent in coastal Senegal.

BAIRD'S SANDPIPER

Calidris bairdii   Not illustrated

Fr. Bécasseau de Baird

IDENTIFICATION 16cm. A small compact-looking wader with a long tapered outline due to the tips of the primaries extending well beyond the tail. This attenuated appearance together with the short black legs produces a low 'horizontal' profile. Predominantly buffish in all plumages. Markedly scaly back recalls immature Curlew Sandpiper. Pale supercilium and white supraloral spot are useful fieldmarks. Wings look long in flight and show a short narrow wing-bar.

HABITS Eclectic in choice of habitats and could be encountered on the coast, on saltpans or on short grassy fields.

VOICE Calls *kreep* or *p-r-r-reet*.

STATUS AND DISTRIBUTION A single record of this Nearctic vagrant in November 1976 in coastal WD. No Senegalese records.

PECTORAL SANDPIPER

Calidris melanotos   Not illustrated

Fr. Bécasseau tacheté

IDENTIFCATION 21cm. Appreciably larger than a Dunlin with a longer neck and a short, slightly decurved bill. Distinguished in all plumages by heavily streaked neck and breast, sharply demarcated from white belly, and narrow faint wing-bar. Base of bill and legs greenish-yellow.

HABITS Muddy shores and marshes with grassy edges.

VOICE Call a loud *krrik-krrik*, often in flight.

STATUS AND DISTRIBUTION Rare but regular Nearctic vagrant to Europe and southern Africa. A single recent sight record from northern Senegal is the only record for Senegambia.

CURLEW SANDPIPER

Calidris ferruginea   Plate 19

Fr. Bécasseau cocorli

IDENTIFICATION 20cm. An elegant wader, taller, larger and longer-legged with a more slender and attenuated appearance than Dunlin, especially when alert, and with an evenly decurved longer bill. Usually seen in non-breeding plumage, when ashy-grey above and white below with upper breast sparsely streaked grey-brown. Long white supercilium contrasts with the dark loral line. Immatures on arrival in September-October have a pinkish-buff wash on the unstreaked upper breast and are attractively scalloped above. In breeding plumage the underparts are warm chestnut-red, the upperparts dark brown, edged chestnut and tipped white. In flight shows a broad all-white rump, a white wing bar, and the tips of the toes project beyond the tail. **Similar species** Dunlin is slightly smaller, with a shorter, less decurved bill and weaker supercilium. In flight, Dunlin lacks a white rump.

HABITS Favours tidal estuarine mud, usually foraging in large flocks, sometimes numbering thousands. Along with Ruff (with which it frequently associates) it is the most abundant medium-sized wader on open wetland habitats..

VOICE A soft rippling *chirrup*.

STATUS AND DISTRIBUTION Palearctic breeder; long-distance migrant to Senegambia. Common to abundant in WD and LRD but with no recent records further inland. Recorded in every month peaking November-February. In Senegal distributed along the whole coast and in the Senegal River delta.

DUNLIN

Calidris alpina   Plate 19

Fr. Bécasseau variable

IDENTIFICATION 18cm. A smallish wader with a round-shouldered dumpy posture, shortish neck and medium-length black legs. Between Little Stint and Curlew Sandpiper in size. In non-breeding plumage upperparts greyish-brown, wing-coverts greyer with white fringing; underparts white with streaky grey-brown breast. Indistinct head pattern a useful field mark with weak whitish supercilium, chin and throat. Bill black, downcurved near tip and about same length as head but variable. Immature shows dark spotting on belly with upperparts scalloped buff, often admixed with grey. In breeding plumage upperparts black, broadly scalloped with chestnut; underparts whitish with black patch in centre of belly. Transitional plumages frequently occur. Various races may differ substantially in measurements and plumage. In flight shows a clear white wing-bar; centre of rump dark with white sides to rump and grey sides to tail. **Similar species** See Curlew Sandpiper and Red Knot. The stints are smaller, shorter-billed and can be separated on tail pattern. Broad-billed Sandpiper *Limicola falcinellus* (Fr. Bécasseau falcinelle) has been recorded inland as a vagrant elsewhere in West Africa and could occur in Senegambia: although similarly plumaged in winter to Dunlin, Broad-billed is smaller; has a distinctive double supercilium.

HABITS Tidal, muddy coastal lagoons with drier fringes are the most suitable sites for small feeding groups, often with other Palearctic waders. Occasionally found on inland expanses of fresh water during the rains. Feeds industriously by picking and probing with a bill held vertically downward. Largest numbers recorded in October; flocks exceeding 30-40 are unusual. Noted in breeding plumage May-July.

VOICE Call a weak though far-carrying, drawn out *treeer* in flight.

STATUS AND DISTRIBUTION Palearctic breeder; long-distance migrant to Senegambia. Frequent to locally common in coastal WD and very occasionally east to CRD. Recorded every month except August. Occurs along the Senegal coast and in the Senegal River delta.

BUFF-BREASTED SANDPIPER
Tryngites subruficollis   **Not illustrated**

Fr. Bécasseau roussâtre

IDENTIFICATION 19cm. Rather plover-like gait but plump. Short black bill, dark eye with narrow white eye-ring all emphasised by pale face. Uniform yellowish-buff below with dark spotted sides to breast, some individuals more whitish; scalloped above, crown spotted. No conspicuous supercilium. Legs yellow. No white (or indistinct) stripe on upperwing, underwing whitish bordered dark with dark wrist. Flight erratic and fast. **Similar species** From larger female Ruff by spotted flanks (plain on Ruff), smaller bill, plain rump with no white on sides of tail and yellow legs.

HABITS Confiding. Prefers dry places, short grassy or sandy habitats. Feeds by picking. Runs with neck extended when alarmed.

VOICE Mostly silent when not breeding, could give a *tik* or a low *chwut* or *chu*.

STATUS AND DISTRIBUTION Vagrant from the Nearctic to West Africa. One record in northern Senegal near Dakar in April 1985; indeterminate in The Gambia with occasional unpublished claims from Bund Road, WD.

RUFF
Philomachus pugnax   **Plate 19**

Fr. Chevalier combattant

IDENTIFICATION Male 30cm, obviously smaller female (Reeve) 23cm. Bill shortish, slightly curved and often orange-based. A hunched-backed variegated brownish wader, with bulky body, small head and thickish neck. Non-breeding male grey-brown above fringed paler, longish legs orange-red but may be green-brown or yellowish, female similar (note size difference). Extreme and highly variable seasonal plumage change in males, infrequently seen in The Gambia. Individuals showing partial ruffs or white necks most likely in late dry months, but birds with upperparts patterned black can be frequent. May return into our area in this phase towards late wet season. Immatures scaly, warmer-toned above with orange-buff; breast golden-buff, whitish below; legs yellow-brown to greenish. Long wings, thin white wing bar (no white trailing edge like Common Redshank). Tail with central dark line and white patches each side (crescent-shaped) is useful in both sexes, at all ages. Feet extend just beyond tail. **Similar species** The two redshanks are the only other species of brown waders with reddish legs, but their legs are usually brighter red. Shape is also a useful distinction, with Ruff being noticeably small-headed and shorter-billed.

HABITS Highly gregarious, in open muddy habitats, rice paddies, ditches and channels. In crowded waterbird assemblages often the commonest wader together with Curlew Sandpiper. Forages slowly at water's edge, wades and swims. Raises hackles and stretches neck in frequent face to face foraging disputes. Other odd body posturing can give rise to problems of identification.

VOICE Mostly silent. When departing or alarmed a *ga-ga* or *gue-gue-gue*.

STATUS AND DISTRIBUTION Palearctic breeder and long-distant migrant to Senegambia. Seasonally abundant throughout. Recorded all months all divisions, with peak November-March; greatest numbers at the coast, much reduced in the wet months. Widespread also in Senegalese wetlands, with records up to one million birds in the north in March.

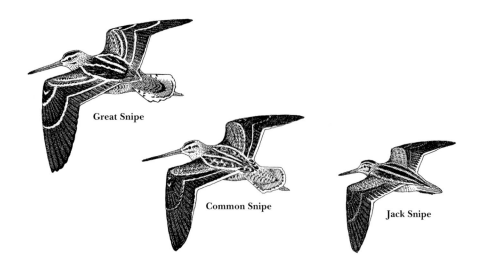

Great Snipe

Common Snipe

Jack Snipe

JACK SNIPE
Lymnocryptes minimus   **Plate 19**

Fr. Bécassine sourde

IDENTIFICATION 17.5cm. A small snipe with medium-length straight brown bill (yellower at base) only slightly longer than head. Cryptically mottled brown and tawny above, showing four contrasting creamy stripes, with a whitish belly. Streaked on neck, breast and flanks. Lacks the pale crown stripe present in other snipes but has well defined double pale supercilium enhanced by the dark cap, land oral and eye stripes. In flight the short, dark tail shows no white, and the white trailing edge to the secondaries is not conspicuous as in Common Snipe. Legs pale greenish. Escape flight is slow and straight, soon dropping back into cover. **Similar species** Common Snipe is much larger with a relatively long bill, single supercilium, white in tail and a conspicuous white trailing edge to the secondaries showing in zig-zag flight.

HABITS Solitary and furtive, skulking in dense freshwater vegetation making detection difficult. Allows very close approach before being flushed.

VOICE Normally silent in winter quarters; a scarcely audible *gah* when flushed.

STATUS AND DISTRIBUTION Long-distance migrant to Senegambia. Rare, recorded in WD and LRD from November to February. More recently in January 1996 in NBD on a drying-up paddyfield. Clandestine habits may be the cause of the dearth of records. Similarly, few records from northern and southern Senegal.

COMMON SNIPE
Gallinago gallinago   **Plate 19**

Fr. Bécassine des marais

Other names: Snipe, Fantail Snipe

IDENTIFICATION 25cm. Cryptic shades of mottled brown and tawny with a whitish belly. Four conspicuous broad longitudinal buff stripes on back. Breast buff with darker streaks, dark barring on flanks. Rounded head, with sloping forehead. Buffy-cream stripe through dark crown, buff supercilium, dark loral stripe broadening at base of bill. Carries bill and head downwards when flying. In flight note clear white trailing edge on secondaries and restricted white on pointed outer tail feathers and along tips. Legs greenish. When flushed flies fast and low in zig-zagging dash then towers to great height before settling at a considerable distance. **Similar species** See Great and Jack Snipes. Flight when flushed is a useful characteristic aiding identification, as is the amount of white in the spread tail – lacking in Jack Snipe, clearly visible in Common Snipe and conspicuous in Great Snipe. Common Snipe has a white trailing edge to the inner wing which is lacking in Great Snipe, the later having a white-edged dark mid-wing panel.

HABITS Found in varied aquatic habitats but with a preference for freshwater swamps and wet fields; also found on coastal mud and saltmarsh. Never far from cover. Usually in twos or threes, sometimes small flocks. Feeds in shallow water with slow methodical probing, food frequently ingested without extracting the bill from the substrate.

VOICE Loud explosive *scaap* or *creech* on being

flushed is diagnostic.

STATUS AND DISTRIBUTION Uncommon to locally common migrant, perhaps under-recorded, with records October-March in all divisions with a peak December-January in URD. In Senegal occurs along the length of the Senegal River, on the coast and along the Gambia River in eastern Senegal.

GREAT SNIPE
Gallinago media   Plate 19

Fr. Bécassine double

IDENTIFICATION 28cm. Bulky, barrel-chested snipe with a long stout bill. Overall appearance dark, heavily and extensively barred on the white flanks and belly. Back not as conspicuously striped as Common Snipe. Prominent broad supercilium and median crown-stripe separated by dark, broad lateral coronal stripes. Long eye-stripe strongest on lores. White spotting on upperwing-coverts frames a dark mid-wing panel, formed by the greater and primary coverts. Outer tail feathers conspicuously white when spread. Legs pale greenish. Does not tower like Common Snipe when flushed but flies directly and slowly; tends to keep the bill horizontal. **Similar species** Common Snipe is slightly smaller and slimmer with whiter belly and barring on flanks; has a slighter and longer bill. Flight characteristics differ: Common Snipe has less white in the tail, Great Snipe lacks a white trailing edge to the wing.

HABITS Most likely to be found in marshy freshwater sedge, rice fields or recently flooded ground.

VOICE A deep hoarse croak, *etch, etch* uttered singly or doubly when flushed.

STATUS AND DISTRIBUTION Rare long-distance migrant. A few records in URD in January-February. Recorded near the coast in northern and southern Senegal in October, February and March.

BLACK-TAILED GODWIT
Limosa limosa   Plate 18

Fr. Barge à queue noire

IDENTIFICATION 43cm. Upperparts uniform grey-brown in the non-breeding plumage normally seen in West Africa. Breast and long neck grey, rest of underparts white or greyish-white. Long whitish supercilium brightest above lores but extending well behind eyes and often giving impression of turning upwards at rear. In breeding plumage head

and breast chestnut, flanks and belly white barred blackish; intermediate plumages regularly seen. Immatures have dull reddish wash on neck and breast. Bill long and straight, blackish with pink base. Legs dark grey. In flight, shows distinctive bold white wing-bar and a black terminal bar on the pure white tail. The legs extend well beyond the tail-tip.

HABITS A wader of tidal mud, coastal lagoons, and also inland rice paddies and swamps. Gregarious, often in flocks of hundreds and sometimes thousands, although occasionally seen singly. Walks gracefully and when feeding wades into deep water until legs completely submerged.

VOICE Flight call a *reeka-reeka-reeka* and softer *tek-tek-tek*.

STATUS AND DISTRIBUTION Palearctic breeder and long distance migrant to Senegambia. Locally very abundant along the estuary of the Gambia River in WD where numbers peak October-January although present in all months with temporary increase in July. Recorded throughout The Gambia. Tens of thousands winter in the Senegal River delta.

BAR-TAILED GODWIT
Limosa lapponica   Plate 18

Fr. Barge rousse

IDENTIFICATION 39cm. Rather dumpier, with shorter legs, neck and bill than Black-tailed Godwit. Dark-streaked brown upperparts typify the non-breeding plumage usually seen in West Africa. Neck and breast tinged pale brown with fine darker streaking, the rest of underparts are white. Whitish supercilium over darker smudgy lores and eye stripe. In breeding plumage entire head, neck and underparts deep chestnut-red with no barring in male; female much paler. Intermediate plumages are frequently seen. Bill long and slightly upcurved, blackish with pink base. Legs dark grey. Immatures are rather more strongly marked than winter adults, with dark centred, buff-edged feathers on upperparts, and neck tinged buff-brown. In flight, white triangular rump extends up the back and lack of a wing-bar is noticeable. The toes only just extend beyond the barred tail.

HABITS A wader of the sandy shoreline but regular also on mud and around rock pools. Often feeds alone although flies in sizeable flocks. Probes deeply while wading in shallow water, but also at the water's edge or on bare mud. Partial to tidal sandbars.

VOICE A seldom heard *kir-ik* or *wik-wik*.

STATUS AND DISTRIBUTION Palearctic breeder

and long distance migrant to Senegambia. Common along the whole coast and occasionally abundant along the Bund Road (Banjul). Highest numbers October-March but present in all months with temporary increase in July when flocks of up to 400 birds regularly occur. Recent records do not extend beyond LRD. Found along the whole of the Senegal coast.

WHIMBREL
Numenius phaeopus　　　　　**Plate 18**

Fr. Courlis corlieu

IDENTIFICATION 46cm. A largish brown wader with a long decurved bill and a distinctive striped crown. A median pale buff crown-stripe is bordered on each side by a broader dark stripe, the pale supercilium and dark eye stripe enhancing the striking head pattern. Upperparts entirely dark olive-brown flecked and edged whitish, underparts paler buff shading to white, with brown streaks. The dark bill is downcurved for the distal third and the relatively short legs are bluish-grey. Underwing white narrowly barred brown. **Similar species** Easily separable from Eurasian Curlew by smaller size, shorter legs and bill, and conspicuous head pattern. In flight Whimbrel looks more uniformly dark and compact and has faster wingbeats. Viewed from behind can be confused in flight with distant Bar-tailed Godwit.

HABITS Favours the tideline on mud in mangrove creeks where it probes for invertebrates, and also chases over sandbars after crabs. Less often seen on sand dunes and even on lawns picking for insects. May be solitary or in groups.

VOICE A characteristic tittering call, typically of about seven notes *pi pi pi pi pi pi pi* often heard shortly after taking flight.

STATUS AND DISTRIBUTION Palearctic breeder and long distance migrant to Senegambia. Common to locally abundant (August-April) along the coastline and along the river in LRD but recorded in every month. Infrequent in CRD and no recent records in URD. Found along the whole Senegal coast.

EURASIAN CURLEW
Numenius arquata　　　　　**Plate 18**

Fr. Courlis cendré

Other names: Common Curlew, Curlew

IDENTIFICATION 60cm with bill measuring up

to 16cm. Our largest wader, generally brown with a distinctively long downcurved bill. Above and below buff-streaked brown, whiter towards the belly and streaked blackish on the flanks. Head uniformly grey-brown with dense fine dark streaking. In flight shows extensive white rump extending as an inverted V up the back. Legs grey-green. Two races occur in West Africa: the smaller and darker western *N. a. arquata* with heavily barred underwing, and the eastern *N. a. orientalis* which is longer-billed, taller and paler, with whiter underwing, and more commonly encountered in Senegambia. **Similar species** Whimbrel has a characteristically striped crown, is smaller, darker and more compact.

HABITS Favours shoreline and sandbars. Normally solitary, walking slowly while feeding, sometimes appearing to over-reach while probing for food. Larger groups occur in creeks at high tide on the elevated banks beside low *Avicennia* mangroves. Flight leisurely and rather gull-like.

VOICE Diagnostic rich, fluty *cour-lee* in flight; bubbling song on breeding grounds.

STATUS AND DISTRIBUTION Palearctic breeder and long-distance migrant to Senegambia. Common along the whole coast; infrequent along the Gambia River up to LRD. Recorded in all months with peak in December-March. Found along the whole coast of Senegal.

SPOTTED REDSHANK
Tringa erythropus　　　　　**Plate 18**

Fr. Chevalier arlequin

IDENTIFICATION 32cm. A tall, long-legged, elegantly slim wader with a long fine bill. In winter pale ashy-grey above and white below. Whitish face and white supercilium above dark loral streak. Bill with blackish upper mandible, red proximally on lower mandible, with a slight droop at tip. Legs deep red. In breeding plumage jet black above and below, upperparts spangled with white. Intermediate plumages may occur. In flight shows large white spindle-shaped patch on upper rump and back, and closely-barred tail; lacks wingbars. Immature browner-grey, finely barred below. **Similar species** See Common Redshank and Greenshank.

HABITS Favours muddy pools, coastal lagoons, paddies, freshwater swamps, mangrove channels. Typically singly or with small groups of other waders. Feeds by probing and picking and being a good swimmer will forage in deeper water. Has been seen to 'up-end' like a duck. Flight is direct and fast.

VOICE A distinctive disyllabic *tew-eet* or *chu-ik* in flight; also *chip*.

STATUS AND DISTRIBUTION Palearctic breeder and long-distance migrant to Senegambia. Uncommon but recorded November-April in WD, CRD and URD. Occurs along the whole Senegal coast and further east on the Gambia River. Several hundred birds may gather in the delta of the Senegal River in December.

COMMON REDSHANK
Tringa totanus   Plate 18

Fr. Chevalier gambette

Other name: Redshank

IDENTIFICATION 27cm. A stridently noisy, medium-sized grey-brown wader with longish red legs and a red and black bill. Non-breeding birds have upperparts mottled brown or greyish-brown and together with the densely dark-streaked white underparts this creates a rather uniformly dark appearance at a distance. Immature warmer brown, more spotted above and with orange-yellow legs. Indistinct short pale supercilium above dark loral line; pale eye-ring. Breeding plumage more boldly and clearly marked with dark spotting on upperparts. In flight unmistakeable, with conspicuous white rump and lower back, blackish-barred tail, and wide white trailing edge to secondaries and inner primaries. **Similar species** Immature Redshank with duller or muddy legs could be confused with feeding Wood Sandpiper but is easily separated in flight. Larger, finer-billed Spotted Redshank is much paler (non-breeding) or predominantly black (breeding) and has no white on the wing. Ruff can sometimes have legs as bright as a Common Redshank, but can be separated by its small-headed appearance and shorter bill.

HABITS Prefers muddy expanses, either tidal or freshwater, but maybe seen in any wetland habitat. Not particularly gregarious, usually feeding alone, but gathers in large roosts. Feeds on the mud or in water by probing, picking or stabbing.

VOICE Very vocal on wintering grounds; a sharp *tew* or *tchew-it* repeated when in flight. Also a characteristic *tew-hoo-hoo* with the first syllable shriller, and a far-carrying mournful *tchioo*.

STATUS AND DISTRIBUTION Palearctic breeder and long- or medium-distance migrant to Senegambia; also breeds North Africa. Common to locally abundant and recorded from every division. Most numerous in coastal WD in July-March with greatest numbers in October-January. Smaller numbers of non-breeders remain all year. In Senegal common all along the coast and east along the upper Gambia River, and on passage in the Senegal River delta.

MARSH SANDPIPER
Tringa stagnatilis   Plate 18

Fr. Chevalier stagnatile

IDENTIFICATION 23cm. An elegant, slim wader with a long neck, small head and black, straight needle-like bill. The greyish-green legs are long, almost stilt-like. In non-breeding plumage pale grey above, with white fringes. Frequently a darker grey patch is evident at the front of the closed wing. Face and supercilium, throat, breast and belly pure white with greyish wash on the flanks. In breeding plumage darker, with blackish spotting on golden-brown upperparts, heavier dark chevron spots on the breast, and legs yellower. In flight resembles a small Greenshank but is slimmer and has a much more obvious leg projection beyond tail-tip. White rump extends up the back. Immature has more conspicuous pale fringes on upperparts with browner wash on head and neck. **Similar species** The very fine, straight, needle-like bill prevents confusion with any other *Tringa* species.

HABITS Favours tidal mud, open saline marshy areas, sewage pools, and occasionally inland on freshwater paddies and swamps. Graceful and delicate foraging technique involves picking and scything across the water surface. Occasionally swims and spins. Usually confiding and approachable.

VOICE A loud *yip* or *tchick* repeated when flushed.

STATUS AND DISTRIBUTION Palearctic breeder and long-distance migrant to Senegambia. Frequent to locally common in WD, October-March though present throughout the year in small numbers. Uncommon but widespread outside WD. In Senegal along the whole coast, including Senegal River delta and on the Gambia River.

COMMON GREENSHANK
Tringa nebularia   Plate 18

Fr. Chevalier aboyeur

Other name: Greenshank

IDENTIFICATION 33cm. A medium-sized, sturdy, pale grey-streaked wader with greenish legs and a long, stout, slightly upturned bill. Non-breeding adults have grey upperparts with white fringes.

Head and neck white, finely streaked grey on crown and nape, forehead white. Underparts white with fine grey streaking sparse and confined to sides of breast. Breeding plumage is darker overall with heavier, blacker streaking and with darker feather centres on back. Immature browner-grey, with more extensive streaking on the head and breast. Greyish-green bill is darker towards the tip. In flight the white of the rump extends characteristically up the back. Wings dark above, pale below; tail white with relatively faint grey-brown barring; toes extend beyond tip of tail. **Similar species** Smaller, slimmer, thin-billed Marsh Sandpiper has marked leg extension beyond tail in flight. Distant roosting groups of Greenshanks can be confused with godwits.

HABITS Favours extensive areas of mud, along mangrove edges, the shoreline and freshwater swamps. Feeds singly or in loose groups at the water's edge or wading deeply probing with the head submerged. Forages beside other waders, herons and egrets to glean disturbed prey. Feeds on fish fry, small crabs, surface insects, insect larvae and worms.

VOICE Characteristic loud, ringing trisyllabic *tew-tew-tew* or softer *tchew-leu-leu*.

STATUS AND DISTRIBUTION Palearctic breeder and long-distance migrant to Senegambia. Common to locally abundant in all divisions. Most numerous in WD in August-April with largest numbers October-February. Some present throughout the year. In Senegal occurs along the whole coast and along the Senegal and upper Gambia Rivers.

LESSER YELLOWLEGS
Tringa flavipes   Not illustrated

Fr. Petit chevalier à pattes jaunes

IDENTIFICATION 25cm. A slim wader between Wood Sandpiper and Common Redshank in size, most resembling a long-legged version of the former. Distinguished by bright yellow or orange-yellow legs, proportionately longer bill and shorter, less conspicuous supercilium. Square white rump.

STATUS AND DISTRIBUTION Two records of this Nearctic vagrant: from coastal WD in January 1976 and the Bund Road in March 1996. In Senegal, one at Dakar-Hann in January 1991.

GREEN SANDPIPER
Tringa ochropus   Plate 18

Fr. Chevalier cul-blanc

IDENTIFICATION 23cm. A stocky dark, olive-brown-backed sandpiper with shortish legs and medium-short straight bill. In non-breeding birds, upperparts only weakly pale-spotted, looking uniformly dark at a distance. Dark-streaked head and breast well demarcated from pure white belly. Dark loral stripe with white supercilium extends back to pale eye-ring but not beyond. In breeding plumage back more profusely peppered with buff-white spots; head, neck and breast appear dark grey streaked white. Immature shows some buffy spotting above. Bill dark brown with olive-green base, legs greyish-green and just projecting beyond tip of tail in flight. Broad black barring on white tail, and white rump contrasting with wholly blackish wings. Flight rapid and irregular, snipe-like. **Similar species** Common Sandpiper is smaller, plumper and browner, with distinctive flickering flight showing white wing-bars and brown rump and centre of tail. Both species bob their tails as does Wood Sandpiper which is paler, sleeker, more profusely spotted on the upperparts and has noticeably protruding legs in flight. Each of the three has its own distinctive flight call.

HABITS Favours sheltered areas along streams, flooded fields, ditches and forest ponds. Often seen running in and out of mangrove root systems or at the edges of grassy puddles. Often solitary.

VOICE A loud, ringing *clewit-wit-wit* or *tweet-weet-weet*, usually when flushed.

STATUS AND DISTRIBUTION Palearctic breeder and long-distance migrant to Senegambia. Common in WD and URD, and frequent in CRD, uncommon elsewhere. Widespread October-March, most numerous in January, but a few non-breeders present all year. Widespread but scattered in Senegal along the coast from the delta and inland to the east and south of The Gambia.

WOOD SANDPIPER
Tringa glareola   Plate 18

Fr. Chevalier sylvain

IDENTIFICATION 21cm. A dainty, long-legged and pale sandpiper. Non-breeding birds have a dull olive-brown back, profusely chequered with white, a useful field-mark, as is the pronounced white supercilium extending well behind the eye. Underparts whitish, lightly streaked greyish-brown on neck, breast and flanks. In breeding plumage upperparts darker with copious pale spotting and fringes; streaking on underparts more intense. Immature has browner back with buff spotting. The dark bill is fine and straight and about the same length as the head. Longish slender yellowish legs. Flight swift and agile, the slim brown wings showing no wing-bar; underwing pale greyish-white, rump white. The legs extend well beyond the tip of the finely barred tail. Regularly flutters. **Similar species** See Green Sandpiper and Lesser Yellowlegs.

HABITS Prefers saline and freshwater habitats with low vegetation, including sewage outlets and drainage ditches. Unlikely to occur on the shoreline. Usually alone or loosely associated with other waders. Often perches on tree stumps. Walks in the shallows picking at food from mud or vegetation, and often bobbing its tail. Stands with head erect looking long-necked when alarmed or in disputes with other birds.

VOICE A quickly delivered, distinctive, shrill *chiff-iff-iff.*

STATUS AND DISTRIBUTION Palearctic breeder and long-distance migrant to Senegambia. Common to locally abundant in WD, CRD, URD and frequent elsewhere, mostly October-March, most numerous November-January. A few non-breeders present throughout the year, with an exceptional June record of thirty birds in breeding plumage.

TEREK SANDPIPER
Xenus cinereus   Not illustrated

Fr. Chevalier de Térek

IDENTIFICATION 23cm. A dumpy pale grey wader with short yellow legs and long characteristically upturned bill. Breast pattern and tail bobbing, as well as shallow flickering wing action, reminiscent of Common Sandpiper. In flight shows whitish trailing edge to wing but distinguished from Common Redshank by grey back and rump. Also has distinctive blackish leading edge to wing and thin dark 'braces' on mantle.

HABITS As well as probing for food has adapted to taking moving prey by adopting a rapid feeding action. Runs in a characterisitc horizontal posture.

VOICE A Whimbrel-like *wit-wit-wit-wit* and Common Redshank-like trisyllabic call *tur-loo-tew.*

STATUS AND DISTRIBUTION Vagrant from the Palearctic with singles from LRD in December 1974 and WD in January 1994. Vagrant elsewhere in West Africa but not yet from Senegal.

COMMON SANDPIPER
Actitis hypoleucos   Plate 18

Fr. Chevalier guignette

IDENTIFICATION 19cm. A medium-sized brown and white sandpiper with a short-billed, short-legged low posture, constant bobbing action and characteristic low flickering flight on bowed wings. In non-breeding plumage, olive-tinged brown upperparts with faint paler feather fringes on the coverts visible only at close quarters. Head, neck and breast brownish and finely streaked, well demarcated from the white throat, lower breast and belly. A diagnostic white pectoral wedge extends upwards in front of the closed wing. In breeding plumage rather brighter, darker streaking on back more prominent, streaked breast forming complete pectoral band. Immature has narrow buff feather edgings on upperparts. In flight the long, dark, white-edged tail and prominent white wing-bar are distinctive. Calls frequently in flight. **Similar species** See Green Sandpiper. Smaller Temminck's Stint is also dully marked and has a dull breast. It too bobs, but is much smaller and lacks the striking white pectoral wedge.

HABITS Frequents a wide range of habitats: attracted by muddy edges along mangroves, shorelines, rivers, rice paddies, swamps and streams. Usually solitary. Feeds by picking and chasing. One of the few waders to regularly eat fiddler crabs.

VOICE A shrill teetering *swee-wee-wee* or *seep-seep-seep* perched or delivered in flight.

STATUS AND DISTRIBUTION Palearctic breeder and long-distance migrant to Senegambia. Common in all divisions and locally abundant in WD, August-April with a November-February peak. Some non-breeders remain all year. Widespread in coastal and inland eastern and southern Senegal as well as the delta of the Senegal River.

RUDDY TURNSTONE
Arenaria interpres   Plate 19

Fr. Tournepierre à collier

Other name: Turnstone

IDENTIFICATION 23cm. A short-necked, thick-set wader with distinctive patchy brown, white and black plumage. Bill short, stout, pointed and black; legs yellowish to orange. In non-breeding plumage upperparts mottled brown and black, underparts white with irregular dark markings on the throat and sides of the breast. Regularly seen in breeding plumage when back is markedly 'tortoise-shelled' in rich chestnut and black, and with complex black and white head pattern; also seen in transitional plumages. Immature browner above with buff fringes. In flight unmistakeably and strikingly patterned, the white wing-bar, braces, mantle and rump contrasting strongly with the otherwise dark upperparts.

HABITS A wader of the beach, tideline, rocky pools and edges of muddy lagoons and mangroves. Rarely alone, mostly in parties of 15-40. On passage congregates along the coast in flocks which may number thousands, often in the company of similar numbers of Sanderlings. Forages by turning over shells, shingle and shoreline debris for hidden prey. Walks and runs in an industrious manner.

VOICE In flight a harsh conversational *kititititit* or *pucka-tuck* when feeding.

STATUS AND DISTRIBUTION Palearctic breeder and long-distance migrant to Senegambia. Common to very abundant along the whole coastline, present in all months with peak in October-March. Regular in LRD and occasional in CRD. Found along the whole coast of Senegal with some inland records in the north.

RED-NECKED PHALAROPE
Phalaropus lobatus   Not illustrated

Fr. Phalarope à bec étroit

Other name: Northern Phalarope

IDENTIFICATION 18.5cm. A slightly built phalarope with small head and very fine needle-like black bill. In non-breeding plumage mottled pale grey on back, with distinctive white V markings, head mainly white with conspicuous dark mark through and behind eye and dark patch on rear of crown; underparts white. Legs dull yellow. When breeding much darker, with scaly brown back and chestnut-red on neck, bright in female, dull in male. In flight pale V on back and white wing-bars are conspicuous. Intermediate plumages occur. Immature has upperparts dark brown. **Similar species** Nearctic Wilson's Phalarope *Phalaropus tricolor* (Fr. Phalarope de Wilson), not illustrated, is a vagrant to north-west Africa but of uncertain status in The Gambia with no substantiated records. It also has a fine bill but is much larger. In winter lacks

prominent dark mark through the eye having a modest eye-stripe, appears very pale overall, and lacks wing-bar.

HABITS Likely to occur in muddy lagoons and coastal waters, and offshore.

VOICE Usually silent in non-breeding season.

STATUS AND DISTRIBUTION First recorded for The Gambia in October 1993; a single adult at Banjul in WD. A group of eight birds in northern Senegal in early December 1983, and singles in October 1991 and April 1992.

GREY PHALAROPE
Phalaropus fulicaria   Plate 19

Fr. Phalarope à bec large

Other name: Red Phalarope

IDENTIFICATION 21cm. Stocky, short-legged phalarope. Non-breeding birds have head and underparts white with variable grey-black cap and pronounced black smudgy mark through eye; otherwise upperparts uniform plain pale grey. Has bold white wing-bar but lacks V markings on back. Broad, thickish yellow-green bill. Legs grey-brown with yellowish lobes along the toes. In breeding plumage: crown, chin and lores dark brown, underparts and much of head strikingly rich chestnut-red with contrasting white patch around eye and on sides of head. Bill yellow tipped black. Immature similar to non-breeding adult but lacks V markings on back. **Similar species** Red-necked Phalarope is smaller with mottled back and finer bill. At sea White-faced Storm Petrel can look similar. A distant Sanderling is also a possible confusion species.

HABITS Normally oceanic when wintering in African waters, likely to be encountered in small groups. Only seen near land after inclement weather, when may occur on muddy pools.

VOICE Usual call is a short sharp *pit*, given in alarm.

STATUS AND DISTRIBUTION Rare but probably under-recorded offshore; coastal records October-March. An unusual inland record in LRD on main river in November and another in flooded rice fields in January in CRD. Senegal records in April, August and September with c 330 off Dakar in September 1990.

SKUAS

Stercorariidae

Five listed for Senegambia but only two from Gambian waters. Mainly dark brown seabirds with gull-like bodies, relatively long wings and slightly hooked bills. Plumage, body shape and behaviour are all useful in identification. Normally associated with mixed flocks of feeding or resting seabirds, and usually first noticed vigorously chasing other seabirds. Size comparison with the victim may assist identification. Their piratical attacks force (mainly) gulls and terns to disgorge their fish prey, but some food is caught by diving or picking from the surface. Identification may be difficult as there are adult and immature plumages and some show polymorphism, with dark, intermediate and pale phases. Until recently, relatively few observations in Senegambian waters, but recent pelagic trips off Dakar have revealed significant numbers offshore.

POMARINE SKUA
Stercorarius pomarinus   Plate 20

Fr. Labbe pomarin

Other name: Pomarine Jaeger

IDENTIFICATION 65-76cm. Gull-like skua with a heavy appearance. Polymorphic: dark phase (all dark brown), pale phase (dark above, yellowish-white below) plus intermediate types (rare). Pale phase birds (uncommon) have distinct dark cap and often a dusky breast-band. Bill rather thick, pale-based with dark tip. Wings broad-based with conspicuous white patches at base of primaries. Tail projection 17-20cm long, broad and twisted with spoon-shaped tips; typically absent in Senegambian waters. Juveniles are generally dark brown above with faint pale scaling; uppertail-coverts prominently pale-barred and all lack pale hind collar. Underbody dark brown, underwings strongly barred. **Similar species** See slimmer, smaller-billed Arctic Skua. All-dark Great Skua is bulkier-bodied with prominent white flashes on broader wings, and short tail.

HABITS Similar to Arctic, but more aggressive. Normally in ones or twos, sometimes more around one site, but greater numbers accumulate after periods of strong winds, often in January. Flight consists of deep wing-beats and rapid pursuits; after a chase often rests on sea.

STATUS AND DISTRIBUTION Locally common October-March at a few sites along coast, around harbours, fish-landing places and seabird roosts; sometimes well up the estuary. Occurs in all seasons. Off Senegal, high numbers recorded in April and September-October; the commonest skua.

ARCTIC SKUA
Stercorarius parasiticus   Plate 20

Fr. Labbe parasite

Other names: Parasitic Jaeger

IDENTIFICATION 46-67cm. Agile, tern-like skua, slimmer and lighter than Pomarine. Like Pomarine it has dark and pale phases as well as intermediates. Pale birds show dark cap and sometimes a dusky breast-band, often incomplete. Bill slimmer than Pomarine and all-dark. Tail is wedge-shaped with elongated pointed central tail feathers, 8-14cm if present. Juveniles are variable appearing slender-winged with a moderately long tail. Head is smallish; slim, bicoloured bill only noticeable at close range. Invariably show pale or rusty hind collar, but lack strongly barred uppertail-coverts of Pomarine. Tail is wedge-shaped with central feathers usually showing slight projection. **Similar species** See heavier, larger-billed Pomarine and smaller, even more tern-like Long-tailed Skua.

HABITS Patrols stretches of coastline, at fish-landing sites and bays, frequently close to the beach, flushing tight flocks of gulls. Flight skilful and elegant; twists, turns and swoops up at acute angles. Glides are followed by powerful acceleration, often low over sea, covering every movement of its victim, and always showing white flashes at base of dark primaries.

STATUS AND DISTRIBUTION Locally common October-March at a few sites along the coast, also well up the estuary; also occurs other months. Off northern Senegal the commonest skua in August, recorded also south of The Gambia off Casamance. Northerly passage noted in April.

LONG-TAILED SKUA
Stercorarius longicaudus Not illustrated

Fr. Labbe à longue queue

Other name: Long-tailed Jaeger

IDENTIFICATION 50-58cm including very long wispy tail extensions of c.18cm in adults (often broken). Smallest, slightly-built and tern-like skua, grey-brown above with blackish cap and primaries, whitish below, often greyish or buffish on belly and flanks. Diminutive dark bill. Slender tern-like wings with relatively inconspicuous white flashes at primary bases. No dark phase. Juveniles rather greyish, lacking warm tones of Arctic but with better contrasing plumage. Bicoloured bill shorter and stubbier; tail longer and more tapered with slight projection of central feathers. **Similar species** See Arctic Skua.

HABITS Flight light and buoyant, often high. Performs astounding, twisting chases and pick-feeds on surface of sea.

STATUS AND DISTRIBUTION Not proven in Gambian waters but to be expected. Seawatching off Dakar since 1990 has shown northerly movements in April; observed also in good numbers in August-September.

GREAT SKUA
Catharacta skua   Plate 20

Fr. Grande Labbe

Other names: Brown Skua, Bonxie

IDENTIFICATION 53-66cm. Largest, heaviest skua always dark brown often with warm tinge to plumage and some coarse pale streaking on mantle. Markedly barrel-shaped appearance with very short tail without obvious projection. Broad wings with striking white patches at bases of dark primaries (above and below). Not as rangy in flight as a gull. Likely to occur singly. Kills and harries; will pursue birds up to size of a gannet. Follows ships, highly pelagic.

STATUS AND DISTRIBUTION. Not proven in Gambian waters. Recorded off northern Senegal, July-February (but see South Polar Skua).

SOUTH POLAR SKUA
Catharacta maccormicki **Not illustrated**

Fr. Labbe de McCormick

IDENTIFICATION 50-60cm. Slightly smaller and less powerful than Great Skua with smaller bill. Plumage greyer and cold-tinged, lacking warm brown colour tones of Great. Three colour phases: dark (uniform grey-brown), intermediate (showing contrast between upper- and underparts), light (sandy-grey head and body with darker wings). Underwing darker than Great Skua.

STATUS AND DISTRIBUTION A bird ringed on King George Island, South Shetlands in February 1991 was found dead on a beach near Dakar in July 1991. In October 1995 and 1996, up to 200 birds (mostly immatures) reported offshore from Dakar indicating regular and substantial autumn passage (T. Marr, verbally). Old records of Great Skua may not have eliminated South Polar Skua as a possibility.

GULLS

Laridae

Fifteen species but few seen regularly. Small to large seabirds usually associated with beaches, harbours and coastal districts. Sexes similar, usually with relatively long wings, rounded short tails and powerful bills. They scavenge for offal and refuse, catch fish by surface plunging, occasionally pirate food from other seabirds. Adults mostly white below with grey or black backs, immatures usually heavily mottled grey-brown reaching adulthood over a period of 2-4 years of transitional stages depending on size. Some are hooded for part of the year; others are white-headed with darker streaking when not breeding. Complex transitional plumages may make identification difficult.

MEDITERRANEAN GULL
Larus melanocephalus   Plate 20

Fr. Mouette mélanocéphale

Other name: Mediterranean Black-headed Gull

IDENTIFICATION 36-38cm. Medium-sized strikingly white-winged, square-headed gull with a thick scarlet 'droop-tipped' bill. Upperparts pearl-grey, underparts white. Breeding adult has jet black hood extending well down neck with broken, thick white eye crescents. Red bill, dark-banded near tip; long legs and feet scarlet. Relatively broad wings appear unmarked in flight with entirely white tips; tail short, white. Non-breeding birds have white on head and streaked crown and nape, with dark eye-patch. Sub-adult has black subterminal markings on outer primaries which at rest look like two or three black dots on very white wing-tips. First-winter much darker-winged with pale mid-wing panel, and a narrow black band on white tail. **Similar species** Non-breeding, paler-backed Mediterranean looks stouter than similar-sized non-breeding Black-headed Gull, which has black primaries. Separated

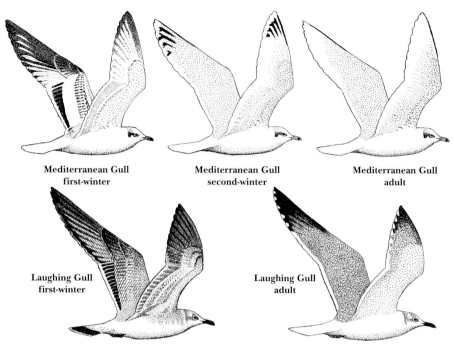

Mediterranean Gull
first-winter

Mediterranean Gull
second-winter

Mediterranean Gull
adult

Laughing Gull
first-winter

Laughing Gull
adult

from much darker, slightly larger Grey-headed Gull on bill and head shape and upperwing pattern. Immature Mediteranean has distinctively different bill structure, a more hooded look and a narrower tail band than immature Common Gull.

HABITS Drawn to gull and other seabird roosts on sandbars and muddy lagoons at low tide. One or two records of individuals lingering for several days at the same locality.

VOICE Low-pitched, shrieking or shrill *eu-err* or *eooe*.

STATUS AND DISTRIBUTION Rare Palearctic vagrant. Eight records from 1975 to 1996, all November-February in coastal WD. Recorded northern Senegal January-March, and once in May. Probably under-recorded.

plumage unlikely to be seen in Senegambia. At rest primary tips black with two or three very small mirrors, which may be absent. Bill dull red looking blackish at distance. First-winter plumage also rather dark, with partial hood and broad tail band. **Similar species** See smaller Franklin's Gull, and also Mediterranean Gull.

HABITS Likely to be attracted to and seen with other gulls at roost on mud, beach or on the sea.

VOICE A *ke-ruh* and a laughing *hah* cry, but not expected to be vocal in West Africa.

STATUS AND DISTRIBUTION Nearctic vagrant to Senegambia. Gambian records in December 1983, February 1984 and January 1991, all in coastal WD.

LAUGHING GULL
Larus atricilla   See above

Fr. Goéland atricille

IDENTIFICATION 36-41cm. Medium, slim-looking gull with a long slightly drooping bill. Larger and darker grey mantle than Grey-headed Gull; underparts white. In flight long wings have a white trailing edge and dark underwing. Long dark sooty legs. Slaty-black hooded head of full breeding

FRANKLIN'S GULL
Larus pipixcan   Not illustrated

Fr. Mouette de Franklin

IDENTIFICATION 32-36cm. Small-medium gull, smaller and darker than Grey-headed, more resembling a scaled-down Laughing Gull in overall appearance. Retains more of a sooty hood in non-breeding plumage but forecrown is white and eye-crescents are retained; primaries prominently

205

tipped white. Conspicuous white trailing edge in flight. In breeding plumage slaty-black hood extends down nape. Wings relatively short and rounded, whitish underwing. Shortish legs dull red to blackish, heavy dark bill tipped red. First-winter birds have darkish upperwings with white trailing edge, black tail-band and a partial hood. **Similar species** Sub-adult Mediterranean Gull is always paler with less prounounced eye-crescents.

HABITS Stragglers may be attracted to flocks of mixed gulls and terns.

VOICE A curlew-like call; unlikely to be particularly vocal in West Africa.

STATUS AND DISTRIBUTION Nearctic vagrant to Senegambia. In The Gambia three records, in February 1984, January 1991 and March 1992 in coastal WD. Two records in northern Senegal in May 1983 and April 1986, at the same site.

LITTLE GULL
Larus minutus
Plate 20

Fr. Mouette pygmée

IDENTIFICATION 25-27cm. Diminutive gull with graceful tern-like flight on rounded wings. Non-breeding adult has white head with blackish eye-crescent and ear-spot, grey crown and hindneck.

Bill short and dull blackish, legs short and dull red. Breeding adult has a black hood, reddish-brown bill and red legs, eye dark with no white eye-ring. Underwing blackish, upperwing pale grey with white wing-tips, giving diagnostic contrast in flight. Immature has distinct inverted dark zig-zag band across wings and back in flight, a feature shared with larger vagrant Black-legged Kittiwake. **Similar species** All other dark-hooded gulls are larger. Larger immature Kittiwake has a dark collar and a stronger zig-zag band on upperwings; Little Gull has browner upperparts compared with greyer-backed Kittiwake and Sabine's Gull has slightly forked tail.

HABITS Outside breeding season spends long periods at sea. In Senegambia most likely to be encountered sitting on tidal mud flats away from other larger gulls and looking tiny and tern-like. Feeding action is similar to marsh terns, picking food items by dipping to the water surface whilst on the wing.

VOICE A repeated, harsh, tern-like *kyek-kyek* or *ki-ki-ki*.

STATUS AND DISTRIBUTION Palearctic vagrant. In The Gambia, recorded in coastal WD in November 1977, September 1984, November 1990 and February 1998. Five records in N Senegal, all singles except 20-30 at Dakar in March 1990.

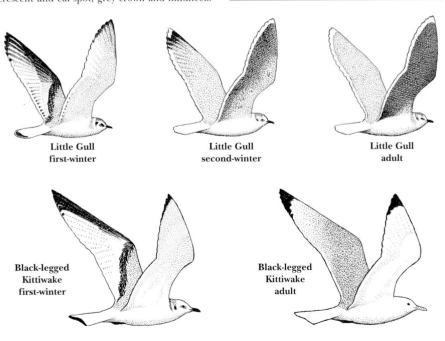

Little Gull
first-winter

Little Gull
second-winter

Little Gull
adult

Black-legged
Kittiwake
first-winter

Black-legged
Kittiwake
adult

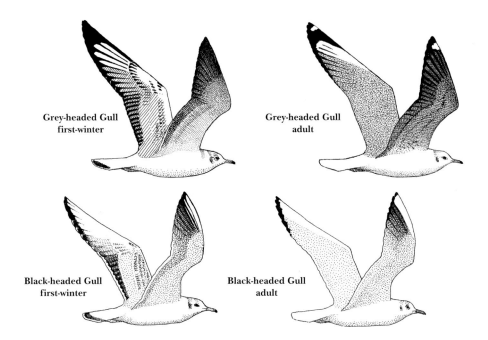

Grey-headed Gull
first-winter

Grey-headed Gull
adult

Black-headed Gull
first-winter

Black-headed Gull
adult

GREY-HEADED GULL
Larus cirrocephalus   Plate 20

Fr. Mouette à tête grise

Other name: Grey-hooded Gull

IDENTIFICATION 42cm. *L. c. poiocephalus.* Me-dium-sized. By far the commonest Senegambian gull. Characteristic pale lavender-grey hood, exten-sive black wing-tips with prominent white mirrors. Upperparts darkish grey; underparts white. Bill and legs red, prominent pale yellowish-white eye. In flight shows grey upperwing with white leading edge to outer wing; underwing diagnostic uniform dark grey with small white mark near tip. At rest, wing-tips extend well beyond the tail. Non-breeding adult lacks the hood and has faint grey crescent behind ear. Immature grey admixed ashy-brown on back and crown; ear-coverts brownish-black; under-parts white, tail white with narrow black subterminal bar, bill pinkish with dark tip, legs and feet brown. **Similar species** In non-breeding plumage Grey-headed and smaller Black-headed are separated by paler back and smaller bill of Black-headed.

HABITS Found along the whole coast especially around estuaries, tidal inlets and docksides; assem-bles also at rubbish dumps. Often congregates in hundreds; sometimes in thousands. Quick to flock inland in pursuit of flying ants at the onset of the rains.

VOICE A harsh and racuous prolonged *caw-caw* or *kraaa* or *ka-ka-ka-ka* or *kruup.*

STATUS AND DISTRIBUTION Abundant to very abundant along the whole Senegambian coast in all seasons, with records inland to LRD in The Gambia.

BREEDING No recent Gambian records. Large colonies June-August in southern Senegal, May in the north.

BLACK-HEADED GULL
Larus ridibundus   Plate 20

Fr. Mouette rieuse

Other names: Common Black-headed Gull; North-ern Black-headed Gull

IDENTIFICATION 38cm. Small-medium. Most frequent of the dark-hooded gulls. Commonly seen in breeding plumage (around February) with chocolate brown hood not extending down rear of neck leaving nape and neck white. Upperparts pale grey, underparts white. Bill slender, pointed and dark red, tipped black, eye with thinnish white incomplete ring, legs red. Non-breeding adult has head white with dark spot behind eye. In flight shows characteristic broad white leading edge, dark wing tips and prominent white outer primaries in all plumages, and pale, not dark, underwing.

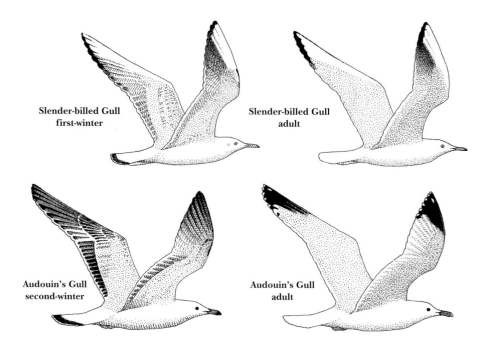

Slender-billed Gull
first-winter

Slender-billed Gull
adult

Audouin's Gull
second-winter

Audouin's Gull
adult

Immature with admixed rusty-brown upperparts and head becoming transitionally grey; tail band dark. Legs dull red, bill yellowish-brown tipped black. **Similar species** Other similarly-sized dark-hooded gulls are rare or vagrants and may be expected singly; all show different bill colour and shape, and greater extent of hood. Slightly larger non-breeding Grey-headed Gull easily separated in flight by its dark underwing. Black-headed is paler grey above at all times and looks slightly more pot-bellied. See Slender-billed Gull for flight separation.

HABITS Frequents the whole coast, especially around estuaries, tidal inlets and docksides. Flocks usually in tens to hundreds, but can occur alone. Groups in breeding plumage often sit on the sea in loose rafts alongside Grey-headed and Slender-billed Gulls.

VOICE A harsh and nasal *kraah* or quieter *kreeaay*.

STATUS AND DISTRIBUTION Seasonally frequent to briefly locally abundant Palearctic migrant at the coast. Recorded in all seasons, mostly November-April, peaking February in most years; rare late rains. Recorded in all months in Senegal.

SLENDER-BILLED GULL
Larus genei   Plate 20

Fr. Goéland railleur

IDENTIFICATION 42-44cm. Medium-sized, very white-looking gull attenuated body shape and long sloping forehead; head all white, never hooded. Upperparts pearl-grey, underparts white with a rosy-pink flush as breeding season approaches. Long slender bill is dark red, looking black at distance. Legs scarlet. Non-breeding birds have smudgy ear-covert spot, orange bill, and orange-red legs. In flight looks hump-backed, showing white wing-tips tipped black, and pale underwing with white trailing edge. First-winter has yellow-orange bill and legs, patchy grey-brown wings, brown smudging over the head and a black terminal bar on white tail. **Similar species** Characteristic long-necked shape and posture are good guides. Always whiter than non-breeding Black-headed, with longer, broader wings, slower wingbeats and longer bill and neck. Grey-headed Gull has dark grey underwing and darker upperparts.

HABITS Gathers in singles, groups or flocks, usually distant from other gulls, on lagoons and river outflows along the whole coast. Often stands or wades in the shallows to feed.

VOICE Deep nasal *ka* and a *kra* calls.

STATUS AND DISTRIBUTION Seasonally common to locally abundant along the whole coast, occasionally along the river to LRD; occurs in all seasons with greatest numbers January-July. Along the whole Senegalese coast year-round.

BREEDING Sizeable colonies in northern and southern Senegal with eggs in May-June in Saloum delta. Display and copulation recorded April-May in The Gambia when numerous birds in full breeding plumage are present.

AUDOUIN'S GULL
Larus audouinii Plate 20

Fr. Goéland d'Audouin

IDENTIFICATION 48-52cm. Sexes similar. Medium-large, pearl-grey and white gull with sloping, long-looking flat head and heavyish angular tricoloured bill. Non-breeding adult has white head slightly streaked grey; tail, rump and underparts white; back, mantle and most of wing pale grey. Coral-red bill with subterminal black band and yellow tip visible at close range appears all-dark at distance. Eye dark brown with red ring, legs characteristically grey. Flight graceful showing black wing tips, tipped white from below, with only minimal white mirrors and narrow white leading and trailing edges. First-winter has a uniform, weakly streaked grey-brown crown and neck, small dark mark behind eye and pale brown upperparts. Legs dark grey. Tail blackish tipped pale, with a distinctive white U shaped mark at base. Later becomes pale grey above, admixed grey-brown, primaries darken to black, and pale bill darkens towards tip. **Similar species** Bill shape and colour, and head profile are characteristic, and separate adult Audouin's from smaller Common Gull. Larger, heavier-built and more hostile-looking adult Yellow-legged Gull has darker upperparts and larger mirrors on broader wing tips. Immatures of larger Lesser Black-backed and Yellow-legged Gulls have a broad dark tail band, with rump, uppertail-coverts and sides all pale with brown barring. Grey legs is the best feature for picking out Audouin's Gulls amongst a mixed flock.

HABITS Pelagic, roosting on sandy spits, coastal lagoons and tidal inlets usually close to the beach. Usually encountered singly or in small groups.

VOICE Likened to cry of human baby or braying donkey *geaak* or *gah-gah ga ga ga*; greets with a *ki-aou*.

STATUS AND DISTRIBUTION Seasonal and locally frequent Palearctic migrant to a few coastal sites e.g. Tanji Bird Reserve. Records in November-

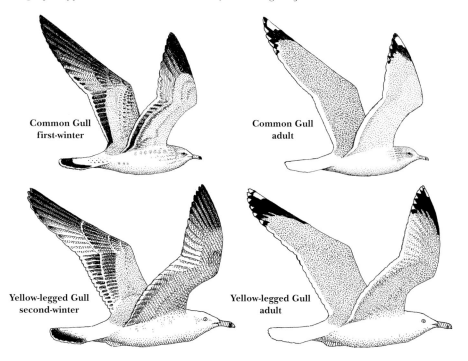

Common Gull
first-winter

Common Gull
adult

Yellow-legged Gull
second-winter

Yellow-legged Gull
adult

Kelp Gull
first-winter

Kelp Gull
adult

Lesser
Black-backed Gull
first-winter

Lesser
Black-backed Gull
adult

March in The Gambia, maximum 400 birds off Tanji in February. An estimated 2.4% of the global population of c. 17,000 pairs winters in Senengambian coastal waters.

COMMON GULL
Larus canus   Plate 21

Fr. Goéland cendré

Other name: Mew Gull

IDENTIFICATION 38-44cm. Sexes similar. A neat, medium-sized gull with a dark eye, round head and relatively weak greenish-yellow bill lacking a red spot. Non-breeding adult pale blue-grey above, white below, with darker streaks on white head (all-white when breeding); tip of bill may be ringed. Legs yellowish-green. Flight looks buoyant on bluntish wings; black-tipped primaries (reaching well beyond tail at rest) with white patches at tips of outer primaries, and white trailing and leading edges. Juvenile has a well defined black band across the tail towards the tip, mottled brown back which soon becomes pale grey, retaining the brown on the wing-coverts. **Similar species** Much larger, bulkier Yellow-legged Gull has more prominent wing-tip mirrors. Ring-billed Gull *Larus delawarensis* (Fr. Goéland à bec cerclé) not illustrated, 43-47cm is slightly larger, heavier and longer-legged. Stouter

yellow bill characteristically banded or tipped dark; wing-tip mirrors small or lacking on broader-based more pointed wing. Not recorded in The Gambia but a single vagrant record of this North American gull from Senegal.

HABITS Most likely to be attracted to mass gull assemblies at fish landing sites, harbours and roosts, or along the beach.

VOICE A *keee-ya*, *kyow kyow* or *kakakaka ka* but not particularly vocal.

STATUS AND DISTRIBUTION Vagrant to Senegambia from the Palearctic with Gambian records from coastal WD in September 1984, January 1986 and a full adult in July 1994. Probably under-recorded and worth searching for in large groups of mixed gulls during the dry season.

KELP GULL
Larus dominicanus   Plate 21

Fr. Goéland dominicain

Other names: Southern Black-backed Gull, Dominican Gull

IDENTIFICATION 60cm. Sexes similar. The biggest and blackest-backed Senegambian gull, robust in appearance, usually in ones or twos. Steep forehead with large heavy bill, olive-green or blue-grey legs with darker joints. White head, underparts and tail, upperparts black. Eye grey to dark brown with orange-red orbital ring. Bill bright yellow with orange-red sometimes scarlet spot and considerable gonys. Immature dark sooty grey-brown with barred rump, tail blackish becoming gradually white with a dark terminal band. **Similar species** Lesser Black-backed Gull is smaller with a slate-grey back and yellow legs, showing a narrower white trailing edge in flight. Yellow-legged Gull is always paler-backed, showing contrasting dark wing-tips on paler wings.

HABITS Usually singly or in small numbers, along the open shoreline near fish-landing sites where large numbers of other gulls accumulate. Scavenges. Stands on the tideline for long periods.

VOICE A loud *ki-ok ko-ko-ko* also a mournful *meew*.

STATUS AND DISTRIBUTION First seen (two adults) in The Gambia in December 1992 at Tanji Bird Reserve, WD; uncommonly since at the same site. Known from Saloum delta, Senegal. No other West African records but possibly overlooked. Widespread in the southern Hemisphere.

BREEDING Isolated record at Saloum, northern Senegal in 1983 of a nest with one egg. Attempted hybridisation with Lesser Black-backed Gull noted in 1980 at same site.

LESSER BLACK-BACKED GULL
Larus fuscus   Plate 21

Fr. Goéland brun

IDENTIFICATION 52-60cm. Sexes similar. The commonest large gull, variable dark grey back, legs usually yellow to yellow-orange (non-breeding). Adult has head, neck, underparts and tail white. Upperparts vary from slate-grey to near-black. Bill yellow with red spot. Wing-tips dark with inconspicuous white spot. Eye creamy-white to pale yellow with red-orange orbital ring. In flight wings dark slate or blackish thinly bordered white. Wings project beyond tail at rest. West European race *L. f. graellsii* is pale slate-grey on back, *L. f. fuscus* of Scandinavian and Baltic origin is much darker. Confusing series of transitional plumages from juvenile to adult; initally dark grey-brown, darkest on the back, with all the flight feathers uniformly brownish-black, dark underwing, uppertail with very narrow white tips and broad dark band, and bill black becoming pale grey with black tip. **Similar species** Even the paler-backed race *graellsii* is darker than adult Yellow-legged Gull which is generally pearl-grey or silver, not slate. Immature Lesser Black-backed generally darker with browner tertials than

immature Yellow-legged which has a less pronounced dark tail band; Black-backs attain adulthoood by turning darker with age whilst Yellow-legged become paler. Rare Kelp Gull is strikingly large and very black-backed.

HABITS Frequents the whole coast in a variety of plumages, particularly numerous near fish-landing sites and in bays with tidal spits. Stands near the shoreline in groups, often slightly separated from other species, or rests on the sea in loose rafts. Visits coastal rubbish dumps. Omnivourous.

VOICE Various deep calls, most commonly a loud *kyow, kokokoko* or *whaa-ow-ow-ow.*

STATUS AND DISTRIBUTION Seasonally common to occasionally locally very abundant Palearctic migrant. Recorded in every month but numbers greatest October-March. Racial composition requires more study. Common along the whole of coastal Senegal, occasional inland.

BREEDING Attempted hybridisation record with Kelp Gull in northern Senegal.

YELLOW-LEGGED GULL
Larus cachinnans   Plate 21

Fr. Goéland leucophée

Other names: Mediterranean Herring Gull; Yellow-legged Herring Gull

IDENTIFICATION 55-67cm. *L. c. atlantis.* A large thick-necked gull with menacing looks, yellow legs (brighter when breeding) and yellow eye with red orbital ring. Bill strong and powerful, orange-yellow with red spot near tip. Head and neck all-white, with little or no streaking in non-breeding plumage (closely related Herring Gull *L. argentatus*, not described, always shows dark streaks on head and neck in winter). Mantle grey, slightly darker than Herring Gull, but always paler than lesser Black-backed. Underparts and tail white. In flight, marked contrast of black wing-tips with white spots and grey wings. Immatures pass through a series of transitional plumages before reaching maturity in four years; typically brownish mottled grey with all-dark primaries and broad dark terminal tail band. Bill usually dark-tipped with paler base. Beware that younger birds have pink legs which may be retained into later stages. **Similar species** Lesser Black-backed Gull is slightly smaller and darker and, in non-breeding plumage, has a streaked head. Herring Gull (unlikely in the region) also has a streaked head but has pink legs and a paler mantle.

HABITS Usually seen singly or in small groups, often with Lesser Black-backed Gulls.

211

VOICE Similar raucous and laughing calls to Lesser Black-backed Gull (sound recordings needed).

STATUS AND DISTRIBUTION Uncommon Palearctic migrant to the coast, recorded in all seasons with highest frequency of observations December-January. Possibly under-recorded. Recorded both to the south and north in coastal Senegal.

NOTE Until recently, Yellow-legged Gull was considered conspecific with Herring Gull and was the southern form of the species. There are now no confirmed records of Herring Gull for West Africa, the nearest vagrants being from Morocco. Gore (1990), however, mentions a presumed Herring Gull "with flesh-coloured legs" at Cape Lagoon in December 1987 but there are no supporting details.

BLACK-LEGGED KITTIWAKE
Rissa tridactyla   Not illustrated

Fr. Mouette tridactyle

Other names: Kittiwake, Common Kittiwake

IDENTIFICATION 38-40cm. Long-winged, slim-built pelagic gull with tern-like buoyant flight. Adult has solid black wing tips with no white mirrors, all white head, tail and underparts, back silvery-grey, bill yellowish, legs black. Immature shows a broad W across wings in a clear 'grey-black-white' format, and showing a dark collar which later fades. In flight also shows short slightly notched tail with a narrow black subterminal bar. **Similar species** Immature Sabine's Gull is slightly smaller and darker with a different wing pattern. Much smaller, red-legged Little Gull has distinctively different wing shape, colour and flight pattern. In flight larger Common

Gull shows white mirror spots on black primary tips and normally streaked head in Africa.

HABITS Pelagic when not breeeding.

VOICE A pugnacious *kek-kek*, also characteristic *kitti-wake* or *kevi-week* normally at colonies.

STATUS AND DISTRIBUTION Palearctic vagrant. Two records of immatures, December 1975 and November 1997 in coastal WD. Senegalese records from north coast March-April 1979 and January 1984.

SABINE'S GULL
Larus sabini   Not illustrated

Fr. Mouette de Sabine

IDENTIFICATION 27-32cm. Strikingly-plumaged, small pelagic gull. Breeding adult has black-bordered grey hood and a yellow-tipped black bill. In non-breeding plumage, has white face and dark rear crown and nape. Distinctive wing pattern of grey mantle and coverts, black primaries and white secondaries, the latter forming a prominent triangle. Tail white and slightly forked. Juvenile shows a similar wing pattern but the grey is replaced by dark grey-brown, concolorous with the head. Tail has a black terminal bar. **Similar species** See Black-legged Kittiwake.

HABITS Breeds in the arctic, in Greenland and North America, spending the winter at sea in the Atlantic off Africa. Highly pelagic and rarely seen from land.

VOICE A harsh *krrr* and various other calls.

STATUS AND DISTRIBUTION Unrecorded in Gambian waters. Recently seen off northern Senegal in substantial numbers, August–November and March–June.

TERNS

Sternidae

Sixteen species. Small to large short-legged, slender birds with narrow pointed wings and elegant flight. Sexes similar. Differences, often subtle, in colour, shape, and size of the bill are often important in identification. In sea terns, seasonal plumage changes mainly involve a loss of black on the forehead and crown, but in marsh terns plumage changes are substantial. Most sea and marsh terns are Palearctic breeders; the noddies are tropical.

GULL-BILLED TERN
Gelochelidon nilotica   Plate 22

Fr. Sterne hansel

IDENTIFICATION 38cm. Robustly-built tern, with more gull-like flight and shorter, heavier all-black

bill than Sandwich Tern. Plumage even at distance appears white. In breeding plumage glossy black cap extends down nape. Non-breeders have forehead and crown white faintly dark-speckled or streaked, sometimes all white, and variable smudgy black patch behind eye. Upperparts and notched (not forked) tail pale grey; underparts white. In flight shows very white underwing with sooty trailing edge to primary tips; upperwing usually uniform pale grey. Relatively long legs black. Immature has grey back with variable amounts of dark brown. **Similar species** Thick all-black bill, notched tail lacking streamers separates it from other black-billed terns. Does not habitually plunge-dive, feeds more by dipping down to surface like a marsh tern, often over bare or fallow land. Occasional birds with all-white crown can cause confusion.

HABITS Regularly found foraging alone, quartering wide expanses of sandbank at low water or on bare mud near mangroves: a useful identification feature. Resting birds appear to stand high compared with other terns, mixing with other terns and sometimes waders. The most likely larger tern to be encountered well inland.

VOICE Harsh *kek-kek-kek* or *kraak.*

STATUS AND DISTRIBUTION Frequent on coast and inland in LRD, less common in CRD and rare in URD. Most records September to March but recorded in all months. Palearctic migrant and regional breeder. Widespread along the whole coast in Senegal, penetrating inland along major rivers.

BREEDING Mostly in the Palearctic. In northern Senegal colony of c. 100 pairs May-July and a few pairs also south of The Gambia.

CASPIAN TERN
Sterna caspia   **Plate 21**

Fr. Sterne caspienne

IDENTIFICATION 55cm. Largest tern, bigger than many gulls. Distinctive huge scarlet-red bill on big head, and strong, powerful, almost gannet-like flight. In breeding plumage glossy black cap with crest cut off square at back of head; non-breeding head is grey with blackish streaks, often sparse; usually darker behind eye. Upperparts including forked tail ash-grey, underparts white. Wings long and broad held relatively straight; white underwing contrasts with large dark wedge on primaries. Legs black. Immature has dark head flecked white, more black on orange bill, copious brown scalloping on grey back, and darker tail. **Similar species** Royal Tern has paler more dagger-like orange bill, longer bristly crest; in flight shows longer, more deeply forked tail and narrower, more pointed wings. At rest Caspian considerably larger than Royal.

HABITS Usually feeds alone or in loose flocks over a wide expanse of water. Gathers to roost in groups of tens to hundreds along the shoreline. Often flies high; hovers frequently but twists and turns with agility despite its size. When hunting flies with head held down watching for prey and makes deep plunges into water, sometimes submerging. Vocal piratical chases of birds with a catch are a regular sight.

VOICE Piercing, loud and raucous *kraaa* or *kak-kak-kak,* sounding heron-like.

STATUS AND DISTRIBUTION Abundant to locally very abundant in all seasons along the coast, frequent along the river to LRD, occasionally further east. Widespread along the whole coast of Senegal and in the delta of the Sengal River.

BREEDING Has attempted to breed in The Gambia April-July.

ROYAL TERN
Sterna maxima   **Plate 21**

Fr. Sterne royale

IDENTIFICATION 49cm. *S. m. albididorsalis.* Commonest big tern of the region and second largest. Bill orange to pale yellow and slightly drooping, less bulky than Caspian Tern but still heavy. Head with smooth black crown and shaggy elongated crest on nape in breeding plumage, when bill is very orange with reddish tinge at base. Non-breeding birds have more yellowish bill; forehead and crown white with black patch through eye merging onto dark nape; head sometimes all white giving a bald look. Upperparts and forked tail pale silvery-grey, underparts white. Rump and tail contrastingly paler than darker upperparts. Wings long, sharply pointed and narrow. Dark primary wedge on underwing less extensive than Caspian Tern. Legs black. Immature has dull yellow bill becoming orange, legs dull yellow becoming black; upperparts and wings scalloped in brown and black. **Similar species** Lesser Crested Tern is slimmer, shorter-legged and thinner-billed. Caspian Tern is bulkier, with a larger darker bill, and has broader wings and only a shallowly-forked tail; Royal also has whiter head in non-breeding plumage.

HABITS Typically flies low and direct over the sea, filing past in continuous streams close to the shore. Spectacular aerial displays involve pairs or squadrons of birds all swooping and stooping in unison. Frequently encountered perched on fish and shrimp-net floats in the river.

VOICE Higher pitched than Caspian; *keer ker ker* and a single *kak* amongst a varied repertoire.

STATUS AND DISTRIBUTION Abundant to locally very abundant in all seasons along the coast (e.g up to 15,000 at Tanji Bird Reserve in April).

Occurs along the river to LRD and occasionally further east. Widespread along the whole Senegal coast.

BREEDING Has attempted to breed in The Gambia. Regular displays and copulation March-April in WD. Large colonies on Banc d'Arguin in Mauritania.

LESSER CRESTED TERN
Sterna bengalensis   Plate 21

Fr. Sterne voyageuse

IDENTIFICATION 38cm. Medium-large tern with body size about that of Sandwich Tern, detectably smaller than Royal. Bill yellow or orange and substantially thinner and straighter than commoner Royal Tern. Head similar to breeding Royal but nape shaggier; non-breeding head pattern also closely resembles Royal. Upperparts pale ashy-grey, slightly darker than Sandwich Tern, underparts white. Ashy-grey rump (not white), uppertail-coverts and tail are only fractionally paler than upperparts. In The Gambia most have yellow bills but orange bills do occur. Legs black. Immature much as non-breeding adult but darker grey with blackish shoulder. **Similar species** Usually conspicuously smaller than Royal but care needed as both species are notably variable. Bill shape, uniform grey upperparts and tail (in flight) of Lesser Crested are important distinguishing features.

HABITS Singly or in small numbers, usually associated with mixed tern groups, but characteristically a few metres away. Visits harbours and fish-landing sites. Often fishes with groups of Sandwich Terns.

VOICE Varied calls including a high pitched *krr-eep* and a twittering *kee-kee-kee*.

STATUS AND DISTRIBUTION Frequent to locally common along the coast e.g. Tanji Bird Reserve but fewer in May-September. Never reported in large numbers. Regarded as a vagrant in coastal Senegal.

SANDWICH TERN
Sterna sandvicensis   Plate 21

Fr. Sterne caugek

IDENTIFICATION 40cm. Medium-large pale tern with diagnostic pale yellow tip to long narrow all-black bill. Non-breeding birds have extensive white forehead, black crown and nape streaked white; changes to all-black crown as breeding season

approaches, developing a short bristly crest, prominent when raised. Upperparts pale pearl-grey, underparts and medium-length forked tail white; rump white. Wings long and narrow with very white underwings, primary tips dark grey; wing-tips extend just beyond tail at rest. Longish legs black. Immature has mottled upperparts and may lack pale tip on shorter bill. **Similar species** In good light conditions the pale yellow tip to all-black bill eliminates all other dark-billed terns. Immature has longer, slender more pointed wings than thicker billed, less forked-tailed Gull-billed Tern.

HABITS Most frequent along the coast, less so on mangrove creeks and on rivers. Plunge-dives, frequently submerges. Attracted to fishermen using hand-cast nets, catching small escaping fish. Frequently encountered resting on groin posts.

VOICE A piercing rasping *kirrink* or *kirr-kitt* and a drawn out *kier-vek*.

STATUS AND DISTRIBUTION Seasonally common to locally very abundant Palearctic migrant; largest numbers mainly November to March but recorded in all seasons. Found along the river to LRD and occasionally further east. Widespread along the whole coast in Senegal.

Identification of Roseate, Common and Arctic Terns

Roseate, Common and Arctic Terns are considered difficult to separate, particularly in the non-breeding and immature plumages prevalent in our area. However, some individuals in more or less full breeding plumage may occur September-October after breeding, or prior to northward departure around April. Non-breeders may be around in West African waters year-round. In general, all three are shades of pale grey above and white below and are usually dark-billed: plumage differences are subtle, and shape is often the best guide.

ROSEATE TERN
Sterna dougallii   Plate 22

Fr. Sterne de Dougall

IDENTIFICATION 38cm. Medium-sized slim-looking tern, much whiter than Common and Arctic Terns and shorter-winged. Distinctively long tail streamers in breeding plumage, extending well beyond wing-tips; slender black bill usually with some red at base, legs red; rosy-tinged below; forehead, crown and nape black. In non-breeding plumage, streamers absent or broken, forehead and crown speckled white. Upperparts pearl-grey, lacking contrast between back, rump and tail. In flight, no dark trailing edge of outer wing at any age and, against light, broad translucent area on secondaries and innermost primaries. Juvenile resembles small juvenile Sandwich Tern; bill and legs dark. **Similar species** Always longer-billed and paler than Common or Arctic in flight, showing substantial areas of mid-grey on primaries above and below.

HABITS Flies with quick stiff wingbeats. Does not hover and plunge-dive with the regularity of Arctic and Common; normally over deeper water flying at 8-15m above the surface.

VOICE A soft *chuwik* and a loud hoarse rasping *raaaach*, lower in pitch than Common or Arctic.

STATUS AND DISTRIBUTION Rare Palearctic migrant, probably under-recorded. Four recent records, all off southern Gambian beaches in September 1992. Older record in July from coastal WD. Frequent on spring passage off northern Senegal, March–June.

BREEDING Curious record in May 1980 in southern Senegal is the only sub-Saharan breeding record.

COMMON TERN
Sterna hirundo   Plate 22

Fr. Sterne pierregarin

IDENTIFICATION 32-38cm. A small-medium tern, slim and elegant with longish legs. Most likely to be seen in non-breeding plumage when bill black, maybe with some red at base, legs dull red, and forehead whitish. At rest, tail does not project beyond wing-tips. In breeding plumage has bright red bill tipped black (no black tip on breeding Arctic), legs and feet red, crown and forehead all black. Slender pointed wings, deeply forked tail. In flight, outer primaries grey (darker than Arctic) with slim black leading edge and broader blackish band along trailing edge; inner primaries paler, giving impression of a paler window in the mid-wing. **Similar species** With experience Common's longer legs distinguish it from shorter-legged Arctic when standing on the beach. Differences in wing shape and translucency are a useful guide in all seasons.

HABITS Fishes offshore, frequently in flocks, usually by typical plunge-diving, but may in some circumstances feed by surface-picking.

VOICE Flight call a *kick kirikiri* or *kree-yair* with emphasis on the drawn-out *yair.*

STATUS AND DISTRIBUTION Palearctic migrant and regional breeder. Frequent to abundant along the coast November-July, uncommon during the rains. Widespread along the whole coast in Senegal.

BREEDING Recorded in northern and southern Senegal.

ARCTIC TERN
Sterna paradisaea   Plate 22

Fr. Sterne arctique

IDENTIFICATION 33-35cm. A small-medium tern, very similar to Common Tern but at rest very short-legged, and tail extends a short way beyond wing-tips. Comparatively slightly shorter-billed and smaller-headed than Common Tern. Non-breeding: legs orange-red or blackish-red, depending on age, becoming red when breeding. Bill blackish possibly with orangey base, becoming uniform coral-red with no black tip in breeding plumage. In flight wings uniformly pale grey, at close range almost translucent, with little contrast between inner and outer primaries. Black leading edge to primaries rarely visible, black trailing edge more precisely defined than Common. **Similar species** See Common Tern.

HABITS Likely to be found at mixed tern assemblages along the shoreline or fishing offshore, associating with similar-sized species. Feeds by plunge-diving, occasionally by surface-picking.

VOICE Higher, clearer and shorter than Common Tern. Di-syllabic *kriii-a* with accent on first syllable; also *kreer* and staccato *kt-kt-kt...*

STATUS AND DISTRIBUTION Rare Palearctic migrant with a few recent coastal records in September 1995. Older records in May-July. Probably under-recorded and may pass by further out to sea. Passage records of substantial numbers off northern Senegal in April and late August-early September.

BRIDLED TERN
Sterna anaethetus   Not illustrated

Fr. Sterne bridée

Other name: Brown-winged Tern

IDENTIFICATION 32cm. Small-medium tern, characteristically grey-brown above and white below, with a deeply forked tail. Shows little seasonal variation. 'Bridle' created by neat narrow white forecrown extending as a short supercilium above and behind the eye (lacking in Sooty Tern). Pale collar below black crown sometimes visible. Crown darker than back and wings. Bill and legs black. Non-breeding birds and immatures browner above, with streaked crown, less defined bridle and shorter

215

tail. **Similar species** See Sooty Tern.

HABITS Likely to be seen out at sea but not truly pelagic. Flight graceful, with quick wing-beats.

VOICE High pitched *kee-yharr*, yelping *wep-wep* and a grating *karr*.

STATUS AND DISTRIBUTION No Gambian records as yet, but to be expected. Several northern Senegal records of singles or small groups, April, June and July.

BREEDING Unsuccessful attempt in northern Senegal, June. breeds in Banc d'Arguin off Mauritania May-July and in Gulf of Guinea October-November. Recently (1996) confirmed at Langue de Barbarie Reserve and Ile des Madeleine.

SOOTY TERN
Sterna fuscata   Not illustrated

Fr. Sterne fuligineuse

IDENTIFICATION 45cm. Medium-sized, appreciably larger and much blacker than Bridled Tern. Very dark sooty-black upperparts with uniform black crown showing no collar; underparts white. White forehead does not extend into a supercilium. Shows little seasonal variation. Wings long and pointed with white leading edge, visible also at rest. Tail dark, deeply-forked, outer feathers edged white; swallow-like when flared. Bill and legs black. Immature all dark above and below, flecked white, reaching adult plumage after two years. **Similar species** Bridled is smaller and browner, with a pronounced white eyebrow. Immature Sooty can be distinguished from noddies by forked, not wedge-shaped tail, and by speckled not uniform brown upperparts.

HABITS More pelagic than most terns; follows ships. Forages mainly at night. Flight buoyant and undulating, sometimes soars on motionless wings.

VOICE Diagnostic raucous, *ker-wacki-wake*, hence alternative name of Wideawake, also *kreaa*.

STATUS AND DISTRIBUTION Single unconfirmed record of two birds off coastal WD in January 1973; but presence likely. Senegalese records largely restricted to nesting reports.

BREEDING Breeds in small numbers on islands in Saloum delta, northern Senegal, May-July, and recorded further north in April. Recently confirmed at Langue de Barbarie Reserve and Ile des Madeleine 1996.

LITTLE TERN
Sterna albifrons   Plate 22

Fr. Sterne naine

Other name: Least Tern

IDENTIFICATION 25cm. Smallest tern of the region, long-winged and short-bodied with a distinctive hurried jerking flight. Non-breeding birds have extensive white forehead with black streak through eye to nape. Straight bill dark when non-breeding but close views show hint of yellow at base, legs brownish-yellow. Breeding adults have neatly demarcated white forehead extending as a narrow V mark just over the eye; rest of head to nape black; bill yellow with black tip, legs yellowish. Upperparts pearl-grey, underparts white; short white forked tail and white rump. In flight shows upperwing pale dove-grey with darker primary tips and white underwing. Immature scalloped grey-brown above with dark shoulder on closed wing. **Similar species** Marsh terns are all shorter-billed, larger and usually darker bodied, with a more relaxed flight.

HABITS Usually feeds close to the beach diving repeatedly into the shallows, occasionally hovering, head down, before plunging. Gathers to roost alongside other larger terns.

VOICE A brisk and trilled *kikikik* or *kree-ick* penetrating but not harsh.

STATUS AND DISTRIBUTION Frequent to locally common, occasionally abundant, along the coast. Occasionally inland to LRD. Most occur November to April but recorded year-round, rare May-July. Palearctic migrant and regional breeder. Recorded along the whole coast in Senegal.

BREEDING Race *S. a. guineae* breeds northern Senegal April-May. Single Gambian record of adults feeding youngsters of unknown age, in coastal WD in August 1966 but no nest found.

Marsh Terns

Three species of small, compact-looking terns compared to the generally larger *Sterna* 'sea terns'. Acrobatic in flight, they have a squarish tail, notched but lacking any pronounced fork. Black Terns occur in vast seasonal feeding flocks, as to a lesser extent do White-winged Black Terns, but Whiskered Terns appear usually in small numbers or alone. They all feed buoyantly over the sea and freshwater e.g. flooded fields, paddies and sewage farms by dipping and surface-picking, but the rare Whiskered Tern may belly-plunge on occasions and tends to fly in direct lines. All three species may appear in any plumage.

WHISKERED TERN
Chlidonias hybridus   **Plate 22**

Fr. Guifette moustac

IDENTIFICATION 25cm. Least frequent marsh tern, plumper, longer-billed and flatter-headed than other two species, in flight more akin to a small sea tern. Distinctive breeding plumage, with lead-grey body separated from black forehead and crown by obvious white cheeks. Underwing and tail white, longish bill red and stout, legs orange. Non-breeding birds have white forehead, underwing and underparts, the latter sometimes patchy grey. Crown speckled black on white, becoming black at nape; rest of upperparts light grey, tail almost white. Immature similar to non-breeding adult but with brownish scalloped back turning grey; bill and legs black. **Similar species** Lead-grey body, darker below with white vent prevents confusion with any small, black-crowned, white-plumaged sea tern in breeding plumage, as do shape and flight. Absence of black patch behind the eye and black pectoral patch on non-breeding Whiskered avoids possible confusion with the other, smaller marsh terns, which rarely fly in a straight line and show more black on the face and head.

HABITS Most likely near or on the coast, occasionally inland on irrigated and extensive rice-growing areas. Dips and swoops but also performs long, straight, patrolling flights.

VOICE Calls *krip-krip*, *kree-ah*, *kik-kik* and a snappy *eeirik*.

STATUS AND DISTRIBUTION Uncommon to frequent migrant at the coast and inland to CRD; probably under-recorded. Recent records of singles and small parties April-June are all from coastal WD and NBD, with a November CRD record. Older records in July-April include flocks of 30 inland in January and 14 in coastal WD in February. In Senegal widespread on the coast and on rice paddies and other wetlands throughout.

BLACK TERN
Chlidonias niger   **Plate 22**

Fr. Guifette noire

IDENTIFICATION 24cm. Commonest marsh tern, unmistakeable in breeding plumage with jet black head and body, uniform dark dove-grey wings, rump and tail. Non-breeding birds have white underparts, forecrown and collar around sides of neck. Hindcrown and nape black, extending as broad mark onto ear-coverts and below eye. Black-

ish smudge on sides of white chest, showing as a small vertical peg in flight, is diagnostic. In flight leading edge of wing appears conspicuously darker than rest of grey upperwing; underwing greyish-white. Bill subtly downcurved, black-tinted red in breeding season; legs red-brown, redder when breeding. Immature closely resembles non-breeding adult but with browner scalloped upperparts, grey rump and paler legs. **Similar species** Slightly longer-billed than White-winged Black Tern, which has shorter rounded wings and slighter stiffer wingbeats. In breeding plumage, White-winged has black and white underwing, not grey. Tail grey in Black Tern but strikingly white in White-winged. In non-breeding plumage, diagnostic black pectoral smudge on side of neck at wing junction of Black Tern is absent in White-winged, but demands close, careful examination. Also, hindcrown of White-winged is speckled and flecked black and white, not all-black as on Black Tern. At rest White-winged stands taller than Black .

HABITS Highly gregarious, sometimes in compact flocks hundreds or thousands strong. Flies low over water with an airy dipping and swooping action, accompanied occasionally by loose fluttery flight. Will hunt high over land to catch flying ants. Used by fishermen as a guide to fish presence.

VOICE A squeaky *krrek* or *ki-ki-ki* and a nasal *ski-aa* but not very vocal.

STATUS AND DISTRIBUTION Seasonally abundant to very abundant offshore along the whole coastline, also along the river up to CRD. Recorded year-round with passage flocks particularly prominent a few hundred metres off the beaches June-October and greatest numbers at the end of the rains in most years. Recorded along the whole coast in Senegal with occasional inland records.

WHITE-WINGED BLACK TERN
Chlidonias leucopterus   **Plate 22**

Fr. Guifette leucoptère

Other name: White-winged Tern

IDENTIFICATION 23cm. A marsh tern unmistakable in breeding plumage, with jet black head, body and underwing-coverts, contrasting strikingly with silvery-white upperwing, tail and underwing flight feathers. Bill crimson, legs vermillion. Non-breeding birds commonly show same breeding season features and look patchy. Otherwise forecrown, underparts and sides of neck white; hindcrown speckled with black (not solid black), large black patch behind and below. Lacks smudgy black pectoral patch seen on sides of chest in Black Tern Rest of upperparts grey (paler than in Black Tern), slightly mottled on wing coverts. In flight, shows white tail (grey in Black Tern). Bill black, legs brownish. Immature as non-breeding adult but with darker back admixed brownish-grey. **Similar species** See Black and Whiskered Terns.

HABITS In most aspects similar to Black Tern. Passage through Gambian waters of White-winged in the first half of the year usually precedes the appearance of Black Terns, while Black far outnumber White-winged in the latter third of the year. The sewage farm at coastal WD is a reliable location to study marsh terns.

VOICE Raucous, brisk *kik-kik-kik-kik*, *kwek* or *kreek*.

STATUS AND DISTRIBUTION Frequent to seasonally abundant passage migrant along the coast and along the river to CRD. Most evident March-May; numbers much reduced in mid-late rains. Recorded along the whole coast in Senegal and inland along the Senegal River with occasional other inland records.

BLACK NODDY
Anous minutus   Not illustrated

Fr. Noddi noire

Other name: White-capped Noddy

IDENTIFICATION 35cm. Much blacker-plumaged and slightly smaller than Common Noddy, about the size of a Common Tern. Top of head very white, not grey, softly merging into greyer nape, sharply demarcated at black lores. In flight shows uniformly dark underwing with no contrasting fringes. Black bill longish and slender, eye dark brown with white crescent above and below, legs brown. Immature similar to adult with buff fringes to body feathers.

STATUS AND DISTRIBUTION Vagrant from southern hemisphere. Two records, October-November 1977 and November 1984, both off coastal WD. Has bred on Gulf of Guinea islands.

BROWN NODDY
Anous stolidus   Not illustrated

Fr. Noddi brun

Other name: Common Noddy

IDENTIFICATION 39cm. A uniformly sooty-brown tern about size of a Sandwich Tern with a wedge-shaped tail that is notched when fanned. Whitish forehead contrasts sharply with black lores, the lavender-grey crown merging to brown at the nape. Small but conspicuous white crescents above and below eye. Pale brown underwing with dark fringes. Bill robust and black, legs brown-black. Immature paler brown with a less well-defined greyish cap.

STATUS AND DISTRIBUTION Vagrant from southern hemisphere. One coastal WD record in August 1974.

SKIMMERS

Rhynchopidae

One species. Unusual bill highly adapted to specialised feeding technique. Despite obvious presence of the species' ecological requirement of a slow watercourse over which to hunt, the lack of substantial sandbanks may conceivably restrict its status in The Gambia.

AFRICAN SKIMMER
Rynchops flavirostris   Plate 22

Fr. Bec-en-ciseaux d'Afrique

IDENTIFICATION 38cm. Unmistakable large tern-like bird with curious orange-red bill. Plumage at distance looks a clean-cut black and white; at closer range lower mandible 2.4cm longer than upper, upperparts and closed wing black-brown, showing a white line across secondaries. Long wings project well beyond white-edged and deeply forked tail at rest. Crown, nape and hindneck blackish, forehead, foreneck and rest of underparts clean white. In flight underwing mostly white with black primaries. Noticeably short bright orange-red legs. Immature duller with mottled darker plumage, shorter bill blackish towards tip. **Similar species** None.

HABITS Gregarious, frequents inland rivers, lakes, creeks and lagoons, occasionally on coastal tidal inlets. Typically rests by day on sandbanks, becoming active at dusk, nocturnally on moonlit nights. Characteristically catches fish by ploughing or skimming with its longer lower mandible inserted below the water surface. During this unique method of fishing the wings are held up in a distinctive position.

VOICE A loud and repeated harsh *kip* and a shrill tern-like *kewee*.

STATUS AND DISTRIBUTION Uncommon to rare; possibly under-recorded on under-watched waterways. Recent main river records are in LRD and CRD in July. Recent coastal NBD records of singles in late September and January at Niumi National Park. Earlier authorities associated its appearance with the very end of the dry season into August, but with occasional records from all divisions in most months. Recorded in Senegal in Saloum delta (c. 300 December-January) and on Senegal River and east on upper reaches of Gambia River.

BREEDING Recorded March-April on Senegal River in northern Senegal. Nest a deep scrape in sand.

SANDGROUSE

Pteroclididae

Three species; one common throughout, one rare in The Gambia but common in northern Senegal and one Saharan vagrant to extreme northern Senegal. Sexes differ, but each shows elaborate and cryptic colour patterns of sandy and brown. Resident of dry habitats; characteristics include very short legs, plump bodies and swift direct flight on longish pointed wings with accompanying flight-whistles. Unless flushed, normally seen at dusk and dawn when flocks gather and flight to drink; highly gregarious. Diurnally furtive, they squat low or shuffle silently away. Entirely terrestrial and all nest on the ground. Adult males are famed for carrying water to their chicks in soaked belly feathers, sometimes over considerable distances.

CHESTNUT-BELLIED SANDGROUSE
Pterocles exustus　　　　Plate 14

Fr. Ganga à ventre brun

IDENTIFICATION 32cm. Sexes differ. Small-bodied sandgrouse lacking white facial markings, both sexes with elongated pin-tailed central tail feathers. Male generally warm khaki with buff barring, sides of head, throat and wing-coverts yellow-buff, more golden on the wings. Narrow black band across chest separates greyish-buff upper breast from dark chestnut lower breast and belly. Female has upperparts thickly streaked and barred with black, crown streaked with black, and blackish-brown belly narrowly barred black, with a very narrow broken black band across chest. Immature shows no sexual differences, lacks breast banding and tail elongations, but shows dark belly. Similar species In flight uniform blackish underwing excludes Four-banded Sandgrouse. Blackish belly, wings with narrow white trailing edge and tapered tail are all useful diagnostic field marks.

HABITS Favours parched scrub and semi-desert. Generally in small flocks, often flying high. Sedentary or nomadic depending on conditions.

VOICE Low-pitched melodic *gutter-gutter*, when drinking a musical *creen*, both distinct from Four-banded Sandgrouse.

STATUS AND DISTRIBUTION Rare in The Gambia, with a few recent records in CRD. Under-surveyed NBD could sustain as yet undiscovered populations. Very common in parts of northern and eastern Senegal.

BREEDING Unlined scrape on dry earth. Season in Senegal extended, with records in most months, most in April-June.

LICHTENSTEIN'S SANDGROUSE
Pterocles lichtensteini　　Not illustrated

Fr. Ganga de Lichtenstein

IDENTIFICATION 27cm. Sexes differ. Distinctively small, about size of European Turtle Dove but with plumper chest and short square tail. Male yellowish-buff with upperparts and belly heavily barred with black and white vermiculations except for plain yellow-buff breast patch, edged above and below by two dark bands. Broad white supercilium behind eye and white forehead crossed with black bars. Female duller sandy rufous, narrowly barred black with no black markings on chest or face, appears greyish at distance. Orange bill brighter on male, legs yellowish-orange. Immature resembles female, with upperparts finely vermiculated. Similar species Four-banded Sandgrouse lacks heavy barring. Chestnut-bellied has a long tail.

VOICE Sharp *quitall* or *quwhetto* and a purring *trrrr*, also a frog-like *kwark-kwark-kwark*.

HABITS Nomadic: movements in north-west Africa little understood. Favours stony and hilly subdesert areas with plentiful *Acacia seyal*, the seeds of which it eats.

STATUS AND DISTRIBUTION Vagrant to northern Senegal, with a few records in July and September.

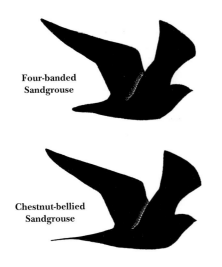

Four-banded Sandgrouse

Chestnut-bellied Sandgrouse

FOUR-BANDED SANDGROUSE
Pterocles quadricinctus   **Plate 14**

Fr. Ganga quadribande

IDENTIFICATION 29cm. Sexes differ, but both show short square tail. Male has conspicuous white and black markings about the forehead. Overall sandy-rufous, back and wings blotched with black, chestnut and white, appearing intricately chequered. Khaki-buff upper breast separated from black-and-white barred lower breast and belly by three bands of chestnut, broad white then black. Female has plain buff head and neck streaked and spotted black, lacking prominent white and black facial pattern. Otherwise plumage as male but duller, with more compact barring, and buff breast lacking bands. Bill dull yellow, legs pale yellow. In flight looks pot-bellied, with characteristic grey underwing and black wing-tips (not uniformly dark as in Chestnut-bellied) and small rounded tail. Immature more rufous, with finer barring. **Similar species** Chestnut-bellied has dark underwings and a long tail. Rare Lichtenstein's is heavily barred.

HABITS Favours a variety of habitats including grassland with scattered trees, coastal thorn-scrub, cultivated land and wooded laterite outcrops. Daytime sightings usually of flushed birds in pairs or singles, otherwise very difficult to locate, relying on good camouflage and skulking behaviour to avoid detection. Evening movement becomes obvious about fifteen minutes before dark when small coveys flying low make their way to water where several hundred birds may congregate, departing abruptly after nightfall.

VOICE Very vocal when coming to drink at dusk and can be heard from far off, a twittering repeated, trisyllabic rising *pip-prrr-trree.*

STATUS AND DISTRIBUTION Seasonally common, recorded in all months; peak in November-March, numbers much reduced during rains when movement north away from The Gambia probably occurs. Less numerous but still frequent at the coast. Common throughout Senegal.

BREEDING Most records in latter half of dry season, early March-June, but one well-feathered chick with adult in February. Nest is a bare scrape in partial shade lined with a few dried leaves.

PIGEONS AND DOVES

Columbidae

Fifteen species, of which one in southern Senegal only, and one in northern Senegal only. Sexes alike in most species. Predominantly graminivorous with two specialised frugivores. Most feed directly on the ground with some (wood doves) having become ground specialists. Very frail platform nest usually with two white eggs; hatchlings (called squabs) are fed on regurgitated 'crop-milk' by the parents. Typically assemble in large flocks to feed and drink. Breeding pairs maintain long-term pair-bonds, nesting primarily in the dry season. The main identification problems arise between the four resident African 'collared doves' and the two wood doves.

AFRICAN GREEN PIGEON
Treron calva   **Plate 23**

Fr. Pigeon vert
Other names: Green Pigeon, Green Fruit-Pigeon

IDENTIFICATION 30cm. *T. c. nudirostris.* Plump heavy-billed pigeon with green breast. Overall plumage bright green and yellow-green. More yellowish-green on the lower underparts, more olive on the back. Wing shoulders tinged mauve, obvious in flight. Bill and cere coral red with a whitish tip.

Feet yellow with fluffy white feathering on thighs. Immature duller overall, mauve on wing absent. **Similar species** Bruce's Green Pigeon has a distinctive grey breast and yellow belly.

HABITS Favours well-wooded coastal areas. Although its range in our area is restricted it is highly mobile within it, appearing and disappearing according to the availability of food. Omnipresent in the few 'forest islands'. Strictly arboreal, congregating at fruiting fig trees to feed. Clambers about fruiting trees like a parrot, often hanging head-down.

VOICE Conversational throaty chuckling, low growling notes and a soft bark. Often heard before being seen.

STATUS AND DISTRIBUTION Frequent to locally abundant resident in WD, slight peak of observations during the rains. Uncommon in LRD and NBD; otherwise largely absent inland. In Senegal along the southern coast and up to 15°N.

BREEDING Latter half of dry season. Frequently nests in swamp date palms and mango trees.

BRUCE'S GREEN PIGEON
Treron waalia   Plate 23

Fr. Pigeon à épaulettes violettes

Other name: Yellow-bellied Fruit-Pigeon

IDENTIFICATION 32cm. Plump, heavy-billed pigeon. Entire head, throat and upper breast ashy-grey. Belly sharply demarcated, characteristic lime-yellow fading to white towards the vent. Wing brownish edged yellow, mauve patch on shoulder visible on closed wing. Bill pale lilac, tipped white, eye blue with red outer ring. Legs and feet orange. Immature duller, mauve on wing absent. **Similar species** African Green Pigeon is uniformerly green below.

HABITS Similar to Similar to African Green Pigeon but favours drier habitats. Also visits mangrove systems and riverine fringes. Nomadic when searching for ripe *Ficus* trees, where they assemble en masse. Frequently forages in association with glossy starlings, orioles, barbets and parrots. Markedly timid and unapproachable.

VOICE A grating creak succeeded by fluting whistles, ending with a growl and a grunt.

STATUS AND DISTRIBUTION Uncommon in WD, becoming frequent to locally abundant in LRD and further east; common in coastal NBD during the rains, but more information needed. Occasional coastal sightings and NBD records are usually

of singletons or small groups and in the company of African Green Pigeons. Widespread throughout Senegal away from coastal districts.

BREEDING Shortly after the rains to late dry season, e.g. late November in CRD on Georgetown Island. Nest well hidden and small.

TAMBOURINE DOVE
Turtur tympanistria   Plate 23

Fr. Tourterelle tambourette

IDENTIFICATION 23cm. Sexes differ. Slightly larger and strikingly distinct from the two other wood doves. Male has dark chocolate brown upperparts with forehead and eyestripe pure white, dark crown and dark patch on ear-coverts. Entire underparts white. Wings brown, with blue-black spots on inner secondaries. In flight chestnut underside of wings, axillaries and flanks are characteristic. Bill purple tipped greenish. Female has the white areas tinged grey. Immature has spotted underparts. **Similar species** The other two wood doves lack the striking white face and underparts of Tambourine.

HABITS Favours dense coastal and riverine forest.

VOICE Call begins with two or three *coos* followed by up to seven shorter notes fading away.

STATUS AND DISTRIBUTION No Gambian records. In Senegal rare in southern Casamance with a few records since 1974, May-July.

BLUE-SPOTTED WOOD DOVE
Turtur afer   Plate 23

Fr. Emerauldine à bec rouge

Other names: Red-billed Wood Dove, Blue-spotted Dove

IDENTIFICATION 20cm. Small, plump, and earthy brown dove with a yellow-tipped red bill. Head pale blue-grey, upperparts earthy-brown, with two black bands across the rump. Underparts pale brown tinged pink, paler towards vent. In flight shows an extensive chestnut flash on axillaries, wing spots change from matt black to shiny metallic violet-blue in sunlight. Immature body feathers edged buff and look scaled, wing spots absent. **Similar species** Black-billed Wood Dove lacks red bill tipped pale yellowish. Typically Blue-spotted is subtly browner not greyer in plumage. Blue-spotted usually occurs in or near heavily wooded areas or on forest floor, uncommon outside WD. Overlap with Black-billed does occur, chiefly at the coast.

HABITS Favours well wooded or thickly vegetated areas; likes moist conditions. Confiding. Usually seen walking and feeding alone or in pairs, often in leaf litter; only gathers in small flocks to drink.

VOICE A series of sombre low-pitched flute-like coos running steadily down the scale, accelerating towards the end; *coo-coo-cu-coo, coo-cu-coo*. Deep, resonant and continuous in forest habitat.

STATUS AND DISTRIBUTION Common year-round in well wooded areas in WD, occasionally up to CRD on both sides of the river. In Senegal found in east but main distribution there is along southern Casamance.

BREEDING Records in October to January.

BLACK-BILLED WOOD DOVE
Turtur abyssinicus   Plate 23

Fr. Emerauldine à bec noir

IDENTIFICATION 20cm. Small, plump, predominantly vinous-grey with black bill. Very similar to earthier-brown Blue-spotted Wood Dove but separable with care. Crown and nape blue-grey fading to whitish on face. Mantle greyish. Two well-spaced dark bars across lower back, and a few large blackish spots on tips of coverts. Wing-spot colour is unreliable. Underparts pale and pinker, fading to white on belly. Immature looks scaly and lacks wing spots. **Similar species** From Blue-spotted Wood Dove by all-black bill and black loral line, and overall paler, greyer plumage.

HABITS Ubiquitous, but less common in closed forest. Grass seeds are principal food. Feeding on charred ground often darkens plumage. Flight rapid, direct and normally low, commonly swerving and banking away at chest height. Does not fly in flocks but congregates in good numbers at waterholes.

VOICE Similar to Blue-spotted but possibly not as persistent. A commonly-heard soporific background sound at all times of day.

STATUS AND DISTRIBUTION Common to abundant, especially in dry savanna woodland, in all divisions year-round. Widespread and common throughout Senegal.

BREEDING At the end and after the rains. Nest is normally low and often in a thorny acacia or mimosa, or on the frond of an emergent palm.

NAMAQUA DOVE
Oena capensis   Plate 23

Fr. Tourterelle du Cap

Other name: Long-tailed Dove

IDENTIFICATION 26cm including long pointed graduated tail. Sexes differ. Male has distinctive black mask over face, throat and breast, rest of head and neck grey, upperparts soft grey-brown. Underparts white. Closed wing rufous and grey-brown with blackish border, up to five iridescent blue-violet spots near tertials. In flight underwing rufous. Bill orange with a purplish base. Female lacks black facial mask otherwise similar but duller. Immature looks paler and scaly. **Similar species** Only dove with long tail in our area; long central feathers brown, graduated outer feathers black.

HABITS Favours drier regions where it forages for minute seeds on patches of bare ground; often comes to roads and tracks. Retreats north at the onset of the rains when fresh growth carpets the ground and interrupts its foraging routine. Flight direct and very fast, usually at no great height. Assembles in numbers at village wells and roadside pools. Perches low, often in groups. Courtship involves a miniature peacock-like display, with the tail cocked up and fanned.

VOICE Unobtrusive; a soft di-syllabic *koo-koo* and a longer quiet *woooo*.

STATUS AND DISTRIBUTION Frequent to seasonally very abundant throughout, largest numbers December-January, mostly in LRD, NBD and east to URD. Locally common at the coast. Disappears almost entirely at height of the rains from August to October. Widespread and common throughout Senegal.

BREEDING Nesting December 1995 in LRD (Batelling, Kiang West). Breeds in northern Senegal in all months.

SPECKLED PIGEON
Columba guinea   Plate 23

Fr. Pigeon de Guinée

Other names: Speckled Rock Pigeon, Senegal Spotted Pigeon

IDENTIFICATION 41cm. Largest of the region's pigeons. Conspicuous elliptical bare red skin patches around eye reaching to ear-coverts. Head and upperparts blue-grey, neck brown speckled pearl-grey. Mantle and wings vinous-chestnut, wings heavily speckled with white spots. Bill blackish with whitish cere, legs purplish-pink. Immature browner with browner eye-skin. **Similar species** Only pigeon in the region with wings peppered with white triangular spots and bare red mask-patches around eyes.

HABITS Prefers open country with nearby rhun palms and baobabs, Has adapted well to urbanisation and congregates in large numbers at grain and groundnut dumps and on harvested ground, often seen sitting on roof tops of town buildings. Gregarious. Flies strongly, sometimes at considerable height.

VOICE Loud, explosive cooing, *doo doo doo* or *whu-whu-whu-whu-WHU* gathering pace, force and volume.

STATUS AND DISTRIBUTION Common breeding resident to locally abundant except in LRD where scarcer. Numbers decrease during rains. Widespread and common through most of Senegal, but scarce in central parts.

BREEDING Mostly in latter half of dry season. Nest a considerable structure for a pigeon, made of twigs, lined with grass, often on the shaded ledge of a building.

ROCK DOVE
Columba livia   **Not illustrated**

Fr. Pigeon biset

Other names: Rock Pigeon, Feral Pigeon

IDENTIFICATION 33cm. *C. l. gymnocyclus*. A largish slate-grey dove, with a white rump and underside of wings. Two conspicuous black lines across the closed wing provide a diagnostic field mark. Neck and throat iridescent violet-green in sunlight. Bill black, legs and feet red. Eye yellowish with red lids. Immature lacks iridescence. **Similar species** Only dove with wing-bars and iridescent neck.

HABITS Favours rocky coastal cliffs and associated habitat not available in The Gambia. Usually but not always gregarious. Note: Rock Dove is the ancestor of the domestic pigeon frequently seen around Gambian villages; many of these show variable plumage characteristics such as white or blotchy feathering; others are 'wild type' and when feral may be impossible to distinguish from wild Rock Doves.

VOICE Call cooing *dooo-roo-dooo* in a display flight that involves a soaring-glide on spread wings.

STATUS AND DISTRIBUTION Not recorded in The Gambia. Inhabits coastal cliffs and caves at a few sites in northern Senegal and in the extreme south-east.

RED-EYED DOVE
Streptopelia semitorquata   **Plate 23**

Fr. Tourterelle collier

IDENTIFICATION 34cm. A largish, dark collared dove. Upperparts umber-brown tinged slate-grey, nape, neck and throat a shade paler; underparts a very deep vinous-pink. Crown and face pale dove-grey, and this subtle contrast with the darker body shows well in flight. Iris brown with a red rim, bare skin around the eye claret. Broadish black collar on hindneck; although fringed pale grey not as well-defined as the very white borders on the collar of smaller paler Mourning Dove. Tail brown with a median dark band, outer tail feathers grey not white. Immature collar indistinct, buff edges to body feathers giving scaly appearance. **Similar species** Largest and darkest of the region's collared doves; lacks white fringes to collar. See African Mourning Dove.

HABITS Ubiquitous, disperses well away from water into arid habitats. Found in pairs and in single and mixed species flocks. Single birds often occupy a high bare branch of a woodland tree to call. Frequent audible wing-clapping displays.

VOICE Loud and far reaching, a repeated series of chugging and chanting coos, *Do-Do du-du do* with slight emphasis on the opening *Do-Do* then a negligible pause before the *du-du-do*.

STATUS AND DISTRIBUTION Common to very abundant resident throughout. In Senegal widespread in the south, rare or absent in the north.

BREEDING Extended season; variety of sites.

AFRICAN MOURNING DOVE
Streptopelia decipiens   **Plate 23**

Fr. Tourterelle pleureuse

Other name: Mourning Dove

IDENTIFICATION 31cm. *S. d. shelleyi*. Medium-large dove. Upperparts grey-brown, narrow black collar with paler edges, head paler, contrasting with darker body. Underparts soft grey. Eye yellow or light orange with contrasting red orbital ring, looking piercing. Collar particularly obvious when neck feathers are puffed up or head is bowed in display. Tail grey, outer feathers broadly tipped white above, all white below. Immature browner, scaly russet edging to body feathers. **Similar species** Larger, darker Red-eyed Dove lacks pale grey head contrasting with grey body; Red-eyed has whitish forehead contrasting with darker, vinous-tinted browner

223

body. In flight Mourning lacks dark distinctive band across the fanned uppertail of Red-eyed.

HABITS Usually associated with flowing water, favouring riparian forest with adjacent acacias, also mangroves in which it nests. Feeds largely on bare ground; often perched on telegraph wires at the coast and feeds on roadsides. Gregarious, mixes with other doves. Display flight involves vertical ascent with fluttering gliding descent. Particularly vocal at first light.

VOICE A fast falling trill or an elated treble gargle, *krrrrrrrrr, oo-OO oo, oo-OO, krrrrrrrrr, oo-OO oo, oo-OO.*

STATUS AND DISTRIBUTION Locally common resident at the coast and abundant to very abundant along the river in CRD and URD. In Senegal, widespread in suitable habitat in the north, restricted or absent in the south.

BREEDING Extended season. Nests are usually near to water.

VINACEOUS DOVE
Streptopelia vinacea Plate 23

Fr. Tourterelle vineuse

Other name: Vinaceous Collared Dove

IDENTIFICATION 25cm. Smallest and by far the commonest of the region's collared doves. Upperparts uniform earthy grey-brown, head and underparts paler and pinker. Palest on forehead, chin and throat. Brown eye lacks eye-ring. Black half-collar of hindneck obscurely fringed grey on the upper edge. Uppertail outer feathers broadly tipped white, uppertail-coverts and middle pair of feathers brown like the back. Undertail with broad white margin. Distinctive slate-grey underwing-coverts, axillaries paler but still dark. **Similar species** Main confusion species is the paler desert-edge associated African Collared Dove. Other common, resident collared doves are larger and greyer.

HABITS Avoids the immediate proximity of human habitation, otherwise ubiquitous where there is an area of bare dry ground upon which to feed. Flight displays commonplace. Congregates in many hundreds in mixed dove flocks at receding water-holes at the start of the dry season.

VOICE Usually four syllables, a tedious *coo-cu-cu-coo,* or 'pieces-of-eight, pieces-of-eight', like the rumbling of a train; incessant even at night and in the heat of the day when other birds are silent.

STATUS AND DISTRIBUTION Abundant resident throughout, occasionally locally very abundant in

some areas e.g. in LRD just after the rains. Widespread throughout Senegal.

BREEDING Extended season. Favourite site is in an acacia sapling. Several observations of nests adjacent to solider-wasp nests.

AFRICAN COLLARED DOVE
Streptopelia roseogrisea Plate 23

Fr. Tourterelle rieuse

Other names: Rose-grey Dove, Pink-headed Dove

IDENTIFICATION 29cm. Similar to but larger than Vinaceous Dove. Head, nape, neck and upperparts uniform sandy-brown paling to grey on the forehead; tail darker than mantle. Underparts pinkish, belly whiter. Black collar on hindneck edged white on upper margin. Pale grey orbital ring surrounding dark red eye. Immature scalloped with less well-defined collar. **Similar species** Separable from ubiquitous Vinaceous Dove on voice and size. Plumage much paler on African Collared. In flight African Collared looks longer-necked. Underwing-coverts are conspicuously white in African Collared, greyer in Vinaceous.

HABITS Principally a species of desert edge, and an indicator of the Sahel. Favoured habitat includes dry thornbush, open desert with annual grass and arid farmland up to 25km from water. Not likely to be found in or near riverine forest. Gregarious, numbers ranging from a few to thousands. Inter-regional migrations and local irruptions recorded.

VOICE Begins with a clear high pitched hiccup, followed after a brief pause by a turtle-dove like croon *ook-crrruuuu.* Also a nasal laughing call of even-spaced notes *wah-wah-wah-wah-wah.*

STATUS AND DISTRIBUTION Uncommon; possibly under-recorded. A few records involving flocks of 25 to 100s mostly from NBD and URD in the early dry months and again in the early rains. Very common in far northern Senegal, some records from east and south.

BREEDING Nest usually quite low in an acacia.

EUROPEAN TURTLE DOVE
Streptopelia turtur Plate 23

Fr. Tourterelle des bois

IDENTIFICATION 26cm. Small slender dove distinguished by rich bronze mantle with black feather centres producing a chequerboard pattern. Head and neck grey, tinged blue, with four diagonal white lines on a black ovoid patch either side of neck. Underparts salmon-pink paling to white on lower belly. Tail dark bluish-grey tipped pure white except central feathers; looks black and white in flight. Bill blackish, legs dark pink. Immature duller, lacks neck patches, brownish underparts scalloped. **Similar species** Only dove with black and white patch on neck, no frontal necklace, and dappled wings. Typically in vast flocks on migration.

HABITS Highly gregarious, migratory flocks estimated at up to one million birds. Amasses to feed in large numbers at riverside peanut landing sites.

VOICE Not recorded on migration. A monotonous, sleepy purred *roor-r-r* on breeding grounds.

STATUS AND DISTRIBUTION Dry season visitor from the Palearctic in annually varying numbers, recorded September-May. Fluctuations in The Gambia may be due to shifting conditions across the Sahel. Some years very abundant in CRD (e.g. Jahali-Pacharr) and URD (near Basse) mostly November-April, with a distinct peak in January in most years. Less frequent in areas of WD and LRD and in mangroves in NBD. In northern Senegal arrival begins in late August, with the southbound passage much less obvious than northbound movement which begins in February. North African race *S. t. arenicola* and the European nominate race are both present in northern Senegal. Grossly overhunted north of the Sahara; protected in The Gambia.

ADAMAWA TURTLE DOVE
Streptopelia hypopyrrha   Plate 23

Fr. Tourterelle poitrine rose

IDENTIFICATION 31cm. Sexes alike. Largish dark dove. Recognised by a solid black half collar (with no white) on either side of neck. Crown grey, forehead, face and throat silver-grey. Mantle and rump feathers earth brown with paler edges. Throat and upper breast grey; lower breast and belly pinkish-cinnamon. Wing feathers black edged tawny and white. Tail black-brown tipped grey except the central feathers. Eye red with darker red fleshy outer ring. **Similar species** Paler European Turtle Dove has a black and white neck patch and lacks the grey breast of Adamawa.

HABITS Reported from tangled riverside vegetation and mango orchards in varying numbers mixed with much greater numbers of European Turtle Dove.

VOICE No Senegambian information. In Nigeria

a series of deep rolling notes, developing sometimes into a four-note *croorr crr-croor coor.*

STATUS AND DISTRIBUTION Poorly known in Senegambia. Detailed reports of small numbers (N. Borrow pers. comm.) from both banks of the river in CRD (near Georgetown) and URD, and in different seasons since 1990. Reported also from Niokolo Koba in southeast Senegal. Formerly regarded as endemic to the Jos-Bauchi mountain plateau from north central Nigeria and northern Cameroon, where it breeds. Although nomadism has been noted, movements are stated as probably only local; vagrant recorded in Togo. It remains to be established if any birds in our area are of the Nigerian form, or a second geographically isolated population. Further records should be submitted to the authors with full information, preferably including photographs and sound recordings if possible.

BREEDING Nearest Nigeria; eggs March-December.

LAUGHING DOVE
Streptopelia senegalensis   Plate 23

Fr. Tourterelle maillée

Other name: Palm Dove

IDENTIFICATION 25cm. Smallish, slender dove with uniform rusty brown upperparts, pale blue-grey wing shoulders and a speckled necklace across the upper breast. Note absence of black collar on back of the neck. Head pale pink. Underparts vinous paling to whitish on the belly. Up to ten rows of black spots across breast (less prominent on female) creates the necklace. In flight shows abundance of white on tail margins. Immature similar but with a rufous wash and less pronounced necklace spots. **Similar species** European Turtle Dove lacks spotted necklace but has dark feather centres on upperparts, and black-and-white collar patches.

HABITS Found in most habitats in The Gambia, including village compounds, fields, bush, woodland and swamp, less common in forest-thicket. Spends much time on the ground feeding and intermittently sunbathing. Often extraordinarily confiding. Distinctive towering display flight, gliding down on open wings. Usually solitary or in loose groups. Assembles in considerable flocks with other doves to drink at receding waterholes after the rains.

VOICE Soft conversational cooing *oo-took-took-oo-roo* with emphasis on *took-took*; effect is more of an extended chuckle than a laugh.

STATUS AND DISTRIBUTION Common resident, locally very abundant, in all divisions. Little seasonal variation. Widespread and common throughout Senegal, but less frequent in heavily forested areas to the south.

BREEDING Extended season. Often nests close to human activity, nest is a very meagre platform of thin twigs and rootlets.

PARROTS

Psittacidae

Three species, two widespread and common, one rare and in need of protection. Noisy birds of wooded country with plumages dominated by shades of green. Bills and feet strong and adapted to climbing and to opening diverse hard seeds and fruits. Flight characteristically swift and direct, usually at tree-top height.

BROWN-NECKED PARROT
Poicephalus robustus   Plate 29

Fr. Perroquet robuste

Other name: Cape Parrot

IDENTIFICATION 30cm. *P. r. fuscicollis*. A large, short-tailed parrot with disproportionately large horn-grey bill and silvery-grey head giving a characteristically top-heavy appearance. Upperparts greenish-yellow, tinged bluish-grey particularly on lower back and rump, lacking any conspicuous bright yellow; underparts greenish-brown. Line of red on shoulder of closed wing; front of thighs scarlet. Female has reddish forehead patch. Immature duller and browner. **Similar species** Senegal Parrot is smaller with yellow underparts and a dark bill.

HABITS Characteristic of the interface between tall riverine mangrove forest and adjacent dry savanna woodland, which provides for both its nesting and foraging requirements. Develops predictable diurnal movements from roosts to various feeding areas and back, when favoured savanna trees are flowering and fruiting; may join mixed parrot flocks at groundnut harvests. Flight is strong and powerful with all wing-beats appearing to be below body-level. Most noticeable on the wing during the first and last hours of daylight, usually in pairs or small groups. Generally wary and unapproachable.

VOICE Deeper, more raucous, but not as persistent as Senegal Parrot: disyllabic *zzk-eek* uttered whilst feeding and in flight.

STATUS AND DISTRIBUTION Very locally frequent, with a modest breeding population in LRD; recorded in most months but infrequent in WD and CRD. Recorded from *Terminalia* woodland across the whole length of Senegal-Gambian border to the north and down to the east and in coastal Casamance. Comparison with earlier records

indicates a decline, hence a need for habitat protection.

BREEDING Feeding fledged young in April in LRD; other regional records February-April, likely sites being holes in tall *Rhizophora* mangroves or baobab.

NOTE This race is under current taxonomic review, and may be elevated to full species level.

SENEGAL PARROT
Poicephalus senegalus   Plate 29

Fr. Perroquet youyou

Other name: Yellow-bellied Parrot

IDENTIFICATION 23cm. Sexes similar. Medium-sized, plump, green-backed, yellow-fronted parrot with a short tail. Head and nape grey, paler around the ear-coverts. Underparts yellow, tinged orange, rump yellow. Upperparts and wings darkish green. Bill grey, eye yellow. Immature duller with a browner head. **Similar species** Brown-necked is larger and lacks the yellow belly.

HABITS Favours savanna woodland, forest thickets and farmland with scattered large trees; also well-established gardens. Nomadic; seasonal food availability affects movements. Normally in pairs or small family parties occasionally in flocks up to 25. Quick to exploit any seasonal fruiting or blossoming tree, but often spends considerable time in the crown of isolated baobabs. Noisy.

VOICE Varied quarrelsome screeches and high-pitched whistles.

STATUS AND DISTRIBUTION Common breeding resident, occasionally abundant, in all divisions in appropriate habitat. Little seasonal variation in numbers but overall more numerous in LRD. Widespread and common in Senegal where lamentably it is exploited for the cage-bird export trade.

BREEDING Normally at the end of the rains, in disused holes of woodpeckers and barbets, often in a rotting oil palm, or in baobabs.

ROSE-RINGED PARAKEET
Psittacula krameri   **Plate 29**

Fr. Perruche à collier

Other names: African Ring-necked Parakeet, Long-tailed Parakeet

IDENTIFICATION 40cm including tail. Sexes differ. A large but slim and elegant long-tailed grass-green parakeet. Male green, yellowish on lower belly and underwing, with blackish flight feathers. Narrow black line extends from cere to eye, chin black. Inconspicuous, rosy-pink collar around hindneck and blue tinge to nape and sides of face. Tail as long as body, green with bluish central feathers. Female lacks black markings and collar, and is shorter-tailed. Bill dark red becoming black on upper mandible, lower mandible black, red towards base. Immature similar to female but duller with shorter tail, bill pink. **Similar species** Other parrots of the region are short-tailed.

HABITS Habitat as for Senegal Parrot. Sizeable flocks may be encountered looking for food, and will descend to ground level to feed on seed heads. Breeding display complex. Birds with curved tails indicate they are currently occupying a nest-hole, and breeding pairs often have predictable daily schedules.

VOICE Vocal in flight and when feeding a variety of shrieks and screams.

STATUS AND DISTRIBUTION Common breeding resident, sometimes abundant, in all divisions in appropriate habitat. Little seasonal variation in numbers. Common and widespread throughout Senegal.

BREEDING Nests in holes; well-weathered baobabs are a favoured site. Season extended.

TURACOS AND PLANTAIN-EATERS
Musophagidae

Three species. A group of frugivorous, highly arboreal birds, with uncertain evolutionary affinities. Agile but look ungainly in flight. Sexes alike. Pair-bonding is strong and involves much mate-feeding even outside the breeding season. Highly vocal. Chicks have a markedly primeval and reptilian look, and are able to clamber away from the nest well before they can fly. Nests are sparse and pigeon-like, building material being gathered by snapping off twigs with the bill. Unique feather pigmentations are turacoverdin (green) and turacin (crimson), both copper-based uroporphyrins.

GREEN TURACO
Tauraco persa   **Plate 29**

Fr. Touraco vert

Other names: Green-crested Turaco, Guinea Turaco

IDENTIFICATION 43cm. *T. p. buffoni*. Upperparts, head and neck bright grass-green, showing a waxy sheen in good light. Wings, tail and back much darker green with a glossy violet iridescence. Primaries and outer secondaries brilliant crimson with black tips, striking in flight. Distinctive facial and head markings include a rounded erectile crest of soft green feathers, an unbroken white cresent from base of bill over the eye, then fragmenting into an inconspicuous slender white collar. Bare scarlet orbital skin is bordered with black. Bill orange-red, legs black. Immature similar but much duller with less crimson in the wing. **Similar species** In flight silhouette, crest gives an arrow-headed appearance and is thinner necked than Violet Touraco.

HABITS A forest species; feeding on a variety of fruits and blossoms in the canopy or the middle storey. May venture from thicket cover to feed in an isolated fruiting *Ficus*, but prompt to return. Usually in pairs or small groups. Flight routes to food are often predictable and occur at the same time each day until the supply is exhausted. Shy and wary, its growling call invariably draws first attention, but despite its size this does not always result in a sighting. Regularly decends to the ground to drink from forest pools in the dry season. Flight undulating, interspersed with glides. Agile on foot, running and hopping swiftly along the gigantic boughs and sweeping branches of forest trees.

VOICE A far-carrying series of *kawr-kawr-kawr* leisurely at first then increasing in tempo, sometimes causing a domino effect through the forest with all turacos in the area joining in.

STATUS AND DISTRIBUTION Restricted to a few forest islands in coastal WD where it is a frequent breeding resident. The Gambia represents the limit of its West African range. Locally common in southern coastal Senegal across to the south-east in Niokolo Koba. Isolated *T. p. buffoni* is restricted to the area from The Gambia to Liberia.

BREEDING Nest building seen in late June and mid-August. Well-grown fledgling records out of the nest in mid and late October.

VIOLET TURACO

Musophaga violacea **Plate 29**

Fr. Touraco violet

Other name: Violet Plantain-eater

IDENTIFICATION 45cm. Unmistakable. Overall a splendid glossy violaceous blue-black, subtly shaded and influenced by light intensities. In flight appears blackish with stunning crimson discs in the primaries and outer secondaries of the wing. Mantle and closed wings purple, showing some crimson, tail and breast tinged green with belly matt black. Distinctive bill, facial and head markings with a characteristic large and raised bright yellow bony shield over the red bill, silky-white line below eye to ear-coverts, bare scarlet orbital skin, and red crown feathering soft and velvety. Immature much duller, lacking head markings of adult. **Similar species** Unique.

HABITS Not as dependent on climax forest as previous species, flourishing in sparser habitat, but requiring tall trees, some creeper clad, or a watercourse with *Ficus*. Occasionally seen in mangroves but probably passing through. The running, tree-hopping habit is regularly employed, but they are mostly seen in flight. Inconspicuous if not on the move. Nearly always in pairs. Food is mainly figs (swallowed whole), the insect larvae within providing animal protein. A West African endemic.

VOICE A resonant rolling series of guttural notes delivered with increasing rapidity *courou-courou-courou*; usually a slightly asynchronous duet becomes a cacophony. Also a single gruff note whilst feeding. With familiarity, readily separable from Green Turaco.

STATUS AND DISTRIBUTION Frequent to locally common breeding resident at the coast and in suitable habitat elsewhere, especially in riparian areas

of CRD and URD. No recent information from NBD but known in northern Senegal up to 14° and right across southern Casamance to the east.

BREEDING Nest a frail platform. One nest at 5m in a dense liana-clad oil palm in the late dry-season and a recently-fledged youngster seen with adults in November, at different localities, in WD suggest breeding synchronous with the rains.

WESTERN GREY PLANTAIN-EATER

Crinifer piscator **Plate 29**

Fr. Touraco gris

Other names: Grey Plantain-eater

IDENTIFICATION 50cm. Overall dull grey with copious dark streaking, shaggy erectile crest often held depressed, large yellow-green conical bill. Head, neck and breast dark brown streaked silver, belly and flanks white, streaked black. Upperparts silver-grey, spotted dark brown. Characteristic conspicuous white stripe on the wing visible in flight from a considerable distance. Long tail grey, darker towards tip. Immature less silvery, lacks crest, dark head looks woolly.

HABITS A conspicuous bird of a variety of habitats, most numerous in dry open savanna with scattered tall trees and *Ficus*. Pair-bonding is exceptionally strong, manifested by greeting displays with bowing and tail-fanning, regurgitated food exchanges and loud calls given on landing in the crown of a tree. Catholic choice of fruit, blossoms and invertebrates; 'plantain eater' somewhat erroneous.

VOICE One of the most familiar avian sounds of the Gambian bush – a loud yelping laughing at all times of the day; *cow-cow-cow* and *kalak-kalak-kalak* carrying over a great distance.

STATUS AND DISTRIBUTION Common to locally abundant resident throughout all divisions year-round, most numerous in WD. Widespread throughout Senegal but less numerous in central northern districts.

BREEDING Season extends through the year. Nest generally more substantial than those of other turacos. Possible cooperative breeding strategy requires further investigation.

CUCKOOS AND COUCALS

Cuculidae

Ten species of true cuckoos, in the subfamily Cuculinae. All are brood-parasites, laying their eggs in the nest of another bird, and most are associated with the rainy season. Six species are regular, the others are vagrant or rare. A knowledge of calls, usually loud and recurring, is of tremendous value in establishing identity, and many are heard before they are seen. Comparatively small-headed, with short, slim slightly decurved bills. Short legs and small feet are distinctive, making them clumsy on the ground. Generally they are long-tailed birds with hawk-like flight patterns. The three short-tailed 'green-backed' species have an undulating flight, and the white in the tail could lead to confusion with honeyguides. All are principally insectivorous. Identification problems mostly centre on immature and subadult plumages. There is marked sexual dimorphism in some species, and brown or rufous (hepatic) morphs are occasionally seen. A non-parasitic subfamily (Phaenicophaeinae) consists of the monotypic African endemic Yellowbill. The non-parasitic, long-tailed and round-winged Coucals (Centropodinae) are represented by three species, one in southern Senegal only.

JACOBIN CUCKOO
Clamator jacobinus   Plate 24

Fr. Coucou jacobin

Other names: Black and White Cuckoo, Pied Crested Cuckoo, Pied Cuckoo

Synonym: *Oxylophus jacobinus*

IDENTIFICATION 34cm. *O. j. pica*. Sexes alike. Slim, long-tailed, pied cuckoo with pointed crest on nape. Underparts pure white; upperparts black with greenish-blue sheen. White wing-bar visible when perched, conspicuous in flight. Tail tipped white. Immature browner above and buffy below, crest reduced and wing-bar less defined but tail tipped white **Similar species** Confusion is possible with commoner, larger and heavier Levaillant's Cuckoo which has characteristic heavy stripes on throat and breast and slightly longer tail. Immature Jacobin has whiter tail tips than Levaillant's. On a poor flight view, plumage similar to rare forest-dwelling Western Little Sparrowhawk.

HABITS Associated with dry wooded savanna and grassy marsh habitats. Usually singly or in pairs, with a low, direct and swift flight. Recent sightings involved birds active in areas of low shrubby acacias. Noted elsewhere to be noisy prior to the breeding season. Possibly under-recorded: additional information sought.

VOICE Coarse gurgled laughing call similar to Levaillant's but not so harsh. Ringing and repeated *pieuu* accelerating before a loud quickfire *tacka-tacka-tacka-tacka* which fades away.

STATUS AND DISTRIBUTION Rare to uncommon, with a few recent records at the onset of the rains, all from WD and LRD. Earlier records in November-December, including a communal roost in CRD. Senegalese sightings are all north and east of The Gambia. Movements not yet understood and may be rain-related.

BREEDING Regional hosts are likely to be the Common Bulbul and Brown Babbler, and in northern Senegal, Fulvous Babbler. Eviction behaviour not proven but host young unlikely to survive.

LEVAILLANT'S CUCKOO
Clamator levaillantii   Plate 24

Fr. Coucou de Levaillant

Other names: Striped Crested Cuckoo, African Striped Cuckoo

Synonym: *Oxylophus levaillantii*

IDENTIFICATION 40cm. Sexes alike. Large black and white cuckoo with shaggy spiky crest and characteristic bold black striping on the throat and breast. Upperparts black with greenish-blue sheen. Prominent white wing-bar in flight evident as white patch on closed wing. Belly variable, whitish to cream. Large white terminal spots on uppertail. Bill black, legs grey. Rare rufous morph has been recorded in The Gambia. Immature dull brown above, underparts buffy with conspicuous black striping on sides of face and on throat. Gambian immatures show white tail spots, not rusty as described on immatures elsewhere. **Similar species** Immature can resemble much rarer and smaller Jacobin Cuckoo but facial and neck striping is distinctive on Levaillant's. Call could be confused initially with Western Grey Plaintain-eater.

HABITS Extremely noisy and active in well-wooded areas, fond of well-grown gardens and found in the few forest-thicket remnants. Flight strong, often seen crossing roads and open spaces, may flare tail on landing or dive headlong into cover. Works in pairs and indulges in chases; often mobbed by parties of irate passerines. Although not shy, often heard but not seen.

VOICE Far-carrying and unusual. Calls are a feature of the wet season and just into the dry period. Series of fluting but sometimes piercing *peee-u peee-u* calls maintained for some time, often followed by a fading quickfire chatter *atack-tacka-tacka*.

STATUS AND DISTRIBUTION Common wet season visitor throughout all divisions, most numerous in WD. Arrives in small numbers late May, peaks July to November. Young may still be fed by host well into January mostly at the coast; occasional late adults also in January. Widespread in Senegal but largely absent from the north-east.

BREEDING Usual host species in Senegambia is Brown Babbler, but also Black-capped Babbler and Chestnut-bellied Starling.

GREAT SPOTTED CUCKOO
Clamator glandarius   Plate 24

Fr. Coucou-geai

IDENTIFICATION 39cm. Sexes alike. A slightly bushy-crested cuckoo with grey wings boldly peppered with white spots in all plumages. Long, graduated blackish tail strongly edged white. Adult has crown and cheeks silver-grey producing a capped appearance in contrast with yellowish throat, face and breast. Upperparts ashy-brown and spotted, underparts yellowish shading to white on belly. Bill black with base of lower mandible pale. Legs grey. Immature has a negligible crest and black cap created by black sides of head and nape; conspicuous chestnut or bronzy primaries may be retained into adulthood. **Similar species** Fleeting views could lead to confusion with Levaillant's and Jacobin Cuckoos but both have glossy black upperparts with no spots, striking white wing-bars and black spiky crests.

HABITS Favours open habitat with scattered trees and low bushy undergrowth, often near water. Most records involve single flying birds. Perches in a horizontal posture with the wing tips drooping below the tail. Wary. Nestling does not evict host's eggs or chicks, and more than one Great Spotted egg may be laid in the same nest.

VOICE A rapid trilled *kittera-kittera-kittera* finishing in a loud *AYEak AYEak*, but not particularly vocal in The Gambia.

STATUS AND DISTRIBUTION Frequent in WD, fewer records in other divisions. Seen year-round, but information lacking on which races involved. November-March reports are fewer, and birds then are possibly of Palearctic origin. Sightings in June-July, including immatures, are probably of African origin; this correspond with breeding dates for the host crow species. However, lack of adult cuckoos prior to these months suggests post-fledging dispersal from elsewhere. Widespread in Senegal in coastal west areas and east of The Gambia to Mali.

BREEDING Host species in our area is Pied Crow *Corvus alba*. However, some *Lamprotornis* glossy starlings that are parasitised in other areas (southern Nigeria and southern Africa) are also found in Senegambia. Egg smaller than crow's but similar in colour. Copulation recorded in May, fledgling food-begging in August in The Gambia. In northern Senegal recorded April-May

RED-CHESTED CUCKOO
Cuculus solitarius   Plate 24

Fr. Coucou solitaire

Other name: Solitary Cuckoo

IDENTIFICATION 30cm. Sexes similar but separable. A grey-backed cuckoo with chestnut breast-band. Underparts pale creamy-white tinged cinnamon, with well-spaced black-brown barring. Clearly defined chestnut-red chest band separates barred belly from dove-grey throat. Female has whiter underparts more extensively barred than male. Upperparts slate, tail darker with outer feathers spotted white, difficult to see from above, conspicuous from below. Bill blackish, horn at the base. Thin yellowish eye-ring, legs yellow. Immature has upperparts, throat and breast dark brown with subtle pale fringes and flecking; closed wings more slaty, may show white nape patch. Underparts boldly barred black on buff. Tail more barred than adult and usually looks shorter. **Similar species** Forest race of Black Cuckoo *C. c. gabonensis* is larger and darker in all plumages, not yet recorded in Senegambia.

HABITS Inland favours escarpment woodland with well-established leafy trees and accumulating ground water as the rains progress. At the coast favours thicket edges with freshwater nearby. Persistence is required to locate it when calling from dense canopies. Calling usually initiated by the first major rain of the season. Most sightings come from following-up calling birds, but beware ventriloquial habits.

VOICE Very loud, distinctive and persistent, carrying on well into the night. A ringing trisyllabic slightly descending *WIP, wip, woe* or *tic-TAC-toe* constantly repeated. Both sexes also communicate with a soft bubbly gurgle.

STATUS AND DISTRIBUTION Uncommon to frequent wet season visitor to WD and LRD, July-October, rare or possibly overlooked elsewhere, peak numbers July-August; occasional sightings after calling ceases. Regular in a few localities in southern Senegal where it begins calling in June.

BREEDING Information sparse in West Africa. Recent record of an immature in LRD in October, and a bird collected in eastern Senegal in July was ready to lay. In Senegambia, hosts unknown but robin-chats, paradise flycatchers, thrushes, wattle-eyes, batises and bulbuls are candidates.

BLACK CUCKOO
Cuculus clamosus   Not illustrated

COMMON CUCKOO
Cuculus canorus   Plate 24

Fr. Coucou criard

IDENTIFICATION 30cm. *C. c. clamosus.* Slender cuckoo, appearing iridescent all-black at a distance; on closer inspection faint and fine reddish-buff barring is visible on the underparts (more so on female). Tail tipped white and in flight underside of wing shows some black and white barring. Bill black, slate-grey orbital skin, legs pink-brown. Immature dark brown, chin to vent profusely but very faintly barred, tail lacking white tip. This migratory savanna race from southern Africa is considered to be the source of the few erratic Senegambian sightings. *C. c. gabonensis* (sedentary forest race with nearest population in Liberia). Sexes similar but female more heavily barred below and duller. Upperparts black with a greenish gloss. Rufous-chestnut breast, poorly-defined and extending onto chin, throat and sides of neck is best field character besides voice. Rest of underparts grey to grey-buff, heavily barred blackish. **Similar species** Melanistic forms of other two regional black-backed cuckoos are always crested. Separating *gabonensis* race from Red-chested poses serious difficulties, but paler adult Red-chested has well-defined breast band, while immature Red-chested has brown breast and whiter underparts. In flight, also beware dark morphs of small raptors, and also male Red-shouldered Cuckoo-shrike.

HABITS Race *clamosus* favours dry acacia savanna and is vocal only when breeding. Race *gabonensis* associates with lowland forests, is largely sedentary and vocal throughout the year. Generally shy.

VOICE Distinctive identification feature: a characteristic trisyllabic mournful drawn-out, rising whistle *too, too, toooooo*, the last syllable often developing into a bubbly note reminiscent of Stone Partridge.

STATUS AND DISTRIBUTION Vagrant. No confirmed recent Gambian records; one during rainy season from end of last century. Two modern records in Casamance, S Senegal August 1971 and February 1980.

BREEDING Elsewhere parasitises bush-shrikes and orioles.

Fr. Coucou gris

Other name: Eurasian Cuckoo; European Grey Cuckoo; Cuckoo)

IDENTIFICATION 33cm. Head, breast and upperparts pale slate-grey, brighter on the face, female overall slightly browner. Underparts white with narrow brown-black barring, well demarcated from grey breast in male; faint barring may be visible on upper breast of female. Darkish primaries contrast with rest of wing. Tail dark with white spotted tiping and irregular spots on outer feathers, reducing towards centre. Eye yellow with neat orange-yellow orbital ring. Bill slighter than African Cuckoo, blackish, with a small but paler base detectable. Legs yellow. Rufous morph is very rare. Immature on first migration brown or grey with a white nape patch. Barred chin, throat and breast, white fringes on upperparts. Underparts more buffy than white. Positive separation from African Cuckoo often difficult and in some instances may not be possible. **Similar species** Differs from the more likely African Cuckoo on several variable factors. Common has narrower yellow-orange orbital ring and less orange gape, more prominent on many African Cuckoos. Flank barring generally finer on African. Uppertail irregularly diamond-spotted near to shaft in Common, as opposed to broad white misshaped bars across the whole feather in African. Undertail completely barred in African, spotted in Common. Distinctive Common Cuckoo call most unlikely to be heard south of the Sahara but African does call 'cuck-oo' albeit in an awkward and rather unpolished hoopoe-like or dove-like manner.

HABITS Likely to favour lightly wooded savanna whilst on passage. Maintains rapid wing beats in flight and does not move the wings above the horizontal, a useful distinction from small raptors.

VOICE Loud disyllabic *cu-coo* call known only in breeding areas.

STATUS AND DISTRIBUTION Very few recent records in The Gambia. Possibly overlooked on passage or overflies Senegambia. Frequent in extreme northern Senegal late July to October. Nowhere common in West Africa with birds ringed in Europe recovered in Togo and Cameroon in October and January respectively.

BREEDING Brood parasitism and host egg or young ejection very well studied in Europe. North African race breeds Morocco to Tunisia April-May, where redstart, warbler and shrike species are listed as hosts.

AFRICAN CUCKOO
Cuculus gularis   **Plate 24**

Fr. Coucou africain

Other names: African Grey Cuckoo, African Yellow-billed Cuckoo

IDENTIFICATION 33cm. The most likely to be seen of the three grey-backed, hawk-like cuckoos. Characteristic yellowish-orange eye with prominent (but individually variable in extent) bare pale orange skin extending from gape to around eye. Bill mainly yellow tipped black. Upperparts and wings slate, paler grey on throat, breast and rump. Barring on whitish underparts well spaced, often diminishing towards lower belly and vent, flanks finely barred. Female shows rufous tinge on the throat. Tail longish, with broad ragged white barring on upper outer feathers, undertail appears completely barred with variability in the spacing. Immature greyish-brown above, extensively speckled and finely barred white, appearing pale overall. **Similar species** See Common and Red-chested Cuckoos, and note possibility of confusion in brief views with small raptors.

HABITS Favours lightly wooded country, often calls from a bare limb. Usually easy to locate when calling and normally first heard in late May a few weeks before the rains break. Exploits the winged termite emergences of the early rains for food.

VOICE Vital for conclusive identification: disyllabic rather weak and forced dove-like or hoopoe-like *ou-ou*, lacking the initial *c* or *k* of the Common Cuckoo; second note is higher.

STATUS AND DISTRIBUTION Frequent to common wet season visitor to all divisions, May to November, most numerous in WD; records peak June-August, calls into September, sometimes later. Northern Senegal is limit of range in West Africa.

BREEDING Not well known. Well grown immature in July (see above). Only recorded hosts elsewhere are Fork-tailed Drongo and Yellow-billed Shrike. Details sought.

AFRICAN EMERALD CUCKOO
Chrysococcyx cupreus   **Plate 24**

Fr. Coucou foliotocol

Other name: Emerald Cuckoo

IDENTIFICATION 20cm. By far the rarest of Senegambia's shiny green-backed cuckoos. Marked sexual differences, but both have green backs. Male has entire upperparts including tail and closed wing, head, neck and chest spectacular glistening emerald green with a golden tinge. Each individual feather has a darker centre than its fringes causing a beautiful scaled effect. Belly and breast are cleanly demarcated and intense sulphur-yellow. In flight, underwing shows white barred green, and undertail is spotted white on outer feathers. Blue-green orbital ring (brighter blue in female). Female has underparts heavily barred green on white, barring narrowest on chin and throat. Entire upperparts including tail, closed wing, head and neck are duller green with a coppery wash and with tawny bars if seen well. Tail bronze-green with white on outer feathers. Immature similar to female with fine metallic barring on the head; male brighter. **Similar species** Adult female Emerald is separable from adult female Klaas's and Diederik by more extensive barring, lack of obvious white marks (not flecks) behind the eye and more bronzy wash on upperparts. Immature and female Emerald both lack white spot behind the eye.

HABITS Not to be expected away from closed canopy and liana-clad forest outside the rainy season months. Perches high and if not calling a sighting is unlikely. Lamentably very heavily persecuted for the feather trade earlier this century; deforestation undoubtedly contributes to its decline in Senegambia.

VOICE Characteristic loud and clearly whistled *choo, choo tow-wee* or *piuu pew-wit*, rythmic and bouncy, slurring quickly down in pitch over first cadence, back up to starting pitch on second.

STATUS AND DISTRIBUTION Very rare wet season visitor. Two recent records from the same forest remnant in WD (near Tanji Bird Reserve) July 1981 and October 1990. Latter record was calling and seen but not relocated subsequently. Suitable tracts of riparian forest in CRD and URD have not been well investigated during the heavy rains in recent years. In southern Senegal heard in a few locations July-October.

BREEDING Several hosts parasitised elsewhere in West Africa are present in Senegambia including weavers and sunbirds.

KLAAS'S CUCKOO
Chrysococcyx klaas   **Plate 24**

Fr. Coucou de Klaas

IDENTIFICATION 17-18cm. Some differences between sexes, but both are green-backed with white mark characteristically restricted to behind the eye only. Males vary from glistening emerald green to bronzy-green above, with conspicuous green half-collar patch on each side of the chest an important fieldmark. Otherwise underparts practically pure white with faint green barring on flanks. Tail metallic green in the centre becoming whiter towards outer feathers, appearing almost white edged in flight. Many females closely resemble the male in West African populations but some have the upperparts mottled coppery and less glossy. Shows half-collar feature (may be greyish) and white spot behind the eye. Flank barring is usually rusty-brown and can be more pronounced than in male, reaching across the belly. Bill colour varies sexually and seasonally from pale green (male) to dark blackish-green (female). Immature has underparts heavily and duskily barred, untidy and variable, usually less on belly and flanks and more on throat. Upperparts show some green iridescence barred with brownish-buff. **Similar species** Diedrick is slightly larger and separated easily on call. Diedrick also has broad white eyebrow before and behind the eye and outer tail feathers with far more extensive green. With a densely-barred immature Klaas's note possibility of larger, more metallic female or immature Emerald Cuckoo (very rare in Senegambia).

HABITS Found in widely varied habitats including savanna woodland, scattered coastal bush, forest-thicket and occasionally mangroves. The only green-backed cuckoo likely to be encountered in the dry-season after December until June when Diederik appears to be absent. Both adult and immatures occur. Individuals may associate with mixed forest-bird feeding parties at mid-storey height where it often forages in the fashion of a sallying flycatcher. Individuals also come to drink at forest pools in the dry season in company with mixed insectivores. Calls from a prominent perch or a leafy tree but is somewhat more timorous than Diederick, although heard practically everywhere during the rains and occasionally during the dry season.

VOICE Call a pining rhythmic and rising disyllabic *whe et-ki, whe et-ki* or *huett jee, huett jee*. Sometimes written as 'who is he ... who is he' repeated in bursts of five or six. Beware possible confusion with the slow call of a solitary Striped Kingisher.

STATUS AND DISTRIBUTION Frequent to common year-round resident especially WD to LRD with a significant wet season increase. In Senegal, well to the north in the rains, but south of 15° in the dry season.

BREEDING Host selection dictated to an extent by habitat. Proven hosts in The Gambia include Grey-backed Camaroptera and Beautiful Sunbird, and in forest Collared Sunbird is a likely candidate along with several others.

DIEDERIK CUCKOO
Chrysococcyx caprius   Plate 24

Fr. Coucou didric

Other names: Didric Cuckoo

IDENTIFICATION 18-20cm. Sexes differ to a degree but both are green above with a coppery sheen on the back (more pronounced in female). Underparts, including chin and throat, white (often washed brownish in female). Sides of chest and flanks are variably barred copper-brown (more so in female). Both sexes show distinctive white marks before and behind the eye, producing 'broken eyebrow'; an important field character, and a narrow dark green moustachial streak. Often distinct white mark over the crown (lacking in female) and rows of white spots (buff in female) on the coverts of the closed wing are other significant fieldmarks. Outer tail feathers green with relatively small white spots. Immature similar to female but more coppery above, with a less-marked eyebrow; underparts buffy-white with brown blotches and streaks, not bars. When recently fledged the bill is diagnostically bright coral-red in good light. **Similar species** Good views of head with eyebrow and moustachial streak showing eliminate Klass's Cuckoo, which is generally more emerald, shows green shoulder patches and near-white outer tail.

HABITS Found in varied habitats, overlapping with Klass's in most facets of behaviour and ecology. Not to be expected in the dry season. Calls regularly from the top of an open tree, often a bare acacia. Early dry season sightings are likely to be juveniles, difficult to find when perched silently.

VOICE Call an onomatopoeic, plaintive, oft-repeated and unmistakable *day-dee did er ick, day-dee did er ick*, first two syllables slower than the rest.

STATUS AND DISTRIBUTION Common wet season visitor July to early December throughout, apparently absent at other times. Calling ceases around October, when records diminish sharply. Widespread in Senegal, where migratory movements are probably rain-related.

BREEDING Parasitises granivorous *Ploceus* weavers especially Village Weaver, and the bishops *Euplectes* may also be victimised.

YELLOWBILL
Ceuthmochares aereus   Plate 24

Fr. Coucal à bec jaune

Other names: Green Coucal, Green Malkoha, Yellowbill Coucal

IDENTIFICATION 32cm including a 20cm tail. *C. a. flavirostris.* Sexes alike. A small, slender, blue-grey coucal with a long graduated tail, large head and curved roller-like yellow bill. Upperparts and wings steel-blue, tinged violet on rump and uppertail-coverts, slate-grey head and neck. Underparts grey, darker on belly and undertail-coverts, paler on the throat. Conspicuous hooked bill chrome-yellow, looks disproportionately large. Eye red surrounded by narrow yellow orbital ring. Immature darker and duller; sole Gambian immature described with blackish bill, head uniform greyish, eye dull grey-brown and feet grey. Similar species Grey plumage and stout yellow bill is unique.

HABITS Restricted to dense liana-clad areas of forest. Agile and furtive, sometimes with mixed forest-bird communities. Sneaks about in creepers in a 'squirrel-like' fashion. Recent sightings in The Gambia all relate to solitary birds in a variety of plumages from immature to adults in both fresh and worn plumages. Flight clumsy, keen to flop into the next thick cover after crossing a clearing.

VOICE A range of calls, from a thinly whistled kite-like *oo sueee sweee*, various clucking or clicking notes similar to those uttered by an irate Gambian Sun Squirrel, and some reminiscent of Western Grey Plaintain-eater *kalack-kalack-kalac.*

STATUS AND DISTRIBUTION Rare to possibly seasonally and very locally frequent; all records in dry season November-June. Recent records all from three forest islands in WD where it is possibly resident in one. Difficult to locate; even intensive studies have revealed a dearth of records. A number of recent records from forest in Casamance, southern Senegal. Threatened by habitat destruction.

BREEDING Not well known; immature in June corresponds with Nigerian dates. An untidy domed or undomed nest built of twigs and debris.

BLACK-THROATED COUCAL
Centropus leucogaster　　　　Plate 24

Fr. Coucal à ventre blanc

Other name: Great Coucal

IDENTIFICATION 50cm. Sexes alike. Unmistakably large, clumsy, black, chestnut and white coucal. Head, throat, chest and mantle black with glossy violet sheen. Clearly demarcated white lower breast, buffy on the vent. Wings and lower back dark rufous-chestnut, long tail blue-black with fine rusty barring more profuse towards the base. Immature 'arrow-marked' with white feather shafts on dark parts, brown wings narrowly barred black, under-

parts buffy-brown. Similar species Smaller Black Coucal lacks white underparts and occupies different habitat.

HABITS A sedentary coucal of furtive habits, favouring forest undergrowth, occasionally appearing in clearings. The largest of the coucals, with a characteristic weak floppy flight. Calls with the head lowered and arched, touching the breast, with the body partially inflated.

VOICE More sonorous than other coucals with low reverberating booms of up to 20 *whoo* calls; said to fill the forest with sound when calling.

STATUS AND DISTRIBUTION Not recorded in The Gambia. Few mid wet-season records from coastal forest in Casamance, southern Senegal.

BREEDING A large domed stucture of grass and leaves in dense forest undergrowth.

BLACK COUCAL
Centropus grillii　　　　Plate 24

Fr. Coucal noir

Other name: African Black Coucal

IDENTIFICATION 35cm. Sexes similar. Small coucal, distinctive in breeding plumage, when all-black with contrasting rufous-brown wings which may show some barring. Black parts with oily sheen, shiny feather shafts creating flecked effect. Tail black, long and bulky, in some birds with narrow buff barring. Non-breeding birds tawny-brown, underparts paler than upperparts, all heavily flecked and streaked. Profuse cream feather shafts on neck, mantle and breast create 'arrow-marked' effect. Tail and wing barring is indicator of age. Immatures are more heavily dark-barred and tawny than non-breeding birds. Similar species Confusion of non-breeders and immatures in poor views with larger immature Senegal Coucal is possible, but Senegal although barred above is more uniform buff not heavily streaked below.

HABITS Restricted to freshwater marsh and wet grassland habitats. Fairly shy but habitually sits well out in the crown of cultivated mango trees in The Gambia so can be watched from a distance. Typically seen with the wings drying-out after a storm, or after having become soaked while creeping about in wet grass. Female controls displays, and may attract several males. Male thought to be responsible for all parental duties once clutch is complete.

VOICE Distinctive once learnt; calls for prolonged periods. Female advertising call *pop op, pop op, pop op* or *kok ock kok ock*; also a gurgling 'water-bottle' call. **STATUS AND DISTRIBUTION** Very locally frequent to common wet season breeding visitor with a restricted range in URD and CRD. Recorded August to late November, then disperses possibly east into Mali. Senegalese records from south-east and both north and south of The Gambia.

BREEDING Polyandrous and monogamous reproductive systems. Bird proffering lizard post-copulation in mid-September, URD 1991, nest building in CRD in August 1963. Nest knitted into low rank vegetation and domed.

SENEGAL COUCAL
Centropus senegalensis   Plate 24

Fr. Coucal du Sénégal

Other name: Bottle-Bird

IDENTIFICATION 39cm. Sexes alike. A medium-sized coucal, white throated and black-capped. Head, nape, upper back and tail black with an olive-green sheen and some 'arrow-marking' caused by white feather shafts. Back and rounded wings rufous-chestnut. Entire underparts creamy-whitish. Immature brown finely barred black above, uniform cinnamon-buff or pale tawny below with fine delicate barring on thighs and flanks. **Similar species** Confusion is possible with immature and non-breeding Black Coucal; see that species. There is also a single but disputed Senegal record for the wetland-associated Blue-headed Coucal *Centropus monachus* (Fr. Coucal à nuque bleue) 46cm; otherwise nearest confirmed records are from Guinea-Bissau and Mali. Blue-headed has head and mantle black, washed glossy blue, and is larger and stockier than Senegal Coucal.

HABITS Ubiquitous, but most favour damper areas. Perches at medium height. Calls from a perch or on the ground with body inflated and the head bowed into the chest. Frequently jerks tail sideways. Spends a considerable amount of its time on the ground creeping amongst low tangled undergrowth.

VOICE A sonorous, guttural, *ouk-ouk-ouk-ouk* or *who-who-who-who* with a pigeon- or cuckoo-like quality. A very familiar sound of the bush, likened to the glugging of water being emptied from a bottle. Sometimes two birds will call in synchrony.

STATUS AND DISTRIBUTION Common to abundant throughout Senegambia in all months.

BREEDING Most records from mid-rains to early dry season. Untidy football-sized nest built of stalks, vines and leaves. Possible reproductive helper/ polygamy strategy needs investigation.

OWLS
Tytonidae and Strigidae

Twelve species. Barn Owls, Tytonidae, with one species. Strigidae with eleven species. Sizes range from small to huge; most are nocturnal, retiring by day to a habitual roost site which may be used for years. Head large and rounded, bill short and hooked and often hidden in a tuft of bristles. Flattened facial disc, large eyes. Soft feathering silences hunting flight on slow wing beats. Prey is swallowed whole and the remains are regurgitated in a pellet, dissection of which allows analysis of diet. Voices distinctive; knowledge of calls of great assistance in identification.

BARN OWL
Tyto alba   Plate 25

Fr. Chouette effraie

Other name: African Barn Owl

IDENTIFICATION 30-33cm. *T. a. affinis*. Medium-sized pale owl. Golden-buff, finely marked grey above. Underparts and underwing white with a varying density of brown flecks on the belly. Distinctly white in flight. Pale facial disc noticeably heart-shaped with dark eyes; lacks ear tufts. Legs longish and white-feathered. Immature darker and more heavily spotted on breast. **Similar species** Could be confused with African Grass Owl *Tyto capensis* (Fr. Effraie du Cap); nearest records Cameroon, but an unconfirmed wet season record in WD.

HABITS Varied habitats used, requiring suitable hunting with a roost nearby. Often in association with man, other sites include woodland, mangroves, swamps and forest. Entirely nocturnal leaving its roost at dark and returning just prior to dawn. Often hunts over roads and tracks, sadly a frequent victim of car strikes. Favoured roosts include tall leafy trees e.g. crowns of palms, cavities in rocks and roofs, and in the nest of the Hamerkop which it may evict. Pellets indicate food is predominantly rodents with occasional bats, birds, beetles, reptiles and amphibians.

VOICE A shrill and screeching *schreeee* often heard just after dark when leaving roost. More vocal before breeding season.

STATUS AND DISTRIBUTION Frequent to common breeding resident in all divisions, possibly most numerous in WD. Widespread throughout Senegal, common in the coastal north.

BREEDING During the dry season in The Gambia, with chicks in February; period more extended in Senegal. Sites include a hole in a baobab, in a bucket of charcoal in an out-house and in the nest of Hamerkop.

EUROPEAN SCOPS OWL
Otus scops   Not illustrated

Fr. Hibou petit-duc

IDENTIFICATION 21cm. Very similar to African Scops Owl but is slightly larger and less boldly marked. Only really separable in the hand when difference in wing shape can be seen. Knowledge of calls assists field identification but European Scops is largely silent in winter.

European Scops Owl

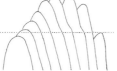

African Scops Owl

STATUS AND DISTRIBUTION Palearctic migrant to the region. Two Gambian records in coastal NBD, January and February 1996. Regular visitor to northern Senegal, September-April.

AFRICAN SCOPS OWL
Otus senegalensis   Plate 25

Fr. Hibou petit-duc africain

IDENTIFICATION 20cm. Smallest owl with ear-

tufts of the region. Overall shades of medium-grey with tawny streaking and fine grey-black vermiculations most pronounced over the head and mantle. Variable rufous tinge to plumage: some appear brown. Prominent feather tufts usually held erect at rest, but head can appear rounded when hunting. Grey facial disc looks unstreaked, with dark border. Iris bright yellow. Immature inseparable from adult. **Similar species** Similar-sized Pearl-spotted Owlet has a characteristic dumpy jizz with well-defined spotting on the back, a mobile tail, and lacks prominent ear-tufts. Larger White-faced Scops has prominent white facial disc with contrasting black fringe and very orange, not yellow, eyes.

HABITS Strictly nocturnal. Begins to call at dusk singly, or at some forest sites several birds responding to each other. Found in wooded savanna, sometimes in mangroves and gardens. Despite continued calling, often very difficult to locate. Far more often heard than seen. Perches during the day close against a tree trunk and if disturbed adopts a body-erect posture relying on cryptic plumage for concealment.

VOICE A diagnostic, sudden, isolated and brief *prrroop* on a single note, moderately high pitched and hollow toned. Sometimes several birds call simultaneously.

STATUS AND DISTRIBUTION Locally common breeding resident; population greatest in WD, but no recent information from NBD. Widespread in Senegal up to 15°, becoming rarer in the north.

BREEDING Apparently mostly towards the end of the rains, coinciding with an upsurge of insect life; immature in WD in August. In northern Senegal late September. Nests in the hole of a tree, fence-post or building.

NOTE This species was formerly considered conspecific with European Scops Owl *O. scops*.

WHITE-FACED SCOPS OWL
Otus leucotis   Plate 25

Fr. Petit-duc à face blanche

Other name: White-faced Owl

IDENTIFICATION 28cm. A smallish, short-tailed owl with characteristically prominent ear-tufts. Overall soft grey with darker streaks and vermiculations and a line of white spots down each side of the back. Upperparts a shade darker than underparts. Distinctive white facial disc contrastingly edged black, eyes large and orange. Immature brownish with greyish-brown facial disc. **Similar species** See African Scops Owl.

236

HABITS Strictly nocturnal; site-faithful, often found in pairs at regular roosts, but communal roosting also known. Adopts body-erect posture if disturbed looking distinctly emaciated. Prefers thorny acacia habitat, but often in proximity to human settlement, even in towns where it will spend the day in holes in buildings. Frequently uses telegraph wires and power cables as hunting stations.

VOICE Most commonly heard call is a double-noted *po-proo*, quiet and understated with accent on second note. Might be mistaken for a small dove cooing softly to itself at dusk.

STATUS AND DISTRIBUTION Locally common breeding resident, most numerous in WD and URD. Recorded in all months. Widespread throughout Senegal but fewer along eastern border.

BREEDING Records in early dry season in The Gambia and in northern Senegal February-April, in a natural or man made crevice. Elsewhere also in the disused nests of other birds. Abundant Buffalo-Weaver nest structures throughout the country are likely sites.

EURASIAN EAGLE OWL
Bubo bubo   **Not illustrated**

Fr. Hibou grand-duc

IDENTIFICATION 65-70cm. *B. b. ascalaphus.* Plumage tawny-brown, with darker streaking, often variable in colour, with prominent dark ear tufts, white throat which inflates when calling, black bill and deep orange-amber eyes.

STATUS AND DISTRIBUTION Vagrant in Senegambia. One record each in northern and southern Senegal; a dead bird near Dakar in March 1974, and a pair heard and seen in atypical habitat in coastal Casamance March 1983. It remains unrecorded in The Gambia, but as an owl of semi-desert its range could be expected to extend as a result of regional desertification.

NOTE The form concerned is sometimes considered to be a separate species, Desert Eagle Owl *B. ascalaphus.*

SPOTTED EAGLE OWL
Bubo africanus   **Plate 25**

Fr. Grand-duc africain

IDENTIFICATION 43-50cm. *B. a. cinerascens.* A largish horned owl. Overall appears mottled

brownish-grey with some bolder white markings on the mantle. Underparts, flanks and feathered legs finely barred, upperparts finely vermiculated. Facial disc greyish-brown, edged blackish. A slightly fluffy white throat patch detectable at close range and when calling. Eye dark, the characteristic feature of *cinerascens* race (nominate *B. a. africanus* has a yellow eye). Immature browner than adult. **Similar species** Verreaux's Eagle Owl is much larger with a stronger face pattern and pink eyelids.

HABITS Favours a mixture of mature woodland, laterite outcrops and nearby open grassland. The scarcity of diurnal records of this owl at roost demonstrates its reclusive nature; attention sometimes drawn to it by parties of mobbing passerines. Most often encountered well after dark, perched beside a road or bush-track waiting for prey, when it may be temporarily 'frozen' in the headlamps of a passing car.

VOICE Either a disyllabic or trisyllabic resonant hoot *hoo-hoo*, or *hoo-whohoo.* Vocal at dawn as well as dusk.

STATUS AND DISTRIBUTION Uncommon but recorded year-round from all divisions; peak sightings in LRD and URD with highest probability in the rains from July to October, but recorded every month. Becomes rare north of 15° in Senegal.

BREEDING Not discovered in Senegambia, but likely to be on the ground in a scrape against a fallen tree, or under a boulder or termite mound.

VERREAUX'S EAGLE OWL
Bubo lacteus   **Plate 25**

Fr. Grand-duc de Verreux

Other names: Milky Eagle Owl, Giant Eagle Owl

IDENTIFICATION 60-65cm. Unmistakable; a huge, powerful owl, grey-brown above with pale milky-grey underparts, stocky horns often readily visible. Looks immense in flight. Underparts very finely vermiculated white on grey-brown. Facial disc dirty white with a broad black border at sides. Prominent black basal bristles around large creamy-horn bill. Eye brown with distinctive pink eyelids. Immature washed browner and lacks ear-tufts. **Similar species** See Eurasian and Spotted Eagle Owls.

HABITS Sedentary, frequents areas with well-developed trees e.g. Grey Plum or mampato *Parinari excelsa* throughout the dry savanna woodlands and sites bordering forest-thicket, occasionally in closed canopy forest. Emerges just prior to dusk from a regular roost onto a prominent branch and typically begins to call. Famed for its habit of 'peeling' hedgehogs and discarding the pelt with the spines attached. Kills large numbers of rats, emphasising its value to agriculture.

VOICE A far-carrying, repeated deep dissatisfied-sounding grunt, usually delivered in twos or threes *whuuc-whuuc-whucc*, also a whine recalling mewing African Harrier-Hawk.

STATUS AND DISTRIBUTION Locally frequent to common breeding resident and site-faithful in all divisions; recorded in every month. Extensively distributed throughout Senegal.

BREEDING Appears to breed once every two years, in the dry season. Known sites are in disused nests of Palm-Nut Vulture and in a deep cavity of a baobab.

PEL'S FISHING OWL
Scotopelia peli
Plate 25

Fr. Chouette-pêcheuse de Pel

IDENTIFICATION 63cm. An awe-inspiringly large ginger-brown owl of riparian habitat. Overall plumage shades of rufous-ginger, upperparts darker with black-brown barring, well spaced on the primaries, underparts paler with arrowmark bars. Raises head feathers when alarmed making the head look huge and bonneted. Large dark eyes dominate the umarked, dusky ill-defined facial disc. Head relatively rounded, ear-tufts not prominent and appear cropped. Characterisitic unfeathered pale yellow-straw legs and feet adapted to fishing. Appears huge in flight. Immature much paler. Pellets reported to be yellowish. **Similar species** Smaller, yellow-cered Rufous Fishing Owl *Scotopelia ussheri* (Fr. Chouette-pêcheuse rousse) is paler with a streaked, not spotted, breast. Nearest verified record Sierra Leone.

HABITS Favours riparian woodland, known from a handful of locations along The Gambia river, both saline and freshwater. Usually located roosting high. May take fiddler crabs and amphibians in addition to a diversity of fish.

VOICE Most likely in the rains. A single deep resonant *hooo* repeated at 10-30second intervals, in bouts of several minutes. This may continue for 2-3 hours. Also a reverberating sonorous wail *whuu-huuuuummm* and a *hmmmm*, audible at a great distance. A penetrating high pitched trill has also been described, and the owlet produces a screaming *miew*.

STATUS AND DISTRIBUTION Rare to uncommon; elusive resident, but precise status difficult to assess. Currently known from locations in LRD, CRD and URD, and recorded in both wet and dry seasons. A single old record from southern Senegal and recent records east along the upper-reaches of the Gambia River in Niokolo Koba.

BREEDING Not yet recorded. Site likely to be a hole in a large riverside tree or possibly the nest of a Hamerkop. Season elsewhere dictated by seasonal lowering of water-levels which concentrates prey.

PEARL-SPOTTED OWLET
Glaucidium perlatum
Plate 25

Fr. Chevêchette perlée

IDENTIFICATION 19cm. Smallest owl in the region. Dumpy shape, with longish tail, well rounded head and characteristic two 'eye-spots' on the nape. Lacks prominent ear-tufts. Plumage chocolate-rufous, sandier on back, spotted white. Tail also appears spotted. Underparts whitish, streaked chocolate-rufous. Facial-disc whitish, edged brown, eyes yellow. Immature paler with more prominent 'eye-spots' on nape which may offer greater protection, tail shorter. **Similar species** See African Scops Owl.

HABITS Favours lightly wooded country with sparse ground cover, also interface of forest and open terrain. When mobbed by small birds the swivelling of its head shows eye-spots on nape well. Flight characteristically fast and deeply undulating. Swings tail, bobs head and arches wings when alarmed.

VOICE A succession of sharp far-carrying piped whistles that rise in a crescendo culminating in a series of longer, falling notes *tu-tu-tu-tu-tu-tu: tuee-tuee-tuee*, alarm call is a repeated querulous *peep peep peep*. Also a diurnal squeaky high pitched note. Frequently duets.

STATUS AND DISTRIBUTION Common to frequent breeding resident in suitable habitat in all divisions; fewer in NBD. Records in every month showing two sharp peaks in November-February and June-July (related to vocal activity). Rare above 15°, otherwise widespread in Senegal.

BREEDING Second half of the dry season, in a tree hole or fence post, often in a disused barbet or woodpecker hole.

AFRICAN WOOD OWL
Strix woodfordii   **Plate 25**

Fr. Chouette africaine

Other name: Woodford's Owl, Wood Owl

IDENTIFICATION 35cm. *S. w. nuchalis*. Unmistakable bulbous-headed, medium-sized forest owl. Upperparts chocolate brown, darker-barred, conspicuous white spots on scapulars. Underparts paler, finely barred brown and white. Facial disc pale chocolate very finely barred, with white crescents on either side of bill; darkest around the eyes which emphasises the size of the dark eyes. Bill yellowish. Immature similar to adult but less well marked. **Similar species** Only medium-sized forest species; no ear-tufts.

HABITS Strictly nocturnal, known only from a restricted stretch of forest-thicket in The Gambia. Elsewhere, including Senegal, partial to riverine forest, a habitat available but seemingly unexploited in The Gambia. A comprehensive nocturnal survey would assist in establishing its distribution.

VOICE Almost a standard *tu-whit-toowho* owl call: a series of rapidly delivered hoots, sometimes in duet, produces a rather high-pitched *whu whu-uh-uh whu-uh*.

STATUS AND DISTRIBUTION Rare, first recorded in The Gambia in 1992; sightings in both dry and wet seasons from same forest-thicket in WD since. Dropped feathers and reports of calls from other forest habitat, all WD. In Senegal, first recorded 1971 in the coastal south; recent records from coastal and central areas, north to 15°30' and to the east where status is often established by call recognition.

BREEDING Breeding suspected in early part of the rains in 1992. Elsewhere, holes in large forest trees preferred but will nest on forest floor.

SHORT-EARED OWL
Asio flammeus   **Plate 25**

Fr. Hibou des marais

IDENTIFICATION 38cm. Pale sandy-brown with boldly dark-streaked underparts. Long barred wings with characteristic dark patch on the underside at the carpal joint conspicuous in flight. Ear-tufts only just discernible. Facial disc pale with thin dark edge. Striking lemon yellow eyes with dark surrounds. **Similar species** Smaller Marsh Owl most likely to be encountered at the coast in the rains; can be separated from dry season Short-eared on eye

colour: lemon yellow in Short-eared, dark brown in Marsh. Also, Marsh is darker and more uniformly coloured, with contrastingly patterned upperwing,

HABITS A diurnal hunter, also active by night. Quarters and glides low over the ground in a harrier-like fashion but with a distinctive jerky bounding flight. Roosts communally on the ground.

VOICE High-pitched nasal yelp, also a continuous resonant hoot, *hoo-hoo-hoo*, not very vocal outside breeding season.

STATUS AND DISTRIBUTION Very locally rare to uncommon trans-Saharan migrant. Re-entry to the Gambian list based on a series of sightings since 1990. Excepting a single November record all others are January-March at a single site in URD. Other suitable and extensive marsh and reedbed habitats remain unresearched. Recorded in northern Senegal near the coast in sugar cane plots in the same months.

MARSH OWL
Asio capensis   **Plate 25**

Fr. Hibou du Cap

Other name: African Marsh Owl

IDENTIFICATION 30cm. Uniformly coloured plain brown owl, with upperparts darker than buffy underparts which are finely paler barred. Thighs, belly and undertail-coverts are unmarked. Long-winged, with some sparse white spotting, pale buff patches at primary bases prominent in flight. Short ear-tufts. Facial disc buff, paler around bill but darker around bulbous brown eyes. Facial disc thinly bordered darker brown. Immature similar to adult but darker on underparts. **Similar species** See Short-eared Owl.

HABITS Crepuscular. Roosts on grassy ground, often in groups. Diet varied, passage through The Gambia correlates with termite availability. Inspected pellets also contained a high proportion of beetle cases. Sensitive to disturbance. Birds appear to leave roost together to visit shared hunting area.

VOICE Various croaking calls: a loud *creeouw* and a more rapid *quick-quark-quark* elsewhere but generally silent in The Gambia.

STATUS AND DISTRIBUTION Seasonal and briefly very locally common. Known to gather at one coastal site which is under serious threat of encroachment. Numbers peak August-September when one recent estimate suggested c. 80 birds leaving roost at dusk. One 1993 mid-rains record of single bird in URD. In Senegal known only from

239

coastal Casamance. Origins of this regional rarity are uncertain, but may be Mali to the east where it breeds January-March. Habitat protection is critical as it appears to be faithful to its probable very restricted moulting grounds in coastal Gambia.

BREEDING Nearest Mali. Nest placed at the end of a short grassy tunnel on the ground.

NIGHTJARS
Caprimulgidae

Ten species, two from northern Senegal only. Only two are regularly seen and widely distributed. Specialised nocturnal insectivores, characterised by cryptic 'dead leaf' patterning and coloration in soft shades of rufous, grey, brown, black and gold. Legs and feet short and weak. They have short bills with a very wide gape, surrounded by a fringe of rictal bristles. Large eyes and soft plumage for silent flight are additional adaptations for aerial pursuit of night flying insects. Crepuscular and nocturnal; songs remarkable including quavering whistles, repetitive 'chucking' phrases and, in some species, extraordinary sustained churring. Flight buoyant. Frequently settle on bare patches of flat ground at night. They do not make nests but usually lay two eggs directly into shallow scrapes or depressions on the ground. Identification of all but the dramatically plumed forms is seldom simple. In flight, look for a combination of size, wing spots and tail shape (rounded, square, indented). For a perched bird, size assessment is again important, and details of white patches (throat, wings) should be carefully noted, as well as general plumage features. Note that white patches on wings and tail of males are often duplicated in buffy-brown or are absent in the corresponding females. Habitat and voice can also be very important.

SWAMP NIGHTJAR
Caprimulgus natalensis Plate 26

Fr. Engoulevent à queue blanche

Other names: Natal Nightjar, White-tailed Nightjar

IDENTIFICATION 23cm. Sexes differ slightly. A medium-sized, squat, short tailed, dusky nightjar. Upperparts heavily spotted with black triangular markings; breast blackish boldly marked with round buff spots; belly to undertail plain pale buff. Dark face mask, whitish throat, buff collar. Blackish crown tapers to a point at the nape and may appear as a cap. Bar across flight feathers (four or five primaries spotted), white in male, rufous in female. Tail rounded, dark grey-brown with five to eight narrow bars. Two outer tail feathers entirely white (male) or white with buffy margins (female). Unusually long legs and toes flesh pink. The darker, greyer and smaller race *C. n accrae* of Liberia to coastal Nigeria is perhaps a more likely candidate than paler brown nominate *C. n. natalensis*, but verification is needed. **Similar species** Common and slim-bodied Long-tailed Nightjar is only other species in the area with outertail completely white and white primary patch, but has in addition a white or buff trailing edge to the wing, white shoulder bar when settled and churrs. Dark but larger male European Nightjar *europeaus* eliminated by limited white in tail.

HABITS Favours the soggy fringes of freshwater swamps. Roosts in rank grassy vegetation during the day. Active before total darkness; likes to hunt low around palms, sits on tracks, fence posts and stumps. Long legs equip it for chasing insects on the ground.

VOICE A characteristic extended, loud, slow *chop-chop-chop-chop-chop* or *chook-chook-chook-chook-chook*. Also liquid, melodious bubbling notes uttered on the wing, *wip wha-hul-hul-hul*, or a tremulous *wuwuwuwuwuwuwuwu* and a slow *chuck-chuck*. The most varied repertoire of all listed nightjars, but beware the vocalisations of tree-frogs and geckos.

STATUS AND DISTRIBUTION Uncertain. No sight records in Senegambia. Inclusion based on vocal identification from two tape recorded birds, March 1975 in swamps near Basse. Extensive swamp areas throughout the country suggest suitable habitat is available. Widely distributed in West Africa but nowhere common.

BREEDING Nearest known in Nigeria, April-May.

LONG-TAILED NIGHTJAR
Caprimulgus climacurus Plate 26

Fr. Engoulevent à longue queue

IDENTIFICATION 23cm or 33cm including diagnostic central tail feathers. Sexes differ. A slim, small-bodied nightjar with a very long straight tail (2 central feathers only; note that moulting birds show only partial elongation). Ground colour varies from greyish-brown to rufous-brown or blackish-brown. Centre of crown streaked dark brown with paler grey-brown over the eyes; tawny half-collar, white throat, conspicuous white wing-bars and pale tawny scapular stripe at rest. Underparts rufous-brown with fine transverse barring from throat to breast, plain buff at vent. Outer web of outer tail feathers completely white; adjoining feathers white at tip only. White bar across flight feathers (4-6 primaries irregularly spotted); white trailing edge to the secondaries and white line along the coverts. Female has buff-brown wing patches, outer tail webs finely barred; central tail feathers shorter. Immature is a dull, plainer version of the female. **Similar species** Swamp Nightjar has more extensive white on outertail, no white trailing edge to wing.

HABITS Favours light bush with bare ground, and notably defunct laterite quarries, where it comes to sit at night. Tamarisk scrub around saltpans with low grassy edges, murrum roads and coastal thorn-scrub behind beaches are other likely sites. Typically seen flying or calling from perches for a brief spell at dusk, or just prior to dawn, but can also be encountered on roads late at night. Hunts close to the ground and around low vegetation.

VOICE An insect-like continuous rapid *churr* or *purr* commencing before total darkness, for several minutes. Often several birds are heard together. The fastest churr of any nightjar in the area. Calls from the ground, also from crown of leafy acacia. May produce a clicking sputtering *k-k-k-k-k* from the ground and a low *chung* or *chow* on the wing.

STATUS AND DISTRIBUTION Locally common throughout all divisions in suitable areas. Records in every month, most frequently November to late-January. Widespread in Senegal, mainly in western and central parts.

BREEDING No proven records in The Gambia but vocalisations and display watched in coastal NBD in January and most full-tailed records occur December-February. In northern Senegal March-September with peak activity May-July.

BLACK-SHOULDERED NIGHTJAR
Caprimulgus nigriscapularis **Plate 26**

Fr. Engoulevent à épaulettes noires

IDENTIFICATION 23cm. Sexes differ slightly. A rich tawny-brown nightjar, associated with forest edges or thickly wooded habitat typically with nearby open terrain. Overall warm orange-brown mixed with black; crown with black and rufous central streaks. Dark rufous and black collar at hind-neck; sides of neck paler. Below dark rufous-brown, closely vermiculated with fine barring. Blackish-brown scapular line contrasts with folded

wing. Bar of smallish spots in outer part of five flight feathers, and last third of two outer tail feathers white in male, buffy in female. Tail square-ended, rufous, barred black. **Similar species** The commoner nightjars are also regularly recorded in the vicinity of coastal forest-thicket. Confirmation of this elusive species in The Gambia without a specimen would at least require good photographs, or a sound recording.

HABITS A little-known nightjar. Habitat requirements best satisfied near to forest thicket edges, which mainly occur in WD. Potential presence in or near tracts of riparian forest in CRD should also be considered. Eastern forms typically hawk insects to and from a tree perch.

VOICE Song characterised by tremulous liquid whistling; *kyeh-oh, kirr-arrh* or *cuiepp-cyrrr* or *t-wip, turr-r-r-r*. Closing notes have quality of African Scops Owl, repeated every 5-8 seconds.

STATUS AND DISTRIBUTION No additional information in The Gambia since the unconfirmed record of two small, dark brown nightjars at a nest containing one egg in CRD, May 1960 (when listed as *C. pectoralis nigriscapularis*). In Senegal, records based on voice recognition in coastal Casamance in January–March and June.

NOTE This species is sometimes treated as conspecific with Fiery-necked Nightjar *C. pectoralis*.

PLAIN NIGHTJAR
Caprimulgus inornatus **Plate 26**

Fr. Engoulevent terne

IDENTIFICATION 22cm. Sexes differ. A pale, smallish nightjar, relatively featureless. Warm mid-brown tones extend from the crown, over mantle and below to the breast. Crown flecked with well spaced, black pin-head spots (sometimes with very light streaking), which extend over faintly streaked upperparts. Lacks conspicuous white throat patch. Ground colour may vary from grey to rufous. Underparts darker at the upper breast, otherwise pale buff with fine barring. Male has white bar (four primary feathers, variable in extent) across darkish outer primaries; one third to half of outer pair of tail feathers white. Female wing spots tawny-buff, outer tail diffusely barred to tip; immature similar to female. **Similar species** Overlaps in northern Senegal with Egyptian Nightjar. Plain is smaller and darker; white in wing and tail of male Plain absent from both sexes of Egyptian. Sandy immature Standard-winged Nightjar has more narrowly and regularly barred wings.

241

HABITS An intra-African migrant. Could occur anywhere in Senegambia, most likely in relatively open bush.

VOICE A prolonged, rapid *churr*, similar to breeding call of European Nightjar; a slight chuckling sound in flight and a low *chuck* whilst on the ground have also been described.

STATUS AND DISTRIBUTION First recorded from two road casualties in CRD in June 1988 and November 1990. On the latter date numerous individuals, almost certainly predominantly of this species, were seen along a tar road just after dark. Probably under-recorded but visits may be both brief and erratic. In Senegal, common east of The Gambia in November, with records north of the Gambia-Senegal border in December.

BREEDING Nearest breeding Mali and Mauritania April-June and Liberia in June.

GOLDEN NIGHTJAR
Caprimulgus eximius   Not illustrated

Fr. Engoulevent doré

IDENTIFICATION 24cm. *C. e. simplicior.* Sexes differ slightly. A thickset nightjar, stunningly coloured golden-buff, spangled with small black and silvery rosettes. Conspicuous white throat patch and large black and white band across the four main flight feathers. Tail has bright white corners to its outer feathers. White in wing and tail of female smaller, more creamy-white. **Similar species** Egyptian Nightjar is uniform sandy-grey not golden-buff, and lacks white wing and tail patches.

HABITS Inhabits semi-desert areas, especially on yellowish soils and stony terrain.

VOICE A low evenly pitched *churr* which may last for up to several minutes, lower-pitched than Long-tailed Nightjar.

STATUS AND DISTRIBUTION No Gambian records. In north-west Senegal, rare in Parc National des Oiseaux du Djoudj (two records 1984-94). Mainly recorded near Richard-Toll.

BREEDING Nest with 2 eggs in northern Senegal April 1965.

EGYPTIAN NIGHTJAR
Caprimulgus aegyptius   Not illustrated

Fr. Engoulevent du désert

IDENTIFICATION 25cm. *C. a. saharae.* Sexes similar. A medium-sized pale sandy nightjar, with limited black flecking and contrasting darker flight feathers. At rest white throat patch inconspicuous; primaries darker than body. There is no white panel or patch in either sex. Open wing shows dark brown and white 'saw-like' barring along the inner webs of the outer primaries. Sandy-buff corners to the outer tail; smaller on the female. **Similar species** Separated from paler forms of Plain Nightjar by sandier, not cinnamon tones; dark primaries contrast more with rest of wing and underwing whiter in Egyptian Nightjar. Thought to migrate with Plain Nightjar.

HABITS In northern Senegal favours scrubby vegetation dominated by tamarisk, where it roosts gregariously. Emerges to forage over nearby rice fields. Confiding by nature and frequently seen around human habitation; flight noticeably slow.

VOICE A repeated knocking *tok tok tok* or *tukl-tukl* alikened to the sound of a distant chugging boat engine. Vocal activity away from breeding grounds is uncertain.

STATUS AND DISTRIBUTION No Gambian records. Considered a locally common non-breeding trans-Saharan migrant to northern Senegal, December-February.

RED-NECKED NIGHTJAR
Caprimulgus ruficollis   Plate 26

Fr. Engoulevent à collier roux

IDENTIFICATION 32m. Sexes differ slightly. A large nightjar with a broad head; colourfully mottled in greys and oranges. Central crown grey, strongly streaked with black and rusty-orange; broad tawny half-collar, rufous scapular region, large white throat patch. Finely streaked with black above, fine dark transverse barring below. Both sexes have white wing spots on both webs of the three longest flight feathers; white outer margin to tail over a quarter its length from corners (wing and tail patches smaller and buff-tinged in female). Open wing barred rusty on inner flight feathers; tail broad, grey with well spaced thin darker bars. Eye brown, bill and legs dark horn-brown. Both the nominate race and greyer, paler, broader-collared *C. r. desertorum* of N Africa may occur. **Similar species** Larger than darker European Nightjar with longer tail, bulkier head and more colourful variegations with stronger rusty tints. With experience, Red-necked has discernibly more powerful and swooping flight.

HABITS Little known away from its breeding grounds. Could appear anywhere but most likely in open dry areas with scattered trees. In flight large size creates impression of small raptor.

VOICE Probably silent on West African passage. On its breeding grounds north of the Sahara a loud, hollow and repetitious *kut-Tok kut-Tok* likened to the striking of an axe on wood.

STATUS AND DISTRIBUTION First recorded in November 1990, a single road-kill in CRD (immature of nominate race; N. Cleere pers. comm.). Fourteen road-kills in CRD, LRD and URD, October 1996-November 1998. Older records in extreme northern Senegal, involving both races, are from November.

EUROPEAN NIGHTJAR
Caprimulgus europaeus   Plate 26

Fr. Engoulevent d'Europe

IDENTIFICATION 28cm. Sexes differ slightly. A large grey nightjar with heavy blackish streaks and speckles. Grey crown strongly marked with black streaks, tawny half collar present but not prominent. Rufous sides to black-freckled face, chin and throat. At rest shows pale moustachial streaks and a series of grey-white spots along shoulders. Male has white spots across the three outer primaries, and white outer margins to tail over one quarter of its length from corners. Female may show buffy tail corners but open wing lacks white spots, having well spaced rufous barring along both webs of the outer primaries. Underparts tawny-buff with narrow barring. Paler north-west African race *C. e. meridionalis* is overall more silvery-grey and male has larger wing spots than nominate. **Similar species** Red-necked Nightjar larger and paler. Female European is separable from Standard-winged Nightjar by less dense barring on primaries, and larger size. Wing shape and heavily streaked grey body separates from vagrant female Pennant-winged.

HABITS Wide range of habitats, some overlapping with Standard-winged Nightjar; also near damp marshy areas well inland and forest edges.

VOICE Probably silent in sub-Saharan Africa. On breeding grounds north of the Sahara two calls are commonly heard, a quiet nasal *coo-uk* in flight and a very long churring that fluctuates in pitch and can be sustained for minutes on end.

STATUS AND DISTRIBUTION Uncommon migrant from north-west Africa and Europe where it breeds. Records (not racially determined) in all divisions south of the river from late October to early February. Considered common earlier this century but no modern records in The Gambia. One Senegalese record from the extreme coastal north in November, was of the north-west African race.

STANDARD-WINGED NIGHTJAR
Macrodipteryx longipennis   Plate 26

Fr. Engoulevent à balanciers

IDENTIFICATION 19-23cm. Sexes differ when breeding. A medium-sized nightjar. Fully-plumed male astonishing, bearing two diagnostic elongated wing feathers, with broad 16cm feather blades trailing at the end of 22cm long bare shafts. 'Standards' sprout in October and are fully grown by February. Partially developed standards can be seen during this period. Non-breeding male and female similar; female slightly paler. Lacks white patches in wing and tail. At rest dumpy, plain mottled greyish; hint of pale arch above eye, weak tawny collar, narrow evenly-spaced rusty and dark brown barring along edge of folded wing. Tail shortish, square-ended. Sandy-buff below with profuse narrow wavy brown bars. Immature resembles the female, but greyer. **Similar species** Larger Pennant-winged Nightjar (very rare and still in need of confirmation) has more widely spaced wing bars and unusaul wing tip shape created by outermost primary (P10) being longer than P9 (see flight silhouettes, Plate 26). Heavier and more regular barring on wings than Plain Nightjar.

HABITS Regularly seen with Long-tailed Nightjar. Found in a range of arid scrub habitats especially lightly wooded savanna with some bare stony ground. In normal flight standards trail flat behind, but displaying males engage in low-level hovering flight, sailing slowly over an open 'arena' with females resting on the ground. Wings then maintain shallow rapid beats, while the standard shafts arch upward and backward, the two 'standards' then flying vertical and parallel above and behind the bird. Generally males with standards are heavily outnumbered by birds without.

VOICE An insect-like, sibilant chirping; a loud shrill churring and an intense low-pitched quiet mouse-like *curr-r-r* during sexual display.

STATUS AND DISTRIBUTION Locally common throughout The Gambia, with two peaks in frequency: November-February and, to a lesser degree, June-September. These mysterious fluctuations in numbers require more fieldwork. Widespread in Senegal.

BREEDING Eggs and chicks June and July in WD; recent observations of full displays in LRD, July and male handled in Kiang West in November prior to breeding with standards beginning to sprout. In eastern Senegal eggs in March.

PENNANT-WINGED NIGHTJAR
Macrodipteryx vexillarius   **Plate 26**

Fr. Engoulevent porte-étendard

IDENTIFICATION 25-28cm. A medium-large nightjar. Sexes differ. Males with full breeding plumes not expected; non-breeding male retains short white wing pennants, and has large white bar across blackish wing, white throat and pale belly. Adult female resembles large Standard-winged female, but wing barring broader and darker. **Similar species** Adult female separable from Standard-winged female on size, though beware some overlap, and wing-bar pattern; in the hand confirm by examination of wing length and relative primary lengths. Female European Nightjar is larger-headed, and greyer.

HABITS Poorly known in West Africa. Gregarious on migration and noted to irrupt sporadically.

VOICE Probably silent on West African non-breeding grounds. In South Africa a low sibilant bat-like or cricket-like twittering, not often heard; in breeding areas *ku-whoop-ku-whoop-ku-whoop* followed by a prolonged *churr.*

STATUS AND DISTRIBUTION Vagrant in need of confirmation in The Gambia. A trans-equatorial migrant, seen on non-breeding grounds north of the equator between February and August. A single specimen examined in the hand near Soma, November 1977. Senegalese Casamance records now considered unreliable. Known from Guinea-Bissau and becomes more frequent in Nigeria and Cameroon but is nowhere common in West Africa.

SWIFTS
Apodidae

Eight species, one only in Senegal. Most are gregarious. A group of largely aerial birds settling only to scurry into crevices in trees or buildings to breed or roost. They feed, sleep, drink, and even mate on the wing. All are fast fliers on long slender curved wings. Voices consist of a variety of high pitched screams. Plumages are predominantly black and shades of grey. Four species have prominent white rumps, two have white on the underparts: the position of these together with tail shape are important in identification.

MOTTLED SPINETAIL
Telacanthura ussheri   **Plate 27**

Fr. Martinet d'Ussher

Other names: Ussher's Spinetailed Swift, Mottled-throated Spinetail

IDENTIFICATION 14cm. Medium-sized. Stout-bodied swift (the only Senegambian spinetail) with a square short tail, pure white rump coupled with a pale patch on the lower belly creating a diagnostic white band around the rear body. Otherwise dark brown, extensive chin and throat patch mottled grey-white, wings extending far beyond short tail. Spines are short barbless extensions of the tail shafts and are almost impossible to detect in flight. Longish, relatively broad wings; shorter secondaries create impression of notched rear margin to wing. **Similar species** Little Swift lacks white belly band and wing notches and has a clear white throat. Spinetails are rarely gregarious and can be best separated from Little Swift on flight jizz.

HABITS Favours dry deciduous woodland with an abundance of aged baobabs. At the coast, regularly seen over forest thickets. Flight is bat-like. Typically forages alone or in very small loose groups, temporarily vanishing then returning; especially active in the last hour of daylight. Swoops in to trawl-drink from standing water in roadside ponds after the rains.

VOICE A rasping and twittering *kak-k-k-k* or *tt-rrit, tt-rrit.*

STATUS AND DISTRIBUTION Frequent to locally common resident throughout all divisions with no significant seasonal changes. Main strongholds in WD to LRD. In Senegal occurs up to 15°, north of which it becomes rare; also found in Casamance and eastern Senegal in pockets of suitable habitat.

BREEDING Groups continously sweeping up into village baobabs with accompanying screaming, mid-July in LRD (near Keneba) where nesting assumed. Display early June, WD (Aboku). Nest a small half saucer of dry leaf pieces and tiny twigs, glued together with saliva.

AFRICAN PALM SWIFT
Cypsiurus parvus   Plate 27

Fr. Martinet des palmes

Other name: Palm Swift

IDENTIFICATION 16cm. Strikingly slender mousey-brown swift with paler throat, chin and sides of face. The slimmest and longest-tailed Senegambian swift. Characteristically deeply forked tail usually held closed looking needle-shaped, occasionally flared wide open; fork reduced in younger birds. **Similar species** Pale coloration and slender shape unique.

HABITS Strongly associated with oil and rhun palms, around which it may congregate in vast flocks. Hunts over a range of heights, coming almost to ground level to feed, unlike most swifts. Regional deforestation of rhun palms may affect long-term population dynamics.

VOICE A shrill, delicate scream, loudest when birds group around a nesting tree or when flock is feeding.

STATUS AND DISTRIBUTION Abundant resident throughout, prone to local short-term population fluctuations. Numbers increase during the rains in most areas. Seasonal intra-regional movements may occur. Abundant throughout Senegal.

BREEDING Solitary, pairs or in groups, season extended. Nest a small flange of plant down and feathers glued with saliva to vertical dangling palm frond. Eggs usually 2, also secured in position with saliva; chick grips vertically to the nest pad during development.

PALLID SWIFT
Apus pallidus   Plate 27

Fr. Martinet pâle

Other name: Mouse-coloured Swift

IDENTIFICATION 17.5cm. Pale brown swift with largish pale throat patch. Palearctic *A. p. brehmorum* or north Saharan nominate *pallidus* may occur. A useful characteristic is that Pallid is comparable in colour to the common resident African Palm Swift. **Similar species** Difficult to separate from Common Swift unless seen well in good light, and preferably at close range. Common is considerably more sooty-grey, especially immature which is even darker but with a more extensive pale throat. Also in Pallid dark outer primaries and leading edge of wing characteristically contrast with paler remainder. Because the mantle is slightly darker than the head, rump

and wing-coverts, Pallid may show a 'saddle' effect. Pallid has a bigger skull thus a stockier head, and the eye patch appears more conspicuous on Pallid because of paler head feathering.

HABITS There appear to be no substantial behavioural or ecological differences in West Africa from those of Common Swift, but Pallid is less agile in flight with more gliding. Occurs almost anywhere in mixed assemblages or in vast single-species flocks.

VOICE A disyllabic *cheeu-eet* or *churr-ic* when breeding.

STATUS AND DISTRIBUTION Frequent to occasionally abundant July-September, and again November-January. Accurate assessment of numbers difficult because of identification problems. Known in extreme northern Senegal at the coast in October and February.

COMMON SWIFT
Apus apus   Plate 27

Fr. Martinet noir

Other name: European Swift

IDENTIFICATION 17.5cm. Largish with characteristic anchor-shaped silhoutte. Adult sooty black-brown, effectively appearing black, except for a white or greyish-white throat patch visible at close range and in good light conditions. Worn plumage after breeding and migration is faded, browner and less glossy, and this is when problems arise with separation from Pallid Swift. Tail slightly forked. Immature generally darker than adult, throat patch usually larger and cleaner white, forehead paler, body plumage edged with pale scaly markings if seen well; immatures often predominate among early returning birds. **Similar species** See Pallid Swift.

HABITS Flight is strong, sometimes erratic with sudden changes of direction. Gregarious at all times, often with other swift species. Often appears shortly after rain storms under low cloud, disappearing once the sky clears. May stay for a few days, sometimes only a few hours. Generally large flocks of all-dark swifts are likely to be Common; conversely large flocks of all-pale swifts are likely to be Pallid.

VOICE Mostly silent on migration; a shrill very high pitched screaming when breeding.

STATUS AND DISTRIBUTION Palearctic migrant recorded in all divisions and in most months. Peak observations on north-south return to Africa from late July to September mostly in WD and URD;

245

fewer records on the south-north return. In northwest Senegal records peak in September. It has recently been suggested that these birds feed along the insect-rich intertropical front that disappears from The Gulf of Guinea at the turn of the year, when the swifts move away from West Africa.

WHITE-RUMPED SWIFT
Apus caffer   Plate 27

Fr. Martinet cafre

Other names: African White-rumped Swift

IDENTIFICATION 15cm. Medium-sized blackish sleek swift. Distinctive features include white throat patch, characteristically narrow white band on upper rump and a markedly forked tail, often looking pointed in flight. White rump may show some faint dark streaks if seen well. **Similar species** The only white-rumped swift with a pointed tail in normal flight, deeply forked if fanned.

HABITS Old bridges on the outskirts of villages are good places to look in the rains. Mostly seen as singles or solitary pairs mingling with other swifts or swallows, where first impression in silhouette is of a chubby African Palm Swift.

VOICE A twittering bat-like *psrrit, psrrit.*

STATUS AND DISTRIBUTION Locally and seasonally frequent in all divisions, with regular stronghold in LRD (Kiang West National Park). An intra-African, possibly trans-equatorial migrant. Wet season visitor to The Gambia with all recent records July-November, but could occur outside these months. Found east in Senegal along Gambia River in same months, also January, and north to 15° along Mali border.

BREEDING Suspected under bridges in the rainy months e.g. August in WD. Nest a pad of feathers glued with saliva in the disused or commandeered flask-shaped nests of a number of the larger swallows, including Mosque, Red-rumped and Rufous-chested. Has bred Niokolo Koba, southeastern Senegal.

LITTLE SWIFT
Apus affinis   Plate 27

Fr. Martinet des maisons

Other names: House Swift, Little African Swift

IDENTIFICATION 12cm. *A. a. aerobates.* Gregarious, small and stocky swift with short square tail and

white rump extending marginally onto the flanks. Chin and forehead white, mantle and underparts blackish. Frequently fans a slightly rounded tail. Immature browner but all white features present, throat patch dirtier. **Similar species** Mottled Spinetail has a complete white band around the lower belly and rump.

HABITS Widespread, numerous and familiar urban and village bird. Crepuscular parties of screaming swifts shortly after the first rain are a regular feature of village life. Flight varied: gliding on motionless wings, particularly in the rain, with some fluttering and sailing with the wings held up in an angled V.

VOICE High-pitched screaming chitter, *chee-ch-ch-ch-ch.*

STATUS AND DISTRIBUTION Common to seasonally abundant throughout; locally very abundant particularly in coastal WD e.g Banjul and Denton Bridge, July-November, but numbers much reduced most years in April-May. Comparable status throughout Senegal.

BREEDING Nest an untidy structure, often in clusters, usually under the eaves of a thatched roof, a bridge or jetty, or sometimes in a rock fissure. Nest built mostly of plant material and a profusion of feathers, cemented with saliva.

ALPINE SWIFT
Tachymarptis melba   Plate 27

Fr. Martinet alpin

IDENTIFICATION 21cm. Big and impressive swift. Adult sandy-brown above with white throat (not always discernible at distance) and characteristic white belly separated by brown breast-band. Wings long, brown, tail dark, held tapered to a point or shallowly forked. Immature similar but scaly above. **Similar species** Confusion possible with similar Mottled Swift *Tachymarptis aequatoralis* (Fr. Martinet marbré), not illustrated: a montane species with barred and speckled underparts. Groups recently reported in south-east Senegal (February) and Guinea (January); nearest stronghold Sierra Leone.

HABITS Long wings and slow deep wing-beats give languid flight impression, almost falcon-like, but direct flight is with great power and speed. Elsewhere forages low around herds of plains game, and may behave similarly around N'dama cattle.

VOICE Silent on wintering grounds.

STATUS AND DISTRIBUTION Vagrant, three records at coastal sites in WD November 1988 (over Kerr-Serign), March 1992 (over Fajara) and 1995 (over Banjul-Barra estuary). In Senegal records in the north in September, south-east (Niokolo Koba) June, December and a large flock in February 1995.

Winters further south-east in West Africa with nearest substantial population in Côte d'Ivoire.

MOUSEBIRDS

Coliidae

One species. Short, thick bill and crested head, with long graduated tail; flies with fast whirring wingbeats. Clambers in tangled vegetation with great agility, scampering mouse-like along branches. Principally frugivorous and may damage crops but also eats insects.

BLUE-NAPED MOUSEBIRD
Urocolius macrourus   Plate 27

Fr. Coliou huppé

Other name: Blue-naped Colie

IDENTIFICATION 38cm including very long stiff tail. Plumage soft, overall grey tinged bluish, underparts buff. Short-cropped crest with turquoise-blue nape-patch. Stubby, powerful bill black, carmine-red at the base of upper mandible. Eye and facial bare skin bright red. Short legs and feet pinkish-purple. Immature lacks blue nape-patch and crest is poorly developed. **Similar species** Colour and shape unique.

HABITS A bird of dry open country and thorn scrub. Highly gregarious moving in small parties. Flight diagnostically low, fast and direct on short rounded wings; dives into scrub at speed on its arrival. Frequently dust-bathes and sunbathes. When perched collectively they adopt a peculiar posture of hanging with their head and shoulders level with their feet. In Senegal, appears to be benefiting from the extensive planting of neem trees, thriving on the fruit.

VOICE A loud whistled *tieee* regularly repeated. Parties call incessantly in flight and when feeding.

STATUS AND DISTRIBUTION Very rare visitor; only two recent records both south of the river in WD. Locally common in northern Senegal above 14°N also recorded just east of The Gambia. More information required from east NBD to clarify Senegambian range.

BREEDING In Senegal, breeds early dry season but season extended. An untidy open nest usually in a thorn bush. May employ cooperative breeding strategies. Breeds also Mauretania and Mali.

KINGFISHERS

Alcedinidae

Nine species, in three subfamilies. Small to medium birds, with one large species. Mostly colourful, with varying amounts and shades of blue in their plumages. Legs short, feet small and weak-looking. Front toes partially fused from base (syndactyl). Bills generally long and dagger-like. None are solely fish-eaters, with insects and lizards contributing substantially to the diet of most species. All nest in holes, often self-excavated, in sand banks, trees or termitaria. Occupy diverse habitats, many associated with dry land.

GREY-HEADED KINGFISHER
Halcyon leucocephala   Plate 28

Fr. Martin-chasseur à tête grise

Other name: Chestnut-bellied Kingfisher

IDENTIFICATION 22cm. Medium-sized woodland kingfisher; rounded head. Breast and mantle light grey paling to off-white on throat. Male slightly brighter and glossy. Belly warm chestnut-brown, back black, rump and tail brilliant azure, especially in flight. Wing coverts black, flight feathers azure-blue tinted violet, white primary bases forming a bold patch in flight and when wings flashed in display. Narrow bill bright red, legs and feet red. Immature duller, with scaly-barred breast, buff belly lacking extensive chestnut-brown, bill blackish. **Similar species** Poor views of immature can lead to confusion with Woodland Kingfisher.

HABITS Usually solitary, often in transit. Prefers a well wooded habitat near freshwater with rank vegetation. Sits unobtrusively in a hunched position when not scanning for prey. Hunts from a variety of tree perches, fence posts and telegraph wires. Predominantly insectivourous.

VOICE A weak vibrant trill.

STATUS AND DISTRIBUTION Frequent to seasonally common throughout, less so in LRD. Recorded in all seasons, peaking June-July and possibly January. Found throughout Senegal except the arid belt south of the Senegal River.

BREEDING Restricted regional information suggested April-June but only recent records are failed attempts in WD in July. In northern Senegal, June-August. Tunnels into banks.

BLUE-BREASTED KINGFISHER
Halcyon malimbica   **Plate 28**

Fr. Martin-pêcheur à poitrine bleue

IDENTIFICATION 25cm. *H.m. torquata*. Relatively large, with a huge red and black bill. Crown and nape dull grey, black patch from bill through eye almost to nape. Cheeks, shoulders and back pale turquoise, tail darker blue. Chin and throat whitish, breast pale blue shading to whitish on belly; flanks lightly barred. Huge bill has red upper mandible, black lower. Legs and feet red. In flight shows striking blue panel in blackish wings. Immature overall greener, bill pale-horn redder towards base. **Similar species** See Woodland Kingfisher.

HABITS Favours aquatic and forest habitats, especially mangroves. Mostly solitary, territorial, and evenly distributed through stretches of suitable habitat. Usually sits in the shade low over water, often drawing initial attention by its song. May gather in some numbers during termite emergences at the onset of rains. Forages by sallying out like a bulky flycatcher to catch insects in mid-air and also to pick insects from the surface of running water and undersides of leaves. Interesting habit of eating palm-oil kernels. An avid bather.

VOICE Vocal throughout the year but more prominent during the rains. Loud and far-carrying, begins with a single emphatic note, hesitating before a mellow sequence of slowing piped whistles, descending mournfully in pitch; *chiu- pu-pu-pu ku ku ku ku*. Usually from deep cover. There is also a raucous flight alarm call *tchup tchup-tchup-tchup*.

STATUS AND DISTRIBUTION Frequent to locally common throughout all divisions, numbers may increase in some areas during the heavy rains. Sedentary at sites where water present year-round, and visits seasonally flooded areas. Throughout Senegal, mostly south of 14°.

BREEDING In mid-late rains tunnels into arboreal termite nests; may re-use sites.

WOODLAND KINGFISHER
Halcyon senegalensis   **Plate 28**

Fr. Martin-pêcheur du Sénégal

Other name: Senegal Kingfisher

IDENTIFICATION 22cm. Medium-sized grey and azure kingfisher. Head grey, tinged blue on ear-coverts and sides of nape, black patch from bill to eye and not beyond. Back and tail brilliant blue with greenish tones, wings black and brilliant azure secondaries conspicuous as a patch in flight. Throat white, breast grey with blue shading, belly whitish. Upper mandible of robust bill red, lower blackish. Legs and feet blackish. Underwing characteristically white in flight. Immature duller blue with very fine barring around face and upper breast, bill blackish. **Similar species** Can be confused with larger and darker Blue-breasted Kingfisher but note that Blue-breasted has diagnostic black mask-patch behind the eye, a larger bill, and a darker ash-grey crown. Woodland has scapulars all blue, black in Blue-breasted.

HABITS Favours light woodland, forest-thicket edges and also gardens. Largely insectivorous, with most prey collected on the ground but will 'flycatch' for winged termites. Strongly territorial. Elaborate aerial display chases accompanied by constant calling are a prominent feature of the early rains. Hasty wing-salutes are given on alighting.

VOICE A loud cascading phrase begins with a sharp *krit* followed by a rapid undulated trill which fades out towards the end *tirrrr-tirrrrr-tirrrrr-tirrrrr*, with familiarity recognisably intermediate between rapid trilling of Striped and slow-piping of Blue-breasted. Also a *kee-kee-kee-kee* alarm call.

STATUS AND DISTRIBUTION Seasonally common throughout, greatest numbers in WD. Arrives in mid-late June and is abruptly almost absent by the end of November; uncommon outside these months. Widespread in Senegal and movements similarly rain-related, but likely also in Casamance during the dry season.

BREEDING Mid-late rains. Usually in an oil-palm in an existing cavity or excavated hole, for which there may be strong interspecific competition.

STRIPED KINGFISHER
Halcyon chelicuti   **Plate 28**

Fr. Martin-pêcheur strié

IDENTIFICATION 17cm. Sexes differ only on underwing features seen during wing-salute display.

A rather small, dumpy and unglamorous kingfisher of the bush. Heavily streaked brown crown with a broad black band through the eye to the nape. Whitish collar, mantle and back brown. Conspicuous turquoise rump patch in flight, largely concealed at rest. Relatively short, stout bill dark red-brown above, red below. Legs and feet dull red. Female browner, lacks black subterminal band across the underside of the opened flight feathers of the male. Immature duller, less streaked. **Similar species** Woodland and Blue-breasted Kingfisher are substantially larger, lacking streaking on crown and underparts.

HABITS Favours dry bush, savanna woodland and farmland. Usually solitary or in pairs and mostly sedentary. Uses regular perches within its territory from which it hunts and sings. Largely insectivorous. Pairs frequently use wing-salute display. Co-operative breeding has been recorded. Vocal throughout the day often causing a 'domino effect' across adjacent territories.

VOICE Loud, high-pitched and far-carrying; one of the first birds to call at dawn in the bush. A repeated disyllabic *keep-kirrr, keep-kirrr*, may sound *pee-hee, pee-hee* at distance. Also a trisyllabic *whi p-whi whi p-whi*. Call could sometimes be confused with either Vieillot's Barbet or Greater Honeyguide and Woodland Kingfisher, which may favour the same habitat.

STATUS AND DISTRIBUTION Common resident throughout, most numerous in LRD. Fewer in WD; becomes less frequent towards URD. No major seasonal population changes. Widespread in Senegal except the eastern region.

BREEDING Little information available but recorded June-July. In northern Senegal early October. Nests in disused holes of woodpeckers and barbets, but a very secretive nester.

AFRICAN PYGMY KINGFISHER
Ceyx picta   Plate 28

Fr. Martin-pêcher pygmée

Other name: Pygmy Kingfisher

Synonym: *Ispidina picta*

IDENTIFICATION 12cm. Minute woodland and savanna kingfisher often seen near water e.g. at a forest puddle. Upperparts ultramarine tinged violet, underparts rufous-orange. Sides of head and ear-coverts pale violet-purple with a white patch on the sides of the neck. Flat-topped head glossy blue, finely darker-barred. Broad chestnut-orange supercilium. Chin and throat white. Short tail darker. Bill, legs and feet red-orange. Immature dark-billed and overall duller. **Similar species** The wetland

Malachite Kingfisher is slightly larger, blue cap reaches down to eye, and has wispy erectile crest. Pygmy has narrow cap restricted to crown, eye fully surrounded by orange feathering. Note possibility of habitat overlap during the rains.

HABITS In the dry season largely restricted to forest-thicket, but during and just prior to the rains can be found in orchards and open coastal scrub. Regularly visits roadside pools and puddles to splash-bathe and may be seen hunting insects around damp grassy margins or picking them off the water surface. Typically perches low and in the shade. Flight is extremely rapid and straight.

VOICE A thin squeaked flight and display call *tseet* or *seet* or *chip-chip.*

STATUS AND DISTRIBUTION Frequent resident in forest-thickets, becoming common in the rains in WD; there is a substantial influx of migrants June to early October. Fewer records in other divisions, restricted to the wet season. In Senegal, south of 14°, seasonal changes correspond with those in The Gambia; recorded in the coastal north during the rains.

BREEDING Excavates burrow in a bank in July, but high speed display chases seen as early as April. Hatching synchronised with termite emergence. Nests frequently lost if heavy rains prevent foraging for prolonged periods.

MALACHITE KINGFISHER
Alcedo cristata   Plate 28

Fr. Petit Martin-pêcheur huppé

Synonym: *Corythornis cristata*

IDENTIFICATION 13cm. *A. c. galerita*. Small blue and orange wetland kingfisher. Upperparts bright glossy ultramarine-blue, underparts and face deep orange-chestnut. The long crown feathers, usually carried flat, are blue strikingly tinged malachite-green with blackish barring; crest when raised or wind blown is extensive and often ragged. Ear-coverts and cheeks rufous contrasting with white chin and throat, with a characteristic prominent white flash on the side of the neck. Bill, legs and feet scarlet-red. Immature has mantle and wing-coverts blackish spangled blue with white markings; bill black. **Similar species** See smaller mostly dry-habitat or coastal Pygmy Kingfisher and larger and rare Shining-blue Kingfisher.

HABITS Practically always seen close to water occupying a range of aquatic habitats. Dives constantly either to feed or bathe. Usually solitary outside breeding season.

VOICE A short, high-pitched, *kweek* or *seet* or *peep-*

peep, normally delivered in flight. Once known, can be distinguished from Pygmy Kingfisher.

STATUS AND DISTRIBUTION Common in suitable habitat throughout, marginally more numerous in WD. Population in all divisions increases dramatically during the rains, suggesting some migration. Numbers peak July-November with a distinct slump February-May. Widespread in Senegal in suitable habitat.

BREEDING Excavates burrow in a sandy bank, September-October.

SHINING-BLUE KINGFISHER
Alcedo quadribrachys   Plate 28

Fr. Martin-pêcheur azuré

IDENTIFICATION 16cm. Similarly coloured but larger than Malachite Kingfisher and lacking latter's untidy crest. Upperparts gleaming bright purple-blue, crown vaguely banded black with glossy ultramarine tips. A small but prominent buff loral spot and blue patches at the sides of the breast are diagnostic field marks. Chin and throat creamy-white, rest of underparts dark chestnut. Bill shiny black, legs and feet dark orange-red. Adult female may have some dark red at the base of the lower mandible. Immature has black bill tipped white, paler underparts, upper breast flecked dusky forming a faint breast-band. Legs and feet orange-pink. **Similar species** Can be confused with smaller and paler, black-billed immature Malachite Kingfisher, separable by shorter slimmer black bill of Malachite, and by obvious barring and crest on Malachite' crown.

HABITS Not well known in The Gambia with a few reports from mangrove and well wooded freshwater streams. Also recorded on temporarily flooded fields, rain-filled ponds and along the banks of the freshwater reaches of the river. Most records involve single birds, occasionally pairs.

VOICE A repeated high-pitched monosyllabic flight call *cheep* or *tschut.*

STATUS AND DISTRIBUTION Rare to seasonally uncommon. A handful of 1990-97 records from WD, LRD and CRD are all early July to late September. May be under-recorded in some sparsely-watched but suitable areas. Older records (1969-86) cover June to January. Senegalese records July-September from southern Casamance, just north of The Gambia in Saloum, and east in Niokolo Koba on upper Gambia River.

BREEDING Little known. Pairs in WD in late September and in LRD and CRD in late July. Nearest confirmed records Nigeria in September-November and Cameroon with nestlings in December. Tunnel nest in a wet bank.

GIANT KINGFISHER
Megaceryle maxima   Plate 28

Fr. Alcyon géant

Other name: African Giant Kingfisher

IDENTIFICATION 42-46cm. *M. m. maxima.* Sexes differ. Large, unmistakable dark kingfisher with massive black bill and stiff shaggy crest. Male has dark grey upperparts with pale edges to each feather creating a barred and streaked effect. Head and face blackish with small white spots, cheeks darker. Chin and throat white; breast rich coppery-brown with faint darker barring; belly white, closely barred grey. White underwing-coverts. Tail brown with several narrow whitish bars. Legs and feet blackish-grey. Female generally similar but breast white, densely spotted black and belly, flanks and underwing-coverts chestnut. Immature male similar to adult male with black speckles on breast. Immature female like adult female but less spotted, resulting in a grey-white band between belly and breast. **Similar species** Huge size is diagnostic.

HABITS Found on coastal inlets, along freshwater streams and beside deep open forest ponds. Despite its size, can be difficult to see when perched in a favoured leafy forest tree. However, also often encountered sitting out on a telegraph wire over a drainage ditch or waterway. Sedentary, ranging over its territory on a regular timetable. Flight is strong and can recall a small duck at a distance; moves freely across areas of dry bush between feeding grounds.

VOICE Call a loud and harsh *kek-kek-kek* or *kahk-kah-kahk*, some sounding crow-like.

STATUS AND DISTRIBUTION Frequent at a limited number of sites in WD, where sightings diminish towards the end of the dry season. Recorded less frequently in other divisions in all seasons. In Senegal distributed thinly across the whole south of the country and up the eastern border to the Sengal River.

BREEDING At the end of the rains in an earth bank, low under a root or high above water, even in a disused quarry.

PIED KINGFISHER
Ceryle rudis **Plate 28**

Fr. Martin-pêcheur pie

Other name: Lesser Pied Kingfisher

IDENTIFICATION 25cm. Medium-sized and unmistakable, Africa's only black and white kingfisher. Sexes differ. Generally blackish above, with copious white barring; wings and tail black with conspicuous white bars. Head black, with untidy crest, broad white supercilium extends from slim dagger-like black bill through to nape. Underparts white: male has two black breast-bands, a broad one above a slim one; female has single incomplete band. Legs and feet blackish-grey. Immature similar to female but with a grey-brown wash. **Similar species** None.

HABITS Found in almost every aquatic habitat. Opportunist feeder, diving conventionally from perch, or uniquely among kingfishers, plunging for prey from hover. Uses this ability to hunt shallow marine surf along sandy beaches as well as a range of estuarine and freshwater environments.. Very approachable. Highly gregarious with occasional roosts of 100 plus birds gathering at dusk in low coastal *Avicennia* mangroves.

VOICE Extremely vocal; a squeaky conversational *kittle-te-ker* chatter delivered whilst perched or in flight and a rapid *chicker-kerker* uttered in a frequent threat or foraging disputes.

STATUS AND DISTRIBUTION Common to locally abundant in all divisions and seasons throughout a diverse range of aquatic habitats; no significant seasonal population changes apart from obvious local increases at breeding colonies. Widespread in all suitable habitat in Senegal.

BREEDING Usually in the rains, sometimes solitary but generally colonially, with a system of nest-helpers. Nests tunnelled into a variety of banks sometimes away from water.

BEE-EATERS
Meropidae

Eight species. Specialised colourful insect hunters, highly sociable with engaging habits and fascinating breeding strategies. Characteristic sailing flight with slim body and triangular pointed wings. Most are grass green with coloured throats and slightly decurved long bills. Some species are hosts to parasitic species of honeyguides. Not normally difficult to identify when seen well, but drabber immatures and moulting individuals lacking tail extensions can cause problems.

LITTLE BEE-EATER
Merops pusillus **Plate 29**

Fr. Guêpier nain

IDENTIFICATION 17cm. Small, sparrow-sized, and always lacking tail streamers, the most familiar of Senegambian bee-eaters. Upperparts green, wings tinged rufous with black trailing edge in flight. Throat yellow with black gorget, upper breast cinnamon, buff belly. Conspicuous dark mask through eye, forehead and eyebrow green, bill black. Notched tail green, rufous at base, with obvious black tip in flight. Immature duller, lacks gorget, breast and throat yellowish-buff shading to brownish on upper breast which merges into a very faintly streaked light green breast. **Similar species** See text for streamerless immature or moulting adult Little Green Bee-eater and blue-tailed immature Swallow-tailed Bee-eater.

HABITS Favours open bushy grassland, often near water, with tall grass or low shrubbery from which to hunt. Often uses telegraph wires. Usually loosely gregarious; family parties tend to stay together until the next breeding season approaches. Confiding, extremely approachable. Ardent sunbathers, adopting a number of contorted body shapes.

VOICE A soft monosyllabic *sip* or *slip* often just on leaving perch. On greeting increases in intensity to *siddle-iddle-ip, d-jeee.*

STATUS AND DISTRIBUTION Common resident throughout; most numerous near marshes and at mangrove-saltpan interface in WD. Rare or absent in drier parts of Senegal, widespread elsewhere.

BREEDING Tunnels into earth banks, often low. Nests May-July, recently fledged young abundant in July at onset of rains. No suggestion of cooperative breeding. Brood parasitism by Lesser Honeyguide has been recorded.

SWALLOW-TAILED BEE-EATER
Merops hirundineus **Plate 29**

Fr. Guêpier á queue d'aronde

IDENTIFICATION 22cm including diagnostic

long, deeply-forked tail. *M. h. chrysolaimus.* Overall emerald-green with brownish on closed wings, blue-green tail white towards tip. Chin and throat yellow with a narrow purple-blue gorget, bluish on vent, rump turquoise. Mask black. Bright blue forehead and green tail with middle feathers blue-green are characteristic of this race. In flight wing looks rufous with conspicuous black trailing edge. Immature a pale duller version of adult, lacking gorget, and with faint streaks on breast, tail only insignificantly forked. **Similar species** Immature Swallow-tailed has a blue-tail, whereas appreciably smaller Little has green/rufous tail. Heavier-looking European has chestnut crown and golden back, Blue-cheeked is also larger and has elongated central tail feathers.

HABITS Gregarious when not breeding. Typically lives in communal roosting and loose foraging parties of up to fifteen birds. Hunts high from lofty shrubbery and bare branches of tall trees, usually approachable. Favours forest-thicket fringes and dry savanna areas with tall trees.

VOICE Easily confused with Little, but more rolling *tip, tip, dip-dip-dip, diddle-diddle-ip;* greeting call an excited *dreee-dreee.*

STATUS AND DISTRIBUTION Locally common breeding resident at the coast and up to LRD, less frequent inland becoming rare or absent URD. Found in the east and across southern Senegal. Local and rare in north, absent north of 15°.

BREEDING Burrows into hard flat ground towards the end of the dry season. Secretive, nest difficult to locate; breeding strategies require further investigation. Fledging recorded before first rains.

RED-THROATED BEE-EATER
Merops bullocki   Plate 29

Fr. Guêpier à gorge rouge

IDENTIFICATION 24cm. Sexes alike. Only Senegambian bee-eater with a scarlet throat and chin, sharply demarcated from a dark buffish breast and belly. Thighs and undertail-coverts bright blue. Upperparts darkish bronzy-green, buffy on the hindneck. Occasional birds have a bright yellow throat. Broad black face mask extends well behind eye. Square-cut rusty-based grey-green tail lacks streamers or elongated central feathers. Buffy under-wing, and green upperwing show broad black trailing edge in flight. Immature much less intensely coloured on red and blue areas, often with a green cheek streak under the mask, sometimes present in adult. **Similar species** Scarlet throat diagnostic within area.

HABITS Highly gregarious. Most frequent along the freshwater reaches of the river, where intense feeding activity takes place just after dawn and towards the end of daylight hours. Forages from high riverside trees and overhanging tangled vegetation. Also visits established gardens and hunts from telegraph wires.

VOICE A *wip* call of varying intensity uttered continuously throughout the day during all activities. Greeting call is an excited trill.

STATUS AND DISTRIBUTION Common to occasionally very abundant breeding resident in CRD and URD, with little seasonal change, some increase in URD during the rains. Only occasional and erratic sightings outside these divisions. Common in parts of eastern Senegal.

BREEDING Tunnels in quarries, sandy riverbanks and storm wash-outs. Breeds in closely-packed 'pepper-pot' colonies in vertical banks. Excavation takes place when rains soften substrate enough to facilitate burrowing; tunnels are returned to for occupation in mid-dry season at some sites. Complex reproductive strategy involves helpers.

WHITE-THROATED BEE-EATER
Merops albicollis   Plate 29

Fr. Guêpier à gorge blanche

IDENTIFICATION 32cm including central streamers. Note that streamers (the longest of all bee-eaters) are frequently absent on otherwise adult-plumaged birds seen in The Gambia. Distinctive bold black and white head and face make for easy identifcation at rest, even duller immatures. Upperparts, wings and tail a soft pale green tinged bluish particularly on the rump, underparts paler green washed silky white. Crown, mask and gorget jet black, contrasting strikingly with white broad superciliary stripe and throat. In flight underwing brownish-green with black trailing edge. Immature washed brown or buff and scaly overall. **Similar species** No other Bee-eater has unique black and white head pattern.

HABITS Appears in various habitats: common requirement seems to be tall trees or mangroves, from the tops of which they can launch foraging campaigns into insect swarms; also feeds low over water. A peculiar trait described from elsewhere and seen in The Gambia is the ingestion of strips of epicarp from the fruit of the oil-palm. Birds catch the discarded peel in mid-air beneath feeding squirrels or monkeys.

VOICE Calls *prrp* and *pruik* in flight and when perched.

STATUS AND DISTRIBUTION Frequent to seasonally common unpredictable intra-African migrant. Flocks most likely in CRD prior to the rains with a distinct peak in May-June when moving north to breed on desert edge. Occasional other records from all divisions and in all months. In Senegal breeds north of 16°; records in the south in late April-May correspond with passage in CRD; also recorded there December-January.

BREEDING In loose colonies, usually tunnels directly into hard flat ground. Helpers assist the rearing of a clutch of up to 7 eggs. Recorded northern Senegal, July and August.

LITTLE GREEN BEE-EATER
Merops orientalis   Plate 29

Fr. Petit guêpier vert

IDENTIFICATION *M. o. viridissimus*. 26cm including tail streamers. Much the smallest of the Senegambian bee-eaters that possess extended central tail feathers. Overall bright glossy lime-green with a coppery-tinge on the upperparts in bright sunlight. Black mask through eye frequently with a narrow turquoise line below it. Throat and breast all-green separated by a narrow black gorget, narrower in female (note: yellow-throated morphs are described from elsewhere in West Africa). In flight shows a dark trailing edge on underside of rufous wing. Immature paler green, particularly on belly, upperparts delicately scaled, gorget absent or ill-defined, streamers absent or showing as a slight projection at centre. **Similar species** Little Green lacking extended tail feathers could be confused with Little, but latter is smaller, browner and dumpier. Less likely is confusion with blue-tailed immature Swallow-tailed, which is infrequent in Little Green strongholds in The Gambia. Observations of attendant adults should assist.

HABITS Favours laterite outcrops with scattered thorn-bush and deciduous woodland; also streams and riverbanks. In pairs or family groups that feed near to each other and huddle along the same branch or stem. Spends alot of time in the air, generally foraging low. Confiding. Frequently mixes with foraging flocks of Red-throated and also on occasions with Northern Carmine Bee-eater.

VOICE A throaty and agreeable waxbill-like trill, *ttree-ttree-ttree* or *tsee-tsee* sharper than Little. Vibrates body and tail when calling. Alarm note, *ti* or *tic*.

STATUS AND DISTRIBUTION Frequent resident in CRD and URD with records in every month, locally common at a few sites; infrequent in WD. Widespread in Senegal.

BREEDING Latter quarter of dry season, April-

June. Tunnels directly into flat earth or a low bank. In The Gambia often solitary or in a small loose colony, with no evidence of helpers.

BLUE-CHEEKED BEE-EATER
Merops persicus   Plate 29

Fr. Guêpier de Perse

IDENTIFICATION *M. p. chrysocercus*. 34-37cm including long central tail feathers, often absent. A large, weighty-looking bee-eater but characteristic in shape and behaviour. Absence of gorget in all stages. Varying shades of burnished green. Harlequin head markings diagnostic: broad black mask through bright wine-red eye, edged with pale blue streaks both above and below; crown green, forehead pale blue, chin yellow and throat rufous. In flight looks long-winged and long-tailed, strong coppery with russet underwing showing only narrow dark trailing edge. Immature duller green with subtle tones of blue, streamers absent or short. Younger birds are scaly on the upperparts. **Similar species** From European by richer underwing colour and more uniform green upperparts lacking any chestnut.

HABITS Gregarious and extremely vocal. Typically perches upright, favouring low coastal *Avicennia* mangroves and nearby power-lines, foraging over the swamp and grassy margins that border this fragile habitat. Particularly agile, with sudden contorted twists and turns in mid-air to catch prey. Benefits from annual butterfly migration through The Gambia in November-Febuary.

VOICE Similar to European Bee-eater but subtly clearer, strident and more detectably di-syllabic, *de-ripp* or *greeip* and a more rolling *pruuk-pruiik*.

STATUS AND DISTRIBUTION Seasonally abundant migrant largely restricted to coastal NBD and WD with occasional sightings elsewhere. Population peaks November-February, but records in every season including records of small flocks moving due north out to sea in March. In Senegal recorded mostly along the entire coastline.

BREEDING In northern Senegal, on desert edge, June-July, in small colonies or alone, nest helping not suspected.

EUROPEAN BEE-EATER
Merops apiaster   Plate 29

Fr. Guêpier d'Europe

Other names: Bee-eater

253

IDENTIFICATION 29cm including central tail feathers. A beautiful multicoloured bee-eater, male brighter than female. Unmistakable, with turquoise blue breast separated by narrow black gorget from a bright sulphur-yellow chin and throat and warm chestnut crown and mantle. White forehead tinged green, black mask. Closed wing chestnut and green offset by golden-yellow scapulars. Back and rump are washed golden-yellow in breeding plumage, greenish at other times. Central tail feathers elongated to blackish point otherwise tail green. In flight, pale rufous wing with a broad black trailing edge appears translucent against a clear blue sky. From above shows obvious golden V, from back to rump. **Similar species** In immature European, dull scaly olive-green replaces chestnut of adult, gorget absent or poorly defined, but dull turquoise belly, pale yellow throat and silvery-green V on back help separate it from greener immature of Blue-cheeked Bee-eater. Habitat and locality will also assist, as Blue-cheeked is primarily associated with mangroves.

HABITS Aerial, highly gregarious and extremely vocal. Preferred habitat is grassland and occasional trees, with dead standing timber from which to hunt, through to dry deciduous woodland on laterite with plentiful baobabs. Often draws attention first by calling at considerable height. Sails and glides elegantly on open wings.

VOICE Distinctive flight-call is a repeated soft yet far-carrying liquid, *kruut* or *kwirri*, detectably di-syllabic to sharp ears. Whole flocks produce a mingled rolling bubbly *touruk, touruk, touruk.*

STATUS AND DISTRIBUTION Seasonal and locally frequent to common Palearctic migrant in dry savanna woodland in LRD, main peak November-January; but recorded in this division in every month. Occasionally recorded elsewhere. Largely absent from north-east and eastern Senegal but more frequent across the south and towards the coast.

NORTHERN CARMINE BEE-EATER
Merops nubicus   Plate 29

Fr. Guêpier écarlate

Other name: Carmine Bee-eater

IDENTIFICATION *M. n. nubicus.* 38cm with impressive central streamers. Unmistakable large bee-eater, strikingly coloured in pinkish-red and blue. Blue tones may range from azure to olive-green depending on the light. Upperparts and belly carmine, lower back, rump and undertail-coverts blue. Adult has characteristic greenish-blue fore-crown, chin and throat, and a well-defined facial hood. Mask is black. In flight underside of relatively long wing is buff with a black trailing edge. Immature has carmine replaced by dusky brown, palest on underparts, looks scaly and lacks tail streamers; chin and throat pale blue. **Similar species** Sheer size may cause confusion with distant Abysinnian Roller or even a falcon. Note that Southern Carmine Bee-eater *M. nubicoides* (Fr. Guêpier carmin) adult has chin and throat carmine; nearest population Zaire.

HABITS Frequents two habitat types in The Gambia. Most frequently encountered in the environs of the freshwater parts of the main river and its adjacent lightly wooded swamps in URD, but also in dry wooded grassland well away from water in LRD. Most numerous in The Gambia during and just after the rains. The habit of following moving livestock to glean disturbed insects is not commonplace but has been noted occasionally.

VOICE Rather unsophisticated short flight call *klunk* or *tunk*, sometimes *chip-chip-chip.*

STATUS AND DISTRIBUTION Seasonally frequent to common in URD, generally August to the early dry season, but recorded in all months. Sometimes present in thousands, e.g. early wet season. Small numbers present in LRD and NBD December to late Febuary. Very occasionally in coastal WD, one recent sighting in December at Tanji Bird Reserve. Absent from dry central Senegal, but present in all border regions.

BREEDING Tunnels near to flowing water, colony size ranging from a few pairs to many thousands. Immatures in URD in August. Occasionally mixes in The Gambia with Red-throated Bee-eater communities. In eastern Senegal more regularly in high sandy river banks with young in the nest in August. Reproductive strategies not well studied.

ROLLERS

Coraciidae

Five species, including one rare Palearctic migrant. A group of thickset, large-headed, medium-sized birds with powerful slightly hooked bills. Sexes similar. Brightly coloured plumages, particularly striking in flight, usually combining shades of blues, purples, greens and browns. *Coracias* species hunt from a conspicuous perch from which they drop to the ground to capture prey. Generally solitary. *Eurystomus* is gregarious and an aerial hunter.

RUFOUS-CROWNED ROLLER
Coracias naevia   **Plate 28**

Fr. Rollier varié

Other name: Purple Roller

IDENTIFICATION 33cm. A stocky, square-tailed roller lacking any prominent areas of blue in the plumage when perched. Overall pinkish-brown, flecked with white on underparts. Prominent broad white supercilium, with distinctive white nape patch, chestnut crown and greenish tinge to mantle. Characteristic bounding flight with a series of rapid shallow wing-beats, when rich dark blue in flight feathers and silvery-white underwings become apparent. Immature browner and duller without pink hues. **Similar species** Other rollers are much brighter.

HABITS Favours dry open grassland with scattered large trees, and burnt or cleared farmland. Usually solitary. Much less active and vocal than other rollers, often seen sitting in an angled posture on a roadside telegraph wire or overhead cable. Display has been described as 'out of control' with the wings appearing to operate independently of each other with the bird appearing to rock on its body axis.

VOICE Not vociferous; a rather subdued *gaa-aaa-aaaaah* in display, also single well-spaced *yaw* notes, sometimes delivered for long periods.

STATUS AND DISTRIBUTION Seasonally common to abundant throughout, reduced numbers in the rains; population strongholds in LRD and CRD. Nomadic and prone to erratic population fluctuations. Widespread in Senegal.

BREEDING Uses a hole in a tree or other natural cavity in mid and late rains.

BLUE-BELLIED ROLLER
Coracias cyanogaster   **Plate 28**

Fr. Rollier à ventre bleu

IDENTIFICATION 36cm including tail streamers. Distinctive thickset roller, largely dark blue with a contrasting pale buffish head and breast and blackish-brown back. Looks caped at a distance. Shortish paler blue tail with wire-like streamers on outer feathers. In flight, striking turquoise and dark blue striping visible in wings, together with unusual combination of very pale front end and dark hind quarters creating a front-heavy, bull-headed impression. Immature duller, lacking tail streamers. **Similar species** None.

HABITS Most numerous near damp ground or watercourses but also found in drier areas. Fre-

quently seen on telegraph wires or on branches of leafy trees. Usually in pairs or small parties. Aggressive, defending foraging and breeding territories with gusto. Rises high above an intruder and stoops vertically on folded wings, providing a spectacular display while doing so. Recorded feeding on oil-palm fruit and treating kernel as an insect by beating and wiping it on the perch before ingestion.

VOICE Highly vocal. A guttural *gah-gah-gah* delivered at varying speed and volume, but typically harsher and more rasping than other rollers.

STATUS AND DISTRIBUTION Common resident throughout, with population strongholds towards the coast. Becomes scarce in northern Senegal, absent north of 15°.

BREEDING In a tree cavity, especially dead oil-palm with vertical entrance into decayed crown. Most activity during the rains; cooperative breeding strategy suspected.

ABYSSINIAN ROLLER
Coracias abyssinica   **Plate 28**

Fr. Rollier d'Abyssinie

IDENTIFICATION 42cm including streamers. Distinctively long, slim roller, with head and body bright azure blue, turquoise and ultramarine, mantle chestnut-brown. Forehead, supercilium and throat whitish. Elongated swallow-like outer tail feathers absent in young and moulting birds, sometimes broken in adults (see European Roller). Flight feathers purple at all ages. Immature paler than adult, browner on the neck. **Similar species** European Roller lacks tail streamers. Abyssinian has paler forehead and purple flight feathers.

HABITS Frequents dry open grassland with scattered trees, also roads and tracks with telegraph wires. Often solitary, occasionally in loose groups. Attracted in large numbers to bush fires. Sits upright and motionless for long periods on a favoured perch, from which it swoops on its prey: termite mounds in open country are favourite perches. Flight swift and direct. Display includes steep, rolling swoops and a very elegant stall turn with wings folded. Aggressive and noisy.

VOICE Crow-like, a raucous and guttural *kra-kra-kra*, particularly in flight.

STATUS AND DISTRIBUTION Common to abundant in all divisions; population peaks November-June, particularly in URD. Present in all months in all divisions, least frequent during heaviest rainfall, when also largely absent from southern Senegal.

255

BREEDING Nest a cavity in a tree, a stump or large termite mound. Records of full displays, visits to suitable nest-hole, and immatures seen with adults, but as yet no complete breeding record in The Gambia. Senegalese records April-July.

EUROPEAN ROLLER
Coracias garrulus　　　　Plate 28

Fr. Rollier d'Europe

IDENTIFICATION 32cm. Square-tailed thickset roller with pale azure-blue head and body, bright chestnut back, azure-blue wings with conspicuous black primaries and secondaries visible both when at rest and in flight. Immature is much duller, also has black primaries, but not recorded in Senegambia. Similar species Rare migrant European can be confused with common Abysinnian Roller, if latter is without streamers; e.g. in moult or in immature plumage, but Abyssinian is paler, with more extensive creamy-white about the forehead; Abyssinian has purple-blue flight feathers, European has diagnostically black flight feathers. Abyssinian is a slimmer, more streamlined bird and jizz of European suggests a heavier-headed bird.

HABITS Favours open grassland with scattered trees, or with overhead wires. Flight easy and buoyant, showing conspicuous blue forewings contrasting with black primaries and secondaries. Not a gregarious bird in Senegambia; most likely to occur singly.

VOICE Typical harsh croak, but relatively quiet on non-breeding grounds.

STATUS AND DISTRIBUTION Seasonally uncommon to rare Palearctic migrant, most recent reports in coastal WD in April with an older January record. A few records in northern Senegal, but considered scarce throughout West Africa.

BROAD-BILLED ROLLER
Eurystomus glaucurus　　　Plate 28

Fr. Rolle violet

Other name: Cinnamon Roller

IDENTIFICATION 26cm. *E. g. afer.* Comparatively small, stocky, square-tailed and dull roller with characteristic yellow bill. Upperparts dark chocolate brown, chin, throat and belly lilac in adults, washedout brown fading to dull pale blue in immature. Tail greenish-blue with darker tip. Bill large, broad and strikingly bright yellow, looks frog-mouthed.

In flight, dark blue flight feathers can be seen in good light. Similar species First glimpses may not suggest a roller but a raptor, nightjar or even a fruit-bat depending on circumstances! Note a disputed record of Blue-throated Roller *Eurystomus gularis* (Fr. Rolle à gorge bleue) near Dakar, northern Senegal; a species not listed for The Gambia, with nearest undisputed records from Guinea Bissau. This is a forest bird with a characteristic azure-blue throat patch, otherwise closely similar but darker.

Broad-billed Roller

HABITS Favours forest thickets and watercourses with tall trees, also well-established gardens. Gregarious in non-breeding season. Feeds aerially like a giant swallow, making sallies from tall forest trees into swarms of termites, dragonflies and other insects. Crepuscular, feeding well after dusk on bright moonlit nights. Flight agile and buoyant, drinks on the wing. An avid bather plunging into still water (even swimming pools) from a high perch. Aggressive.

VOICE Characteristic croaks and cackling, an ascending *crik-crik-crik-crik* whilst feeding, usefully learnt as especially vocal when foraging.

STATUS AND DISTRIBUTION Seasonally abundant in suitable habitat throughout all divisions; much reduced January-May. Birds begin to appear at the coast a few weeks prior to the rains. Gambian population may only be moving into southern Senegal in the dry season but intra-regional migrations not yet fully understood.

BREEDING Nests in a tree cavity (baobabs particularly favoured); also holes in masonry.

WOOD HOOPOES

Phoeniculidae

Two species. Largely arboreal, with fine decurved bills for probing. Both have long tails and shortish rounded wings, barred white. In addition there is a disputed record of White-headed Wood Hoopoe *Phoeniculus bollei* (Fr. Moqueur à tête blanche) from southern Senegal (Cap Roxo, December 1976) whose nearest other records are from Liberia.

GREEN WOOD HOOPOE
Phoeniculus purpureus   Plate 27

Fr. Irrisor moqueur

Other names: Red-billed Wood Hoopoe, Senegal Green Wood Hoopoe

IDENTIFICATION 44cm including immensely long, graduated tail. Sexes similar but female slightly shorter-billed. Overall dark green with sheens of black, green, purple and blue, with conspicuous white barring on wings and white spots on long tail. Variable bill colour ranging from black to red, which may be age- and sex-related; bill long and decurved. Short legs reddish. Immature duller, bill straighter. Six recognised races, difficult to separate in the field, two in Senegambia: *P. p. senegalensis* throughout The Gambia and southern Senegal and *P. p. guineensis* in northern Senegal. Distinctions are based primarily on plumage sheen and adult mandible colour: *guineensis* is brighter than *senegalensis* from crown to mantle and *senegalensis* is also slightly larger and has darker mandible coloration. **Similar species** Other gregarious species with long tails and dark plumage which might cause initial problems are Piapiac and Long-tailed Glossy Starling, but neither has any white barring in its plumage.

HABITS Highly sociable and found wherever there are sufficient trees. Typically announce their presence vocally well before they are seen. Move in a follow-my-leader style, generally in flocks numbering 4-15 birds. They clamber around collectively and inquisitively on the trunk of a tree, on a termite mound or on the bare ground, probing for food in cracks and crevices, busily looking over and under likely sites.

VOICE Abundant communication and conversational calls, *kuk-uk-uk-uk* and *keek* culminating in a frenzied choral cacophony. Old regional name 'Kakelaar' is derived from its voice. One of the noisiest birds of the bush.

STATUS AND DISTRIBUTION Common to locally abundant throughout all divisions, with no significant seasonal population changes. Widespread and common throughout all districts of Senegal.

BREEDING Nest-helper system well studied in East Africa. Nests in a natural tree cavity, apparently subject to high levels of predation. Eggs 2-4, blue.

BLACK WOOD HOOPOE
Rhinopomastus aterrimus   Plate 27

Fr. Irrisor noir

Other name: Lesser Wood Hoopoe

IDENTIFICATION 22cm including long tail. Sexes similar. Small and dark, a half-sized version of much commoner Green Wood Hoopoe. Black overall with a subtle violet sheen, bluer about the head and face. Wing with broad white bar conspicuous in flight, but appearing as a small patch on the closed wing. Tail squarish and only slightly tapered. Decurved black bill proportionately shorter than Green Wood Hoopoe. Legs and feet greyish-black. Immature duller; lacking iridescence and with shorter and less decurved bill. **Similar species** Northern Black Flycatcher has a short straight bill, is uniformly black and with no white in the plumage, and does not clamber. Similarly, the two Senegambian drongos have no white in the plumage.

HABITS Predominantly a bird of dry Guinea-savanna dominated by locust bean *Parkia biglobosa* trees. Behaves like a giant tit clambering among the dry pods probing for its food. Characteristically makes a rapid vertical dive onto a lower branch to continue feeding. Not gregarious, usually in pairs, but associates with travelling mixed feeding groups.

VOICE Vocal. A mournful *poui-poui-poui* or rising *whee-wheep-wheep* increasing in volume; also a cracked *wha-wha-wha-wha*.

STATUS AND DISTRIBUTION Frequent to very locally common in WD and LRD, smaller numbers in other divisions. Peak sightings in the early rains. Widespread throughout Senegal, but nowhere common.

BREEDING Nests in a tree-hole. Fledged young being fed in late May and in the late rains, young in the nest in mid-October; in WD and LRD.

HOOPOE

Upupidae

One species: a monotypic worldwide genus with unmistakable plumage. Characterisitc long crest, usually held flat, long decurved bill and butterfly-like flight on broad, rounded wings. Further study of racial variation is needed.

HOOPOE
Upupa epops Plate 27

Fr. Huppé fasciée

IDENTIFICATION 26cm. Generally cinnamon-buff, with black and white barred back, wings and tail. Long black-and-white-tipped ginger crest, stunning when fanned and erected in aggression or on landing. Long fine blackish decurved bill. Immature duller with shorter straighter bill. Distinctive floppy flight on rounded wings. Races occurring are Palearctic *U. e. epops*, warm buff to a sandy washed-out orange with a browner mantle, and African *U. e. senegalensis*, overall rufous-sandy, belly to vent whitish. African forms are often darker and show more white in the wings (broader bars). **Similar species** None.

HABITS Open wooded savanna, coastal scrub and irrigated gardens with sparse ground cover are all favoured habitats. Mostly alone or in pairs. Feeds by probing on bare or short grassy ground; approachable. When disturbed takes flight usually only for a short distance.

VOICE A monotonous far-carrying trisyallabic *hoop-hoop-hoop* or *hoo-poo-poo*, sometimes maintained for long periods; beware possible confusion with some doves and African Cuckoo.

STATUS AND DISTRIBUTION Frequent throughout all divisions, especially LRD and CRD; recorded in all seasons but most records December-March. Recorded throughout Senegal

BREEDING Not yet proven, but two adults with a pale straight-billed juvenile in January in WD. Nests in holes.

HORNBILLS

Bucerotidae

Seven species; four found throughout Senegambia, one vagrant to The Gambia only and two forest residents restricted to southern Senegal only. Medium-small to huge, with massive bills usually with casques or ridges, and with far-carrying calls. Characteristic slow and undulating flight with intermittent flaps and glides. Predominant plumage colours are greys, blacks and white.

ABYSSINIAN GROUND HORNBILL
Bucorvus abyssinicus Plate 30

Fr. Calou terrestre d'Abyssinie

IDENTIFICATION 100cm. Huge and unmistakable. Terrestrial hornbill. Incomplete open casque looks broken. Prominent white primaries in flight are mostly unseen when wing closed, and walking bird looks wholly black. Sexes differ: female is slightly smaller and has inflatable dark blue bare skin about the face, chin and throat. Male has a red patch at base of lower mandible, blue bare skin around eye and upper throat, lower throat and neck red. Slimmer, duller immature has poorly developed casque on noticeably smaller bill. **Similar species** Unique.

HABITS Usually encountered in pairs or small parties in open country. Moves at a leisurely pace, only rarely flies unless disturbed. Usually extremely wary but occasionally confiding.

VOICE A deep, sonorous and repeated booming *umm-hum-umm* or *uuh-uh-uh*, usually delivered from a prominent bare branch.

STATUS AND DISTRIBUTION Frequent to seasonally locally common, evidently site-faithful e.g. near Bijolo Forest Park WD. Most likely in LRD (Kiang West) becoming uncommon in URD. Records peak April-July. Widespread throughout Senegal.

BREEDING Adults seen leaving nest cavity early July, LRD. Recently fledged chick with adults in November in LRD.

RED-BILLED HORNBILL
Tockus erythrorhynchus Plate 30

Fr. Petit calao à bec rouge

Other names: Red-beaked Hornbill

IDENTIFICATION 42cm. Smallest Senegambian hornbill, immediately recognisable by its characteristic orange-red mandibles with no casque. Sexes similar but female has a smaller bill; base of lower mandible always blackish in male. Sooty-black stripe from forehead extends over crown, tapering to a point on the mantle. Underparts all white, white line down centre of back, closed wings dark brown with bold white patches on shoulders. Tail dark brown edged white. Legs and feet brownish-black. Immature similar to adult but duller, with shorter and straighter bill. **Similar species** Grey Hornbill has a darker head with a prominent white supercilium.

HABITS Favours open woodland and dry savanna, but an opportunistic feeder, exploiting any abundances of insects, grain and fruit. Found in pairs and in feeding flocks of 40-80 birds, especially over burnt ground and upland rice paddies after harvest. Terrestrial, enjoys sunbathing.

VOICE Clucks and chuckles, a chanted *kok-kok-kok-kokok-kokok-kokok* rising in tempo and volume.

STATUS AND DISTRIBUTION Common to abundant breeding resident throughout all divisions; no significant seasonal population changes. Ubiquitous throughout Senegal.

BREEDING Paired display involves shoulder-shrugging and body-swaying with opened wings and head raised. Nest is usually a plastered-up tree cavity; breeds during the rains with most records in the latter part.

AFRICAN PIED HORNBILL
Tockus fasciatus   Plate 30

Fr. Calao longibande

Other names: Black-and-White-tailed Hornbill, Allied Hornbill, Pied Hornbill

IDENTIFICATION 53cm. *T. f. semifasciatus.* Largest of the three regularly seen arboreal hornbills in The Gambia. Overall black with a contrasting clear-cut white belly. Black tail with white-tipped rectrices 3 and 4. Cream-yellow bill with a modest casque, with substantial but erratic black markings. Looks emaciated because of oversized head and tail on a slim body. Immature duller, with a smaller bill, lacks casque. **Similar species** Piping Hornbill has a shorter, thicker bill with obvious crinkles, and prominent white tail tips. The other two *Ceratogymna* hornbills are much larger with pronounced casques.

HABITS A forest species, wandering into adjacent woodland, especially when the fruits of oil-palm or an isolated *Ficus* are ripe. In the forest, a canopy

bird. Extremely vocal; far-carrying yelps often draw attention to its presence. Gregarious when not breeding.

VOICE An echoing querulous whistle that rises and falls, *Pii-pii-pii pii-pii-pii pii-pii-pieeu*; could be likened to human laughter.

STATUS AND DISTRIBUTION Locally common breeding resident in scattered coastal forest islands in WD showing no seasonal changes. Infrequent in LRD, two recent March records in CRD, no recent information from NBD. Recorded in Senegal in coastal north below 14° and from forested Casamance to the south.

BREEDING During the rains in a tree hole; adults with well grown fledgling in February.

AFRICAN GREY HORNBILL
Tockus nasutus   Plate 30

Fr. Petit Calao à bec noir

Other name: Grey Hornbill

IDENTIFICATION 45cm. Medium-sized hornbill readily recognised by overall grey-brown appearance and long dark mandibles. Head, neck and cheeks dark grey with prominent white stripe from eye to nape. Upperparts smoky grey-brown with central white stripe down back. Tail dark and tipped white, except central feathers. Wings dark grey-brown dappled white. Underparts whitish tinged grey, darker on upper breast. Sexes distinguishable on bill colour: male black with a small area of cream at the base of upper mandible, lower ridged with white; female dark red mandibles with the base half of the upper yellow and yellow ridges on the lower. Poorly-developed casque is slightly more pronounced in male. Immature more uniform grey with shorter bill. **Similar species** Red-billed Hornbill has whiter head and underparts and brighter bill.

HABITS Widespread in a diversity of habitats preferring open dry savanna with tall trees, less frequent in closed forest. Flight is undulating and buoyant with flaps and glides. Mostly arboreal, seen capturing large-winged insects like a cumbersome flycatcher. Can be nomadic, probably associated with food supplies. Vocal shortly after daybreak.

VOICE A sorrowful and far-carrying piping disyllabic *pee-o pee-o pee-o* at varying speeds; also a hawk-like *ki ki ki ki* during the rains.

STATUS AND DISTRIBUTION Common breeding resident throughout all divisions, no seasonal population variation; numbers greatest in WD. Ubiquitous throughout Senegal where populations may fluctuate seasonally.

259

BREEDING In tree holes, usually in the early rains; season may be extended in Senegambia.

PIPING HORNBILL
Ceratogymna fistulator　　　Plate 30

Fr. Calao siffleur

Other names: White-tailed Hornbill, Laughing Hornbill

Synonym: *Bycanistes fistulator*

IDENTIFICATION 53cm. Similar to Pied Hornbill but with relatively short casqued bill and a rough crest. Head, neck, upper breast and upperparts are glossy black. Rump white. Wings all black except white-tipped central secondaries. Underparts below breast white. Tail black (except central feathers) broadly tipped white. Bill dusky brown with a cream base and tip; casque forms a low protuberance on the heavily ridged bill, more crinkled and smaller in female. Immature duller, lacks casque. **Similar species** African Pied Hornbill has a different bill shape and pattern, and lacks white rump and prominent white tail tips.

HABITS Forest-adapted and generally similar habitats to those of African Pied Hornbill. Gregarious when not breeding.

VOICE Raucous laughing *kah-k-k-k-k* and a pentrating piped *peep-peep-peep*.

STATUS AND DISTRIBUTION Not listed for The Gambia; sole record in 1961 not now acceptable. Locally common in thick forested areas in southern Senegal.

BREEDING Not recorded for Senegal, but likely. Elsewhere nests in a tree hole.

BLACK-AND-WHITE-CASQUED HORNBILL
Ceratogymna subcylindricus　　Plate 30

Fr. Calao à joues grises

Synonym: *Bycanistes subcylindricus*

IDENTIFICATION 75cm. Large heavy-looking black and white forest hornbill with a huge bill. Secondaries and inner primaries white, rest black, resulting in strikingly characteristic half black, half white wing. Tail black with broad white edges. Sexually dimorphic bill and casque: female smaller-billed than male with blackish-grey bill, casque reduced and rounded, and marked wrinkling at bill base. Male has dark brown bill with a long high

casque, pale at the base. Head appears immense. Immatures resemble adults with much reduced bill structure. **Similar species** Piping Hornbill is considerably smaller and lighter-billed with mainly black wings.

HABITS Favours dense forest and forest margins sometimes emerging into open woodland. Recent Gambian records well authenticated, evidently not involving escapes from captivity.

VOICE Slow hooting and quacking *kakaka-kawack* or *aahkaaak-kaaak* in flight.

STATUS AND DISTRIBUTION First recorded in The Gambia in November 1993: single female seen by two independent parties at Tanji Bird Reserve and in *Avicennia* mangroves (Old Cape Road near Bakau) in WD. A single bird in November 1994 again in WD at Kotu. Single bird in October 1995 in NBD (Niumi National Park) on Gambia-Senegal border. A record from Banjul in March 1969 rejected as an escape. There are no records from southern Senegal.

BREEDING The nearest significant breeding population is in Côte d'Ivoire c. 1,000km to the south.

YELLOW-CASQUED WATTLED HORNBILL
Ceratogymna elata　　　Plate 30

Fr. Calao à casque jaune

Other names: Yellow-casqued Hornbill

Synonym: *Bycanistes elata*

IDENTIFICATION 100cm. Immense forest hornbill with distinctive head and neck pattern. Overall iridescent black, tail white with central feathers black. Heads of sexes differ. Male head and bushy crest blackish-brown, white flecked on the neck. Bare skin around a crimson eye and gular pouch cobalt-blue, central line of buff feathers down the throat tipped brown. High, pale-yellow casque ends half-way along the grey-black bill. Female is smaller with crest and hindcrown rufous, facial feathering and central throat rufous-buff flecked white, bare skin pale blue. Bill yellowish, casque much reduced compared with male. Immature resembles female, young males show black in the crest at an early stage. **Similar species** Black-and-white-casqued Hornbill lacks blue skin around face and shows much white in the wings.

HABITS Little known. Forages in the high canopy of lowland primary forest.

VOICE A sonorous bray *aa-a aa-a*; whooshing of wings in flight clearly audible.

STATUS AND DISTRIBUTION No Gambian records. A fragile population known from one forest reserve in southern Senegal, where it was first observed in 1962. Range extends to Nigeria: now sparsely distributed throughout but it was common in Sierra Leone in the 1930s.

BREEDING Nests in a natural hole, high in a forest tree.

BARBETS AND TINKERBIRDS

Lybiidae

Seven species, three in southern Senegal only. A colourful group of very small to medium-sized birds closely related to the woodpeckers, sharing the feature of zygodactyl toe structure and the climbing agility that this confers. Frugivorous. Nest in self-excavated holes or ones they have enlarged. Calls often attract attention; 'popping' tinkerbirds can be frustrating to locate. Barbets usually perch prominently. No conspicuous sexual dimorphism in any species.

RED-RUMPED TINKERBIRD
Pogoniulus atroflavus Plate 30

Fr. Barbion à croupion rouge

IDENTIFICATION 12cm. Not as diminutive as other tinkerbirds and the only tinkerbird in the region with a scarlet rump. Upperparts, crown and cheeks glossy blue-black, wing coverts and secondaries edged yellow. Three fine yellow stripes on side of head, (two white stripes on Yellow-rumped Tinkerbird). Throat yellow with black malar stripe. Rest of underparts yellowish-green. Legs blue-green. Immature duller with less brightness on the rump. **Similar species** Other tinkerbirds have yellow rumps.

HABITS Little-known forest species. Usually encountered in tall creeper-clad trees sometimes in small parties. Forages high and joins in travelling mixed feeding groups. Feeds on berries and insects. Whilst calling and displaying it puffs out the feathers of its throat, spreads its tail vertically downwards splaying the uppertail-coverts to display its red rump whilst twisting and turning its head from side to side.

VOICE Popping call is lower pitched than very similar sounding Yellow-fronted Tinkerbird and is only to be expected in forest, where Yellow-fronted is generally absent. Frequent *oop... oop... oop* delivered in a long series, or a nasal *onk... onk... onk*; also trills *peet-peet-peet*.

STATUS AND DISTRIBUTION No records from The Gambia. Known from coastal lowland forest in southern Senegal only. West African range extends to Gabon, but nowhere common.

BREEDING Excavates nest hole in a dead stump.

YELLOW-THROATED TINKERBIRD
Pogoniulus subsulphureus Plate 30

Fr. Barbion à gorge jaune

IDENTIFICATION 10cm *P. s. chrysopygius*. Very small, similar to slightly larger Golden-rumped Tinkerbird, upperparts jet black. Plumage differences between these two tinkerbirds are subtle, throat tinted very pale yellow similar to coloration of the breast in Yellow-rumped, therefore less obvious contrast between white throat and yellow underparts as Yellow-rumped. Two facial stripes whitish but not as well defined and clear as Yellow-rumped. **Similar species** Yellow-rumped Tinkerbird.

HABITS Little-known forest species, found in thicker and denser vegetation than Yellow-rumped and expected high in the canopy.

VOICE Call *pyop*, much faster in tempo and with a sharper quality than other tinkerbirds, delivered with urgency; could be likened to a rapid burst of the popping call of Yellow-fronted Tinkerbird.

STATUS AND DISTRIBUTION No records from The Gambia. Little known but possibly locally frequent in Senegal in the coastal forest of Casamance.

BREEDING Practically unknown, information sought.

YELLOW-RUMPED TINKERBIRD
Pogoniulus bilineatus Plate 30

Fr. Barbion à croupion jaune

Other names: Golden-rumped Tinkerbird; Lemon-rumped Tinkerbird; Yellow-rumped Tinkerbarbet

IDENTIFICATION 11cm. *P. b. leucolaima*. Very small and boldly marked. Crown and upperparts uniform glossy black. Two white stripes above and below the dark eye; lower stripe is broader than the supercilium. Characteristic bright yellow rump, tail and wings black and edged bright yellow. Underparts yellowish-white with white-tinged grey throat. Immature duller with a paler bill. **Similar species** In southern Senegal see Yellow-throated Tinkerbird.

HABITS Favours liana-entangled forest-thicket. Difficult to find if not actively on the move through the trees or calling, partial to *Parkia* locust-bean trees, and fruiting figs. Sightings normally of singletons. Appears most vocal just at the onset of the rains but heard at other times, calling records outnumbering sightings. Bold and approachable when calling.

VOICE Clinking *tonk-tonk-tonk-tonk ..tonk-tonk-tonk-tonk* delivered in short phrases of four to seven notes, sustained for long periods. Pauses between each sequence are momentary, but clear when attention focused; the characteristic criterion for separating from unbroken rhythm of otherwise similar sounding Yellow-fronted Tinkerbird.

STATUS AND DISTRIBUTION Very locally frequent, sedentary, with all known sites close to a few forest thicket sites in WD (e.g. inland of Tanji) where sightings peak in the late dry and early wet season when most vocally active. Single recent record in CRD. In Senegal disjunct in coastal Casamance then again in the south-east in Niokolo Koba.

BREEDING Second half of the dry season towards the onset of the rains. Pair behaving as if incubation in progress lost their nest to woodpecker predation in late April. Excavates in dead or decaying wood.

YELLOW-FRONTED TINKERBIRD
Pogoniulus chrysoconus   Plate 30

Fr. Barbion à front jaune

Other name: Yellow-fronted Tinkerbarbet

IDENTIFICATION 11cm. Very small. Distinctive black and white head and facial markings. Forecrown with yellow-golden circular spot. Upperparts black, heavily striated yellow and white, short tail brownish-black. Underparts lemon-yellow, paler on belly and on undertail-coverts. Immature has crown blackish and lacks well-defined crown-spot. **Similar species** Other tinkerbirds have uniform dark upperparts.

HABITS Favours dry habitat with tall trees and low scrub. In flight looks stumpy with noticeably short tail. Foraging rather warbler- or tit-like, picking and hovering from the underside of leaves. Ventriloquial nature of popping calls often creates difficulty in pinpointing the whereabouts of the calling bird.

VOICE By the far most regularly heard tinkerbird in the region, a continuous background sound of the Senegambian bush. Repetitous *tink-tink-tink-tink-tink* sustained in unbroken rhythm for long periods, lacking regular momentary pauses of Yellow-rumped Tinkerbird.

STATUS AND DISTRIBUTION Common resident throughout all divisions in appropriate habitat, no significant seasonal changes. Largely absent from the central arid part of northern Senegal, otherwise widespread.

BREEDING Synchronises breeding with the insect surges associated with the rains. Tiny hole excavated in a dead branch, usually quite low.

HAIRY-BREASTED BARBET
Tricholaema hirsuta   Plate 30

Fr. Barbican hérissé

Other name: Hairy-breasted Toothbill

IDENTIFICATION 19cm. Largish barbet with head, crown, throat and chin all black with contrasting white eye and moustachial stripes. Upperparts blackish flecked yellow, underparts golden-yellow streaked and spotted, probably brighter in female. Breast yellow with very fine black hair-like tips to lower throat. Underwing greyish-white with black spots and bars. Toothed bill black, eye red-brown, legs black-grey. Immature similar with a fluffy quality to the plumage. **Similar species** None in Senegambia.

HABITS Forest species, likely to be seen high in the canopy. Difficult to locate, attention attracted by calls. Joins mixed foraging parties.

VOICE Principal call is a low, slow series of brief *oork* calls delivered for 20 seconds or more at a time. Rapid *poop-poop-poop-poop*, recalls the resonance of a wood dove.

STATUS AND DISTRIBUTION No records from The Gambia. Known only from lowland coastal forest in Casamance, Senegal.

BREEDING Nest and eggs remain undescribed. Nuptial display involves the flaring of hairy-tipped breast feathers into an apron whilst elongating and stretching the body.

VIEILLOT'S BARBET
Lybius vieilloti   Plate 30

Fr. Barbican de Vieillot

IDENTIFICATION 15cm. Sexes similar. Small-medium, stumpy sparrow-like bird with a heavy bill and neck. Scarlet head and face with black-spotted pale yellow body. Three poorly defined races with Senegambian race *L. v. rubescens* overall darker and redder than northern sub-Saharan desert form. Upperparts, wings and tail dark brown. A prominent broad yellowish-sulphur stripe down the central back as the bird flies is diagnostic. Throat and breast whitish with broad scarlet tips, diffuse streak of scarlet down the centre of the belly. Toothed bill blue-black, eye chestnut-red to yellow. Legs and feet grey-black. Immature duller, particularly on red areas. **Similar species** Beware confusion with Red-headed Weaver and queleas.

HABITS Frequents dry Guinea-savanna and thorn scrubland, preferably with some *Ficus*. Confiding and approachable when calling, often perches in the crown of an acacia. Mostly found in pairs or occasionally small parties. Vocal throughout the year, heard at all times of the day.

VOICE Main call a trumpeting, repeated, slightly out-of-time duet *whu-oupe whu-oupe* as if being sucked and blown; this is often initiated with a chattering purr. Until learnt, could be confused with Striped Kingfisher.

STATUS AND DISTRIBUTION Common resident throughout all divisions in appropriate habitat; most numerous in WD and LRD. Widespread throughout Senegal.

BREEDING Usually just prior to and into the rains. Excavates hole in dead wood, often on the underside; also associated with termite hills. Parasitism by Lesser Honeyguide could occur.

BEARDED BARBET
Lybius dubius   Plate 30

Fr. Barbican à poitrine rouge

IDENTIFICATION 25.5cm. The largest Senegambian barbet, thickset black, red and white bird with enormous yellowish-horn bill which is both grooved and double-toothed. Conspicuous bare eye-patches. Close up views show a thick tuft of bristles on each side of bill base. Upperparts and shortish tail black-blue with a prominent large white rump patch apparent as it flies or hops. Red throat and belly suffused with specks of white separated by a broad black band. Legs reddish-orange. Immature similar but duller. **Similar species** Similar Double-toothed Barbet *Lybius bidentatus* is not within our range (nearest Guinea-Bissau) and lacks the black breast band of Bearded.

HABITS Favours well wooded areas with an abundance of figs and regularly enters gardens to eat ripe papaya fruits. Flight heavy and direct and usually accompanied with a growl.

VOICE Not especially vocal; raucous growl or bark uttered whilst feeding or in flight, *toa rk* or *sccrawk*, or a more crow-like *caw*.

STATUS AND DISTRIBUTION Common resident throughout appropriate wooded habitat in all divisions with no seasonal change; most frequent in WD. Ranges upto 14°30' in northern Senegal, common to the south.

BREEDING Excavates in both live and dead wood, May-August.

HONEYGUIDES

Indicatoridae

Four species, one vagrant from northern Senegal only. Small to medium-sized birds. Soberly coloured in greens, olives and greys, tails are characteristically patterned in brown and white. Interesting habits include brood parasitism, mostly on hole-nesters. Chicks hatch with a hooked appendage on the bill which can be used to wound or kill the young of the host. All feed on beeswax, and one species actively guides man to bees nests. Thick-skinned, which may protect them against insect stings. Apart from wax, they are mainly insectivorous with a little fruit also being taken. If not calling, initial attention is usually attracted by a flash of the white outer tail feathers. This important field mark can cause possible confusion with the 'green-backed' cuckoo species.

CASSIN'S HONEYBIRD
Prodotiscus insignis   Plate 31

Fr. Indicateur pygmée

Other names: Cassin's Honeyguide, Western Green-backed Honeybird

IDENTIFICATION 12cm. Small, with flycatcher/warbler jizz, showing characteristic honeyguide tail pattern of all-white outer tail feathers which are frequently flared. Blackish narrow pointed bill on a small head, small greenish area at corner of gape. Upperparts dark yellowish-green with an olive-grey tinge, underparts paler. Important field marks are two white patches from rump to flanks (not always visible), tail grey-black, with outer two pairs immaculately white with no dark tips. **Similar species** Other honeyguides are larger with more swollen bills.

HABITS Favours forest-thicket habitat where it flits about tit-like whilst feeding in high canopy; joins mixed feeding parties.

VOICE Weak *whi-hihi* also a chatter; more information needed.

STATUS AND DISTRIBUTION No Gambian records. Known from a single sighting in northern coastal Senegal close to The Gambia. Predominantly a forest bird in the areas where it is resident. Northern limit of West African race *P. i. flavodorsalis* generally regarded as Guinea-Bissau.

BREEDING Parasitic on small cup-nesting insectivorous passerines including apalis, wattle-eye and flycatcher species.

SPOTTED HONEYGUIDE
Indicator maculatus   Plate 31

Fr. Indicateur tacheté

IDENTIFICATION 16.5cm. Greenbul-sized and overall olive-brown tinged greenish with distinctive dirty yellowish-white spotting which is denser on the upper breast, sparser towards lower belly. Distinctively marked honeyguide with a rather chubby jizz, dense feathering and a thickset neck. Brownish crown is speckled buff-white and there is a hint of olive-yellow streaking in the closed wing if seen well. Upperparts slightly more olive-greenish than more olive-brown underparts. Blackish conical bill. Eye brown, legs olive-green. Immature more streaked than spotted. **Similar species** Underparts unique amongst Senegambian honeyguides.

HABITS Little known. Recent sightings are from thickly wooded areas usually with freshwater ponds and streams where it comes to drink. Difficult to find when perched, remaining motionless for long-periods. Range overlaps at most sites with forest dependent Buff-spotted Woodpecker to which it bears a remarkable plumage resemblance; possibly the Spotted Honeyguide's main host. Could be expected in tracts of riverine forest along the freshwater region of the Gambia River.

VOICE Little known in The Gambia, but vocal similarity to or mimicking of Buff-spotted Woodpecker worthy of investigation. Strange extended raptor-like mewing *peeeoo-peeeoo* in the first hour of light. Also described is a loud *woe-woe-woe* similarly alikened to the cry of a hawk or falcon; also a trill-

ing *brrrr* uttered with tail fanned and throat and body feathers puffed out.

STATUS AND DISTRIBUTION Rare; sporadic sightings since 1985 from a handful of locations, all in WD within 50km of the coast. Probably sedentary. Older records from riverine forest in CRD. In Senegal known from southern coastal Casamance and in riverine forest east of The Gambia only.

BREEDING Practically unknown, nearest proven breeding Liberia.

GREATER HONEYGUIDE
Indicator indicator   Plate 31

Fr. Grand Indicateur

Other name: Black-throated Honeyguide

IDENTIFICATION 20cm. Sexes differ, immature looks completely different from adult. Bulbul-sized, obvious white outer tail feathers in flight at all ages. Male has upperparts dark grey; contrasting facial markings with obvious black throat and white ear patches, dark crown. Bill pale pink. Wing-streaked whitish, a bright golden-yellow shoulder patch is not always seen. Female has bill blackish-horn, plumage overall browner and paler and lacks facial markings, otherwise similar. Eye dark brown, legs grey. Immature, seen in all seasons, has upperparts and sides of face uniform olive-brown, underparts from chin to breast a pale yellow wash. Eye with thin bluish eye ring. **Similar species** Beware confusion of forest greenbuls with immature.

HABITS The most frequent honeyguide in our area, favouring tall scattered trees in dry terrain, normally absent from closed forest but also in mangroves. Intensely site-faithful. Despite the persistent calls often difficult to pinpoint; look high on outer branches near to clusters of leaves. Calls throughout the day, even in the hottest hours. Comes to water where bees drink with a view to following them back to the hive.

VOICE Loud far-carrying disyllabic *WHit-purr, WHit-purr, WHit-purr, WHIT* or *VIC-tor;* guiding call is an excited rattling *chitik-chitik-chitik.* Records of calls far outnumber sightings. Also a drumming *whurr* noise made with either the flared tail or wing feathers, or both, in a lowish circular flight display; noted in dry and wet seasons.

STATUS AND DISTRIBUTION Frequent throughout, less numerous in URD and NBD. Seasonally common when calling, particularly in the savanna woodlands of LRD. Extensive in Senegal but rare north of Dakar up to Senegal River.

264

BREEDING Elsewhere parasitises woodpeckers, barbets, wood-hoopoes, kingfishers, bee-eaters and glossy starlings (all hole-nesters); also the larger swallows with a flask-shaped nest.

LESSER HONEYGUIDE
Indicator minor   Plate 31

Fr. Petit Indicateur

IDENTIFICATION 15cm. *I. m. senegalensis*. Sexes similar, size and jizz of a largish and active sparrow-like bird. Head, nape and breast grey tinged olive paling to whitish on breast and lower belly; poorly-defined moustachial stripe and conspicuous white loral spot. Upperparts and wings olive with darker streaking and in good light subtly tinged golden. Tail dark with white outer feathers tipped brown. Black bill markedly stubby, pinkish at the base, eye dark brown, legs grey. Immature has loral mark and moustachial stripe absent and is overall darker grey. **Similar species** For any 'Lesser-type' showing strong facial markings note possibility of Western Least Honeyguide *Indicator exilis* (Fr. Indicateur minule) on southern coastal Senegalese border with Guinea-Bissau; this species could occur elsewhere in forest habitat in Senegambia. Western Least has a prominent black malar stripe below a

white eye and is slightly smaller with bolder markings than Lesser Honeyguide.

HABITS Well-wooded freshwater-courses are favoured habitats, frequently found around isolated fruiting *Ficus* trees. In forest-thicket most likely at the edges or in the canopy, comes also to well established gardens. Feeding behaviour is bee-eater- or flycatcher-like, sallying out from and back to the same perch. Does not guide to bees nests but gives a sharp *chik* when near a hive. Usually solitary and sedentary. Circular undulating flight display is accompanied by drumming feather reverberations.

VOICE Not often heard. Series of trilled notes *krrr* or a trisyllabic *pew-pew-pew* or *chee-chee-chee* sounding similar to the hurried notes of a bulbul with up to half-minute pauses. Drumming *whurr* noise made with either the tail or wing feathers or both.

STATUS AND DISTRIBUTION Locally frequent in WD in suitable habitat, becoming rare elsewhere with occasional records from LRD and CRD. Recently frequent in coastal NBD. Limited in Senegal with records in south Casamance north to Dakar and through to the east.

BREEDING Vieillot's Barbet is a candidate host; a pair of honeyguides was pestering this species over a period of weeks in June in WD near Brufut. Brood parasitism proven with Little Bee-eater.

WOODPECKERS AND WRYNECKS
Picidae

Eight species, one from Senegal only. Three are commonly seen, one regularly, the others are infrequent or rare. Ecologically specialised and morphologicaly adapted arboreal birds ranging from very small to medium-sized. With the exception of the cryptically-coloured wryneck they rarely descend to the ground. Characteristics include strong, straight chisel-tipped bills, sturdy necks and stiff tails (except wrynecks) used for support when climbing and foraging. Usually solitary, sometimes in pairs or family parties. Strongly undulating flight pattern sometimes accompanied with shrill excited calls. Few drum in Senegambia but some regularly tap. All woodpeckers excavate their own nest hole and chamber, usually in a tree trunk or branch of either live or dead wood, and a few are forest termitaria specialists.

EURASIAN WRYNECK
Jynx torquilla   Plate 31

Fr. Torcol fourmilier

Other names: Northern Wryneck, European Wryneck

IDENTIFICATION 17cm. Sexes alike; well camouflaged in cryptic nightjar-like plumage. Upperparts, from crown to the longish broad tail and including wings, are brown tinged grey to rufous and finely speckled and vermiculated in black and

white, with line of black from centre of crown to back. Distinct whitish, brown and black striped markings around the face and sides of head and neck. Tail conspicuously barred, underparts paler with some barring especially prominent on the flanks. Short pointed bill. North African race *mauretanica* overall darker with paler throat. **Similar species** Spotted Creeper has long decurved bill and different habits.

HABITS Favours dry open woodland with very occasional records from inside forest or forest edge.

Elusive, probably under-recorded in what is suspected to be a brief transit period when migrating north or south through The Gambia. Mostly seen on the ground feeding on termites and ants, or in slightly undulating weakish flight when the wings are characteristically flicked and the tail looks long. Capable of climbing like a woodpecker but does not use the soft tail as a prop for support. Common name acquired from a series of unusual snake-like threat postures of head and neck if disturbed at the nest.

VOICE Probably silent on migration. When breeding in Europe and N Africa, a plaintive repeated *tu-tu-tu-tu-tu*, reminiscent of a small falcon.

STATUS AND DISTRIBUTION Uncommon to frequent Palearctic migrant, mostly recorded in dry sandy areas with scattered trees in WD, NBD and LRD; recorded November–April. Records from extreme north and less frequently from southern coastal Senegal.

FINE-SPOTTED WOODPECKER
Campethera punctuligera Plate 31

Fr. Pic ponctué

IDENTIFICATION 22cm. Medium-sized woodpecker; sexes differ. Shows slightly crested nape when excited. Female with bicoloured crown, hind crown and crest crimson, forehead black flecked white. Male entire forehead to crown,and prominent moustachial stripe crimson (absent in female), narrow white supercilary stripe. In both, yellowish-white chin, throat and sides of neck finely peppered with black dots. Upperparts and wings overall green-olive, tinged-yellow, covered in paler yellowish spots and bars. Rump delicately barred, tail barred yellow-brown with obvious yellow shafts. Underparts lighter yellow with black spots, more profuse on breast and flanks. Bill dark, eye reddish, legs green-grey. Immature: sexes alike, no forehead differences, overall darker green, facial stripes blacker. **Similar species** See rare, heavily streaked and blotched Golden-tailed Woodpecker.

HABITS Favours scattered woodland with plentiful stands of oil-palms, on which it often feeds. Adjacent open land with termite mounds also attract, periodically seen in closed forest. Will forage on the ground. Especially vocal in the first hour of daylight when leaving thicker bush on its way out into the open to find food. Frequently perches on prominent dead branches. Adults with immatures stay together for some time after fledging and are regularly encountered in noisy group gatherings, typically held on an eroded termite mound.

VOICE Loud far-carrying ringing calls, often delivered in unison in flight and especially on landing with crest raised *wik-wik-whew-wee-yeu, wee-yeu*; also a harsh *chew-chew-chew*.

STATUS AND DISTRIBUTION Frequent to common in favoured habitat throughout, seasonally stable in most divisions but reduced numbers in URD in the late dry season. Widespread and common throughout Senegal.

BREEDING Small number of records in the first half of the rains in The Gambia.

GOLDEN-TAILED WOODPECKER
Campethera abingoni Plate 31

Fr. Pic à queue doré

IDENTIFICATION 22cm. *C. a. chrysura*. Heavily streaked woodpecker with a barred green back. Sexes differ and separable on crown, forehead and malar differences. Female has forehead and crown black with fine white spotting, malar area streaked black and white. Male has forehead and crown grey-black, feathers tipped red. Tail barred with very yellow shafts but not as significant a field mark as common name suggests. Eye reddish, bill slate, legs greenish. Immature more heavily streaked and greener on the back. **Similar species** From common Fine-spotted Woodpecker by darker greenish olive-brown back more barred than spotted; underparts very conspicuously streaked and blotched (not pin-head spots); boldly so on the upper breast and this is the critical field mark. At rest Golden-tailed is more squat and heavy-shouldered. See text to eliminate substantially smaller, widespread Cardinal Woodpecker.

HABITS Much as Fine-spotted. In some areas reported to favour freshwater-courses and to assemble in mixed feeding parties. Likely to stay close to areas with abundant ant and termite nests. Probably but inexplicably under-recorded.

VOICE No details recorded so far in Senegambia. Elsewhere a loud nasal *yaooaakk-yaaaaak* and a fast rattled *weet-weet-wit* or *wit-wit-it-it*.

STATUS AND DISTRIBUTION Rare. Data suggest a rare sedentary or occasional breeding species in WD; modern reports from coastal NBD, single records from URD and northern and south-eastern Senegal. West African population is geographically isolated.

BREEDING Solitary modern record in WD with a female feeding a fledgling in June 1984.

266

LITTLE GREEN WOODPECKER
Campethera maculosa   Plate 31

Fr. Pic barré

Other name: Golden-backed Woodpecker

IDENTIFICATION 16cm. Sexes differ. Small-bodied, small-headed woodpecker. Upperparts practically uniform yellow or bronzy-green, may show some sparse pale spotting on upper back, rump indistinctly barred, tail blackish, shafts with some yellow. Underparts greenish-white with complete olive barring, breast and throat additionally tinged buffish. Sexes separable by red nape patch on male, absent on female. Male has crown and forehead olive-black, crown feathers tipped red. Both sexes have brown spotting, most profuse on female's buff sides of face, neck and chin. No prominent malar or supercilium features. **Similar species** Only woodpecker in the region showing combination of green (not brown) back and barred (not streaked) underparts.

HABITS Not well known. Most likely to be encountered close to forest edges or beside creeper-tangled freshwater-courses, feeding on ants.

VOICE Repeated series of three or four rising notes *teeay, teeay*.

STATUS AND DISTRIBUTION No records in The Gambia. Rare in Senegal but a West African endemic with little information from anywhere in its restricted range. Reports from both northern coastal and southern Senegal and to the east are all close to the borders with The Gambia, and all within the last 30 years. Main range is from Guinea-Bissau to Ghana and Cameroon.

BREEDING Sole record in southern Senegal in early August 1971 with nest excavated in a tree termitarium.

BUFF-SPOTTED WOODPECKER
Campethera nivosa   Plate 31

Fr. Pic tacheté

IDENTIFICATION 15cm. Medium-small, dark brown forest woodpecker. Sexes differ. Female shows blackish-olive nape patch, red in male. Both sexes have distinctive small rounded head, blackish-olive with fine paler streaks. Underparts dark olive-green liberally but subtly spotted yellowish-buff, becoming barred on the sides and flanks. Wings and back dark brownish-green with some barring when seen well, and if caught in a shaft of sunlight showing bronzy-green. Tail blackish. Short bill slaty-black. Immature lacks bronzy sheen, is

greener and barred not spotted. **Similar species** From Little Green Woodpecker by spotted not barred underparts, and note superficially similar pattern of associating brood parasite, the Spotted Honeyguide.

HABITS Favours forest thicket, most often encountered in the company of a passing mixed species foraging party. Any woodpecker-type tapping heard in a forest habitat is worth investigation. Elusive, but normally confiding. Feeds with agility like a tit on flimsy canopy twigs.

VOICE Not very vocal, most often a rattled trill *dee-dee-dee* and a rolled and repeated disyllabic *preuuu preuuu* sounding feebly raptor-like.

STATUS AND DISTRIBUTION Locally frequent in coastal forest-thickets of WD and NBD, with a disjunct population in swamp forest patches on both banks of the river in CRD. In Senegal found in southern coastal forests.

BREEDING Adults seen feeding fledged young in mid-dry season and after the completion of the rains which suggests an extended breeding season: more information sought. Excavates into tree termitaria and possibly rotten wood. See Spotted Honeyguide for note on brood parasitism and plumage resemblance.

LITTLE GREY WOODPECKER
Dendropicos elachus   Plate 31

Fr. Petit pic gris

Other name: Least Grey Woodpecker

IDENTIFICATION 12cm. Very small, sexes differ. Male has hindcrown and nape red, which can be erected as a small crest; female all-brown. Forehead and crown brown in both sexes, with a fine white supercilium and line beneath the eye. Slender moustachial stripe brown, throat and sides of face whitish with very fine brown streaking, prone to fade. Both sexes have upperparts barred greyish-brown and white. Important field mark is that rump and uppertail-coverts are conspicuously red; tail brownish barred dirty-white. Wings brown barred white with hint of yellow on shafts. Underparts whitish, spotted or barred, densest on breast faintest on belly and sides. Immature duller. **Similar species** Red rump distinctive, overall pale brown and whitish plumage (may look bleached) and weak facial markings are characteristic, but see text for Brown-backed Woodpecker.

HABITS A bird of Sahel and semi-desert, currently best known in Senegambia (predictably) from the extreme north of Senegal. Older Gambian records are from areas that are not compatible with these

267

habitat requirements. To be expected in belts of acacias and dry stream beds with scattered trees where any tiny (tinkerbird-sized) woodpecker should be scrutinised for a red rump. Possibly overlooked. All ecological and behavioural details eagerly sought.

VOICE Main call a very fast 1-2 second harsh rattle at 8-9 notes per second, delivered in a series of 3-10 phrases, *skree-eek-eee-eee-eeee-ee-eee-eeek*; recalls abbreviated call of Cardinal Woodpecker.

STATUS AND DISTRIBUTION No recent records in The Gambia; a few older sightings from WD, LRD and CRD. Sparse in northern Senegal and to the east, no records in the south.

BREEDING Recorded in Senegal, January-February; one nest in thin acacia branch.

CARDINAL WOODPECKER
Dendropicos fuscescens   Plate 31

Fr. Pic cardinal

IDENTIFICATION 14cm. *D. f. lafresnayi*. Sparrow-sized woodpecker. Sexes differ. Forehead brown; crown and nape scarlet in male, uniform brownish-black in female. Raises crest in excitement. Initial impression is of a scaled-down Fine-spotted Woodpecker. Broad pale superciliary stripe, sides of face, cheeks and chin with very fine streaks or tiny spots, moustachial stripes brownish-black. Upperparts green with multiple but insignificant yellow-green-laddered spots and bars, underparts yellow-buff with well spaced brown streaks on breast, denser on flanks. Eye red-brown, bill black-grey, legs lead-grey. Immature duller green above; both sexes have dull red crown. Note: West African race significantly different in plumage to some other races, initially suggesting another woodpecker species. **Similar species** Cardinal is common throughout, with green back faintly barred, and yellow-buff underparts streaked brown. Brown-backed is frequent mostly in dry savanna, with well-defined facial markings, white and brown streaked (not red) rump, and no yellow in tail. Little Grey frequents subdesert, and conspicuous red rump, pale facial markings, and greyish-brown back. Golden-backed (unrecorded in The Gambia) is completely barred (not streaked) below with a nearly unmarked yellow-bronzy-green back.

HABITS Favours similar habitat to larger Fine-spotted, and often seen near to it. Not expected in thick forest or very dry areas but may occur. By far the commonest of the small woodpeckers, usually in pairs. Agile on thin branches and leafy twigs; joins mixed feeding parties, may cling motionless to the trunk of an oil-palm for long periods.

VOICE Main call, often delivered by pair in unison on arrival at a perching site to feed, is a fast rattling *kweek-eek-eekik-ik*.

STATUS AND DISTRIBUTION Frequent to common throughout in all divisions in all months, most frequent in WD where sightings are reduced in the rains. Rare north of 14°N in Senegal.

BREEDING The few records suggest activity is initiated by the rains and thus coincides with termite emergence.

GREY WOODPECKER
Dendropicos goertae   Plate 31

Fr. Pic gris

IDENTIFICATION 20cm. Sexes differ. Male has scarlet on the hindcrown and nape, female uniform grey. Both show pale markings on the face. Plumage overall lacks conspicuous spots, streaks or barring; wings show some barring at close range. Upperparts green with a bronzy tinge in strong sunlight, rump and uppertail-coverts red. Underparts grey with a small patch of orange-red on the belly. Rather fine, sharp bill slaty-black and slightly longer in male, eye red-brown, legs dark grey. Immature similar to female, reds paler, with some faint barring on sides. **Similar species** Lack of obvious spots or barring is diagnostic.

HABITS Commonest and most widespread woodpecker in Senegambia, occurring anywhere where there are trees or tall mangroves. Stands of oil-palms are favoured for feeding and breeding. The most regularly heard 'tapping' wooodpecker, also the most often sighted in flight, across open land between stands of trees. Also drums. Often in pairs. Feeds low and occasionally descends to the ground.

VOICE Loud and typically rapid, shrill and far-carrying *peet-peet-peet-peet*, often answered by mate.

STATUS AND DISTRIBUTION Common and widespread in all divisions in all months. Population greatest in WD with gradual but slight decrease eastward to URD. Widespread throughout Senegal.

BREEDING Most records in latter part of dry season; most sites in a rotten palm trunk.

BROWN-BACKED WOODPECKER
Picoides obsoletus   Plate 31

Fr. Petit Pic à dos brun

Other name: Lesser White-spotted Woodpecker

IDENTIFICATION 13cm. Sexes differ, nape patch scarlet in male, brown in female. Both have a very pale brown forehead, hindneck brown, crown darker. Well-defined facial markings: white stripe from eye down side of neck, broad brown stripe over ear-coverts with a further white line below, moustachial streak also brown. Upperparts unmarked dull brown, rump white streaked brown; wings brown, barred and spotted white, tail brown with white bars, tinged yellow underneath. Underparts dirty white with faint and fine brown flecking. Immature more barred than flecked below. **Similar species** Beware occasional faded individuals; Little Grey Woodpecker of semi-desert areas has red rump.

HABITS Frequents dry savanna woodland and a few sites of degraded agricultural land with scattered acacias. Silk-cotton saplings are a favourite place to feed and caterpillars are a favoured food item. Gleans with an industrious intensity with far-carrying tapping. Often a member of mixed foraging parties. Easier to find in the dry season.

VOICE Weak trill in flight, *kweek-week*, with a shrill note on arrival at a perch, sometimes prolonged.

STATUS AND DISTRIBUTION Locally frequent to common in WD and LRD; may be under-recorded. Rare in other divisions possibly overlooked in URD. Widespread in Senegal to 14°30'N.

BREEDING Little information; one adult with beak crammed full of insect larvae at nest hole in mid-March in LRD 1995; less well-documented records from WD in March and an immature in June.

LARKS

Alaudidae

Ten species recorded, three only from Senegal. A group of largely terrestrial birds, feeding and nesting on the ground. Typically, larks are cryptically patterned in shades of brown, variously streaked and blotched. Sexes alike except for Sparrow-Larks. Identification is often difficult, complicated by racial variation and wear. Songs and calls are often helpful. Some species are migratory or nomadic, their movements related to climate and vegetation growth.

SINGING BUSH LARK
Mirafra cantillans Plate 32

Fr. Alouette chanteuse

IDENTIFICATION 13-14cm. Small plain lark, lacking a crest. White outer tail feathers distinctive once seen, but often awkward to confirm. Sandy brown above, lightly streaked darker brown on head and mantle; moderate to weak supercilium. Narrow white crescent on side of head caused by pale tips of ear-coverts. Folded wing feathers slightly more rufous, tertials with narrow pale buff fringes; central retrices of tail match tone of back and wings. Underparts whitish, with weak streaking on chest; belly white. Short, stout pink-yellow bill with slightly curved culmen; legs flesh. Immature more contrasting; speckled with dark brown and buff on crown and mantle; heavier sandy fringes on flight feathers; more heavily streaked on gorget than adult. **Similar species** Flappet Lark is darker and streakier above with more rufous on tail and a less finch-like, horn-coloured bill; Sun Lark is more boldly patterned with a heavy supercilium and much more distinct dark brown streaking on the whitish underparts; Kordofan Bush Lark, which occurs rarely in northern Senegal but has not been recorded from The Gambia, has a distinctive tail.

HABITS Usually found scattered in areas of medium to longish grass on dry ground; creeps around, often flushed unexpectedly from underfoot and usually settles again nearby after brief hesitant flight. Breeding males sing from a low shrub or in a low fluttering and gliding flight.

VOICE Typical lark phrases with rapid jumbled notes interspersed with distinct thinner ones.

STATUS AND DISTRIBUTION Local migrant following the intra-tropical convergence. Recorded from all areas of The Gambia, mainly in the dry season, generally local and thinly scattered.

BREEDING Not proven in The Gambia, but a few birds present on the north bank of the river in CRD (July), male giving half-hearted display flight with song. Common wet season breeding bird July-September in northern Senegal.

KORDOFAN LARK
Mirafra cordofanica Not illustrated

Fr. Alouette du Kordofan

IDENTIFICATION 14cm. A sandy-cinnamon lark, basically similar to Singing Bush Lark, except that the tail shows a three-coloured pattern, being mainly black, with rufous central and white outer tail feathers. Bill whitish, legs flesh. **Similar species** See Singing Bush Lark.

HABITS Favours arid areas with thin grass cover and scattered low bushes. Males sing from bushes or low tree tops.

VOICE No information.

STATUS AND DISTRIBUTION Not recorded in The Gambia, very rarely seen in northern Senegal.

FLAPPET LARK
Mirafra rufocinnamomea Plate 32

Fr. Alouette bourdonnante

IDENTIFICATION 14-15cm. *M. r. buckleyi*. Medium-sized lark, warm mid-brown to rufous brown with pale cinnamon wash to upperparts. Blackish streaking on crown, and mantle. Closed flight feathers and wing coverts dark brown, narrowly fringed buff; in flight may show broad rufous cinnamon fringes to secondaries. Tail narrow, often scruffy; prominent rufous-cinnamon central feathers, otherwise dark brown with buff outers. Underparts orange-tinted buff, speckled and streaked dark brown on the upper breast. Bill dark brown, slightly decurved; legs flesh. **Similar species** Singing Bush Lark is paler, has characteristic bill. From Sun Lark by lack of crest, cinnamon tones, much weaker supercilium and finer speckling on breast. Flight display diagnostic.

HABITS Favours open bush and light woodlands near the edges of lateritic outcrops. On the ground easily overlooked, but display flight of the male, most prominent in the early morning, attracts attention. Rising steadily from the ground, the bird breaks into bursts of extra rapid wing-beats, which

270

are maintained at regular intervals as it circles, the bursts accompanied by a dry, rattling *brrrp..brrrp*. As it calls the bird characteristically rises momentarily, then partially stalls. Song flight is often terminated by a rapid plunge toward the ground.

VOICE Reported as a thin short song phrase between bouts of wing-clapping, not easily heard, and a soft two-note call.

STATUS AND DISTRIBUTION Rarely recorded, status poorly known, but displaying birds seen regularly in limited areas of CRD at the end of the dry season suggest it is a rare breeder at least. In Senegal restricted to the drainage system of the Gambia River.

BREEDING Males displaying May to October in The Gambia, but no confirmation of breeding. Nearest confirmed breeding Côte d'Ivoire.

black, white outer web to outer feather. Upper breast blotched and streaked, sometimes very heavily. In flight the wings reveal a strong black and white pattern hence the name. Legs long and flesh-coloured. **Similar species** Large size and long decurved bill preclude confusion with any other species.

HABITS Prefers arid scrub, largely terrestrial and a swift runner. Characteristic display, visible at long range, rising with loose jointed wing beats to make a rolling stall turn and dropping swiftly to the ground.

VOICE A mellow ringing series of discordant whistles, often delivered from the top of a low bush, occasionally accelerating and rising to a more emphatic trill.

STATUS AND DISTRIBUTION Not recorded in The Gambia, but reported from northern Senegal since the 1970s.

RUFOUS-RUMPED LARK
Pinarocorys erythropygia   Plate 32

Fr. Alouette à queue rousse

Other name: Red-tailed Bush Lark

IDENTIFICATION 18cm. Distinctive large lark with whitish face marked by an elaborate dark brown pattern. Dark crown and mantle, characteristic rufous rump and tail. Heavily blotched breast and whitish underparts. **Similar species** Rump and tail pattern are diagnostic.

HABITS Favours savanna woodlands, attracted to burnt areas.

VOICE Song includes whistles and buzzing trills.

STATUS AND DISTRIBUTION Vagrant. One bird at the Bund Road in The Gambia in 1961 is the only record for The Gambia or Senegal. An intra-African migrant in West Africa between the woodlands (breeding) and the Sahel.

HOOPOE LARK
Alaemon alaudipes   Not illustrated

Fr. Sirli du désert

IDENTIFICATION 20cm. Large slim lark, immediately distinguished on the ground by its unstreaked sandy grey-brown upperparts and long, slim, black decurved bill. Head patterned with blackish lores and eye-stripe below a pale supercilium, and black moustachial stripe. Closed wing shows varying amounts of black and grey; central tail feathers grey-brown, the rest dark brown to

SHORT-TOED LARK
Calandrella brachydactyla   Plate 32

Fr. Alouette calandrelle

IDENTIFICATION 14cm. *C. b. rubiginosa*. A small upright lark with a noticeable rufous cap above a partial supercilium and short narrow darker mark behind dark eye. Cap finely streaked with brown; nape and back brown with dark streaking on mantle; the background colour tone is variable being distinctly richer in some individuals. Tertials long, covering primaries, darker brown, with narrow pale edging, (pronounced in juveniles). Underparts plain whitish, though with smudgy buff upper breast usually showing two ill-defined darker patches on either side. In flight whitish underparts contrast with short dark undertail. Juveniles may show some light streaking on upper breast. Bill pale horn, legs flesh. **Similar species** Singing Bush Lark slightly larger with heavier bill, showing weak necklace of smudgy streaks on upper breast. Female Chestnut-backed Sparrow-Lark is smaller and plainer, but shares similar crouching behaviour.

HABITS Gambian records mainly singletons, but wintering birds may occur in flocks in dry open habitats with low vegetation. Usually crouches low, easily overlooked, but stands upright when alert, and makes short fast dashes over the ground.

VOICE Various calls reported; typically a short, throaty *trrlp* flight and contact call.

STATUS AND DISTRIBUTION Rare, formally added to The Gambia list in 1994 (one observation at Jahali Pacharr, CRD) after previous rejections. Common dry season visitor to northern Senegal from Western Palearctic.

271

SUN LARK
Galerida modesta   Plate 32

Fr. Cochevis modeste

IDENTIFICATION 13cm. Smaller, streakier version of Crested Lark, but crest less prominent and more often relaxed, though readily raised. Crown heavily streaked dark brown; very prominent buff-white supercilia extend to join at nape, with narrow blackish eyestripe below. Adult shows distinct narrow dark line running diagonally upward from below eye to join eyestripe behind the eye. Upperparts streaky brown. Tail dark brown with narrow buff edges to outer retrices. Throat and upper breast heavily streaked with dark brown on whitish background, lower breast, belly and undertail-coverts white. Pinkish legs and feet, bill dark brown with pale base to lower mandible. Immatures show pale buff spotting on mantle, broad fringes to coverts and flight feathers, car-coverts strongly defined by downward extension of the supercilia to form a pale collar and dark crown peppered by small off-white feather tips. **Similar species** Partial crest, more distinct streaking on the underparts and strong supercilium all separate from Singing Bush Lark, Flappet Lark and larger Crested Lark; bill more dagger-shaped than yellowish bill of Singing Bush Lark, not so heavy as that of Flappet Lark.

HABITS Found in characteristic clearings of short vegetation occurring as habitat 'islands' in savanna woodlands on laterite. In pairs or groups; in larger groups in the wet season. May settle on dead trees if flushed.

VOICE Soft contact calls, including a three note rattle; typically elaborate lark flight song.

STATUS AND DISTRIBUTION Accepted onto The Gambian list in 1994. Rare year-round resident known from a few locations on the north bank of the river in URD. Almost certainly breeds. Found in the Niokolo Koba area of Senegal and once near the coast .

BREEDING Newly-fledged birds seen in December, immatures present through late dry season.

CRESTED LARK
Galerida cristata   Plate 32

Fr. Cochevis huppé

IDENTIFICATION 17cm. *G. c. senegallensis*. Sexes similar. Medium-sized pale brown lark, immediately distinguished by prominent upright crest. When depressed, crest usually still apparent as a small spike pointing backwards from crown. Gambian birds have a pale supercilium and darker edges to ear-coverts giving distinct 'face', separated by narrow pale zone from a necklace of heavy streaks across upper breast. Rest of underparts plain off-white. Mantle pale brown with darker mottling, flight feathers pale brown with darker median covert bar on folded wing. Tail appears dark brown, with whitish outer edge and pale brown central retrices. Pale orange tinge apparent on underwing-coverts in flight; broad based wings. Strong bill horn brown. Immature speckled with small creamy spots on mantle and broad pale edging to coverts and flight feathers. **Similar species** Sun Lark is smaller, with a heavier supercilium and a partial crest, not the diagnostic spike of Crested; also has different distribution and habitat. Nearest accepted record of the marginally smaller-billed, but otherwise very similar, Thekla Lark *G. theklae* (Fr. Cochevis de Thékla) is a vagrant in Mauritania.

HABITS Associated with areas of short scrubby vegetation, tracks and roadsides, often close to water, and beach scrub in pairs or family parties. On flushing, flutters and moves only a short distance before settling. Often calls before dodging into cover.

VOICE Mellow *doo-lee-oo* call; song in flight a sustained throaty warbling cascade given at moderate height.

STATUS AND DISTRIBUTION Resident, frequent to locally common along the coast. Present in one or two localities along the lower estuary, for reasons not apparent does not occur regularly further inland. Occurs well inland along the Senegal River and elsewhere in northern Senegal.

BREEDING Ground nester with extended breeding season.

CHESTNUT-BACKED SPARROW-LARK
Eremopterix leucotis   Plate 32

Fr. Alouette-moineau à oreillons blancs

Other name: Chestnut-backed Finch-Lark

IDENTIFICATION 11cm. *E. l. melanocephala*. Sexes differ. Very small front-heavy lark usually encountered when suddenly flushed from the ground. Adult male distinctive, underparts all-black, striking bright white ear-covert and hind neck patches contrast with all-black head. Mantle and wings chestnut-brown with slightly darker mottling and pale buff-white patch at bend of wing. Female and immature duller, uniform mottled brown above, head showing paler patches in poorly-defined repetition of male head pattern. Underparts white, heavily speckled brown on throat and upper breast. Unlikely to be encountered without males. Bill finch-like and whitish in both sexes, legs pink-brown. **Similar species** Female distinguished from Black-crowned Sparrow-Lark by darker mottled head and mantle, hint of white collar and dark brown mottling on upper breast.

HABITS Frequents open ground with bare soil and short scrubby vegetation, usually in flocks. Attracted to recently burnt ground; often seen rising from murrum tracks ahead of vehicle. Squats low, shuffling around with belly and tail almost on the ground; characteristically holds folded wings stiffly away from body.

VOICE Song a short scratchy phrase followed by two clearer notes.

STATUS AND DISTRIBUTION A common dry season visitor, becoming locally abundant, particularly to the north bank of CRD and URD, but seen in all areas in most years. In Senegal found mainly to the north of The Gambia; typically erratic in numbers and distribution.

BREEDING Recently fledged young in February in The Gambia; recorded in all seasons in Senegal.

BLACK-CROWNED SPARROW-LARK
Eremopterix nigriceps   Plate 32

Fr. Alouette-moineau à front blanc

Other names: Black-crowned Finch-Lark, White-fronted Sparrow-Lark

IDENTIFICATION 11cm. Sexes differ. Similar in general shape to Chestnut-backed Sparrow-Lark but adult male immediately distinguished by pale grey-brown back and wings. Underparts, throat, sides of neck black; lores, face and hind-crown black; white ear-coverts merging into a grey-brown nape and an additional white forehead patch. Upperparts grey-brown, with pale area along carpal joint. Tail dark. Female sandy-brown above, finely streaked dark brown on crown, faint collar pattern, weakly mottled and streaked on mantle. Sandy cheeks merging into pale creamy underparts. In flight both sexes show characteristic black underwing-coverts. Bill finch-like, pale grey with bluish tint. Immature resembles female. **Similar species** Female less variegated and paler than female Chestnut-backed; separated from other small nondescript larks by combination of extremely small size, plainness and deep finch-like pale bluish bill.

HABITS When not breeding found in parties of variable size, running over ground in areas of short vegetation.

VOICE Breeding male in song flight gives a short high-pitched phrase, finishing with a characteristic clear whistle, slurring down over two notes, *p-tik, p-teeuu*. Various shorter flight calls including a soft *trrp*.

STATUS AND DISTRIBUTION Not so far recorded from The Gambia, but currently thought to be increasing and spreading southwards in Senegal, where it is widely distributed through the north.

BREEDING Dry season records from Senegal.

SWALLOWS AND MARTINS
Hirundinidae

Seventeen species, six of them rare or very rare in The Gambia, one recorded only from Senegal. A familiar family of medium-small, largely aerial birds with streamlined bodies, relatively long slender curved wings and swooping flight. Tail shapes vary from short and almost square, to notched and ultimately to deeply forked with long to extremely long streamers; Barn Swallows have proved productive subjects in studies of streamer symmetry and mate choice. Legs short, feet small. Most build nests of mud pellets, others excavate tunnels in banks.

FANTI SAW-WING
Psalidoprocne obscura   Plate 33

Fr. Hirondelle de Fanti ou Hirondelle hérisée

Other name: Fanti Rough-winged Swallow

IDENTIFICATION 17cm. The only all-dark swallow in Senegambia. Male is blackish, lightly glossed green except characteristically dull underwings. In practice the greenish sheen is hard to see and the bird appears black against the sky. Forked tail is much deeper and longer in the male, broad but tapered streamers producing a characteristic 'set of dividers' shape. Immature browner, less glossy. Occasional individuals with cleanly demarcated pale grey lower belly have been seen. **Similar species** No other hirundine is all-dark, but beware shorter-tailed females and immatures.

HABITS Typically found in flocks, in wooded areas, or over forest pools, wheeling in loose-knit groups, at low level with a wispy, quiet flight, the wings frequently briefly folded to the body in a bounding action; individuals often settle briefly on adjacent trees.

VOICE A quiet *wheep* and a soft, rising, nasal *shooroop* (bee-eater-like but much softer).

STATUS AND DISTRIBUTION Considered to have spread into the region since the end of the 1960s and now established in favoured sites in the coastal area e. g. Abuko, year-round. Recorded in all inland divisions, mainly in the wet season. Confined to Casamance and the south-east in Senegal.

BREEDING Not yet proven in The Gambia; young have been seen in southern Senegal in September. A tunnel-nester in sand banks.

SAND MARTIN
Riparia riparia   Plate 33

Fr. Hirondelle de rivage

Other names: Bank Martin, European Sand Martin

IDENTIFICATION 12cm. The common small brown and white martin of wetland habitats. Upperparts, including head, mantle, rump and tail plain brown, with the tail only slightly notched. Chin and throat are whitish, with a plain brown breast-band. Note that this may be less well-defined in fresh plumage when feathers of breast-band show grey-brown tips. Rest of underparts white. Wings brown, flight feathers darker. Juveniles scaly due to buff feather edges. **Similar species** Separated from Banded Martin by smaller size, uniform underwings and absence of white mark over eye; from Brown-throated Martin by brown breast band across white underparts.

HABITS On migration often seen in large flocks over riverine or marshy areas, occasionally over farmland and light woodlands. Busy flight with little gliding.

VOICE A harsh contact call *tschr* and twittering song.

STATUS AND DISTRIBUTION Frequent to abundant winter visitor, mainly September to March in CRD and URD, though seen in all divisions. Roosts in very large numbers in reedbeds of Senegal River and may also do so in the middle reaches of the Gambia River.

BROWN-THROATED MARTIN
Riparia paludicola   Plate 33

Fr. Hirondelle paludicole

Other names: African Sand Martin, Plain Sand Martin, Brown-throated Sand Martin

IDENTIFICATION 12cm. A small brown martin with fluttering flight. Dark brown upperparts, with plain brown tail, only slightly notched. Chin, throat and upper breast grey-brown, fading into a white belly and undertail-coverts; thus lacking a distinct breast band. Flight feathers brown, underwing grey-brown. Juveniles similar but with buff tips to the feathers. **Similar species** Separated from other brown swallows and martins by combination of small size and absence of breast band.

HABITS Normally associated with aquatic habitats. Usually in small groups and may mix with other species such as Sand Martin.

VOICE A *svee* flight call and soft twittering song.

STATUS AND DISTRIBUTION Rarely recorded in The Gambia, November to January, but also June in Senegal, where it also seems to be rare, with a few records from both the Gambia and Senegal Rivers. May be overlooked among large flocks of Sand Martins.

BREEDING No breeding information in The Gambia or Senegal.

BANDED MARTIN
Riparia cincta   Plate 33

Fr. Hirondelle à collier ou Hirondelle de rivage à front blanc

IDENTIFICATION 17cm. Large, brown and white square-tailed martin. Brown upperparts, with a short narrow white superciliary streak above black lores from base of the bill to the eye, ear-coverts brown. Lower back and rump brown, rump paler. Chin and throat white; broad brown breast-band, remainder of the underparts white. Wings long, brown flight feathers contrasting sharply with white underwing-coverts. Juveniles like adults, but mottled due to pale tips on brown feathering. **Similar species** Separated from Sand Martin by larger size, white supercilium and characteristically marked underside.

HABITS Usually seen near water in small groups; noted for languid flight; mixes with other hirundines.

VOICE A *chip* flight call, and squeaky warbling song.

STATUS AND DISTRIBUTION Rare in The Gambia, the very few records over the past twenty years were possibly overshooting individuals from wet season movements into Nigeria and Ghana. Reported to have been frequent early in the century.

GREY-RUMPED SWALLOW
Pseudhirundo griseopyga   Plate 33

Fr. Hirondelle à croupion gris

IDENTIFICATION 14cm. *H. g. melbina*. A small swallow with contrasting dark upperparts and white underparts. Top of head and neck, brownish, darkening slightly around the eye. Mantle and back glossed dark blue; rump grey-brown. Tail distinctively deeply forked, blackish, with very narrow pale outer fringes if seen well, but generally appearing all black. Underparts uniform whitish from chin to vent; underwing-coverts dirty white contrasting with dark flight feathers. Immatures paler with shorter tail streamers. **Similar species** Separation of adult from Common House Martin, which shows grey-tinted rump in Africa, straightforward if deeply forked tail seen well, but note immature has shallower fork.

HABITS Associated with open grasslands or savanna woodland, often near water and attracted to burnt areas. Flight weak and fluttering.

VOICE Generally silent. A grating *chraa* described.

STATUS AND DISTRIBUTION Rare: recorded sporadically from The Gambia, November–February, and once in Senegal in February 1993. The situation is confused and all sightings merit detailed notes.

BREEDING Nearest breeding Liberia. A tunnel nester, burrowing into flat sandy ground.

RUFOUS-CHESTED SWALLOW
Hirundo semirufa   Plate 33

Fr. Hirondelle à ventre roux

IDENTIFICATION 18-19cm. *H. s. gordoni*. A large blue and orange swallow, similar to Mosque Swallow, but being slimmer, appears slightly smaller. Head blue-black from lores, below eye to ear-coverts, giving distinctive masked appearance; neck and mantle blue-black, rump rufous. Tail blackish, forked, with very long outer streamers in breeding adult with white patches on inner webs. Underparts from chin to vent variable, mainly pale rufous-orange, sometimes darker. West African race generally paler below than the more orange-tinted

birds of eastern and southern Africa. Underwing-coverts buffy-orange. Females have shorter tail streamers. Juveniles similar to adult, but duller and browner, with very pale chin and throat, and short tail streamers. **Similar species** Identification in good conditions straightforward, but in poor conditions requires care: look for extension of dark cap onto sides of head and ear-coverts. This plus slimmer proportions, uniform underparts and extreme elongation of tail feathers distinguishes from other three rufous-rumped swallows; Red-rumped also distinguished by black undertail-coverts. Size differences apparent when flying together, but may be difficult to judge in isolation.

HABITS Associated with wooded habitats in The Gambia, though also present over farmland and villages. Usually in pairs, but larger numbers near water especially at the beginning of the rains. Flies with steady deep wing-beats interrupted by long glides on depressed wings, similar to Mosque Swallow.

VOICE Calls include a *wik-wik-wik*; and a bubbling song in flight, sometimes accompanied by a stalling wing-shimmer.

STATUS AND DISTRIBUTION Variable in numbers, mainly April to December; seen most frequently June to November, when in good years it may outnumber Mosque Swallow. Most frequent in LRD through CRD to URD, but also regularly encountered near the coast. Moves northward with the rains. Probably under-recorded in the past owing to confusion with Mosque Swallow.

BREEDING Nesting in a culvert in URD in August and feeding juveniles near the coast in November. Mud pellet nest with grasses, with a tunnel entrance.

MOSQUE SWALLOW
Hirundo senegalensis   Plate 33

Fr. Hirondelle des mosquées

IDENTIFICATION 21cm. A large swallow, relatively bulky and deliberate in movements compared to other species, but still elegant. Crown and nape blue-black down to the eye; lores whitish; ear-coverts uniform with buff-orange partial collar. Mantle and wings blue-black; prominent rufous-orange rump and deeply forked tail which in nominate race of West Africa lacks white patches. Chin and throat creamy-buff, darkening progressively to rufous-orange on the lower belly. Underwing-coverts pale buff; flight feathers blackish. Juveniles browner on upperparts, with whiter throat and collar and more patchy orange tint to underparts. **Similar species** Separated from Rufous-chested by larger size and bulk, dark cap limited to top of the head, contrast between pale throat and darker lower belly, and proportionately shorter tail streamers; from Red-rumped by size and orange undertail-coverts; from both by characteristic call.

HABITS Usually in pairs. Flight with typical measured wing-beats and long glides, wing tips below body and tail flared, resembling a small falcon. Spends long periods perched on favoured vantage points. Favours all kinds of woodland and farmland habitats, also around buildings and radio masts.

VOICE A single-note call with a distinctive reedy whining quality often gives away its presence overhead; equally characteristic slow descending *peeooo*.

STATUS AND DISTRIBUTION Found throughout The Gambia and present in all months.

BREEDING Nesting July and August, recorded in WD and URD; a mud pellet and grass structure with tunnel entrance under an overhanging surface.

LESSER STRIPED SWALLOW
Hirundo abyssinica   **Plate 33**

Fr. Hirondelle à gorge striée

IDENTIFICATION 15-18cm. *H. a. puella*. A slim, long-tailed swallow; the only Senegambian swallow with streaked underparts. Pale rufous-orange head, slightly darker lores; cleanly separated at nape from blue-black mantle; rump pale rufous, matching head. Forked tail black with long streamers (female shorter), white patches of variable extent on inner webs. Underparts white, clearly streaked with dark brown from chin to lower belly; apricot-buff undertail-coverts. Overhead, pale buff underwing-coverts contrast with blackish flight feathers. Immatures show blackish speckling on crown, and buffer wash underneath. **Similar species** The only other blue and rufous swallow with heavily streaked underparts, the Greater Striped Swallow *H. cucullata* (Fr. Hirondelle à tête rousse), has not been recorded closer than Zaire.

HABITS Generally in pairs or small groups, usually associated with woodland. In the few Gambian observations, usually associated with other hirundines.

VOICE A sequence of notes with a distinctive creaking, wheezing quality.

STATUS AND DISTRIBUTION Rare visitor to The Gambia and Senegal, almost all records in the wet season. Near the northern limit of its range in this area, thought to be a partial migrant, pushing north with the rains. The local race shows the least heavy streaking of all the forms of Lesser Striped Swallow.

BREEDING No information from The Gambia, but seen in small colonies at bridges in Niokolo Koba,

eastern Senegal. Mud pellet nest with tube entrance under overhanging surface.

RED-RUMPED SWALLOW
Hirundo daurica   **Plate 33**

Fr. Hirondelle rousseline

IDENTIFICATION 16-18cm. *H. d. domicella*. The smallest and palest of the rufous-rumped, blue-black and rufous swallows in Senegambia. Crown and mantle glossy blue-black, separated by a chestnut collar. A large patch on the rump area is cleanly demarcated pale rufous. Long forked tail feathers are blue-black, lacking white mirrors. Underparts whitish, very lightly stained buff, with no streaking. Undertail-coverts diagnostically jet black Underwing-coverts buffish, and contrast with darker flight feathers. **Similar species** Four other swallows share rufous rumps in the region, Mosque, Rufous-chested (both larger) and Lesser Striped. Red-rumped is separated from all by sharply defined black undertail-coverts. European race *H. d. rufula* is slightly larger, and more buff-tinged underparts may show narrow dark streaks in close views. Rufous stripe over eye broader and rufous on rump fades towards uppertail-coverts. Preuss's Cliff Swallow has a pale buffy-toned rump and underparts, and a relatively square-ended tail.

HABITS In The Gambia occurs in a wide range of habitats, often in flocks of mixed hirundines. Occurs over the river, light woodlands or farmland. Perhaps most predictably encountered around rural villages, where its steady flight pattern low over the thatched roofs is characteristic. Note that the larger rufous-rumped species often share habitats and behave similarly.

VOICE A quiet twittering song when perched distinctively musical; frequent *djuit* contact call.

STATUS AND DISTRIBUTION Resident African race less common at the coast, but common inland from LRD through to URD, seen around most villages inland. Birds of European origin reported in northern Senegal.

BREEDING Mainly in the dry season, often using man-made structures such as the eaves of thatched huts for constructing a retort-shaped mud nest.

AFRICAN ROCK MARTIN
Hirundo fuligula   **Plate 33**

Fr. Hirondelle isabelline

IDENTIFICATION 12-13cm. Small, compact all-brown martin. The races found in tropical West Africa are the smallest and darkest of a widespread variable species. Upperparts uniform dark brown with hint of gloss. Tail brown, with a scarcely noticeable fork and characteristic white patches making a row of distinct 'windows' either side of the midline when flared. Blackish lores with a plain dull pinkish-brown throat fading into the rest of the underparts, which are plain brown, paler than upperparts. Flight feathers dark brown, with brown underwing-coverts tinged rufous. Juveniles similar to adults but with buff fringing on upper body feathers. **Similar species** Crag and African Rock Martins readily distinguished from other hirundines by diagnostic tail and leisurely flight style, but separating the two requires care; darker brown races of Rock Martin are sometimes similar to Crag Martin in colour but smaller, being closer to a bulky Sand Martin in dimensions. Pale North African and Saharan races of African Rock Martin (e.g. *H. f. spatzi* recorded as close as Mali and *H. f. presaharica* in Mauritania), should be borne in mind especially during the northern winter when some individuals are thought to move southward. These are distinguished from Crag Martin by smaller size, and by the contrasting pale grey patch on the lower back and rump; pale underwings have less contrasting underwing-coverts. African Sand Martin lacks white spots in tail and is whitish on the undertail-coverts, with a fluttering flight.

HABITS Normally associated with rocky outcrops; occasionally around buildings. Flight pattern steady, level and unhurried, with frequent long glides on very flat wings, lending a distinctive dignified air, though still capable of speed.

VOICE Generally quiet; a soft twittering song and *wit* flight call.

STATUS AND DISTRIBUTION Single record from The Gambia not well substantiated, but specimens attributed to the dark race *H. f. bansoensis* have been collected in south-eastern Senegal. Representatives of pale races may occur as rare winter visitors in northern Senegal. All observations merit careful and detailed notes.

CRAG MARTIN
Hirundo rupestris   Plate 33

Fr. Hirondelle de rochers

IDENTIFICATION 15cm. A bulky, medium-sized all-brown martin. Brown head, mantle and rump similar to or slightly darker in tone than Sand Martin. Tail nearly square with distinctive row of white patches either side of mid-line apparent when

flared. Lores slightly darker than crown, throat and cheeks weakly speckled and streaked (not easily seen), underparts pale brown, with darker brown undertail-coverts. Flight feathers and tail slightly darker than the body from above, underwing-coverts noticeably blackish. Juveniles slightly darker, with buff fringes to feathering. **Similar species** Separated from dark forms of African Rock Martin by size, and blackish underwing-coverts, and from pale forms by these features plus uniform brown mantle to rump (see African Rock Martin).

HABITS Similar to African Rock Martin, associated with cliff faces and rocky outcrops, around which it cruises at low level with a steady, direct flight, including long glides.

VOICE Soft *prrt* and twittering song.

STATUS AND DISTRIBUTION One definite record in The Gambia, an individual around lateritic outcrops on the north bank of the river in CRD, November. Considered a rare migrant to northern Senegal.

WIRE-TAILED SWALLOW
Hirundo smithii   Plate 33

Fr. Hirondelle à longs brins

IDENTIFICATION 14cm. Small noticeably blue and white swallow. Crown from forehead to nape, down to the top of the eye bright chestnut, often less obvious than might be expected. Black lores and dark blue ear-coverts; white chin and underparts. Neck, mantle and rump dark glossy blue; tail glossy blue with white patches; outer two feathers elongated into filamentous extensions, the 'wires'. Wires present but shorter in females and young; at all times wires hard to see while the bird is on the wing. Wings blackish, with blue iridescence more marked than in most other swallows. Underwing-coverts white. Juvenile has brownish cap and less iridescent upperparts, but the basic pattern of the adult is present. **Similar species** Only swallow with clean white underparts and a chestnut cap. Bright colours and flight pattern often attract attention first.

HABITS Usually in pairs or small groups; strongly associated with waterside habitats and bridges. Quick, twisting and purposeful flight pattern, mostly at low level, is distinctive. May also occur away from water when easily overlooked .

VOICE Quiet, with a soft *chip* call and a twittering song.

STATUS AND DISTRIBUTION Resident in The Gambia; most often encountered in the coastal area, but thinly scattered in all divisions. Seldom recorded much further north in Senegal, though present to the south and east.

BREEDING Recorded breeding in nearly all months, possibly more frequent January-March and July-August. Builds a shallow bowl of mud pellets and dry grasses, often fixed to man-made structures such as bridges.

PIED-WINGED SWALLOW
Hirundo leucosoma   **Plate 33**

Fr. Hirondelle à ailes tachetées

IDENTIFICATION 12cm. A small, compact, blue and white swallow, with a variably prominent white line across the base of each wing. Upperparts from top of head glossy blue-black; tail shallowish-forked. Wings blackish apart from a line of white across the wing base. Underparts bright white from chin to vent, with hint of partial breast-band. Immature similar to adult, but duller with a grey-brown tinge. **Similar species** White wing bases diagnostic.

HABITS Singly, pairs, sometimes small parties, over a range of habitats including farmland, and coastal and savanna woodlands. Hawks with other hirundines low over the freshwater reaches of the river.

STATUS AND DISTRIBUTION Resident in low numbers throughout from the coast to CRD; not recently recorded in URD though almost certainly occurs there. Patchily recorded in Senegal. West African endemic, found eastward to Cameroon.

BREEDING Probably mainly April-June, immatures seen in The Gambia in July and October. Cup-shaped nest of mud pellets and grass, often on buildings.

ETHIOPIAN SWALLOW
Hirundo aethiopica   **Plate 33**

Fr. Hirondelle à gorge blanche ou Hirondelle d'Ethiopie

IDENTIFICATION 13cm. Resembles Red-chested Swallow but smaller, with chin and throat pale creamy-buff and partial bar projecting either side of breast; paler underwing-coverts and brighter upperparts. Immature has brownish forehead and crown, brown tinges on upperparts. **Similar species** Separated and from Red-chested and Barn Swallows by creamy-buff, not red, throat and partial breast-band; from Wire-tailed by restricted patch of chestnut on forehead.

HABITS Favours open savanna habitats and around villages, not usually in heavily wooded areas.

VOICE Typical swallow twittering.

STATUS AND DISTRIBUTION Not recorded in The Gambia, but recorded in Niokolo Koba Senegal in April and February. Thought to be spreading westward.

BREEDING Recorded in Senegal February and April; mainly March to August in rest of range in West Africa, a cup-shaped nest of mud pellets mixed with grass, built against a vertical surface.

RED-CHESTED SWALLOW
Hirundo lucida   **Plate 33**

Fr. Hirondelle à gorge rousse

IDENTIFICATION 15cm. Commonest swallow in The Gambia. Upperparts dark blue-black, small patch of rufous on forehead. Blackish flight feathers lacking bluish gloss. Tail dark, forked, outer feathers less elongated than Barn Swallow, with extensive white patches; perched adults show lower surface of tail all white with narrow black margin. Variable but generally narrow blackish breast-band, often scruffy and sometimes incomplete, surrounding rufous throat. Rest of underparts clean white, merging with the mainly white undertail. Underwing-coverts grey-white, lacking buff tones of the Barn Swallow. Immature similar to adult, browner about the head with narrow buff line from bill to above eye. **Similar species** Main points of separation from Barn Swallow are slightly smaller size, with relatively large rufous throat patch, smaller rufous spot on forehead, narrower breast band, shorter tail streamers and especially in flight the whiter appearance of the underparts. The latter, in combination with the narrow breast-band, also assist in the more difficult separation of juveniles. In flight the Red-chested is less languid and fluid its its movements than Barn.

HABITS Frequent over nearly all habitats, usually in flocks often around villages and towns.

VOICE Maintains a continuous twittering, similar to Barn Swallow.

STATUS AND DISTRIBUTION Abundant resident throughout The Gambia and southern Senegal, only occasional in the extreme north of Senegal.

BREEDING Mainly in the dry season, February-July; a cup-shaped nest of mud pellets and grasses, often placed on buildings.

NOTE This species was formerly treated as a subspecies of Barn Swallow.

BARN SWALLOW
Hirundo rustica   Plate 33

Fr. Hirondelle de cheminée

Other names: European Swallow, Swallow

IDENTIFICATION 17-19cm. Sleek swallow, with glossy blue-black upperparts, deeply forked tail and elegant movements. Outer feathers (streamers) longer in males. The dark tail shows a white subterminal band when flared. Flight feathers blackish, washed with glossy dark blue, fading towards wingtip. Chin and throat rufous-chestnut, rufous patch immediately above the bill. A broad blueblack breast band separates the rufous throat from creamy underparts. Underwing-coverts pale buff. Immatures washed-out version of adult, retaining broad breast-band. **Similar species** See Red-chested Swallow.

HABITS Hawks at low level in groups over almost any habitat. Very much a transit migrant in The Gambia.

VOICE Maintains a prolonged twittering with various rattling sounds and a number of other chirps.

STATUS AND DISTRIBUTION Occurs throughout the country, mainly on passage October-November and again April-May, when it may be briefly frequent in any area, but is seldom very common.

COMMON HOUSE MARTIN
Delichon urbica   Plate 33

Fr. Hirondelle de fenêtre

Other name: House Martin

IDENTIFICATION 12.5cm. A typical martin, holding wings in more widely spread gliding position than swallow. Distinctive clean-cut separation of dark blue-black upperparts, wings and shallowforked tail contrasting with white underparts and rump. Adults in Africa (and immatures) show slight greyish tinge to the rump. **Similar species** See Greyrumped Swallow. Preuss's Cliff Swallow *Hirundo preussi* (Fr. Hirondelle de Preuss) occurs in nearby Guinea-Bissau and Mali, and has been reported recently from coastal WD in The Gambia. It is dark blue above with brownish wings, squarish indented tail, and pale buff rump and underparts.

HABITS Typically gregarious in milling groups. Flight a mixture of glides and fluttering stalls and tumbles. Regularly associates with other swallows around roadside pools not associated with urban areas or buildings.

VOICE Commonly a distinctive throaty *chirit*.

STATUS AND DISTRIBUTION A Palearctic migrant, frequent to temporarily very abundant; seen sporadically from October to June in all parts of The Gambia, most obvious November to January. Numbers variable from year to year; may be absent for long periods, appearing for brief spells in large flocks.

PIPITS AND WAGTAILS

Motacillidae

Four wagtails (two rare), one longclaw and seven pipits, three of which are rare or of uncertain status. A group of fast-running largely terrestrial birds with relatively strong legs and feet and short slim bills. Wagtails are generally strikingly coloured with long, white-bordered tails continually wagged up and down; pipits in contrast have shorter tails and are well-camouflaged in streaky buff and brown plumage, presenting greater identification problems. Calls may be a useful identification guide. The single longclaw is like a large, brightly coloured pipit.

YELLOW WAGTAIL
Motacilla flava   Plate 34

Fr. Bergeronnette printanière

IDENTIFICATION 17cm. Several sometimes reasonably distinct subspecies recorded. Smallest and commonest of the wagtails, with proportionately the shortest tail. Extremely variable with subspecies, age and moult, but as a general rule any wagtail showing yellow on underparts and a greenish or brownish back will be this species. In breeding plumage (after about March) head patterns are as below, but all subspecies show green mantle, brown wings with paler fringes, and blackish tail with white outer margins. Underparts uniform yellow. In flight shows weak double wing-bar. Non-breeding (after July-August): browner wash above, buffier yellow below, sometimes with mottled brownish necklace. Many first-winter birds very washed out but buffy tones usually apparent. Legs black. At least four subspecies, all from western Europe and Scandinavia, of which the nominate *M. f. flava* and the British subspecies *M. f. flavissima* are the most frequent. They may be separable on the head patterns of males in breeding plumage, which are described below, but in non-breeding plumages are usually indistinguishable.

M. f. flava The breeding male has a blue-grey head and nape; grey lores, ear-coverts and cheeks are demarcated by a white supercilium above and a narrow whitish streak immediately below, with a restricted patch of white on the chin. Breeding range in western and central Europe.

M. f. flavissima Has an olive-green crown, pronounced yellow supercilium with greenish-yellow lores and cheeks and solid yellow underparts. Breeding range mainly in the British Isles.

M. f. thunbergi A dark slate-grey crown with very narrow white supercilium, if present, but more often only a trace or lacking; blackish lores and dark grey cheeks merge with greyer top of head. Chin and throat mainly yellow, sometimes with vestigial white patch. Breeding range in northern Europe and Scandinavia.

M. f. iberiae Grey-crowned with a narrow white supercilium which is sometimes absent in front of the eye, dark grey to blackish lores and cheeks, with a white chin and throat. Small in size. Breeding range in Spain and the western Mediterranean.

M. f. cinereocapilla Grey-crowned with blackish lores and cheeks and very fine white supercilium usually present behind eye; throat whitish. Breeds central Mediterranean. Not yet recorded in the Senegambian region.

Similar species Yellow, brownish or buffy tones separate first-winter from same plumage of White Wagtail.

HABITS Prefers open areas, feeding on the ground, often associating with herds of grazing animals. Wintering birds may also occur singly or in small groups; larger assemblies may be of mixed races. Very large flocks assemble at roost sites in extensive reed beds.

VOICE Distinctive *tsweet* call.

STATUS AND DISTRIBUTION Passage migrant and winter visitor to all parts of the country from September to April, with the majority arriving November, most numerous January to March, stragglers remaining as late as June. Common to abundant in suitable habitat; individual flocks likely to be transient. Similar status in Senegal.

GREY WAGTAIL
Motacilla cinerea　　　　Plate 34

Fr. Bergeronnette des ruisseaux

IDENTIFICATION 18-19cm. Sexes differ. A long-

tailed, slim wagtail, with strong yellow undertail. Breeding plumage (after February-March): male grey head and ear-coverts, white supercilium, grey mantle, relatively long black tail with white margins. Wings dark with pale fringes. Black bib, white moustachial stripe, sulphur yellow underparts. Non-breeding (from August): similar but throat white, underparts pale yellow, apart from rich yellow undertail. In flight shows white wing bar. **Similar species** Very long tail, slim body, whitish supercilium, grey mantle and yellowish rump eliminate all other wagtails. Voice also distinctive.

HABITS Strongly associated with waterside habitats, where it patrols in typical wagtail fashion, frequently dipping the very long tail up and down; sallying upward to take insects overhead.

VOICE Contact call a clear *tchee*. Flight call a loud *tip-tip* or *chiss-ick* with accent on second syllable.

STATUS AND DISTRIBUTION Rare winter visitor from Palearctic; the very few sight records in The Gambia are from October to March. Regular as far south as Mauritania, only a few scattered records from northern Senegal.

WHITE WAGTAIL
Motacilla alba　　　　Plate 34

Fr. Bergeronnette grise

Other name: Pied Wagtail

IDENTIFICATION 18cm. Sexes differ. Slim, with black or grey and white plumage variable according to sex, season and age. Breeding plumage from February-April: male has white forehead, face and ear-coverts, black crown and nape, black bib. Mantle plain grey; wings dark; tail black with white outer margins. Female white-throated with grey gorget. Non-breeding (from August): greyer on crown, white throat, bib reduced to smudgy dark grey gorget across white underparts; female greyer over face, loses capped effect on crown. Immature similar to non-breeding female. **Similar species** Distinctive, grey mantle and absence of supercilium always separates it from African Pied Wagtail. Immature distinguished from other wagtails by absence of buff or yellowish tones.

HABITS Found singly or in flocks, usually in open areas, runs with characteristic bustle, pumping head back and forth in time with footsteps, making short undulating flights forward to resume feeding.

VOICE Flight call a crisp *chizzick*.

STATUS AND DISTRIBUTION Palearctic migrant to The Gambia in good numbers, with large flocks on passage October and again March; wintering individuals regularly encountered in the interim. Recorded in all divisions. Widespread in all but the most arid districts of Senegal, September-April.

AFRICAN PIED WAGTAIL
Motacilla aguimp   Plate 34

Fr. Bergeronnette pie

IDENTIFICATION 19cm. Immediately distinguished by bright white supercilium over black face mask and very clean black and white appearance. Head and mantle black; tail black, outer margins white. Prominent, broad white supercilium above black lores and ear-coverts; white patch at the side of the neck. Chin and throat white separated from the rest of the white underparts by a cleanly defined black breast band. Black wings show a large white patch at rest and in flight. Immature echoes adult pattern in dark grey-brown and white above, washed buffy below. **Similar species** Adult unlikely to be confused with any other wagtail once supercilium seen well; prominent white in wings distinguishes adult and immature from all others, though latter can otherwise resemble White Wagtail.

HABITS In West Africa associated mainly with freshwater riverside habitat, especially where water levels are low enough for small islets and shingle banks to form, a factor which may explain its limited penetration in the eastern section of The Gambia. Usually seen in pairs or small groups, working its way along shoreline.

VOICE Flight call *chizzit* more liquid than White Wagtail; an elaborate song.

STATUS AND DISTRIBUTION Very few records in The Gambia from URD, once May and once December; reports of wanderers in coastal area unsubstantiated. A small, isolated population is resident along the Gambia River in Senegal, from which some birds may occasionally disperse downstream.

BREEDING Builds a cup nest, usually in a crevice; recorded once in Senegal, March.

TAWNY PIPIT
Anthus campestris   Plate 32

Fr. Pipit rousseline

IDENTIFICATION 16.5cm. Sexes similar. Slim, plain, sandy-coloured pipit. Forehead and crown sandy with fine brown streaks; pronounced white

supercilium, lores dark, a short moustachial stripe and longer, thin black malar stripe give distinctive head pattern. Mantle and back similarly sandy, appear plain but brownish streaking and mottling visible in good views. Rump unstreaked sandy-brown; tail brown with creamy-white outer webs to the outer feathers. Chin throat and underparts creamy-white with faint streaking on sides of breast. Wing often appears dark, though strong buff fringing in fresh plumaged birds may reduce this. Legs flesh or yellowish. Immature more streaked on head, mantle and upper breast mottled on tawnier background, head may show faint yellow wash. **Similar species** Sandier colouring and slimmer build than more frequent Plain-backed Pipit. Richard's Pipit *Anthus richardi* (Fr. Pipit de Richard) is a potential confusion species; although not yet recorded in Senegambia, it is known from Mali and Mauritania.

HABITS Typically in ones and twos or small groups, in flat open country, including dry harvested croplands and rice stubble. Walks with head jerking and tail bobbing pattern of a wagtail. Often confiding and slow to flush, possibly preferring to use cryptic coloration to avoid detection.

VOICE Characteristic *tsheup* call on rising from the ground; metallic *che-vee* or *chireeo* notes.

STATUS AND DISTRIBUTION Uncommon passage migrant in The Gambia, recent records from the coastal area, once in LRD (December-April). Possibly under-recorded. Frequent in arid scrub habitats in northern Senegal in the same period.

LONG-BILLED PIPIT
Anthus similis   Not illustrated

Fr. Pipit à long bec

IDENTIFICATION 17cm. *A. s. asbenaicus*. Marginally larger than Tawny Pipit with dark streaking on snady-buff upperparts, indistinct streaking on the underparts, and buff outer tail feathers. It gives a simple rising whistle note, often when perched in a small tree.

STATUS AND DISTRIBUTION Not recorded in The Gambia, but recently reported in northern Senegal; considered to be uncommon and elusive in Djoudj, November-March. Has been recorded from Mali and Niger.

PLAIN-BACKED PIPIT
Anthus leucophrys   **Plate 32**

Fr. Pipit à dos roux ou Pipit à dos uni

IDENTIFICATION 17cm. The only resident pipit, rather featureless but identified by combination of upright stance and strong moustachial stripes. Top of head brown with very faint streaking; mantle unstreaked brown; dark brown tail showing buff outer webs to outer feathers, not easy to see well. Broad creamy-white supercilium above brown lores and eyestripe, with characteristic well-developed moustachial stripe; malar stripe indistinct. Chin and throat are whitish, merging with uniform buff underparts, finely streaked with dark brown in a poorly-defined belt across the upper breast. Wing-coverts and flight feathers dark brown, with buff edging; underwing-coverts buffish. Bill horn-brown with a yellowish base to the lower mandible. Legs pinkish or yellowish-flesh. Immature browner with heavier streaking on breast and whiter edges to flight feathers. **Similar species** Distinguished from Tawny Pipit, by bulkier darker body, darker bill, and tendency to a more upright posture, lacking wagtail impression of Tawny Pipit.

HABITS Typically in pairs, sometimes small groups, in short grass areas, cultivation after harvest, and burnt ground, especially rice fields. Not directly associated with water and not often seen in arid areas. Usually does not fly far when flushed. Sometimes flies up into trees and fence posts.

VOICE Characteristic *ssissik* flight, repetitive slow song of simple notes.

STATUS AND DISTRIBUTION Thinly scattered breeding resident, recorded in all divisions of The Gambia. Sometimes locally common and typically site-faithful. A similar pattern applies in Senegal, though largely absent from central arid regions.

BREEDING A ground nester, cup-shaped nest typically at the base of a tussock; mostly April to July in The Gambia, once November.

OLIVE-BACKED PIPIT
Anthus hodgsoni   **Not illustrated**

Fr. Pipit d'Hodgson

IDENTIFICATION 14.5cm. Size and habits similar to Tree Pipit. Can be separated on head pattern difference; Olive-backed shows striking thick creamy-white supercilium (usually buffish on Tree) highlighted by a thin black lateral crown stripe and thin dark eye-stripe. A pale and a dark spot on rear ear-coverts (more marked than on Tree) also use-

ful. Greenish-olive mantle when fresh, becomes more streaky and greyish when worn.

STATUS AND DISTRIBUTION Uncertain. Vagrant from Asia. A single record at Cape Point, WD, in December 1984 was accepted onto the Gambian list, although later considered questionable by Gore (1990).

TREE PIPIT
Anthus trivialis   **Plate 32**

Fr. Pipit des arbres

IDENTIFICATION 15cm. Sexes similar. Most frequent migrant pipit. Upperparts brown, streaked over the head and mantle, plain on rump. Tail brown, with mainly white outer feathers. Neat yellow-buff supercilium, blackish lores and eyestripe. Fine whitish eye-ring, more prominent than other pipits. Submoustachial area plain yellow-buff, set off by black malar streaks. Breast yellow-buff, marked by a band of heavy dark brown streaks. Lower belly whitish and unstreaked; whitish-buff flanks weakly striped. Legs distinctively pink. Immature similar to adult. **Similar species** From Red-throated Pipit by paler colour and less streaky overall appearance and absence of streaking on rump.

HABITS On migration usually encountered in pairs or small parties moving stealthily on the ground in semi-open woodland habitats. On disturbance likely to rise and settle in the branches of a nearby tree.

VOICE Characteristic flight call a short *tzeep*, distinct from longer call of Red-throated Pipit.

STATUS AND DISTRIBUTION Palearctic migrant. Thinly scattered throughout The Gambia in suitable habitat September–April; larger numbers in April on passage. Similarly in Senegal, though slightly more frequent in the south.

RED-THROATED PIPIT
Anthus cervinus   **Plate 32**

Fr. Pipit à gorge rousse

IDENTIFICATION 15cm. Sexes differ to variable degree. A dark heavily streaked pipit with marked seasonal plumage variation. In all plumages, check for heavy dark streaking on the breast, flanks and rump. Non-breeding birds sometimes show reddish tinge on supercilium of male and margins of gorget; buff chin and submoustachial stripe with distinct dark malar stripe. Crown mantle and rump, brown with black streaking, forming black-buff-black 'tramline' stripes either side of midline. Tail blackish-brown, whitish outer webs to outermost feathers. Diagnostic reddish-pink face and throat of full breeding male rarely seen early April. Legs flesh, brightening to pinkish in April. Immature like non-breeding adult, though with whiter background to heavy breast streaking. **Similar species** From Tree pipit by darker appearance, heavier streaking, which continues on flanks and rump. Breeding male unmistakable.

HABITS Favours areas of short vegetation, notably hotel and camp lawns, also rubbish dumps. Usually singly or in small parties, often with Yellow Wagtails.

VOICE Characteristic penetrating, sibilant drawn out *pseeep* when flushed and in flight.

STATUS AND DISTRIBUTION Palearctic migrant uncommon in The Gambia, perhaps underrecorded. Sporadic September-May, most records March-April at the coast, less frequently seen inland. Thinly recorded in Senegal, mostly to the north of The Gambia in the same months.

YELLOW-THROATED LONGCLAW
Macronyx croceus   **Plate 34**

Fr. Sentinelle à gorge jaune ou Alouette sentinelle

IDENTIFICATION 20cm. Unmistakable robust, brown and yellow bird of damp ground. Upperparts tawny-brown streaked with black. Tail brown with prominent white corners. Yellow supercilium. Entire underparts bright yellow, intensified by extension of the black moustachial stripes to form a complete black band across the upper breast, with black streaks along the lower edge of this necklace. Heavy-looking bill blue-grey, sometimes tipped black. Legs brown with greatly elongated claw on hind toe. Immature dull buffy yellow on underparts; only a hint of the black necklace. **Similar species** The only longclaw found in West Africa; not likely to be confused with any other species.

HABITS Associated with grassy areas near water, with scattered low bushes used as vantage points and singing posts. Typically in pairs or small parties, often attracts attention by calling. Fluttering flight, interrupted by brief, stiff-winged glides when white patches in the tail become apparent. Frequently sings on the wing.

VOICE Variable; in The Gambia includes a far-carrying repeated sequence of plaintive upward whistles, followed by a rapid series of staccato notes on a monotone *shoe-ee ti-ki-ti-ki-ti.*

STATUS AND DISTRIBUTION Thinly scattered resident in suitable habitat in all divisions of The Gambia, though not recently recorded in LRD. Most likely near the coast. In Senegal mostly confined to southern regions.

BREEDING Recorded June-August in The Gambia; cup-shaped nest, usually on or very close to the ground at base of vegetation in damp locations.

CUCKOO-SHRIKES

Campephagidae

Two species. Often solitary, medium-sized, rather secretive and largely arboreal birds, in some ways shrike-like, with relatively powerful drongo-like bills, favouring open forest and wooded savanna.

RED-SHOULDERED CUCKOO-SHRIKE
Campephaga phoenicea   **Plate 34**

Fr. Echenilleur à épaulettes rouges

IDENTIFICATION Length 20cm. Sexes differ. Adult male glossy blue-black with bright scarlet shoulder patch (occasionally orange); female greenish with yellow flight feather edgings and dense black scalloping on breast and underparts creating unusual barred appearance. In good views a small yellow wattle may be seen at the gape of the male, and small yellow patches at base of legs. Female greyer toned on head, more olive tones on lower back and rump; tail green-brown. Immatures similar to the female, though darker. **Similar species** Female might suggest immature green-backed cuckoo species, but all are ruled out by cuckoo-shrike's yellow edging on flight feathers; barred underparts eliminate female orioles and Amethyst Starling. The rounded tail of the male eliminates both drongos and is broader and shorter than tail of the slimmer Northern Black Flycatcher.

HABITS Usually in pairs, sometimes several birds loosely associated, in thickets and woodland, usually with large trees, where they search unobtrusively and methodically through leaves and outer branches picking off caterpillars and insects. Occasional sallies flycatching. The male indulges in short display flights in the breeding season in which the shoulder patches are puffed up.

VOICE Not vocal. Contact notes between a feeding

pair soft and crisp, *tchit* etc. and a soft flight call, *chi-pit...chi-pit*. A soft waxbill-like *sweee-sweee* in display to females.

STATUS AND DISTRIBUTION Resident in all parts of The Gambia in wooded habitats, generally uncommon though occasionally locally frequent. Seen much more frequently in the wet season months July–September, especially in the inland divisions, suggesting local movements. In Senegal found to the north mainly in the wet season, though not penetrating to the extreme north; resident in Casamance.

BREEDING In The Gambia seen carrying nesting material and in display flight in August-September. Builds a cup-shaped nest on a fork of a tree.

WHITE-BREASTED CUCKOO-SHRIKE
Coracina pectoralis   Plate 34

Fr. Echenilleur à ventre blanc

IDENTIFICATION Length 25cm. Sexes similar. Medium-sized tree-top species, unmistakably elegant in powder blue-grey with a clean white waistcoat. Entire upperparts plus chin and upper breast mid-grey, underparts pure white. Flight feathers and outer tail are blackish. Male dark grey in front of the eye. Adult female similar but lacks darker lores, usually paler on chin and upper breast. Bill black, eye brown with pale eye ring, legs grey. Immature very pale at fledging, marked with dark brown spots and bars above and below, with white edging to the flight feathers. **Similar species** Unlikely to be confused with any other West African species.

HABITS Found in Guinea savanna and dry savanna woodland areas with high canopy. Usually in pairs, moving unobtrusively amongst upper leaves; may spend longish periods sitting quietly at a high vantage point at the canopy edge, easily overlooked. Mostly encountered making short flights from tree to tree, using a flap and glide action, and may indulge occasionally in aerial pursuit of insects. Faithful to particular sites, but always difficult to find.

VOICE Not vocal, but gives a long wheezy downward whistle, *chreeeee*, and various thin and high pitched contact calls.

STATUS AND DISTRIBUTION Rare to uncommon resident in WD and LRD, with records evenly scattered through all seasons. In Senegal, found almost entirely to the south of The Gambia, rare but widespread.

BREEDING Recorded breeding in The Gambia once in June and once nest building in late dry season March-April. Builds a small cup nest of stems, leaf skeletons and lichen usually in the fork of a tall tree, also on extreme terminal twigs, at a good height.

BULBULS
Pycnonotidae

Twelve species, three rare, six only in Senegal. Widespread family of small–medium birds, most with a rather upright perching posture; largely arboreal and frugivorous. Some species are notoriously difficult to identify but a knowledge of songs and calls is helpful.

LITTLE GREENBUL
Andropadus virens   Plate 35

Fr. Bulbul verdâtre

Other name: Little Green Bulbul

IDENTIFICATION 16-17cm. *A. v. erythropterus.* Sexes similar. A medium-small bulbul with uniform dull olive-brown head, breast and mantle, with only weak greenish tones, flight feathers and tail russet-brown; often scruffy. Bill horn (yellow-tipped in the immature), legs variable yellowish-flesh. Immature browner, fresher looking; may show prominent orange gape marks. **Similar species** Slender-billed Greenbul recorded in Casamance, but not The Gambia; has a pale grey underside, darker brown legs, and a finer bill. Immature Yellow-whiskered Greenbul (Casamance only) very similar, but shows dark malar streaks.

HABITS Found in forest and thicket habitats, tends to remain in thick cover. Typically seen alone or in pairs. Remains mainly in low vegetation, but may sing from high point. Thought to be strongly territorial in the breeding season, though aggregations of singing birds at this time also noted. Where found, easily heard, but requires patience to observe well.

VOICE Song may emanate from dense thickets throughout the year, but is most prominent June–December. Consistent and diagnostic phrase lasting 1-2 seconds, composed of chortling notes rising steadily up the scale, and culminating in a trademark rising slurred whistle. Repeated from the same perch over long periods. Also a repetitive chuckling, *ch-ch-ch-prrrro*.

STATUS AND DISTRIBUTION Threatened by habitat loss in The Gambia, though still a locally common resident within the few remaining suitable areas. All recent records confined to small and mainly diminishing forest and thicket patches in the coastal region south of the river, e.g. Abuko. In Senegal restricted to coastal Casamance.

BREEDING Seen gathering nest material May, most breeding commencing at the onset of the rains, June-July in The Gambia; cup-shaped nest in low vegetation.

SLENDER-BILLED GREENBUL
Andropadus gracilirostris   Plate 35

Fr. Bulbul à bec grêle

IDENTIFICATION 20-21cm. Sexes similar. Southern Senegal only. Medium-sized two-toned bulbul, olive above, grey below, and strictly arboreal.. Upperparts uniform olive-brown, dividing cleanly below eye and along line of wing from uniform pale grey underparts. Underwing-coverts pale yellow. **Similar species** The only typical bulbul with plain grey underparts, all others showing tones of olive or yellowish; the forest-skulking Western Nicator is also grey on the underparts, but is larger, with a fearsome bill and bold yellow spots on the olive-green upperparts.

HABITS Favours a range of forest and thickets from full primary forest to dense woodland patches in savanna and abandoned cultivation, often making use of thicket edges. Found alone, in pairs or family parties, typically spending long periods in the canopy. Generally unobtrusive.

VOICE Call is a drawn out soft whistle, *tseeeu*. The song, heard infrequently, is a phrase of three or four notes with the accent on the highest pitched second note *chip-cheeo-wu*.

STATUS AND DISTRIBUTION Not recorded in The Gambia, but was first collected in Casamance in 1981; since then seen regularly in the vicinity of Parc National de Basse Casamance, year-round.

BREEDING Nearest recorded breeding Sierra Leone.

YELLOW-WHISKERED GREENBUL
Andropadus latirostris   Plate 35

Fr.Bulbul à moustaches jaunes

IDENTIFICATION 17cm. *A. l. congener*. Sexes

similar. Plain dull olive-toned bulbul, similar to Little Greenbul but adult distinguished by narrow lemon yellow moustachial feathering and voice. Dull olive-grey on the upperparts, tail very dark grey-brown tending to black, underparts more or less uniform greyish olive, save for a pale yellowish patch in the middle of the belly. Legs orange or yellow-brown. Immature may show hint of pale whiskers. **Similar species** Adult distinctive; immature separable from Little Greenbul by dusky malar stripe.

HABITS A bird of dense primary and secondary forest that will also spread into thickets regenerating in deforested areas. Often alone, lurking in dense cover near the ground. Sings for prolonged periods, sometimes throughout the day. Ringing evidence indicates that it can be unusually nomadic for a forest bulbul.

VOICE Monotonous sequence of crisp throaty notes, becoming louder through each phrase, which often ends with a forceful sequence of three hard notes; repeated, sometimes for very long periods, from dense cover. Also a simple call, *chip*.

STATUS AND DISTRIBUTION Not recorded from The Gambia but present in the forests of southern Casamance, where more localised than the Little Greenbul. This population appears to be isolated from the next known area in Guinea.

BREEDING In low-density populations conventionally territorial and monogamous with the pair sharing parental duties. Nest typically at low level in dense cover.

YELLOW-THROATED LEAFLOVE
Chlorocichla flavicollis   Plate 35

Fr. Bulbul à gorge jaune

IDENTIFICATION 25cm. Sexes similar. A babbler-sized plain olive-brown bird, with a characteristic broad, cleanly defined bright yellow chin and throat patch. Yellow streaking on underside not a useful field mark. Bill dark grey, eye grey-buff, paler in the female, legs dark grey. **Similar species** Generally distinctive but at a distance in dense foliage may be confused with Brown Babbler; more uniform dark body tone should alert observer to look for throat patch. Simple Greenbul *C. simplex* (Fr. Bulbul modeste) occurs in Guinea-Bissau and has been seen once in southern Senegal. It is dark olive-brown above and dark buff below with a white chin and throat.

HABITS Sociable, usually in groups of 3-6 and highly vocal as parties move through shrubs and trees at varying heights, searching industriously for food. Associated with narrow patches of thicket habitat, often along watercourses or drainage line margins, but also in creeper-strewn forest thickets or dense patches alongside mangrove margins. Throat patch distended when calling. Generally shy and remains in cover when not on the wing; typically glimpsed only briefly. Best located by calls.

VOICE Noisy: bouts of calling when several birds repeat a series of hoarse squawks, creating an effect similar to babblers. Distinguishable by the slightly cleaner tone of the notes, which are more individually enunciated and less slurred than those of the babblers, phrases often open with a hesitant stutter; *eh-h-h, quoik.quoik - kwi - quoik-quoik-quoik* etc. Bold *chak-chuk-chuk-chik*.

STATUS AND DISTRIBUTION Resident throughout the country where there is reasonable broadleaf tree cover, including riparian woodlands, but not dry savanna woodlands. Particularly frequent near the coast and along the banks of the river in CRD. In Senegal widely distributed at the latitude of The Gambia and southward; not in the north.

BREEDING Seen carrying food and nesting material late April and May, but most nest records in the wet season, July to December. Builds a cup-shaped nest in leafy cover usually well off the ground, and pairs may raise brood with cooperative assistance from related adults.

SWAMP PALM GREENBUL
Thescelocichla leucopleura Plate 35

Fr. Bulbul à queue tachetée

Other name: Swamp Palm Bulbul, White-tailed Greenbul

IDENTIFICATION 25cm. Sexes similar. A large bulbul with uniquely patterned tail. Head greyish with dark olive-brown mantle and wings. Chin whitish; weak pale and dark striations on cheeks, grey upper breast, pale yellow-buff lower breast and belly. Long tail olive-brown, prominent white tips to outer rectrices distinctive. **Similar species** Tail markings distinctive once seen. Call, coming from thick cover similar to Leaflove but harsher.

HABITS Favours dense forest patches particularly associated with swamp palms near damp ground. Moves in parties, given to coordinated bouts of noisy calling in the style of babblers and some bulbuls.

VOICE Harsh, grating scolds, given by several birds at once; basic phrase a group of several notes moving tentatively up the scale in chattery, very nasal tone, *quek, quek...qui-quek-qui-quek, qui-quock*

and a harsh *kek-kek-kek-kek*, also very nasal.

STATUS AND DISTRIBUTION Extremely rare in The Gambia, with no recent records. Recorded from Abuko October 1968, and one other possible occurrence, based on voice recognition only, Pirang October 1986. Seen regularly in favoured locations in coastal Casamance.

BREEDING No information for Senegambia; nearest confirmed breeding Liberia.

LEAFLOVE
Pyrrhurus scandens Plate 35

Fr. Bulbul à queue rousse

IDENTIFICATION 23cm. Sexes similar. A large elongated pale pastel-toned bulbul, shading from pale grey head to pale fawn-brown mantle and wings. Distinctive pale rufous-brown tail. On underside white throat merges to uniform pale buff underparts. Bill dark brown or bluish-grey upper mandible, pale grey on lower; whitish cutting edges and tip; eye variable reddish to pale grey or yellow; legs blue-grey to olive. **Similar species** Pale female Northern Puffback has similar pattern, but is much smaller, lacks the long rufous tinted tail, and usually accompanies the black and white male.

HABITS Favours dense habitats and thickets, usually shy and elusive, most likely to be located when pairs or small family parties move about noisily, mainly in upper and mid levels. Frequently indulges in calling bouts with accompanying tail fanning and posturing.

VOICE Simultaneous calling by two or more birds, similar in tone and quality to the Yellow-throated Leaflove (see that species for distinction from babbler calls). A series of clucks, rising slightly in pitch and speed *..kyuk-kyok..kyuk-kyuk-kyuk*. A diagnostic sharper *kyik-kyok...kyik-kyok*, and harsher sounds, may break out in syncopated rhythm over the background squawking and is a useful recognition point.

STATUS AND DISTRIBUTION A rare resident in The Gambia, with all records from coastal forest and thickets. Occasionally reported in Abuko, but recently seen regularly only in two very small populations at unprotected locations in NBD and WD; seriously threatened by habitat loss. In Senegal restricted to coastal Casamance.

BREEDING No breeding information from The Gambia, but a wet season breeder in Casamance.

WHITE-THROATED GREENBUL
Phyllastrephus albigularis   Plate 35

Fr. Bulbul à gorge blanche

IDENTIFICATION 16cm. Sexes similar. A slim agile greenbul combining olive and yellow body plumage with a reddish tail. Dark grey loosely feathered crown. Grey lores and ear-coverts with hint of pale ring around eye. Upperparts olive, greener on mantle; rufous-olive tail, with greenish edges to outer feathers. White chin and throat, breast grey-olive finely streaked pale yellow, merging into pale yellow belly and undertail; underwing pale yellow. Bill characteristically straight and slim, variably black, dark grey or brown with yellowish cutting edges and tip. Immature similar to adult, greener on head, extensive yellow on bill. **Similar species** Slimmer-billed than other bulbuls. Smaller and darker than Leaflove. Red-tailed Greenbul is larger, bright yellow below, bare skin over eye.

HABITS In pairs or small family parties in primary and secondary forest, uses tangled areas at forest edge. Usually stays at low levels and moves easily and swiftly through tangled vegetation.

VOICE A busy nasal song phrase, rising in intensity and pitch to a central peak before diminishing. Various additional calls, including a harsh *jerr-it* and a stuttering chatter, reminiscent of Little Greenbul chatter.

STATUS AND DISTRIBUTION Not recorded in The Gambia; infrequently encountered resident in forests of south Casamance; first observed there 1980.

BREEDING Appears to breed mainly in the rainy season in West Africa, constructing a nest low in the vegetation. Nearest confirmed breeding Ivory Coast, but resident in several countries in between.

GREY-HEADED BRISTLEBILL
Bleda canicapilla   Plate 35

Fr. Bulbul moustac à tête grise

IDENTIFICATION 20-21cm. Sexes similar. *B. c. moreli*. A skulking, low-flying bulbul, green above, lemon yellow below, with a grey head and stout wedge-shaped grey bill. Head and cheeks pale grey, cleanly separated at collar from greenish back, wings and tail. Neat creamy patch on chin and throat seen in good views; belly lemon yellow. Pale yellow tips to outer tail feathers useful when seen, either in flight or when tail is fanned, but often hidden. Immature shows pale chestnut patch on

the carpal. **Similar species** Senegambian subspecies pale, with thicker, shorter bill than nominate. Full feathering around eye eliminates Red-tailed Greenbul (Casamance only) and extralimital Green-tailed Bristlebill *Bleda eximia* (Fr. Bulbul moustac à queue verte), nearest records Guinea, which both show bare blue orbital skin, though of similar coloration. The throat patch is smooth, creamy and neat, never puffed up and white like Red-tailed Greenbul.

HABITS Confined to thickets and forest, usually at low level. Often glimpsed flying swiftly through dense cover to a low perch, where it sits quietly or drops to the leaf litter to forage. May scold noisily if disturbed. Elusive, but not necessarily shy once located.

VOICE Characteristic song of clear, mellow whistling tones given from deep cover opens with 3-4 hesitant pure notes rising in pitch, accelerating to a more fluid descending sequence progressing evenly back down the scale. A loud alarm *pew-pew-pew-pew* and a rapid chattering scold is also used, reminiscent of Little Greenbul.

STATUS AND DISTRIBUTION A resident confined to a few small forest and thicket patches in coastal areas and southern borders of WD in The Gambia. Locally frequent within such places, such as Abuko, but threatened by habitat loss. In Senegal confined to coastal Casamance, with a few sightings just to the north of The Gambia in Delta du Saloum.

BREEDING No conclusive evidence, though seen carrying food in June. Young birds observed in Casamance in January.

YELLOW-BEARDED GREENBUL
Criniger olivaceus   Not illustrated

Fr. Bulbul à barbe jaune

IDENTIFICATION 18cm. A medium-sized, mainly uniform olive-green forest bulbul with a bright pale yellow throat patch and rufous-tinted tail. Much smaller than Yellow-throated Leaflove, and a forest bird.

STATUS AND DISTRIBUTION Has been recorded as present in southern Casamance on the basis of single 19th century specimen, which may have been erroneously labelled. All recent records are from Guinea eastwards to Ghana.

RED-TAILED GREENBUL
Criniger calurus **Plate 35**

Fr. Bulbul huppé à barbe blanche

Other name: White-bearded Bulbul

IDENTIFICATION 19cm. Sexes similar. *C. c. verreauxi.* Local race not strongly red-tailed. A forest bulbul, olive above, bright yellow and olive below with prominent white throat patch and grey head. Grey cheeks finely streaked with white; ring of bare blue-grey skin around eye; elongate bristles project from nape. Upperparts dark olive, tail mainly olive, with dark red stripe down the centre of each feather. Throat white, often puffed out, contrasts with head and mainly bright yellow centre to belly, olive tones on breast and flanks. **Similar species** Separated from other bulbuls by bold white throat, bright yellow belly and bare skin around eye; see Grey-headed Bristlcbill.

HABITS Favours primary and secondary forest, also makes use of large gallery forest patches at the edge of the savanna, generally preferring to stay in the lower vegetation levels. Typically in family parties, sometimes quite large, (up to 12 individuals). Noted for maintaining close proximity to wood-peckers when participating in mixed species foraging groups.

VOICE A vocal species, although not particularly noisy. Commonly gives a whistled call, *preep* or a quick *pu-ee-u*, with a rise in pitch and emphasis on the middle syllable. The typical song is a 4-5 note phrase on a throaty rising sequence, *chup-chup-chwirulup,* or *chit-chiro-chiro.*

STATUS AND DISTRIBUTION Not recorded in The Gambia but regular, though infrequently seen, in the forest reserves of southern Casamance.

BREEDING Nearest confirmed breeding Sierra Leone; breeding season not definitely established.

COMMON BULBUL
Pycnonotus barbatus **Plate 35**

Fr. Bulbul commun

IDENTIFICATION 18cm. Sexes similar. *P. b. inornatus.* An ubiquitous, sombre mousey-grey crested bulbul. Confiding manner, always on the move and always calling. Dark grey, nearly black head, with crown feathers giving a characteristic domed head shape. All-brown body and tail, with dirty grey underparts and whitish vent. Heavy bill. Immature paler, with poorly developed crest giving fluffy cropped look; gape creamy-white. **Similar species** Distinctive, but flycatcher sallying habit is potential pitfall and black cap may lead to possible confusion with Black-capped Babbler.

HABITS A prominent and active species in a wide range of habitats from town and garden to field, woodland, forest and mangrove. Often present in numbers near fruiting trees. Displays with arched wings and fanned tail; sways side to side.

VOICE Many calls and phrases, often loud, mainly involving a confident liquid chortle of full throated notes; a regular background sound in many areas, and usually the first bird to call at dawn, when extremely emphatic notes characteristic around farms and hotel gardens. Variable, but choppy character summarised *doctor-quick doctor-quick be-quick be-quick* sometimes followed by a more complex group, rattling up the scale in cheerful confident style; *whip-chip-chop - whip-chip-chop, chirrop- chirrip* etc.

STATUS AND DISTRIBUTION Common to abundant throughout the country. Common throughout Senegal with the exception of the most arid parts of the north-east.

BREEDING Breeds in all seasons, increased activity in the wet season. Nest a flimsy shallow cup, in a small shrub or tree.

THRUSHES AND CHATS

Turdidae

Twenty five species, seven rare in The Gambia, eight confined to Senegal. Includes Song Thrush *Turdus philomelos* (Fr. Grive musicienne), known in the region from a single vagrant in northern Senegal in November 1958. Small to medium-small, generally upright, often relatively long-legged birds which though primarily arboreal feed mostly on the ground. Several are renowned for their fine songs. Some species show strong sexual dimorphism, but in many others the sexes are similar.

COMMON NIGHTINGALE
Luscinia megarhynchos **Plate 37**

Fr. Rossignol philomèle

IDENTIFICATION 16.5cm. Sexes similar. Small,

plain brown thrush, with rufous tail, lurking in dense thickets. Uniform brown upperparts, with bright chestnut uppertail-coverts and plain rufous-chestnut tail. Slightly paler on the lores, with a pale eye-ring and some faint pale streaks on the ear-coverts, otherwise pale creamy on throat and belly, with grey-brown sides to the neck merging over the breast. Undertail-coverts pale buff. **Similar species** Bulkier than female Common Redstart, with plain rufous tail, skulking habits and usually first detected by voice.

HABITS Favours dense thickets, regenerating old cultivation, lush vegetation alongside water courses, and forest edges; also dense thorny thickets growing close to the coastline. Remains in dense cover nearly all the time, usually solitary. Makes short dashes between cover at low level, when rufous hind parts obvious. Moves with long hops; stands with head and tail high, wings drooping.

VOICE Often gives testy grating *krrrr* from dense cover when disturbed. Regularly sings in The Gambia, rich, throaty voice, characterised by full-toned, emphatic repeating notes given variously with accelerating crescendos and abrupt changes of pace, usually mixed with harsher sounds. In Africa seldom sings at night.

STATUS AND DISTRIBUTION A locally frequent Palearctic winter visitor and passage migrant, first appearing in September, and most prominent around November and December. Less obvious on spring migration, though recorded through to March. Appears in suitable habitat throughout, most commonly recorded at the coast and in URD. Records from Senegal suggest winter distribution influenced by coastline and major river valleys.

BLUETHROAT
Luscinia svecica   Plate 37

Fr. Gorgebleue

IDENTIFICATION 14cm. Seasonal and racial variation, and sexes differ. Always secretive. All plumages share diagnostic combination of olive-brown upperparts, strong, buffy-white supercilium, basal two-thirds of outer tail feathers rufous-chestnut. Wintering males of the most commonly reported race in West Africa, *L. s. cyanecula*, have whitish throat and neck, surrounded by a gorget of dappled black, linking the blue malar regions across the breast; the gorget is variably suffused with blue and chestnut, which may in this plumage encroach onto central spot area. Lower breast and belly grey-white, merging to buffish undertail. Moult to summer plumage occurs February to April, when *L. s. cyanecula* male develops all blue throat with a clean white central spot (chestnut in nominate form) bordered below by narrow bands of black, white and chestnut. Females are whitish on the neck and throat, with a dappled necklace of blackish feathers across the breast linking the black malar stripes, which may occasionally show

limited blue and chestnut suffusion, especially in breeding plumage. First-winter birds resemble female, but retain buff spotting on greater coverts. **Similar species** Tail pattern is diagnostic.

HABITS Wintering birds strongly associate with damp places. Spends most of the time near or on the ground, moving easily through very dense vegetation, but also coming out to feed in the open, with upright stance, drooped wings and cocked narrow tail, frequently bobbed.

VOICE An elaborate songster, more restrained on migration. Produces 10-20 second phrases characterised by a hesitant start and tinkling and ringing sounds. Also a throaty *tchick*, with tail flick.

STATUS AND DISTRIBUTION Rare-uncommon winter visitor, with a few records from the coast and URD, February-March, and in November. *L. s. cyanecula* annual in northern Senegal along the Senegal River basin, mainly September-March.

SNOWY-CROWNED ROBIN-CHAT
Cossypha niveicapilla   Plate 37

Fr. Petit Cossyphe à tête blanche

IDENTIFICATION 20cm. Sexes similar. The smaller of two boldly marked black and orange robin-chats found in The Gambia. Top of the crown sports a rectangular patch of pure white, offset by jet black front and sides of the head. Upperparts slaty-grey; rufous-orange rump and tail, save for brownish-black central rectrices and outer webs to outer feathers. Wings brown, primaries edged blue-grey. Entire underparts clean uniform rufous-orange, extending to form a complete rufous collar across hind neck. Immature a washed out echo of this pattern, heavily spotted with rufous on the upperparts and scalloped with dark brown underneath. **Similar species** From White-crowned Robin-Chat by smaller size, squared shape of white crown patch; contrast between black sides to head and grey mantle, orange collar and voice.

HABITS A bird of dense thickets in forest patches and regenerating bush, favouring even heavier vegetation than White-crowned. Occasionally in mangroves. Usually in pairs. Remains low, moving quietly through vegetation and feeding in ground litter. Elusive, but cautiously curious, often one of the first species to be seen by a quiet observer waiting patiently near a thicket. Frequent soft calls best indicator of presence.

VOICE Persistent, soft, monotone whistle, *tooo*, given at all times of the day. Sometimes extended to two notes, the second of higher pitch, though distinct, not slurred like White-crowned. Otherwise a great songster, especially towards sunset, with sustained and variable fluty warblings. May commence with repeated base phrase, rising and falling *too-too-li-oo*, but an accomplished mimic, freely integrating calls and phrases from a wide variety of sources.

STATUS AND DISTRIBUTION Frequent resident breeder in patches of dense habitat near the coast, becoming uncommon further inland, though sporadic individuals seen along the river banks through CRD and URD. Spreads a little way to the north of The Gambia in Senegal, though commoner to the south, widespread across the rest of West Africa.

BREEDING Builds a cup-shaped nest at low level in the wet season months in The Gambia.

WHITE-CROWNED ROBIN-CHAT
Cossypha albicapilla   **Plate 37**

Fr. Grand Cossyphe à tête blanche

IDENTIFICATION 27cm. Sexes similar. Robust orange and brown bird, the larger of the two robin-chats. Capped with silvery white, finely scalloped with black when fresh, from the base of the bill, over the width of the crown, to neat point on the nape. Lores, sides of the face down to line of malar stripe brownish-black, extending over the neck, back, wings and central tail feathers. Rump dark rufous, outer tail feathers bright rufous-orange, save black-brown outermost web. Entire underparts are clean rufous-orange, except narrowly black chin. Immature birds are similar, but paler on the upperparts, with some rufous spotting on the wing-coverts. **Similar species** From Snowy-crowned by more extensive, scaly white cap, absence of orange collar, less contrast between cheeks and upperparts, much larger size; see also voice.

HABITS Strongly associated with forest thickets and dense vegetation where there are trees, including areas of good canopy cover with more open interior beneath, and river bank vegetation where this is thick. Likes thickly planted gardens, including hotels, where it becomes tame, otherwise elusive. Usually in pairs, sometimes small parties; at such times may be very active and prominent, birds chasing each other rapidly through and around bushes and trees, with much flashing of tail colours and raucous cries. Otherwise mostly quiet, flitting with characteristic, bouncy, gather-and-hop action.

VOICE Characteristic repeated, piercing, upward slurred two-note whistle *tooo-eeee* separates from Snowy-crowned Robin Chat. Elegant song of melodic throaty whistles and some grating notes, sustained, but not as varied as that of Snowy-crowned;

mimics other species. Rising *tsueee* followed by quicker *tsu, tsu-tsu-tsu-tsu-tsu* on a monotone, all in one phrase. Loud and raucous cries when excited, especially rasping rattles, *krrr-krrrr-krrr* and nasal monotone *sweeee....sweeee*.

STATUS AND DISTRIBUTION Overall infrequent because of habitat restriction, but within this constraint frequent to locally common breeding resident throughout, being found in thickets close to the river bank, especially CRD and URD. The Gambia River marks the northern limit of the species' range; elsewhere found in an unusually narrow belt in the Guinea savanna eastward to central Africa.

BREEDING Nests during wet season months, June-September.

FIRE-CRESTED ALETHE
Alethe diademata   **Plate 37**

Fr. Alèthe à huppe rousse

Other name: White-tailed Alethe

IDENTIFICATION 18cm. Sexes similar. An elusive forest floor species, best detected by distinctive song. Adult dull brown above, save for a broad, partially erectile, orange crown. Wings brown, tail blackish, marked by large white spots on the three outer feathers. Sides of face grey, underparts whitish from throat to belly, tending to greyer on the flanks. Legs slaty-blue to grey. Juvenile blackish on the upperparts, streaked with orange on the crown, spotted with large orange blotches on the mantle; throat pale orange with a band of black and orange streaking and scalloping, merging to a whitish belly. **Similar species** White tail-spots distinguish immediately from smaller *Illadopsis* species that share these habitats.

HABITS Favours dense habitats, including regenerating forests and forest-savanna mosaic, as well as primary forest. Territorial, mostly solitary except when following ant columns and joining mixed species feeding parties, which it does frequently. Perches low, flicking wings and tail. Aggressive towards others of its species.

VOICE Song a remarkable mellow whistle of three clear notes, given slowly in a rising cadence - *toooo-leeee-eeeeii*; tends to slur up in pitch on the last note. May raise interval between each note by an erratic and unmusical semi-tone in successive sequences. Also a complex subsong; when agitated a harsh *chit-chit-chit*.

STATUS AND DISTRIBUTION Not recorded in The Gambia, but a common and widespread resident in south-west Casamance.

BREEDING Simple cup-shaped nest on or under a fallen log or stump, always close to the ground. In Liberia nests June-September.

RUFOUS SCRUB ROBIN
Cercotrichas galactotes Plate 37

Fr. Agrobate roux

Other name: Rufous Bush Robin

Synonym: *Erythropygia galactotes*

IDENTIFICATION 15cm. Sexes similar. *E. g. minor* resident, augmented by migrant birds of the nominate European form. Active but secretive medium-sized chat; black and white tail-tip markings catch the eye against a rufous tail and rump. Resident birds are pinkish-brown on the upperparts from crown to mantle; with creamy-white supercilium, narrow black eyestripe, and narrow dark malar stripe. Bright foxy-rufous from rump to fan-shaped tail. Wings brown. Underparts creamy-white, with faint rufous-brown wash on breast, undertail buff. Immature similar to adult. Migrant birds appreciably larger with darker rufous-brown upperparts and broader black tail-bar; these differences are not easily discerned in the field. **Similar species** Skulking birds if glimpsed recall Nightingale, but movements quite different, head and tail markings obvious.

HABITS Favours dry bush, woodland and edges of farmland. Usually low in shrubs or on the ground. Fans and flirts mobile tail, particularly raising and lowering by stages with rather mechanical precision from horizontal to arched sharply over back; also droops wings and intermittently may flick them forward. Usually singly or in pairs; males sing from exposed tops of bushes or small trees.

VOICE Thin repetitive series of twitters and more liquid whistles, sustained with little variation in pitch. Common *tek tek* call.

STATUS AND DISTRIBUTION An uncommon to locally frequent species north of the river, mainly CRD and URD; occasional south of the river. Observed January-July. Singing males, conforming to description of *E. g. minor*, present in the late dry season. This race is widespread in central Senegal, and *C. g. galactotes* seasonally present in northern Senegal.

BREEDING Untidy low cup nest in dense vegetation, April to June. Not proven in The Gambia, but probable north of the river.

BLACK SCRUB ROBIN
Cercotrichas podobe Plate 37

Fr. Merle podobé

Other name: Black Bush Robin

Synonym: *Erythropygia podobe*

IDENTIFICATION 18cm. Sexes similar. Distinctive slim all-black bird, elongated for a chat, with white tips to very long tail which is waved and fanned conspicuously. Entirely matt black, marked only by bright white tips to the outer tail feathers and narrow white chevrons on the tips of the undertail-coverts. In flight, dark brown inner webs to the inner part of the flight feathers create noticeable rufous wing panel. Legs rather long and black. Immature generally duller and smaller with white marks in tail. **Similar species** Unmistakable.

HABITS A bird of dry scrub areas of the Sahel and desert, associated with low dense vegetation and small trees, spending much of its time at low level in cover or on the ground. Usually singly or in pairs, territorial and generally sedentary. Active, much in the habit of waving its long tail up and down to an exaggerated degree, displaying the white markings of the underside. Characteristic flight with slight arch to back and head-high posture.

VOICE Song given from top of low bush includes a short phrase with first and third note on a falling pitch, separated by a harsh trill. Also a sustained quieter babbling and when disturbed a testy *chew-ch-ch-ch-chew-chew*.

STATUS AND DISTRIBUTION Not recorded in The Gambia, but widespread over most of Senegal not far to the north.

BREEDING Breeds February-July in Senegal, loose cup nest in the fork of a dense thorn bush or palm, usually only 1-2m from the ground.

COMMON REDSTART
Phoenicurus phoenicurus Plate 37

Fr. Rougequeue à front blanc

Other names: European Redstart, Redstart

IDENTIFICATION 14cm. Sexes differ, also seasonal variation. A small chat, conspicuous in flight with distinctive deft movements, especially rapid shivering of rufous tail. Breeding male blue-grey over crown, mantle and back, bright orange-rufous rump and tail, save two brown central feathers; cleanly defined black face mask offset by wedge of pure white from the forehead to above the eye. Underparts bright orange rufous. In first-winter this pattern initially obscured by brown feathers, but weak echo of mask and white forecrown usually discernible. Female plain light brown above, pale buff ring around the eye; underparts whitish on the throat, merging to pale buff breast and belly; rufous tail similar to male. Immature resembles female, dappled with buff spots. **Similar species** See Familiar Chat. Black Redstart *Phoenicurus ochruros* (Fr. Rougequeue noir) is a vagrant to northern Senegal (three records in Djoudj, December-April). Males are grey-black with a red tail, females similar to Common Redstart but plumage is greyer brown.

HABITS Found in a wide variety of habitats on migration, notably the shrub and tree parkland on vegetated coastal sand dunes; generally associated with leafy vegetation, regularly using interior perches in shrubs and small trees. Flies out from concealed perch to procure insect food on the ground, occasionally in the air, and moves with agility, sweeping back into cover, leaving an impression of the flashing orange tail.

VOICE Uses slightly tetchy *tsip* call frequently, rarely sings.

STATUS AND DISTRIBUTION Frequent to locally common migrant to all areas, seen most often from December to April, particularly in LRD and CRD. Occasional records in September-October. Recorded throughout Senegal.

STONECHAT
Saxicola torquata　　　Not illustrated

Fr. Traquet pâtre

IDENTIFCATION 12cm. Slightly dumpier than Whinchat. Sexes differ. Male has black head and throat, reddish breast, white half-collar and wing patches, and black tail. Duller in non-breeding plumage but usually retaining a partial head mask. Female similar to female Whinchat but lacks prominent supercilium.

STATUS AND DISTRIBUTION Not recorded in The Gambia, but a few pairs of *S. t. moptana* breed in northern Senegal. Some winter records may relate to vagrant *S. t. rubicola* from the Palearctic.

WHINCHAT
Saxicola rubetra　　　**Plate 36**

Fr. Traquet tarier

IDENTIFICATION 12.5cm. Sexes differ, also seasonal variation. Small streaky brown chat, white markings create variegated impression, especially

in flight. Breeding male (January to June) brown, heavily streaked on crown and mantle; blackish sides to the head defined by broad whitish supercilium above and white border below. Wings dark brown; short white bar on the coverts and a white dash at the edge of the folded wing. Tail dark brown, two white patches either side of base. Underparts plain orange-tinted buff. Adult female a paler echo of this pattern, retaining a broad buff supercilium, but less dark on the sides of the head; white markings on the wings nearly absent, and on the tail much less prominent. Wintering birds of both sexes develop buffish overall tone, pale throat and underparts, with well-defined dark centres to mantle feathers creating spotted effect on back; buff supercilium prominent, white wing and tail patches as for female. **Similar species** See Stonechat.

HABITS Typically found in open bush. Perches prominently and upright on low but exposed vantage points; drops suddenly to the ground, pouncing on small insect prey. Often bobs and flicks wings on perch. Usually singly, though forms larger groups when migrating.

VOICE Usually quiet in winter but may give crisp *tuc-tuc* call.

STATUS AND DISTRIBUTION Winter visitor and passage migrant, present September-April; peak numbers January-February. Recorded throughout The Gambia; generally uncommon but may be locally frequent to common for short periods. Mainly coastal and near water in Senegal.

NORTHERN WHEATEAR
Oenanthe oenanthe　　　**Plate 36**

Fr. Traquet motteux

Other names: Common Wheatear, Wheatear

IDENTIFICATION 15cm. Sexes differ. Three races described from Senegambia, primarily *O. o. leucorhoa*, also *O. o. oenanthe* and *O. o. seebohmi*. Alert chat of open fields, readily identified as a wheatear in all plumages by white rump and tail with black T-bar, (note tail pattern in all wheatears in flight). Breeding male (January-March), grey over crown and mantle, black eye mask highlighted by white line above and below, dark wings, black and white tail, warm orange-buff suffusion over breast; female similar but browner, weaker mask, brown-fringed flight feathers. Non-breeding male (September-December): grey-brown upperparts, diffuse ghost of mask with buff supercilium, may show darker lores, wings narrowly edged with buff; female and first-winter brown above, buffy supercilium, wings fringed with buff, buff breast. **Similar species**

Greenland race *O. o. leucorhoa* large, darker, more uniform buff-orange below; breeding *O. o. seebohmi* black-throated, may show speckling on throat in non-breeding plumage. First-winter and non-breeding female of all races from Isabelline by slimmer build, less heavy bill, primaries project longer than exposed tertials; tail tip near ground when perching up; more contrast between upperparts and underparts; T-bar on tail less deep, supercilium strongest behind eye; from same plumage of Desert and Black-eared by tail pattern.

HABITS Favours relatively open habitats, notably harvested fields and fallow with emergent low shrub cover; perches, often not particularly high, watching around; bobbing action with short tail flick.

VOICE Call hollow *tuc*, sometimes *tuc weet*; rapid creeky warbling song.

STATUS AND DISTRIBUTION Winter visitor to all parts of The Gambia in variable numbers, tending to be less common in recent years. Seen September-March, peaking December to February. Found throughout Senegal, but not frequent south of The Gambia. Senegambia noted as a major wintering area for Greenland race *O. o. leucorhoa*.

BLACK-EARED WHEATEAR
Oenanthe hispanica   Plate 36

Fr. Traquet oreillard

IDENTIFICATION 14.5cm. Sexes differ. Relatively slim wheatear with black and white tail in T-bar pattern; males seen with two differing face patterns. Breeding (January-June): male sandy-buff crown, mantle and breast, contrasting black wings and black mask through eye, which is continuous with full black throat in some, otherwise throat whitish. Female similar, more brown-fringed wings. Non-breeding (July-December): male reduced mask and browner fringing on dark wings. Female and first-winter paler brown above, dark wings fringed buff, no mask. **Similar species** Breeding male distinguished from Desert by tail pattern, black of wings extending onto scapulars and gap between wings and face mask. Non-breeding and first-year extremely similar to same stages of Northern and Desert; slighter with less grey-toned upperparts than Northern; good view of tail in flight, not always easily obtained, is best distinction.

HABITS Usually solitary. Characteristically favours areas with more trees than other wheatears and selects higher vantage points, but otherwise similar to other wheatears in general behaviour.

VOICE Does sing on migration; quiet scratchy song

mixed with whistles and a sharp *tack* call note.

STATUS AND DISTRIBUTION Rare in The Gambia with three records in last thirty years. Winter visitor to northern Senegal, September-April, but less frequent than Northern Wheatear.

DESERT WHEATEAR
Oenanthe deserti   Plate 36

Fr. Traquet du désert

IDENTIFICATION 14-15cm. Sexes differ. *O. d. homochroa*. A black and brown wheatear; distinctive white rump and all-black tail in all plumages. Breeding (January-June): male distinctive, whitish supercilium above black mask and throat, linking to black wing at side of neck, pink wash to brown body; female similar but duller, more fringing on flight feathers. Non-breeding and first winter (September-December): loses face mask, wings buff-fringed but retains tail pattern. **Similar species** Non-breeding and first-winter resembles same stages of Northern and Black-eared when perched but tail pattern diagnostic. Blackstart lacks white rump, with black tail coverts as well as tail.

HABITS Favours arid open habitats and found in true desert in much of its range. Usually solitary, flicks wings, wags tail more rapidly and persistently up and down than others. Characteristically dodges behind objects to get out of sight.

VOICE Not particularly vocal; sharp *tuc* call and a slow descending *chee-you*.

STATUS AND DISTRIBUTION Three sight records from the coastal strip area in WD in 1970s and 1980s, one of which was rejected by Gore (1990), others remain unsubstantiated. Rare winter visitor in Senegal River delta December-February. Limited to the Sahel and desert margins.

ISABELLINE WHEATEAR
Oenanthe isabellina   Plate 36

Fr. Traquet isabelle

IDENTIFICATION 16cm. Sexes similar. Relatively bulky, plain, buff-coloured wheatear. White tail base with broad black terminal T-bar. Little seasonal plumage variation, but beware general similarity to non-breeding plumages of other wheatears. Uniform sandy-brown upperparts, pale buff breast, whiter on throat and belly, short buffy-white supercilium (less prominent in winter), dark lores sometimes noticeable. Folded inner wing does not contrast greatly with mantle. **Similar species** Locally frequent Greenland race of Northern Wheatear approaches Isabelline in size but Isabelline heavier-billed, paler overall, especially below and on throat, less contrast in wing, black alula stands out against coverts, gives impression of less white on tail; Isabelline stands more upright, long-legged and dumpy bodied, deep measured bobbing action and looks shorter tailed. Most of these points also distinguish from non-breeding Desert and Black-eared which also have more distinctive tail patterns.

HABITS On migration usually solitary. Favours open habitats with low shrubs. One of the more vocal wheatears in non-breeding range, with some individuals singing regularly; hostile and dominant over congeners.

VOICE Call an emphatic *chack* also a descending whistle *wheew*. Song a jumbled phrase of whistles and chacking notes with nasal twanging.

STATUS AND DISTRIBUTION Very rare winter straggler to The Gambia. One accepted record from the coastal region near Yundum 1984; possibly under-recorded. Regular winter visitor to northern Senegal from October, with an isolated record from the south-east in March.

FAMILIAR CHAT
Cercomela familiaris   Plate 36

Fr. Traquet de roche à queue rousse

Other name: Red-tailed Chat

IDENTIFICATION 14cm. Sexes similar. A plain brown chat of rocky habitats, distinguished by rufous rump and tail. Head and upperparts brown, marginally darker on the wings. Rump and tail rufous-orange, save for a narrow dark terminal bar and two central tail feathers dark brown throughout their length. Greyish-brown below, slightly paler on throat and belly, generally uniform. Immature with buff spotting over the upper body, otherwise resembles adult. **Similar species** Rump and tail not so bright as that of female Common Redstart which also lacks terminal tail bar, shivers tail repeatedly and does not flick wings in style of Familiar Chat.

HABITS Favours woodland habitats, strongly associated with rocky outcrops or boulder-strewn patches along watercourses. In pairs or small parties. Often raises and lowers tail on landing, accompanied by wing flicking.

VOICE Song a quiet series of whistles and harsher notes *tsweep-tsweep-tsweep-cher-cher-tsweep* etc. and harsher chat-like alarm notes, *chak chak* or *cher-cher*.

STATUS AND DISTRIBUTION Not recorded in

The Gambia but thought to be a scarce resident in extreme south-east Senegal; most observations in the dry season.

BREEDING Cavity nester in boulder-strewn areas often establishing a platform of small pebbles on which to construct a more conventionally woven cup nest.

BLACKSTART
Cercomela melanura   Plate 36

Fr. Traquet à queue noire

Other name: Black-tailed Rock-Chat

IDENTIFICATION 14cm. *C. m. ultima*. Sexes similar. A slim plain chat with uniform grey-brown upperparts fading indistinctly to whitish on the lower breast and belly. Narrow dark area through lores; distinguished at once by uniform matt black tail coverts and tail. Immature resembles adult. **Similar species** First-winter Desert Wheatear shows white rump, greater contrast between wings and mantle. Lead-coloured Flycatcher smaller, favours woodland, has bright white outer margins to prominently flared black tail.

HABITS A desert bird of arid rocky hillsides, usually with some low scrub. Perches prominently. Given to swift but deliberate and elegant simultaneous fanning of wings and tail, especially when on the ground. Usually alone or in pairs in the breeding season.

VOICE A thin querulous song of rapid notes making a short phrase, repeated infrequently. Call a throaty *chirrrip*, rising in pitch.

STATUS AND DISTRIBUTION Vagrant. Included on the basis of a single record of three-four birds together, photographed near Fatoto 1974. Not recorded in Senegal and normally a bird of desert; closest populations in rocky parts of the southern Sahara in Mali and Niger.

NORTHERN ANTEATER CHAT
Myrmecocichla aethiops   Plate 36

Fr. Traquet-fourmilier brun du nord

Other name: Ant Chat

IDENTIFICATION 20cm. Sexes similar, male marginally larger. A dark, upright chat, looking black at distance, immediately distinguished in flight by the presence of two large white panels in the primaries. At close range brown colouring seen to be dappled by tobacco-brown fringes of feathers about the forehead, chin and throat; otherwise uniform dark brown over rest of body, folded wings and tail. In flight extensive white on inner webs of primaries create large white patches visible above and below the wing. Immature darker brown than adult. **Similar species** Distinctive short-tailed stocky shape, obviously bulkier and larger than female White-fronted Black Chat, and usually seen in much more open habitats.

HABITS A bird of open habitats, associated with areas of termite mounds, quarries and wells, often around periphery of villages. Usually in pairs or family groups, which may defend local territories using joint displays; said by one authority to engage in 'what looks and sounds like a swearing match'. Perches prominently on posts, termite mounds or small shrubs.

VOICE Call a clear rising whistle. Song a series of piping notes, mixed with harsher sounds; may sway and droop wings during delivery.

STATUS AND DISTRIBUTION Resident breeder through NBD and CRD on the north bank of the river. Discontinuously scattered so not particularly common, though readily found where it occurs. Sporadic south of the river in WD and CRD.

BREEDING Digs a deep hole in soft earth, using old wells, quarries and termitaria only in the wet season in The Gambia. Young birds July-September.

SOOTY CHAT
Myrmecocichla nigra   **Not illustrated**

Fr. Traquet-fourmilier noir

IDENTIFICATION 16.5cm. Distinctive all-dark chat, glossy black in the male, distinguished by a bright white shoulder patch; plain dark brown female. Close to White-fronted Black Chat in size, lacks the Northern Anteater Chat white wing panel. Found in lightly wooded savannas.

STATUS AND DISTRIBUTION Reported twice in south-eastern Senegal, the only records west of Nigeria, and far outside its normal range.

WHITE-FRONTED BLACK CHAT
Myrmecocichla albifrons   **Plate 36**

Fr. Traquet noir à front blanc

IDENTIFICATION 15cm. Sexes differ slightly. Medium-small chat, characteristically seen silhouetted against the sky on projecting branch of a dead tree. Both sexes are uniform slate-black, save for a bright white blaze on the forehead to forecrown of the male, very occasionally partially developed on the female as well. Wing feathers slightly browner tone. White underside of flight feathers may create a slight silver-grey sheen in the spread wing, but not an important field mark. Immature very dark black-brown with pale mottling over the body. **Similar species** White forehead not always obvious. Smaller than any of the other all-dark chats of the region and associated with denser woodlands; shorter, square-ended tail, more horizontal stance and slimmer bill distinguish from Northern Black Flycatcher and Fork-tailed Drongo.

HABITS A bird of savanna woodland and the periphery of cultivation where there are still reasonable numbers of trees and shrubs. Usually in pairs, sometimes singly, spends considerable time perched on conspicuous vantage points. Flicks tail intermittently and male said to indulge in butterfly-like display flight.

VOICE Usual note a monosyllabic *tweet*. Song a short tumbling phrase of cleanly whistled short notes, not loud, but carries well through the woodland.

STATUS AND DISTRIBUTION An uncommon to locally frequent resident breeder in wooded areas throughout the country, most readily encountered in the larger forest parks and extensive woodlands of LRD. No recent records from NBD. Thinly spread through central and south-eastern Senegal.

BREEDING Builds a cup nest in a crevice under boulders. Seen displaying at the beginning of the rains in The Gambia, but elsewhere a dry season breeder.

MOCKING CLIFF-CHAT
Myrmecocichla cinnamomeiventris **Plate 36**

Fr. Traquet de roches à ventre roux

Other name: Mocking Chat

IDENTIFICATION 20.5m. *M. c. bambarae*. Sexes

differ. Large slim chat of rocky hillsides and gorges, boldly coloured in black, white and rufous-orange. Dull uniform black on crown, upperparts and tail; narrow rufous-orange rump patch. Small wing-covert patches white, mixed with black. Chin to breast black, changing directly to dull rufous belly. Female dull uniform grey on upperparts, breast and tail, lacks white wing patches and is dull rufous on the belly. Immature resembles female. **Similar species** Unmistakable.

HABITS Closely associated with rocky habitats on hillsides and gorges, where it moves swiftly and inquisitively over rocks, dodging through crevices, nearly always in pairs. Particularly favours locations with fig trees. Perches on prominent boulders, and characteristically pumps tail slowly up and down.

VOICE Both sexes give a loud song of clear ringing whistles, pair maintain close contact by answering and duetting. Accomplished and accurate mimic.

STATUS AND DISTRIBUTION Not recorded in The Gambia, but an established local population in south-eastern Senegal. In West Africa generally fragmented, according to habitat availability, with distinct forms developing in these isolated populations.

BREEDING Usually in crevices in boulders and rockfalls, and often appropriates nest of Lesser Striped Swallows.

BLUE ROCK THRUSH
Monticola solitarius   Plate 37

Fr. Merle bleu

IDENTIFICATION 20cm. Sexes differ. A long-billed, rangy thrush of uniform dark coloration. Male in spring dark slaty blue-grey above and below, tending to blackish on the wings and tail. In fresh autumn plumage this is overlaid by a scaled appearance created by narrow buff fringes to the body feathers, which wear away through the winter. Female plain dark brown with scarcely any blue tones, lightly speckled on the upperparts; plain dark brown tail; spotted brown-buff around the face, closely scalloped on the underparts with pale buff feather fringes, creating a finely barred effect. Immatures are dark brown, heavily spotted and scaled with buff above and below. **Similar species** Separated from Rock Thrush in all plumages by absence of orange on tail, which is also longer, giving a more attenuated shape.

HABITS On migration travels in small groups, sometimes in association with Rock Thrush, but becomes solitary, seeking out rocky hill top habitats

in the interior, and cliffs near the coast; individuals also take up residence around buildings, usually isolated in farmland. Perches on rocks or roofs with upright stance, peering over up-tilted bill, returning regularly to vantage point; will often disappear into crevices, under eaves etc.

VOICE Mainly silent on wintering grounds, but gives a liquid *uit-uit* and has a piping fluent song.

STATUS AND DISTRIBUTION Rare-uncommon winter visitor and passage migrant, spring passage said to be more prominent. Tends to be reported regularly over some weeks at favoured locations. All recent records from WD and LRD, October-January. Similarly infrequent in Senegal, reported from a handful of widespread locations but also suspected of becoming resident near Dakar.

EUROPEAN ROCK THRUSH
Monticola saxatilis   Plate 37

Fr. Merle de roche

IDENTIFICATION 18.5cm. Sexes differ. A large headed, short-tailed thrush; variably speckled and mottled over the body (except breeding male), some orange-rufous below the tail in all plumages. Non-breeding birds most likely to be encountered: grey-brown over head, breast and mantle, extensively speckled with black and creamy buff; wings darker brown with buff edging to primaries; pale throat, plain orange-chestnut tail with brown centre, orange wash to lower belly and undertail. Males show limited amount of white on mid-back. Breeding males (January to March) cobalt blue-grey upperparts and throat, white back, brown wings; bright orange-chestnut underparts and tail. Immatures resemble washed-out non-breeding birds, heavily mottled, but retaining strong hint of orange-chestnut in tail. **Similar species** Distinguished by orange-tinted short tail. Similar to female Blue Rock Thrush, which lacks rufous undertail, is larger, darker, less mottled.

HABITS Favours open bush and light woodlands, also burnt areas, coastal rubbish tips etc. Travels in loose small groups on migration, but on wintering grounds tends to be solitary and shy, skulking around bushes; perches with distinctive upright posture and long bill held up at an angle. Flies swiftly and direct at low level. May shiver tail like a redstart.

VOICE Quiet on migration, but utters a *chack-chack* call.

STATUS AND DISTRIBUTION Rare-uncommon passage migrant, recorded from coastal, central and interior parts of The Gambia, December-March.

Not reported every year. Similarly infrequent in Senegal, with records from all parts of the country.

AFRICAN THRUSH
Turdus pelios   Plate 37

Fr. Grive grisâtre

Other name: West African Thrush

IDENTIFICATION 23cm. Sexes similar. Typical thrush; quick movements and furtive habits, distinguished by a prominent bright yellow bill. Upperparts soft pale brown, darkening on wings and tail. Whitish throat marked with rows of black flecks, plain pale brown breast merging to whitish belly and undertail-coverts. In flight warm orange buff tone of underwings apparent. Robust bill uniform yellow, brightening in the breeding season. Immature diffusely freckled dark brown on a buff-washed

throat and upper breast, whiter below, with narrow buff tips to greater coverts; bill dark brown.

HABITS Found mainly in or near well-wooded habitats and generally shy. Has been observed to smash snails on stones. Also takes fruit in season.

VOICE Sings well; sustained confident warbling, confusable with Snowy-crowned Robin Chat, but tends to repeat each sub-phrase several times in any sequence; *too-lioo, toolioo, toolioo, wee-tutu, wee-tutu, wee-tutu* etc. A very high-pitched, squealing whistle, frequently given in flight.

STATUS AND DISTRIBUTION Frequent to common resident in coastal areas, becoming less frequent inland, though recorded regularly from all parts of The Gambia. Movements not well understood, but a significant increase in records June-September. Recorded a short distance north of The Gambia into Senegal.

BREEDING Cup-shaped nest, with all records from the wet season, June-October.

WARBLERS, CISTICOLAS, PRINIAS, EREMOMELAS, CAMAROPTERAS, HYLIA AND CROMBECS
Sylviidae

Fifty three species. Large, diverse family of small to tiny birds from a variety of habitats, largely arboreal, slimly-built, many as readily identified by song and call as by plumage, which is often nondescript. In several subfamilies the sexes are similar in plumage and the immatures closely resemble the adults; in others (e.g. *Sylvia*) there are distinctive age/sex plumage differences. A substantial number of resident species are joined by migrants during the Palearctic winter.

AFRICAN MOUSTACHED WARBLER
Melocichla mentalis   Plate 38

Fr. Fauvette à moustaches

IDENTIFICATION 18cm. Sexes similar. Large plain brown warbler, distinctive broad dark tail and patterned face. Upperparts plain warm brown, with chestnut forecrown, longer fluffy feathers on rump rufous-tinted; darker wings show faint panel caused by rufous edges to flight feathers; very broad, rounded, plain dark brown tail. Lores and cheeks whitish, extending to a white supercilium; ear-coverts pale brown; narrow black malar stripe extends from lower corner of the bill. Underparts pale brown, whitish in centre of belly, merging to pale brown on flanks and undertail. Legs greyish. Young birds lack the chestnut forecrown and are mottled below. **Similar species** Comparable size to Great Reed Warbler and Greater Swamp Warbler, but distinguished from both by malar stripe and contrasting dark brown tail.

HABITS Favours tall grass habitats, usually beside water, typically alone or in pairs. Most readily encountered morning and evening, remaining silent in dense cover through much of the day. Flicks tail when nervous and quickly dives out of sight. Flies low between clumps of vegetation, but generally most readily detected by characteristic scolding alarm coming from dense cover.

VOICE Bright warbling song starts with two slow notes followed by a more hurried chortle, rendered with great uniformity, *tip-tip-twiddle-iddle-ee*. Alarm call a rasped *ti-ti-ti-ti*.

STATUS AND DISTRIBUTION One old but unsubstantiated record from the Bund Road in The Gambia; otherwise resident through south-eastern Senegal from the mid-Casamance eastward to Niokolo Koba.

BREEDING Grass cup near to ground in dense tussock of large grass species; recorded over a wide range of months in West Africa, though mainly associated with the rains.

GRASSHOPPER WARBLER
Locustella naevia   **Plate 38**

Fr. Locustelle tachetée

Other name: Eurasian Grasshopper Warbler

IDENTIFICATION 12.5cm. Sexes similar. A small, variably streaky, Palearctic visitor to long grass habitats. Upperparts are mainly olive-brown, sometimes tinged yellower, sometimes greyer, with fine dark streaks on the crown, becoming broader and blotchier on the mantle to rump, though not always easily seen. The rounded broad brown tail may be lightly barred. Smudgy supercilium and narrow pale eye-ring. Underparts buffish or yellowish-white, with a faint gorget of buff which shows very fine streaking across the breast. Heavier dark streaks commence on buffy flanks and continue onto long, brown undertail-coverts. Comparatively fine bill, noticeably pink or orange legs. First-year birds yellower on underparts. **Similar species** Proportionately large rounded tail, regularly dipped, distinguishes from Aquatic and Sedge Warblers; streaks on breast and undertail, plus small bill, separates from all cisticolas.

HABITS Exceptionally furtive, preferring rank grass habitats, including stands of tall *Andropogon*, *Mimosa* thickets and tamarisks around marshy areas. Hops and scrambles through dense vegetation, moving with a wriggling scuttle close to the ground, but a high stepped loping run also described. Nearly always difficult to see, best located by sound. Flights short, whirring and low, broad tail apparent.

VOICE Commonest call, in Africa, a crisp, *pitt* or *tswit*, resembling an isolated element of the reeling song. Has been recorded singing in Sierra Leone in March, but apparently not a common event.

STATUS AND DISTRIBUTION Very limited information in The Gambia. Regular March-May in the Senegal River basin, and a handful of records from surrounding countries, to the south and east, mainly in spring, sufficient to imply that it is underrecorded in the entire region. Likely to occur in The Gambia more regularly.

SAVI'S WARBLER
Locustella luscinioides   **Plate 38**

Fr. Locustelle luscinioïde

IDENTIFICATION 14cm. Sexes similar. Plain brown medium-sized warbler of dense reedbeds. Upperparts dark rufous-brown, uniform on mantle and rump; slightly darker on the broad, strongly graduated tail. Wings brown, though a whitish outer web to the outermost primary may be visible. Indistinct buffy supercilium fades rapidly behind the eye, which has narrow pale ring; ear-coverts lightly flecked buff. Plain underparts are paler rufous-brown along the flanks, merging to whitish along the centre line from throat to belly. Buffy brown undertail-coverts are long and may show some white scalloping. Legs brownish or flesh. **Similar species** To separate from plain *Acrocephalus* species sharing same habitat, note broad and rounded tail, great elongation of undertail-coverts, and small-headed look. Reed Warbler is smaller, Greater Swamp Warbler much larger.

HABITS A skulking species of freshwater reed swamps, though sometimes also recorded in thickets near water in Africa. Will sing and call from exposed reed stem, but drops if disturbed, scuttling away unseen, like a rodent.

VOICE Main song a dry, buzzing, trill on a monotone, usually starting up with a few hesitant ticking notes; ventriloquial, variable in volume, lower in pitch and of shorter duration than typical Grasshopper Warbler. More often a simple *tswick* or *pstink* call is heard on wintering grounds.

STATUS AND DISTRIBUTION Palearctic migrant not yet recorded in The Gambia, but note a very large area of suitable habitat, very little explored, in the lower reaches of CRD. Regular winter visitor December-February in Senegal River basin, recorded once in southern Senegal.

AQUATIC WARBLER
Acrocephalus paludicola   **Plate 39**

Fr. Phragmite aquatique

IDENTIFICATION 13cm. Sexes similar. A small streaky warbler, patterned with pale yellowish and black. Strongly striped on the head, with a pale buff central-crown stripe, long blackish lateral crown stripes, broad, long, pale-buff supercilium, and plain pale lores. Strongly streaked on the back with sandy-yellow and black, black streaking continues onto the rump area. Rounded brown tail shows pointed individual feathers when flared. Underparts pale whitish-buff, with necklace of dark streaks, narrow and obscure streaking on flanks. Bill comparatively short and thick, legs pink. Juveniles similar, but with stronger contrast to pale and dark streaking of mantle, buffer below. **Similar species** Beware first-winter Sedge Warbler's tendency to tawny median crown stripe and possible remnant speckled gorget; Aquatic has longer lateral crown stripe, plain lores, cleanly streaked upperparts with continuation of streaks on rump. Small bill and strong crown pattern unlike any cisticola, which all have black and white tips to tail feathers.

HABITS Skulking bird of wet grassy and shrubby marsh habitats, with shallow water. Clambers low in vegetation; flight short, low, and whirring. Not especially associated with reedbeds in breeding range, situation in wintering range in Africa poorly known.

VOICE Simple *tack* call, song of short phrases of more simple structure than Sedge and other *Acrocephalus*. Winter calls not recorded.

STATUS AND DISTRIBUTION Not recorded in The Gambia, but known to occur sporadically in Senegal River delta, mainly from ringing captures. Winter range poorly known, but sparse records suggest swampy areas, in or close to the Sahel in West Africa, may be important.

SEDGE WARBLER
Acrocephalus schoenobaenus   **Plate 39**

Fr. Phragmite des joncs

IDENTIFICATION 12.5-13cm. Sexes similar, juvenile distinguishable. Small streaky warbler distinguished by prominent supercilium and rufous rump colour. Adult grey-brown or olive-brown above, densely streaked with fine black lines on the crown, more lightly on the mantle. Rump plain rufous-brown, contrasting with the rounded dark brown tail. Buff-white supercilium bordered on lower margin by a narrow black line running across lores and extending behind the eye. Underparts white, stained rufous on flanks. Legs pale grey-brown. First-winter birds buffy-yellow ground colour, often showing a pale wedge through the centre of the crown and gorget of darker speckles. **Similar species** Juvenile features can suggest Aquatic Warbler: separate by plain rump, narrow black line on lores, shorter supercilium and lack of spiky tail feathers. Less black than on streaky cisticolas, with brighter supercilium; also lacks black and white in tips of tail.

HABITS In winter range found primarily in freshwater swamps, or in reedbeds, but is noted for using drier sites on passage. Mainly solitary. First arrivals skulking, but may establish winter territories; by February chasing may occur, singing not unusual. Generally secretive but reacts to disturbance with testy alarm call to which song burst may be added.

VOICE Alarm call a rasping *trrrr*, also a soft *tic...tic*. Song an elaborate non-repetitive warble, starting with harsh notes, but adding in a mix of clear whistles and trills, sometimes with mimicry.

STATUS AND DISTRIBUTION Regular winter visitor, mainly November-March, to the coast, CRD and URD; uncommon, but likely to be under-recorded. Earliest record in August. In Senegal regular in large numbers in the Senegal River delta, with scattered records further inland.

REED WARBLER
Acrocephalus scirpaceus   **Plate 38**

Fr. Rousserolle effarvatte

IDENTIFICATION 12.5-13cm. Sexes similar. Small brown warbler, plain save for a slight rufous-tinge on rump. Plain brown on upperparts from crown to square-ended tail, tending to greyer tone in worn post-breeding plumage; hint of rufous on rump may show between the short, folded wing tips, or in flight. Pale buffy-white short supercilium and pale eye-ring present, but indistinct. Underparts whitish on throat to belly, rufous-brown on flanks. Undertail-coverts long, pale buff. Legs brownish to grey-tinged. First-winter birds richer brown than worn adults, with buff wash especially strong on flanks. Note flat forehead, peaked crown and spiky bill. **Similar species** Lack of black and white tips to tail feathers eliminates all plain cisticolas, long undertail-coverts eliminates *Hippolais* warblers. Greater Swamp Warbler is much larger, with greyish flanks; separation from slightly smaller, shorter-winged resident African Reed Warbler is the main difficulty, sometimes regarded as conspecific. In the hand, longest primary P3. Marsh Warbler *Acrocephalus palustris* (Fr. Rousserole verderolle) is also very similar to Reed Warbler. It is more olive-brown with pale-fringed dark tertials and long primary projection. A vagrant from the Palearctic it has been seen once in northern Senegal in January 1994; two previous rejected records from The Gambia.

HABITS A warbler of long grass habitats, mainly wetlands, though on migration freely uses dry grass and bush, sometimes in numbers in mangroves; clambers on single vertical stems, or grasps one in each foot. Tilts entire body to axis of movement, whether, up, horizontal, or down. Flies neatly and buoyantly, usually low. Spends much time hidden in cover, where tetchy alarm calls are best indicator of presence.

VOICE Short dry and raspy emphatic alarm, *skurr*, or *tchar*. Sings sporadically in winter range using harsh scolding notes, given with measured, repetitive delivery, *chara-chara-char*, *krik.krik.krik*, *chirruc-chirruc*, *jag-jag-jag* etc.

STATUS AND DISTRIBUTION Winter visitor, mainly November-April, to all parts of the country, with highest frequency of records in January. Not commonly recorded, but elusive habits may exaggerate impression that it is uncommon. Abundant on passage in northern Senegal, September-October and again March-May.

AFRICAN REED WARBLER
Acrocephalus baeticatus   Plate 38

Fr. Rousserolle africaine

IDENTIFICATION 12-13cm. Sexes similar. Small brown warbler, perhaps best identified by season of observation; in the field scarcely separable from Reed Warbler on appearance alone and sometimes regarded as conspecific. Plain brown above, with pale sides to face and short pale supercilium. Whitish below with buffish-tinge on flanks. Legs brown. **Similar species** Long undertail-coverts and relatively indistinct face markings distinguish it as an *Acrocephalus* warbler, size eliminates Greater Swamp and Great Reed Warblers. Slightly smaller and shorter-winged than Reed Warbler, but other than in the hand, probably not safely distinguished. Greyish birds from the Senegal River basin distinguished as *A. b. guiersi*; a single specimen from Casamance has been attributed to the more widespread rufous form *A. b. cinnamomeus*. Affiliation of birds in The Gambia as yet unknown. In hand, longest primary P4.

HABITS Typical reed warbler, associated with dense, freshwater marsh vegetation. Mostly out of sight, but ventures briefly into view on upper stems to investigate source of disturbance, before flitting quickly back down. Sings mainly from deep within reeds.

VOICE Dry scratchy song, given at measured pace, very similar to Reed and similarly lacking fluty or whistled notes. Gives harsh *churr* alarm.

STATUS AND DISTRIBUTION Poorly known, mainly due to field identification difficulties. Presence originally established by collection of 5 nests in July 1945 near the coast; in recent years singing birds observed regularly in freshwater swamps of CRD and URD in June and July, where it is likely to be frequent. Situation in winter and spring confused by presence of Reed Warbler.

BREEDING Nesting in the early rains, June-July, building deep cup slung between several vertical stems.

GREAT REED WARBLER
Acrocephalus arundinaceus   Plate 38

Fr. Rousserole turdoïde

IDENTIFICATION 19-20cm. Sexes similar. Relatively large warbler plain warm-brown, size and shape more that of an attenuated small thrush than a warbler. Upperparts brown, slightly darker on the crown, paler on the rump, more marked in female. Wings and tail brown. Pronounced pale creamy supercilium in front of the eye, variably extending a little way behind the eye where it is narrower, and this contrasts with duskier lores to give a distinctly patterned face. Throat and central belly white, indistinct streaking on throat smudgy and not noticeable; flanks and undertail-coverts buffy. Worn postbreeding adults greyer above, whiter below. Bill heavy and long, dark brown above, prominent pink side of lower mandible; legs greyish. First-winter bird warmer above and flushed patchily with orange-buff on underparts. **Similar species** Size alone separates from all other migrant warblers; larger than Greater Swamp Warbler, which lacks face pattern; plain tail and wings separates from smaller Rufous Scrub Robin. Lores darker and lacks rufous-toned rump and tail of Nightingale.

HABITS Strongly tied to reed swamps in breeding range; similar habitats in winter but may also be encountered in drier areas, for example *Papyrus* near water or stands of long *Pennisetum* grass in savanna. Heard more easily than seen, but less shy than most *Acrocephalus*. Flies low and heavily with spread tail and extended body.

VOICE Loud harsh song frequently heard in winter: harsh clicking and grating notes delivered in rhythmically repeated sequences, interspersed with thin, sometimes strangulated, squealing notes; *karra-karra-karra*, *sqee-ee*, *krit-krit-krit*; variable and long-winded. Call a forceful *chack*.

STATUS AND DISTRIBUTION Rare in The Gambia, with a handful of records October-May from the coast, CRD and URD. Rare in northern Senegal, with most seen on spring migration. A widespread visitor to sub-Saharan Africa, generally more numerous to the east.

GREATER SWAMP WARBLER
Acrocephalus rufescens   Plate 38

Fr. Rousserolle des cannes

Other name: Rufous Cane Warbler

IDENTIFICATION 18cm. Sexes similar. A large,

plain, dark brown and grey warbler of retiring habits, much more readily heard than seen. Upperparts plain brown, slightly darker on top of head, with a plain face lacking a supercilium, though with paler lores. Chin, throat and belly white, merging to grey-brown on sides of breast and flanks. Bill fairly heavy, dark brown above, noticeable yellowish patch at base of lower mandible; legs grey. **Similar species** Approaches Great Reed Warbler in size, but darker, with greyish, not rufous-tinted sides, plain face, and very rounded wings and tail if seen in flight; voice distinctive.

HABITS An elusive reedbed skulker, remaining in deep cover. Favours emergent vegetation at the edges of fresh water.

VOICE Short phrases of mellow resonant warbling with reedy contralto and yodelling notes; given with measured confidence and volume, less garrulous manner than most reedbed warblers: *churr-churr, chirrup, chuckle* etc.

STATUS AND DISTRIBUTION Singing birds first identified in the extensive but little explored freshwater swamps of CRD in 1991, within which it is likely to be frequent to common. All records to date July-November. Subspecies *A. r. senegalensis* is frequent in the swamps and sugar cane plantations of the lower Senegal River, and in the upper reaches of the Gambia River valley near Tambacounda.

BREEDING Substantial cup nest suspended on thick vertical stems; nests from late dry season in April through the rains to August.

OLIVACEOUS WARBLER
Hippolais pallida　　　　　Plate 38

Fr. Hypolaïs pâle

IDENTIFICATION 13cm. Sexes similar. Plain pale grey medium-sized warbler with a flat rather angular head, persistent tail-dipping action, broad-based bill. Upperparts plain grey-brown, usually quite pale, from crown to tail. Short wings may be slightly darker, and in fresh-plumaged birds may show a weak pale panel though this quickly wears away. Pale eye-ring and pale lores; weak supercilium seldom extends behind eye. Throat white, entire underside very pale buffy-white. Bill long, broad-based, dark brown above and pale flesh-yellow below, often catching the eye on birds foraging overhead. Legs variably brown to grey. First-winter birds similar, pale fringes to flight feathers may be more prominent. **Similar species** From *Sylvia* warblers, especially Garden, by angular crown and long bill, and from *Acrocephalus* by short undertail-coverts and lack of rufous tones. From Melodious by lack of

yellow tones, flatter crown, and especially regular downward tail-pump accompanied by persistent contact call. Individuals of the north African race *H. p. reiseri* are relatively small, with a shorter finer bill, and may cause confusion with *Phylloscopus* warblers.

HABITS Usually encountered singly in well wooded areas, particularly broadleaf riparian habitats, sometimes mangroves; less commonly in drier thorn bush. Apt to take up residence in canopy of *Acacia* tree, singing and feeding there over several days. Less abrupt in movements than *Phylloscopus* species.

VOICE High-pitched vigorous song, scratchy in quality with a few fluty notes; regularly heard in The Gambia. Characteristic call a persistent, rich *tchik*.

STATUS AND DISTRIBUTION Migrant from Europe and North Africa, but recorded year-round in The Gambia, with a January peak, when it is frequent especially in CRD and URD. Most are likely to be the large west European race *H. p. opaca*, including wet season records, but smaller north African *H. p. reiseri* have been captured in Senegambia.

MELODIOUS WARBLER
Hippolais polyglotta　　　　　Plate 38

Fr. Hypolaïs polyglotte

IDENTIFICATION 13cm. Sexes similar. Well-built, yellowish and green warbler, often moving quietly in leafy cover, revealing yellowish wedge along side of the bill. Upperparts pale greenish-brown, with yellow supercilium and eye-ring. In fresh spring plumage, attained by January, narrow pale edges to inner flight feathers may create hint of pale wing panel, but in worn autumn adults this is plain; tail square-ended, matching rest of upperparts. Underparts usually washed pale yellow in most cases, occasionally bright yellow, especially in spring, and some first winter birds pale creamy, though yellow traces nearly always detectable around throat and breast. Bill dark brown above, side of the lower mandible noticeable orange or yellow. Legs grey brown. **Similar species** Green and yellow tones distinctive; Olivaceous lacks yellow and gives regular downward tail flick. Bulkier than *Phylloscopus* species, with heavier bill and legs, less patterned face, and less active movements. Most similar to larger, long-winged and more vocal Icterine Warbler *Hippolais icterina* (Fr. Hypolaïs ictérine), from which best separated by shorter wings. Icterine not recorded in Senegambia, and generally rare in West Africa.

HABITS Found in well wooded savannas, also

301

mangroves and along the forest edge, though not apparently in densely forested areas. Less abrupt in movement than most small warblers. Usually solitary but may also be found joining in mixed species feeding parties in open woodland.

VOICE Sings quite frequently in winter quarters, around time of arrival and departure, sustained song of mixed scratchy and more tuneful notes, reminiscent of some *Acrocephalus* species. Sparrow-like *chrrrt* call.

STATUS AND DISTRIBUTION Frequent winter visitor to all parts of The Gambia, September-May, with a rapid build-up to a peak around November, reducing gradually through to April, possibly surging briefly again in May. Strong passage in northern Senegal in August-November, though less obvious in following spring. Entire population winters between Senegal and Cameroon.

Cisticolas

A diverse group of small specialized warblers, 42 species currently recorded from Africa, 12 in Senegambia. Small brown birds, with or without black streaking and varying degrees of rufous; broad, usually rather rounded tails with distinctive pattern of subterminal black and terminal white spots; slightly decurved bills, characteristic black gape displayed while singing. Habitat and song are often key supporting features in identification in a generally difficult group.

Often use prominent perches when singing on territory, otherwise skulking and awkward to observe, though not necessarily shy. The general similarity of many species, with breeding and non-breeding plumages creating variation, in addition to variation due to plumage wear all combine to give them this reputation. In absence of singing birds positive identification may not always be possible.

Birdwatchers encountering this group for the first time can minimise confusion with the abundant Tawny-flanked Prinia by familiarising themselves with narrow elongate-tailed jizz of the Prinia which waves its tail side to side. Cisticola tails are broader, more rounded and less mobile. The tail spotting in conjunction with the subtly decurved bills distinguish the larger plain cisticola species from all migrant Palearctic warblers.

The English nomenclature has been complicated by the publication of more than one name for several species; beware especially 'Black-backed Cisticola', which is often applied to *C. galactotes* in southern Africa, but is reserved for *C. eximius* in west and east Africa.

RED-FACED CISTICOLA
Cisticola erythrops Plate 39

Fr. Cisticole à face rousse

IDENTIFICATION 12-13cms. Sexes similar. Medium-sized cisticola, unstreaked brown above with strong gingery-rufous wash over face and underparts, favouring rank vegetation close to fresh water; intermittent bursts of loud ringing song from dense cover. Adept at remaining out of sight. Upperparts and wings uniform warm olive-brown (darkening with wear), unstreaked above, plain rufous-orange around eye, face and cheeks, fading to apricot-tinted creamy throat, breast and belly. Rufous-orange suffusion over underparts may extend over throat and upper breast as well as face. Tail brown; black subterminal spots below. Immature lacks rufous-orange tones, plain brown above, diffusely pale around eye, white throat, tawny-buff on flanks, extensively yellow on lower mandible of bill. **Similar species** A distinctive cisticola; orange-buff tones on face and underparts, waterside habitat and loud ringing song render identification straightforward.

HABITS Closely associated with rank grass and broad-leaved vegetation immediately beside freshwater, usually in pairs; not often encountered away from such places. Occurs in reedbeds in other parts of Africa, though this has not yet been noted in The Gambia. Presence most likely to be revealed by a burst of song.

VOICE An arresting burst of disproportionately loud and spirited song, rather untypical of cisticolas generally. Commences with rhythmic repetition of crisp ringing notes at constant pitch, *whip.whip. whip.whip.whip - too-tee-teee- tee - tiptiptiptip* etc. leading into more complex phrases. Typically pauses for some minutes between each song bout. Unlike most cisticolas sings from within dense cover. Occasional sings in the non-breeding season.

STATUS AND DISTRIBUTION Frequent resident along freshwater streams in WD, occasionally in thick scrub behind beaches e.g. Sanyang. Regular along the banks of the main river in freshwater and swampy areas of CRD and URD, but not reported from mangrove and saline stretches; not recently recorded from either LRD or NBD. Found through Casamance to Niokolo Koba; not reported north of The Gambia in Senegal.

BREEDING Like Singing Cisticola, constructs grass cup inside pocket formed by binding opposing margins of living leaves on a low growing broad-leaved herb using spider's silk – with beautifully camouflaged results. Recorded September, fledglings being fed by adults December.

Cross section of Gambian habitats, illustrating cisticola distribution and typical song display locations.

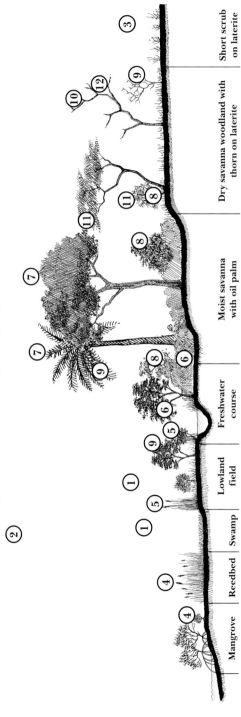

| Mangrove | Reedbed | Swamp | Lowland field | Freshwater course | Moist savanna with oil palm | Dry savanna woodland with thorn on laterite | Short scrub on laterite |

AERIAL SONG FLIGHT DISPLAYS

1 Zitting Cisticola *C. juncidis* – usually low-level damp areas and field margins (Note: may sometimes be close to Desert Cisticola habitat)

2 Black-backed Cisticola *C. eximius* – high-level

3 Desert Cisticola *C. aridulus* – low-level, dry open areas

PERCHED SONG DISPLAY – STREAKED SPECIES

4 Winding Cisticola *C. galactotes* – exposed; reedbed

5 Croaking Cisticola *C. natalensis* – exposed; long grass (sometimes calls on wing)

PERCHED SONG DISPLAY – UNSTREAKED SPECIES

6 Red-faced Cisticola *C. erythrops* – dense cover near freshwater

7 Siffling Cisticola *C. brachypterus* – usually exposed (very weak streaking may be apparent in best view)

8 Singing Cisticola *C. cantans* – usually from edge or slightly within canopy; widespread

9 Whistling Cisticola *C. lateralis* – exposed; edge of canopy in bush, palm or savanna

10 Rufous Cisticola *C. rufus* – usually exposed; dry savanna woodland

11 Plaintive Cisticola *C. dorsti* – exposed; dry savanna woodland

12 Red-pate Cisticola *C. ruficeps* – exposed; open dry savanna woodland

SINGING CISTICOLA
Cisticola cantans **Plate 39**

Fr. Cisticole chanteuse

IDENTIFICATION 11-12cms. *C. c. swanzii.* Sexes similar. A medium-sized cisticola, unstreaked, favouring wide range of shrub and woodland habitats, simple repeated clinking song. Rufous crown and wing patch contrast with greyer mantle. Pale, ill-defined supercilium created by dark lores, throat creamy fading to dirty buff on flanks. Tail grey-brown with smudgy dark subterminal spots. Immature similar but less contrasting between grey and rufous, bill paler. **Similar species** From Whistling Cisticola by smaller size, contrast between grey mantle and rufous crown, which also separates from smaller Siffling and Rufous Cisticolas. Red-faced Cisticola is much more orangey-buff around face and underparts; larger than breeding Red-winged Warbler which has narrower tail, cleaner creamy underparts and all-black bill.

HABITS Found in a range of habitats, from farmland and scrub to savanna woodlands; tends to avoid thorny vegetation. Typically skulking, but sometimes curious, approaching through dense low vegetation to scold observer; may join parties mobbing e.g. Pearl-spotted Owlet. Singing males may perch in the open, but more frequently remain partially concealed within topmost foliage. Sings most intensively through the breeding season, but may be heard sporadically in other seasons.

VOICE In The Gambia most often detected by measured, repeated, disyllabic *p-tink*, or more *pi-choo*. Superficially resembles some calls of the Grey-backed Camaroptera and Tawny-flanked Prinia, but steady tempo of Singing Cisticola distinctive, as is more metallic and emphatic quality. May give a single call, or a burst of five or six; occasionally more prolonged bouts.

STATUS AND DISTRIBUTION Common breeding resident, though elusive. Found in all parts of The Gambia with suitable rank vegetation. Mainly in southern Senegal with populations stretching north to Dakar near the coast and slightly north of The Gambia in the extreme east.

BREEDING Recorded breeding in late dry season, April-May in The Gambia, males most vocal in the wet season. Builds low in vegetation, binding two or three broad green leaves along their margins using spider's web to conceal more conventional grass lining within, reminiscent of elaborate stitched nests of tailorbirds in India. Shares this distinctive habit with the Red-faced Cisticola.

WHISTLING CISTICOLA
Cisticola lateralis **Plate 39**

Fr. Cisticole siffleuse

IDENTIFICATION 12-14cm. Sexes similar; males larger. Large, plain brown unstreaked cisticola; of savanna woodland to field margins with trees; confident rising song phrase, more warbling than whistling in character. Adult grey-brown on head nape and mantle, uniformity the main feature. Narrow pale buff wedge at lores stands out. Weak rufous wing panel; tail brown with black subterminal spots and whitish tips, indistinct on upperside, clear on underside. Chin and throat white; washed out grey-brown smudgy areas form either side of breast; rest of the underparts very pale buff. Non-breeding birds more diffusely stained buff below. A bright foxy-rufous phase and a nearly black phase recognised in West Africa, neither prominent in The Gambia. Immature dull brown above, yellow stained below, diffuse tail spotting. Bill strong and black (flesh tinted in female), flesh legs stout. **Similar species** Distinguished from Singing Cisticola, with which it is often found, by lack of contrast between head, nape and mantle, larger size and song; from Nightingale by lack of contrast between tail and body; from Garden Warbler by richer body colour, bill shape and leg colour.

HABITS Found in open woodlands. Skulking and elusive in the dry season, when it remains near the ground in dry herbaceous layer; sings and perches prominently July-November. Especially fond of areas of low dense bushes dotted with emergent larger trees, as often occurs in and around old cultivation. Individuals move regularly between perches giving repeated bouts of song at each place, typically ending with a shallow, fluttering, descent to disappear into the surrounding vegetation. In the breeding season usual to find several singing males scattered in a favoured area.

VOICE Loud throaty song of confident warbled notes, lasting one to two seconds, usually rising steadily in pitch in the latter half. Usually a leisurely pause between repetitions in each song bout. Alarm calls of loud, bright, scolding notes on a repeating monotone; delivered from cover when disturbed *tswink, tswink, tswink-tswink-tswink.* Soft conversational twittering prior to full song, especially at start of rains.

STATUS AND DISTRIBUTION Locally common breeding resident near the coast e.g. Yundum, present in Kiang West National Park in LRD, becoming progressively less frequent inland. No recent records from CRD or URD. Present but seldom recorded in the dry season. In Senegal recorded patchily from Casamance and the south-

304

east, not north of The Gambia.

BREEDING Wet season breeder, June-October, typical dome nest with side entrance in grass clump close to the ground

RED-PATE CISTICOLA
Cisticola ruficeps   Not illustrated

Fr. Cisticole à tête rousse

IDENTIFICATION 10-11cm. Sexes similar, some seasonal variation. Medium-sized cisticola with distinct rufous cap, plain brown body; favours thinly wooded areas; whistling and warbling song. Short white line in front of eye, underparts whitish on chin, grey-buff smudging at sides of breast and flanks, whitish belly. Non-breeding birds lightly mottled on mantle. **Similar species** Separated in breeding plumage from Plaintive Cisticola by drier habitat preference, whiter vent and larger brighter tail spot pattern on shorter tail (38-40mm); see also Voice.

HABITS Similar to Plaintive, but Red-pate is associated with drier, grassland habitats with scattered bushes, a habitat shared with Desert Cisticola.

VOICE Main song opens with brief rising note with clean whistled quality, followed by falling warble and a secondary rising warble. Calls buzzy or rattling.

STATUS AND DISTRIBUTION Uncertain. Hitherto the Red-pate Cisticola was considered to be thinly scattered in savanna woodlands throughout the Gambia (Gore 1990), but there have been no confirmed recent records and it cannot be assumed that all previous Red-pate records should be referred to Plaintive Cisticola (see below). It remains possible that both species are present and further observations are necessary. Separation in the field relies primarily on song. Similar problems may apply to all recent records of Red-pate from eastern Senegal, where Plaintive Cisticola has not been considered in recent field work (G. Morel pers. comm.).

BREEDING No records.

NOTE The Red-pate Cisticola *C. ruficeps* has recently been split into two, Red-pate and Plaintive Cisticola *C. dorsti*. The split is based on the fact that differing song patterns are stable and appear to coincide consistently with habitat separation and small morphological differences (Chappuis 1985, Chappuis & Erard 1991). Recent voice recordings from Kiang West National Park have been identified by C. Chappuis as *C. dorsti*.

PLAINTIVE CISTICOLA
Cisticola dorsti   Plate 39

Fr. Cisticole de Jean Dorst

Other name: Dorst's Cisticola

IDENTIFICATION 10-11cm. Sexes similar. Medium-sized cisticola, with well formed rufous cap, plain milky-tea upperparts; favours savanna woodlands; quavering monotone song. Breeding (June-November): rufous cap cleanly delineated at the front and sides, merging at the nape into light brown mantle; rufous-brown rump and tail, subterminal spots and narrow greyish tips most marked on the underside. Wings brown. A thin pale line from lores over front of eye. Chin and throat white; buffy wash over flanks and vent. Legs pink. Non-breeding plumage not described. **Similar species** From Red-pate primarily by voice and habitat plus buffier undertail-coverts, with longer tail (42-43mm) but smaller, duller undertail spotting. From Singing Cisticola by more pronounced rufous cap contrasting with brown, not grey, mantle. From female Blackcap by warmer brown body, broad tail, decurved bill, more rufous, smaller cap; from all by voice.

HABITS In The Gambia known from broad firebreaks with dense secondary shrub growth cut through extensive savanna woodland. Elusive and seldom observed, especially in dry season. During rains July-September, while breeding, indulges in song displays from high perches, particularly large dead trees. Several regular song posts, using each for several song bursts before characteristic fluttering descent into the surrounding bush. Elsewhere associated with grassland with thicket clumps or extensive shrub layer, including growth in old fields.

VOICE Singing males produce a clearly articulated, slow quavering monotone call, with marginally accelerating delivery, *twe-we-we-we-we-we-we*, lasting 1-2 seconds. This is followed by a pause of variable length, often half a minute or more, before repetition. Rhythm more urgent in territorial disputes, when it may occasionally give plain, descending call-note.

STATUS AND DISTRIBUTION Not well known; perhaps under-recorded. Locally common in a limited area of savanna woodlands in LRD, where records are of singing birds found annually June to November. Should be looked for in open shrub areas of Kiang West National Park and similar habitats throughout our region. No confirmed observations in Senegal, or anywhere closer than Chad and Nigeria at present, though this can be expected to change.

BREEDING Presence of singing males in July to

November the best evidence of breeding in The Gambia.

WINDING CISTICOLA
Cisticola galactotes **Plate 39**

Fr. Cisticole roussâtre

Other names: Black-backed Cisticola, Greater Black-backed Cisticola, Rufous Grass Warbler

IDENTIFICATION 12-13cm. Boldly-streaked cisticola of swamp and reedbed; song a rising, rasping, nasal rattle. Markedly larger than the Zitting Cisticola; shows contrast between breeding and non-breeding plumage. Breeding (June-November): rather dark crown dull rufous-brown, becoming greyer with wear; untidy dark streaks on grey-brown mantle, rufous patch on primaries, grey rump; tail grey brown, black subterminal spots, white tips. Throat pale creamy, dirty buff on flanks. Non-breeding in The Gambia: much brighter with longer tail; top of head orange-rufous, boldly patterned mantle of broad black stripes on a tawny-buff background contrasts with grey-mottled shoulder and rufous wing panel; black tertials (the black 'back') with narrow rufous-buff edging may be more or less prominent depending on feather positions; tail rufous-brown with diffuse dark subterminal spots, white tips below; dirty-white throat merging to pale buffish underparts. Immature similar to non-breeding, but with pale lemon yellow wash on throat and upper breast. **Similar species** Distinguished from Zitting Cisticola by larger size and contrasting plumage; from Croaking Cisticola by smaller size, finer bill and absence of strong black continuous streaking over crown, nape and mantle.

HABITS Mainly found in freshwater swamps, marshes and reedbeds; also brackish sandflats with young mangroves or tamarisks at the coast. Prominent in the breeding season when readily located indulging in singing bouts from prominent perches, typically an emergent stem above a reedbed. Regularly flies between a series of such perches, but quickly drops down into vegetation and out of sight when not singing. Skulking.

VOICE Dry rasping ratchet song, rising slightly in pitch. English name derived from similarity to slow winding clockwork mechanism. Forceful delivery with emphasis on snap cut-off: *zrrrrt*; given individually or repeated at a steady, unhurried pace when in full song. Song carries well and is characteristic in marshy areas from the start of the rains through to November.

STATUS AND DISTRIBUTION Frequent eastward from LRD, at Farafenni ferry crossing NBD, becom-ing locally common in the extensive marshlands of CRD to URD. Uncommon nearer the coast, where it occurs in some freshwater sites close to the beach on the north bank near Jinnack. In Senegal mainly in Casamance and western regions; also reported from suitable habitat on the Senegal River.

BREEDING Breeds August-November; domed woven grass structure with side entrance, low in reeds.

CROAKING CISTICOLA
Cisticola natalensis **Plate 39**

Fr. Cisticole striée

Other name: Striped Cisticola

IDENTIFICATION 12.5-16cm. Largest cisticola; females 20% smaller than males. Dark-streaked over crown and upperparts, heavy bill; grasslands and swamp margins; percussive, plopping, calls. Breeding (June-November): grey-brown above, smudgily streaked with dark brown on crown, nape and mantle. Tail brown with dark subterminal spotting and pale tips; rufous wing panel. Sides of face brown merging into white chin, throat and underparts, which are variably stained buff-orange on lower flanks. Non-breeding: evenly streaked with black on tawny background. Tail longer with tawny edges to the feathers. Immature like non-breeding adult, but yellow-stained on breast. **Similar species** Distinguished from Winding Cisticola by large size, very heavy bill and uniform streaky appearance on crown and nape as well as mantle; frog-like voice.

HABITS Found in a range of habitats with rank grasslands, usually but not exclusively associated with moist areas and swamps. Perches prominently when calling and indulges in high display flights when breeding.

VOICE Variable loud bubbly repetitive croak *breep- breep- breep*; a *pee-chuck*, rising to an explosive second note, plus more elaborate variable sequences often with frog-like or dripping water-drop qualities, or harsher stridulating sounds..

STATUS AND DISTRIBUTION A rare bird in The Gambia, historically recorded near the coast and in CRD swamps on both banks of the river e.g. Sapu. Similar paucity of records from Senegal, from central and southeastern regions. Widespread through the rest of sub-Saharan Africa.

BREEDING August-October in the Gambia, family party at Sapu, October 1974. Nest ball-shaped with side entrance, often close to the ground in a grass clump, and using soft green grasses as the dome.

SIFFLING CISTICOLA
Cisticola brachypterus **Plate 39**

Fr. Cisticole à ailes courtes

Other name: Shortwing Cisticola

IDENTIFICATION 9.5-10cms. Tiny, plain cisticola, favouring open broadleaved woodlands delivering repetitive whistling song phrase from exposed high point. Appears virtually unstreaked in the breeding season, with faint mantle streaking at other times not easily visible. Crown, nape and mantle brown, sides of face grey-brown, short pale superciliary mark to above eye only, hint of an eye-ring, mainly when viewed from below. Throat creamy-white, rest of underparts pale buff. Primaries show weak rufous-tinge, contrasting with mantle. Upperside of tail matches mantle, ill-defined dark subterminal tail spots. Non-breeding shows more rufous-tinge on crown, rump and tail, paler cheeks, and in good views faint darker streaking on mantle can be made out; tail slightly longer. **Similar species** Main identification problem is to separate from Rufous Cisticola, which has slightly less upright posture when singing, more even-toned rufous-tinge to all the upperparts, and favours drier bush on lateritic soils (see also notes on voice). Singing Cisticola is noticeably larger, and more contrasting in pattern.

HABITS Associated with farmland and degraded woodlands, generally on reasonably well-watered soils, often adjacent to aquatic habitats though not dependent on them. Persistent bouts of singing, mainly July-November, occasionally sallying out in a loop flight display. Otherwise skulks low in tangled vegetation, surprisingly seldom seen relative to frequency of wet season calling males.

VOICE A wispy series of 3-4 whistled notes on an energetic rising and falling cadence, with emphasis typically on the second note, *ti-twee-oo....ti-twee-oo*. In The Gambia song usually less complex than typical *mississipi-ing* recorded further east; may lapse to a two-note call rather similar to that of the Rufous Cisticola, though usually Siffling notes are slightly fuller in tone, and more hurried.

STATUS AND DISTRIBUTION Frequent to locally common July-November, mainly WD, but present in suitable habitat throughout. Not clear whether infrequency of records in remainder of year reflects a reduction in numbers by movements or is caused by secretive behaviour once males stop singing. In Senegal mostly to the south of The Gambia, with a few observations to the north near the coast.

BREEDING Breeds in the rains, July-September, ball-shaped woven nest low in vegetation, with side entrance.

RUFOUS CISTICOLA
Cisticola rufus **Plate 39**

Fr. Cisticole rousse

IDENTIFICATION 9.5-10cm. Sexes similar. Tiny plain cisticola of savanna woodlands; song a soft, two-note whistle. Plain rufous-brown upperparts, uniform from crown to rump and tail, which has dark subterminal spots weakly formed on the upperside, stronger on the underside, narrow white tips. Lores show weak buff patch; underparts pale, buffy wash strongest on flanks. Non-breeding birds may be slightly foxier-brown above. **Similar species** From Siffling Cisticola by more rufous body colour and complete absence of streaking. Immediately distinguished from Reed Warbler by dumpy shape, broad tail lacking projecting undertail-coverts.

A few very bright rufous individuals have been seen in The Gambia. The Foxy Cisticola *C. troglodytes* (Fr. Cisticole grisâtre) is poorly known and not recorded near our area. Similar in appearance but bright foxy-red with a blackish aspect to tail. Known from Mali, its identification would need to be confirmed by inspection of the alula in the hand, blade-shaped in *rufus*, pointed in *troglodytes*.

HABITS Prefers dry savanna woodland habitats, with numerous emergent trees, though may also be found around cultivation. Typically forages low amongst shrubs and tall grasses, usually alone or in small parties. In the wet season, breeding males sing from exposed points, usually high on bare branches of a dead tree, or clearly in view at the top of a live tree. Even in these positions, difficult to locate and easily overlooked.

VOICE Song a thin, simple whistle of two or three notes; a small sound from the tree-tops which typically comes faintly to the observer at first, clarifying only as attention is focused. The main part of the call is an unhurried two-note *tsi..eee*, or *tsee- wip*; the first note of higher pitch and breathy quality, there is a momentary pause before dropping to the more cleanly whistled final note. May sound similar to some calls of Siffling Cisticola but latter usually more hurried and less thin in tone. Song repeated methodically for several minutes; volume rises and falls with the breeze and movement of the bird's head.

STATUS AND DISTRIBUTION Frequent to locally common breeding resident found in all parts of The Gambia with more extensive woodland; most frequently encountered inland in the dry woodland areas of LRD through CRD and URD. Present near the coast in WD on both sides of the river, but uncommon, e.g. Yundum and Jinnack. Recorded more frequently in the wet season because of song behaviour. In Senegal poorly known, with scattered

307

records mostly through central and eastern regions.

BREEDING Singing males recorded June-November, but no nests definitely identified. Nearest confirmed breeding Ghana.

ZITTING CISTICOLA
Cisticola juncidis   Plate 39

Fr. Cisticole des joncs ou Cisticole commune

Other names: Common Fantail Warbler, Fan-tailed Warbler

IDENTIFICATION 9.5-10.5cm. Tiny streaky cisticola, found in rank vegetation in more open habitats throughout The Gambia; short buzzy song coordinated with dipping flight action. Predominantly buffish upperparts, strongly streaked on crown and mantle with very dark brown; plain buff-orange rump; brown tail marked by a row of dark subterminal spots and white tips, broadest on outer tail feathers, contrast stronger in the male than female. Underparts dirty buff, merging to smudgy buffy-orange on flanks. Non-breeding plumage: tail longer, slightly increased contrast in streaking, especially mantle and tertials. Immatures pale yellow-stained below. **Similar species** Song and flight display the most useful guide. From Desert Cisticola by darker colouring, especially the underparts, and clearly contrasting subterminal tail spots. Winding Cisticola is larger, with rufous wing panel, and likely to be in centre of marsh while Zitting would be mainly at edges and in adjacent scrub. From non-breeding Black-backed Cisticola which can share habitat, by proportionately smaller head, less contrasting colouring, but more patterned tail, less capped appearance.

HABITS Spends much of its time in low rank vegetation in a variety of habitats, such as field edges; margins of swamps, scrubby grassland and low bushland. In wet season present in rank field areas that become more suitable for Desert Cisticola by subsequent dry season. Frequently seen flying just above the ground between clumps of vegetation, otherwise creeps stealthily amongst the stems; may occasionally feed on the ground, usually remaining within cover when it does so.

VOICE In the breeding season sings in a persistent bouncy display flight rising a few metres above the vegetation and coordinating short bursts of wing flapping with simple, brief, raspy song-notes to produce a rhythmic display of sound and movement; *zip...zip...zip...zip* etc. This may be continued for a minute or more before the bird drops back to the vegetation, but at the height of the season displaying birds may be in the air for prolonged

periods in favoured habitats. Characteristically also calls while perched along telegraph wires where available, e.g. at Bund Road. May occasionally emit a single *zip* when disturbed in the non-breeding season.

STATUS AND DISTRIBUTION Common breeding resident in suitable habitats throughout The Gambia; readily seen at all times of the year, especially when displaying July-November. Found widely through Senegal, cosmopolitan.

BREEDING Wet season breeder, July-October. Elongated oblong nest with entrance at the top, using young grass stems bound with cobwebs, differing from all other cisticolas.

DESERT CISTICOLA
Cisticola aridulus   Plate 39

Fr. Cisticole du désert

IDENTIFICATION 9.5-10cm. Tiny, pale streaky cisticola with pale supercilium; favours dry open habitats; simple tinkling song delivered on the wing, plus wing-snap display. Breeding (June-November): dark streaking on the pale sandy upperparts tends to break into a spotted pattern on the mantle; wings brown with broad sandy edges. Tail rather uniform black-brown obscuring subterminal spots, narrow white tips; black spotting visible below. Chin and throat white, lightly stained with faded pale dirty yellow on flanks. Non-breeding: tail shorter with buff-fringed feathers, mantle more streaked. Immature similar but yellow-washed below. **Similar species** Differs from Zitting Cisticola by generally paler-toned body, uniform whitish underparts with only trace of staining around lower flanks and under the tail; upperside of tail uniformly dark. Positive identification secured by a combination of voice and behaviour (see below) as well as habitat, since Zitting Cisticola uses field areas in the wet season that subsequently become very harsh and dry later in the year.

HABITS Generally in dry, open places with short vegetation and scattered small shrubs; not associated with water. Feeds extensively on the ground; may actively run about giving appearance of tiny, dark-tailed pipit. In breeding season, male employs a swooping low display flight which incorporates audible wing-snaps, diagnostic in separating this species from Zitting Cisticola.

VOICE A repetitive monotone *ting...ting...ting* given from a perch or in cruising flight, when intermittent wing-snaps also heard. In alarm agitated *tuc..tuc* followed by swoop towards the ground in which sharp mechanical snapping also given.

STATUS AND DISTRIBUTION Not well known, but likely to be present in e.g. CRD north of the river, where resident birds in dry open woodland to field margins are suspected but not yet confirmed by records of song and display. Reasonably common in dry plains to the north of The Gambia in Senegal, becoming more widespread further north.

BREEDING Wet season breeder, July-October.

BLACK-BACKED CISTICOLA
Cisticola eximius   Plate 39

Fr. Cisticole à dos noir

Other names: Black-backed Cloud Cisticola, Nigerian Cloud-scraper

IDENTIFICATION 9-10cm. Sexes similar; some seasonal variation. Tiny, dark streaky cisticola, bright orange-buff and black tones; favours damp grassy locations; flies high in song display. Breeding male (not illustrated) shows a plain brown crown (lightly streaked in female), merging into a brown back streaked with black; orange rump; wings black-brown edged with buff; tail short and dark brown, narrow white terminal spots below. Chin to throat dirty white, merging to strong buff on lower flanks and undertail-coverts. Non-breeding plumage more contrasting; heavy black streaks confined to the top of the crown create a capped effect, plain brown sides to face and orange nape, heavy black streaking on orange-buff mantle; tail slightly longer, blackish. Legs pink. Similar species Much smaller than Winding Cisticola; from similar-sized Zitting Cisticola by subtly larger-headed, capped appearance with plain nape (non-breeding but varies with posture), shorter and less contrasting tail; overall usually brighter coloured, especially rump and flanks. Song and display diagnostic.

HABITS Associated with open grassland areas, often near water or salt-flats. Secretive. The wet season song display is dramatic, but being a small bird, is easily missed: ascends to a considerable height, all but out of sight, where it circles and undulates giving a thin wispy call, the 'cloud-scraping' habit, before descending 'like a thunderbolt' swooping into low vegetation.

VOICE Sharp dissonant *tchereet-tchereet* and during flight display a thin *tsree-tsree-tsree-tsree, preep, preep, preep*.

STATUS AND DISTRIBUTION Rare resident in the extreme south of the coastal area of The Gambia, but possibly more widespread. Locally frequent in southern Senegal and extends eastward to Nigeria and Cameroon.

BREEDING No breeding records. Likely to be a wet season breeder.

TAWNY-FLANKED PRINIA
Prinia subflava   Plate 40

Fr. Prinia (ou Fauvette-roitelet) commune

Other name: West African Prinia

IDENTIFICATION 10.5-12.5cm. Sexes similar, slight seasonal changes. One of the commonest small birds of the savanna throughout the region. Wet season breeding period: greyish above, with narrow, moderately long, graduated tail showing subterminal dark bar and whitish tips. Pronounced whitish supercilium over slightly darkened loral stripe; flight feathers narrowly edged brighter brown. Throat and breast pearly-white, flanks and undertail-coverts stained tawny-buff. Dry season: upperparts paler tawny-brown, supercilium remains, tail up to a centimetre longer; more extensively buff below. Iris pale brown, red-ringed in breeding season. Juveniles pale yellow on underparts with yellowish bill. Similar species Long narrow tail, waved side to side and bold supercilium eliminates cisticolas; from Red-winged Warbler dry season plumage by supercilium, brownish on wing not a solid chestnut panel, smaller and finer bill, calls. Separate from possible River Prinia on habitat and voice.

HABITS Found in a wide range of mainly dry habitats, where it spends most of its time at low level in amongst grass tussocks and shrubs. Buzzing alarm calls are characteristic background sound in the bush. Commonly found in pairs, or small groups. Frequently cocks and swings tail side to side. Singing males may take up a prominent post well off the ground in an exposed location, like many cisticolas.

VOICE Short persistently repeated *bzzzt....bzzzt* call, wheezy in tone, diagnostic and frequently heard when disturbed, sometimes a more extended version with increased anxiety. Song a rhythmic monotone *cli-cli-cli-cli-cli-cli-cli-cli*, reminiscent of Grey-backed Camaroptera, from which it is separated by usually quicker delivery, more liquid, less mechanical quality, lacking accompanying clicks delivered by camaroptera.

STATUS AND DISTRIBUTION Abundant resident in most of The Gambia, though seldom penetrating the densest thickets. Found throughout Senegal.

BREEDING Purse-shaped nest of woven grass at one to two metres off the ground, August-December in the Senegambia. Eggs stippled pink-white sometimes sprinkled with reddish spots. River Prinias lay blue eggs.

RIVER PRINIA
Prinia fluviatilis   Not illustrated

Fr. Prinia aquatique

IDENTIFICATION 11-12cm. First described from the Senegal River valley in 1974, closely resembles the Tawny-flanked Prinia. **Similar species** Has greyer upperparts than Tawny-flanked and much whiter underparts, while the tail is proportionately even longer than that of Tawny-flanked when compared to the wing length.

HABITS Closely associated with rank waterside vegetation, along river banks, irrigated rice fields, and marshy areas.

VOICE Interval between calls is subtly but consistently shorter.

STATUS AND DISTRIBUTION Not yet recorded in The Gambia, but probably present; nests have been found in Guinea-Bissau as well as along the Senegal River in northern Senegal. Recorded from Niger, Chad and Cameroon.

BREEDING Deep flask- or hammock-shaped nest, rather resembling that of Zitting Cisticola, in rank waterside vegetation often just above water, September. Recently fledged chicks seen up to January.

RED-WINGED WARBLER
Heliolais erythroptera   Plate 40

Fr. Fauvette à ailes rousses.

IDENTIFICATION 14cm. Sexes similar, but marked seasonal variation. Small and long-tailed, recalling a robust Prinia. In breeding plumage, June-December, grey head and mantle contrasts with chestnut wing; throat white, clean creamy lower underparts develop pale apricot bloom. Tail long, graduated, rufous-brown at base, becoming greyer with subterminal black bar, white tips showing below. In strong light may appear slightly darker on lores to ear-coverts. Dry season birds: head and mantle pale coppery pink-brown, the chestnut wing does not contrast so strongly and the underparts are duller pale buff. Immature resembles pale non-breeding plumage. **Similar species** Wet season plumage distinctive, the bird being heavier-billed, larger and more contrasting than Tawny-flanked Prinia. In dry season it lacks prinia's clear supercilium. Structure helps separate non-breeding plumage from all plain cisticolas, especially Singing Cisticola, which have a proportionately broader and shorter tails.

HABITS Found singly or more often in small parties, usually moving with industrious activity at low level through dense vegetation in areas where small bushes and shrubs are abundant, notably fallow fields. May perch in open, but difficult to observe clearly or for any length of time. Intermittently noisy, regular contact calls often being the first feature to draw attention to the bird's presence. When giving alarm flicks wings.

VOICE A loud down-slurring *tseeu-tseeu-tseeu-tseeu*, at medium pitch, similar to alarm calls of some cisticolas, but less throaty. Sustained, loud *drreee-drreee-drreee* and an array of other sounds, including a thin higher pitched *psee-psee-psee*.

STATUS AND DISTRIBUTION Uncommon resident WD and LRD, has been recorded in all divisions but only rarely in CRD and URD. Most records from late dry season and start of rains; seen much less frequently once wet season vegetation established. In Senegal mainly limited to Gambia River valley, marking the north-western limit of its range.

BREEDING Thought to be mainly a wet season breeder in The Gambia, seen with fledglings in December; a domed nest of grasses fixed between leaves in the manner of some cisticolas.

SCALY-FRONTED WARBLER
Spiloptila clamans   Not illustrated

Fr. Prinia grillon ou Fauvette à front écailleux

Other names: Cricket Warbler, Cricket Longtail, Scaly-fronted Prinia

IDENTIFICATION 11.5cm. Shaped like a prinia, but with an additional pair of tail feathers, and distinguished by black and white feathers flecking the crown and wing-coverts. Otherwise pale sandy-brown mantle, white tips and black subterminal markings to graduated tail, especially prominent on the underside; male grey-naped, female brown nape merges with mantle. Often in parties, very active; repetitive, rapid contact call.

STATUS AND DISTRIBUTION A very small inhabitant of acacia scrub areas associated with the southern edge of the Sahara desert; moderately common in northern Senegal but not found in The Gambia.

YELLOW-BREASTED APALIS
Apalis flavida   Plate 40

Fr. Apalis à gorge jaune

IDENTIFICATION 11.5cm. Sexes similar. Small inhabitant of forest thicket canopies, broad yellow breast patch prominent. Plain grey from top of head to below eye, cleanly separated at nape from greenish mantle; greenish fringes on brown flight feathers and greenish wash on tail maintains impression of uniform olive-tinged upper body. Narrow whitish tips to two central tail feathers seldom show in field. Throat and belly bright white, separated by broad, well-defined bright yellow breast band. Gambian birds do not show any sign of a black mark on the lower edge of this yellow band; recent inspection in the hand confirms this. Juvenile paler yellow on the breast. **Similar species** Neat contrasting head pattern, distinct yellow patch and tail length in proportion to body all separate from Olive-green Camaroptera. Individual lacking tail separated from Green Crombec by much more prominent and well-defined yellow breast-patch, no supercilium.

HABITS Primarily found in forest thickets and strips of riverine and swamp forest, sometimes in dense coastal thickets and mangroves. Usually in pairs or small parties, frequently in mixed species feeding parties. Moves with agility through branches and leaves of canopy, and lower in dense vegetation.

VOICE Sings by steady repetition of a wooden-toned double note, repeated with metronomic regularity at constant pitch; *tik, k-tik k-tik k-tik k-tik k-tik* (emphasis on *tik*); maintained for prolonged bouts of around 45 seconds, with irregular pauses every few seconds. Especially vocal at onset of rains. When agitated may accelerate call, becoming very similar to camaroptera, but basic rhythm still discernible. Duetting behaviour reported.

STATUS AND DISTRIBUTION Resident, uncommon overall, though locally frequent in a few small areas of suitable habitat including remnant forest thickets and secondary thickets near the coast in WD, e.g. Abuko, Tanji; also NBD, and present in riverine and swamp forest near water in CRD. Not recorded from Senegal.

BREEDING Adults with fledgling in Banjul December 1996. Builds half-domed nest of lichen and moss, and has been recorded using old weaver nests.

GREY-BACKED CAMAROPTERA
Camaroptera brachyura   **Plate 40**

Fr. Camaroptère à dos gris

Other name: Bleating Warbler

IDENTIFICATION 13cm. Sexes similar, limited seasonal variation. A compact bird, dark grey and mossy-green, short cocked tail; frequent whining call. Entire head, mantle and back grey, contrasting

with olive-green of folded wings. Greyish tail shows hint of dark subterminal bar in certain lights, not a field mark. In non-breeding dry season plumage grey areas more brown-toned, but retain contrast with wings. Underparts grey over throat to upper breast, paling to a whitish patch in middle of belly, yellowish underwing-coverts show at wing edge and bright tawny thighs contrast with body if seen. Bill longish and black, eye red-brown, legs flesh-brown. Juveniles may show a variable wash of yellow on the underparts, and more olive-brown tone to the upperparts, otherwise similar. **Similar species** See Olive-green Camaroptera.

HABITS Found in a wide range of habitats, typically in dense cover usually close to the ground. Flies direct and low with rapid wingbeats, and may perch briefly in open with tail cocked and wings slightly drooped. Presence usually made obvious through persistent bleating calls.

VOICE Highly vocal. Most frequently a whining nasal bleat from thick cover, *zbeeee.....zbeeee*; song a vigorously delivered repetitive phrase of crisp snapping notes *twik.twik.twik.twik-twik.twik.twik.twik.twik.*, continuous for prolonged sequences with brief irregular hesitations, each note articulated simultaneously with underlying rapid doublets or triplets of clicks, creating a distinctive hollow echo effect. The rhythmic phrase is on constant pitch; hard tonal quality and mechanical clicking distinguishes from more fluid Tawny-flanked Prinia and more measured, wooden and (usually) double note of Yellow-breasted Apalis. Also a variety of additional whistles and nasal buzzes.

STATUS AND DISTRIBUTION Common to locally abundant resident throughout The Gambia wherever dense vegetation, even in quite small patches, can be found; does not penetrate mangroves to any significant extent. Found throughout Senegal, sparser in the north.

BREEDING Wet season, June-November. Binds several large leaves together close to the ground, and constructs nest from dry grasses within protection of the leaves.

OLIVE-GREEN CAMAROPTERA
Camaroptera chloronota   **Plate 40**

Fr. Camaroptère à dos vert

Other name: Green-backed Camaroptera

IDENTIFICATION 11.5cm. *C. c. kelsalli*. Sexes similar. Small, short-tailed skulker in secondary thicket. Breeding male uniform dull olive above, hint of brown on crown and distinctive rusty-buff patch around eye and ear-coverts; yellow patches at edge of folded wing betray yellowish underwing-coverts. Below uniform grey from chin to undertail-coverts, with small patch of tawny feathers around the base of each thigh. Female similar, but grey underparts quickly give way to whitish belly. Legs yellow-brown. Juveniles pale yellow-buff from chin to belly, marginally more yellow-tinged on upperparts. **Similar species** From Grey-backed by lack of contrast between wings and mantle, rusty patches on sides of head, greyer below; juveniles from Green Crombec by larger size, no real contrast between crown and body and absence of supercilium.

HABITS Usually singly or in pairs in dense secondary vegetation, thickets etc. Vocal and more easily heard than seen.

VOICE Soft contact call *wheet..wheet*. Song a remarkably sustained and quickly repeating series of clear whistles, very slightly descending the scale, slowing and diminishing in volume as if tiring, *tu-tu-tu-tu-tu-tu-tu-tu-tu-tu.*

STATUS AND DISTRIBUTION Rarely recorded species; two published sight records from The Gambia, at Abuko and Pirang, and seven records from Senegal, in Casamance and Niokolo Koba.

BREEDING Nest of down and silk within folded leaves of bushes; not recorded in Senegambia.

GREY-BACKED EREMOMELA
Eremomela icteropygialis **Not illustrated**

Fr. Erémomèle gris-jaune

Other name: Yellow-bellied Eremomela

IDENTIFICATION 10cm. Similar in size and shape to Green-backed Eremomela, but slightly shorter-tailed, pale grey over crown and mantle and grey-brown on the wings, not greenish above; also more extensive white to buff on the underparts, with yellow (paler than Green-backed Eremomela) restricted to a smaller area on the belly only. Quick moving, arboreal and usually in small parties.

STATUS AND DISTRIBUTION This tiny warbler replaces the Green-backed Eremomela in the drier acacia-steppe habitats that lie to the north of The Gambia. Widespread in northern Senegal, with a zone of overlap around 15°.

GREEN-BACKED EREMOMELA
Eremomela pusilla **Plate 40**

Fr. Erémomèle à dos vert

Other name: Senegal Eremomela

IDENTIFICATION 10-11cm. Sexes similar. Common, tiny, very pale greyish and yellow warbler. Head to nape pale grey; rest of upperparts soft greenish, slightly yellower-toned on the rump. Throat and upper breast white to pale buff, changing to bright yellow over lower breast and entire belly. Small pointed bill dark, eye brownish and legs flesh-brown. **Similar species** Smaller than any migrant warbler (e.g. *Phylloscopus*), open-faced expression, and combination of pale grey and white front half, and green and yellow rear half distinctive.

HABITS Found wherever there are reasonable numbers of broadleaved trees, notably *Combretum* and *Terminalia*, including open woodland, cultivation and savanna woodland. Usually in industrious parties flitting rapidly from tree to tree, searching busily through branches and leaves.

VOICE Soft squeaking *tuuu*, also distinctive soft chattering trill on one note during the breeding season.

STATUS AND DISTRIBUTION Common resident throughout The Gambia, but particularly prominent in LRD, recorded most frequently April-August. Widespread in southern Senegal, largely replaced by Grey-backed Eremomela in the north, though a few far northern populations exist where *Combretum* is present.

BREEDING Recorded in The Gambia in April, thought to breed February-August elsewhere, though few observations. Nest a very small compact cup of leaves and silk at no great height in trees or shrubs.

GREEN CROMBEC
Sylvietta virens **Plate 40**

Fr. Crombec vert

IDENTIFICATION 8-9cm. *S. v. flaviventris*. Sexes similar. A tiny, dumpy, practically tailless bird of thickets, easily overlooked. Head and nape brownish grey, upperparts olive; yellowish-olive edges on flight feathers. Smudgy grey-white supercilium ill-defined but noticeable. Buff-tinged chin and brownish throat with variably developed pale yellow smudge in centre of breast, sometimes quite extensive and bright, on otherwise pale greyish-brown underparts, save for small clean white area on lower belly. Relatively long straight bill is dark brown on the upper mandible, fleshy on the lower mandible. **Similar species** Distinguished from Lemon-bellied by supercilium; two-toned, not all-black, bill which takes good views to determine; generally less extensive yellow on underparts.

HABITS A bird of dense thickets and tangled secondary forest vegetation, also venturing into surrounding bush where cover is heavy; particularly associated with creeper-clad trees. Often in small family parties and mixed species feeding groups, active and restless, frequently flicks wings. Maintains regular contact calls, though at longish intervals, and sings from within thick cover near top of tree, where it is difficult to observe.

VOICE Principal contact call a soft but crisply defined and sibilant *prrt...prrrt*, given at long intervals, usually the first indicator of the presence of Green Crombec in dense thickets. Distinctive twittering song carries well; a cheerful, hurried, downscale tumble; in The Gambia noticeably followed by a brief turn back up the scale over the last two notes *diddle-i-dee di-dee di-di.*

STATUS AND DISTRIBUTION Frequently encountered in a few small thicket areas near the coast, where it has been recorded in all months. Like all forest and thicket birds in The Gambia, rare overall and at risk from tree felling. Known from the Casamance region of Senegal.

BREEDING Seen gathering nest material in July in The Gambia; suspended nest of woven grass and silk.

NOTE Once regarded as a separate species, *Sylvietta flaviventris,* with the English name White-bellied Crombec (Bannerman 1953), now recognised as a western subspecies of Green Crombec, *S. v. flaviventris* by Hall & Moreau (1975). Confusingly both Gore (1990) and Morel & Morel (1990) use White-bellied Crombec as an alternative English name for *Sylvietta denti hardyi,* although this is more usually known as Lemon-bellied Crombec. This leads to some doubt over reliability of Lemon-bellied Crombec records from The Gambia (see below).

LEMON-BELLIED CROMBEC
Sylvietta denti   Not illustrated

Fr. Crombec à gorge tachetée

IDENTIFICATION 8-9cm. *S. d. hardyi.* Sexes similar though female generally duller than the male, and note that local race of Green Crombec also has yellow belly patch. A tiny forest warbler of typical crombec tailless shape. Greyish head and olive-green upperparts, pale on the throat, turning to

dull yellow underside from breast to belly. Bill comparatively short for a crombec and both mandibles are black; eye orange-brown and legs described as grey-brown or dusky pink. Young birds not described. **Similar species** Western form rather brighter green above and greyer-headed than nominate; all-black, shortish, bill and absence of smudgy supercilium important features to separate from Green Crombec; see that species for further discussion.

HABITS Similar in most ways to Green Crombec, though possibly even more closely associated with dense secondary forest, where it may also prefer to stay higher in the canopy.

VOICE Clear squeaky whistle, breaking to a clear sequence of even notes sliding slightly down the scale: *creek, p-p-p see-see-see-see-see-see.*

STATUS AND DISTRIBUTION Uncertain. Three published sight records from The Gambia, the first from Brufut in 1964, where Green Crombec is now regularly seen. Not recorded from Senegal. In view of ambiguity in English name in published records (see note above) and modern observations of Green Crombec, possibility of confusion with Green Crombec *S. v. flaviventris* must be considered probable. Elsewhere *S. d. hardyi* known from Guinea east to Nigeria.

BREEDING Purse-shaped nest suspended from a twig, using petioles and silk; decoration with litter makes effective camouflage.

NORTHERN CROMBEC
Sylvietta brachyura   Plate 40

Fr. Crombec

Other name: Nuthatch Warbler

IDENTIFICATION 9cm. Sexes similar. Minute, with a unique combination of tailless shape, pale grey above, soft buff-orange below. Entire upperparts soft pale blue-grey with a brownish wash. Narrow darker brown line through lores to earcoverts is not strong, but picks out creamy supercilium effect where colour fades at edges of crown. Underparts pale gingery over breast and flanks, merging imperceptibly to white in centre of throat and belly. Bill moderately long, dark pinkish with horn tip. Juvenile birds show brown tips to wingcoverts. **Similar species** Unlikely to be confused given reasonable views, tailless dumpy appearance and characteristic climbing activity distinctive.

HABITS Active and quick moving, though surprisingly unobtrusive, mainly encountered in pairs and small groups. Climbs up and head-first down bark with agility, holding body away from the surface.

313

VOICE Maintains a soft but emphatically delivered contact call, *prrrt....prrrt* or soft very high pitched *tsweet-prrt*, noticeable only at close range. Song composed of thin notes, lisping and sibilant, soft in character, basic phrase first rising in pitch and volume, before falling away down the scale in a single smooth phrase, *psee-see-psi-psi-si-suuu* and variations. In early rains may maintain sustained song, more twittering in character but with recognisable elements of the above.

STATUS AND DISTRIBUTION Frequent breeding resident throughout The Gambia wherever there are reasonable numbers of trees, including scattered small trees on old cultivation, though often unnoticed. Particularly frequent in LRD. Widespread throughout Senegal, less so in the far north, where like the Green-backed Eremomela it is associated with *Combretum*.

BREEDING Purse-shaped nest attached to a twig, outside camouflaged with litter; recorded breeding in all months except Jan-Feb and Oct-Nov., with an apparent peak June-July.

ORIOLE WARBLER
Hypergerus atriceps   Plate 35

Fr. Moho à tête noire ou Noircap loriot

Other name: Moho

IDENTIFICATION 20cm. Sexes similar. Distinctive green and yellow medium-sized bird with curiously 'silvered' black head; found in stream-side thickets and heavy undergrowth. Entire head to nape black, the stiff feathers narrowly fringed with white, creating an unusual silvered impression over the crown, nape and cheeks. Rest of upperparts and graduated tail uniform light olive green, sometimes untidy. Chin, throat and upper breast plain black, cleanly divided from bright yellow underparts which may be lightly tinged with rufous-brown on flanks and undertail-coverts. Bill long, slightly decurved, greenish black; eye red-brown; legs brown. Duller juvenile is grey-eyed. Similar species Distinctive, superficial resemblance to some weavers completely overridden by bill shape. Absence of black on wings and body eliminates all true orioles.

HABITS Skulking bird of tangled thickets, usually but not exclusively near water, sometimes mangroves. Remains mainly out of sight at low level. Though silent for long periods, breaks out into startlingly loud bursts of arresting duetted song, particularly at dawn and dusk. Pairs display with weird simultaneous sinuous swaying action, usually in vicinity of nest.

VOICE Remarkable and very characteristic. Usually the male begins with a series of three to five, (typically four) clear-toned whistled notes, smoothly slurred in a (usually) rising arpeggio sometimes switching to downscale or other variation, adding subtle embellishment to individual notes. In simultaneous accompaniment the female cycles through continuously repeating sequences which open with a stuttering staccato, accelerating to a pulsing vibrant churr, sustained for as long as the male keeps going. A pair in full territorial display duet vigorously. Partial versions of a few arpeggios from one bird also frequently heard; may resemble similar, but thinner, and less persistent rising whistle produced by Western Bluebill, which shares similar habitat.

STATUS AND DISTRIBUTION Restricted to dense and damp freshwater habitats, making it vulnerable in The Gambia. Locally frequent resident in Abuko, and in mangroves around Kotu. Found at most sites near fresh water in WD, few records from LRD, becoming locally frequent mainly in denser riverbank vegetation and small patches of swamp forest through CRD and URD. In Senegal not found to the north of The Gambia, confined to Casamance and the extreme south-east.

BREEDING Large tangled nest of creepers, grass and leaves, usually suspended from underside of palm frond, often near or directly over water. Decorated with various items of litter and twigs resulting in untidy appearance successfully creating impression of fallen forest debris; the side entrance not obvious to casual inspection. Breeding in the Gambia, June-November.

NOTE DNA analysis suggests links with *Apalis* spp. and *Camaroptera* spp. in Sylviidae. Probable nearest relative is Grey-capped Warbler *Eminia lepida* of East Africa.

YELLOW-BELLIED HYLIOTA
Hyliota flavigaster   Plate 41

Fr. Hyliote à ventre jaune

Other name: Yellow-bellied Flycatcher

IDENTIFICATION 13cm. Sexes similar but distinguishable. Beautifully marked in blue-black above, peach below, with a white wing-bar; jizz resembles sunbird or warbler. Male upperparts glossy blue-black, snow white bar across the wing-coverts, sometimes slightly extended along full length of secondaries; whitish rump usually does not show; tail black. Below clean soft pale peachy-yellow. Female echoes the same pattern, but dull slaty above and duller pale buff below. Slightly extended and narrow bill black above, bluish-grey at the base. Immature resembles female, but with buffy tips to upperpart feathers creating a washed-out appearance. **Similar species** From vagrant Violet-backed Hyliota by white wing bars and white, not black, underwing-coverts. May resemble Violet-backed Sunbird, but white wing-bars, yellow-buff underside and shorter bill distinguish it.

HABITS Favours well-wooded dry savanna, particularly areas with a good diversity of trees. An agile and restless gleaner from foliage and branches, climbing hastily, looking over and under likely sites and moving on quickly to the next tree. Dumpy short-tailed appearance in flight. Usually in pairs or small family parties, often associating with mixed species foraging parties.

VOICE A slight two-syllable whistle described, and excited chattering calls. In flight may produce a rapid *tiki.teeep...tiki.teeep* second note with a slight rising inflection.

STATUS AND DISTRIBUTION Uncommon resident in all divisions of The Gambia, locally frequent at a few favoured sites near the coast and in LRD. Thinly scattered and probably affected by loss of woodlands, though historically noted to be found at low density in most of its range. Found a little way to the north of The Gambia near the coast, but mostly confined to Casamance and southeastern Senegal.

BREEDING Pair with two recently-fledged young in October is the only breeding record in The Gambia.

VIOLET-BACKED HYLIOTA
Hyliota violacea   Not illustrated

Fr. Hyliote à dos violet

IDENTIFICATION 13cm. Rare forest species, similar to the Yellow-bellied Hyliota, but lacks white on wing-coverts in West African *H. v. nehrkorni*. Male has dark violet gloss to upperparts, greyish rump; underparts whitish, faintly washed tawny-yellow. Female was for a long time not recognised since she is dull slaty on the upperparts and much darker tawny (darker than Yellow-bellied Hyliota) on the underparts. Immature not described. **Similar species** Male from Violet-backed Sunbird by shorter bill, pale chin and throat. Both sexes distinguished from Yellow-bellied Hyliota by black, not white, underwing-coverts.

HABITS Very poorly known, but generally regarded as the forest equivalent of the Yellow-bellied Hyliota.

VOICE Not recorded.

STATUS AND DISTRIBUTION Uncertain. A single sight record from a mango orchard near a heavily vegetated stream in coastal south Gambia is the only record; otherwise known from a few records from Sierra Leone east through the thick forest belt.

GREEN HYLIA
Hylia prasina   Plate 38

Fr. Hylia verte

IDENTIFICATION 12cm. Sexes similar, male larger. Compact dull olive bird, with heavy whitish supercilium, stout bill for its size, difficult to observe in dense habitats. Entire upperparts, including the broad tail, dark olive-green; male richer green than female. Broad off-white supercilium arcs from base of the bill over eye, with a solid olive-black line through the lores to ear-coverts. Underparts contrasting pale grey. Legs olive-grey. Juvenile paler above, suffused with yellowish-green below, including the supercilium (which may remain yellow-stained as rest of plumage matures); yellow-horn bill and yellowish legs. **Similar species** Broad, arched supercilium very distinctive once seen.

HABITS A bird of canopy and dense vegetation in coastal forest patches and thickets, where it remains within cover for most of the time. May fly out from thicket to canopy of leafy adjacent tree, but never seen away from thick cover. Best detected by loud voice.

VOICE Suddenly emits two forceful clear-toned notes at measured pace, of equal length, emphasis, and pitch; *tseee-tseee* or *tsooo-tsoooo*, usually followed by longish silence before repetition. May be heard with regularity and at considerable distance. Characteristic insistent scolding dry rattles, *trrit, trrrit, trrrrrit*.

STATUS AND DISTRIBUTION Rare overall in The Gambia due to a shortage of required habitat; within a few small areas frequent to locally common breeding resident of WD. At the northern limit of its range in The Gambia, known only from coastal Casamance in Senegal.

BREEDING Extended breeding season; in The Gambia mating in August, feeding newly fledged young in August and January. Builds a large, rough, globular nest on low branches using leaves and fibres, with a narrow off-centre entrance near the top.

WILLOW WARBLER
Phylloscopus trochilus   **Plate 38**

Fr. Pouillot fitis

IDENTIFICATION 11cm. Sexes similar. Slim, plain leaf warbler, uniform greenish or yellowish-brown above (varies with moult and wear); striking narrow pale yellowish supercilium to behind eye, emphasised by narrow olive-brown eyestripe. Wings and tail brown; wing relatively long. Underparts whitish with variable yellowish wash from throat, weakening towards belly. Migrant adults arrive on completion of full moult, rather strong yellow-green above, subsequently fading. First-year arrivals extensively yellow-washed on underside. Legs usually yellow-brown, sometimes dark. All moult again January-December, creating second wave of birds strongly tinted with greenish and yellowish. **Similar species** From Chiffchaff by combination of longer primary wings, more pronounced supercilium, slimmer build, occasional not frequent wing-flicking, and pale leg colour; see also voice; from Wood Warbler by smaller size, lack of sharp transition from yellow breast to white belly; Western Bonelli's Warbler has 'open face' with indistinct dark stripe through eye, more uniform whitish below. First-winter Willow Warbler smaller, quicker moving, more patterned head than Melodious Warbler.

HABITS Favours canopies and shrub cover at a variety of heights, alert and agile with occasional sallying flights. Attracted to flowering trees where several may be found hunting abundant insects.

VOICE Call a rising, near disyllabic, *hooeet*; song, heard on arrival and prior to departure, a genuine warbled phrase of clear, thin-toned notes tumbling rapidly down scale.

STATUS AND DISTRIBUTION Passage migrant to all parts of The Gambia, locally frequent in season, dispersing widely further south. Most numerous October-November and again March-April. Main arrivals precede Chiffchaff and some present in early August. Mainly a passage migrant through most of northern Senegal, probably with smaller numbers overwintering in the south.

CHIFFCHAFF
Phylloscopus collybita   **Plate 38**

Fr. Pouillot véloce

IDENTIFICATION 10-11cm. Sexes similar. Most frequently encountered of the three regularly recorded small greenish Palearctic leaf warblers seen in The Gambia through the dry season. Similar to

Willow Warbler, but generally more olive-brown above, but subtle variations in precise tone common. Short smudgy supercilium with narrow darkish eyestripe. Wings and tail brown, primaries relatively short. Below buffy over throat and breast, some light yellow wash, paler towards belly; small yellow patch at bend of wing notable but not unique. Legs usually dark to blackish, occasionally yellow-brown. First-winter birds similar, but yellower-washed. **Similar species** From Willow Warbler by shorter wings, relatively stronger eye-ring, weaker supercilium and stubbier blackish bill; with familiarity also by subtly dumpier, rounder-crowned shape, duller appearance; see also voice. From Western Bonelli's Warbler by smaller size, dumpier shape, smudgier underparts, head pattern, usually less green and white body colours.

HABITS Lively and mobile, encountered in bushes, notably coastal mangroves, as well as areas of scattered trees. Often several birds seen working same tree alongside other canopy-feeding species. Regularly flicks wings and wags tail.

VOICE Call a simple *hweet*, more monosyllabic than Willow Warbler but still with 'enquiring' rise in pitch. Song diagnostic, recorded sporadically in The Gambia in December and January, but do not expect to hear it; a monotonous run of crisp, evenly weighted and unhurried notes *chiff chaff-chiff-chiff-chaff-chiff-chaff* etc.

STATUS AND DISTRIBUTION Frequent winter visitor to all parts of The Gambia, September-May, peak numbers December-March. Widely recorded throughout Senegal, less frequent in the south. Primarily a migrant to the southern Sahel, with virtually no dispersal further south.

WOOD WARBLER
Phylloscopus sibilatrix   **Plate 38**

Fr. Pouillot siffleur

IDENTIFICATION 12cm. Sexes similar. Largest and brightest of the migrant Palearctic leaf warblers, but very rarely recorded in Senegambia. Brown upperparts brightly washed greenish-yellow when fresh (first-winter and spring adults), otherwise duller. Pronounced long yellow supercilium, long pointed wings. Face, throat and breast yellow, cleanly separated from pure white belly and undertail. Legs yellow-brown. **Similar species** From Western Bonelli's Warbler by strong yellow supercilium, contrast between yellow and white on underparts, larger and more elongated overall.

HABITS A canopy-loving species, frequently droops wings.

VOICE Call a quick *whit*; song an accelerating sequence of silvery notes on a monotone, breaking to a shivering trill.

STATUS AND DISTRIBUTION Very few positive coastal records, from NBD and WD. Two netted in northern Senegal, with a handful of possible sight records, including one from Casamance. Bulk of population spend northern winter in the Guinea forest belt of West Africa; it seems possible that small numbers may briefly transit through our region each year, but are seldom detected.

WESTERN BONELLI'S WARBLER
Phylloscopus bonelli   Plate 38

Fr. Pouillot de Bonelli

IDENTIFICATION 11.5cm. Sexes similar. A plain-faced warbler, with weak supercilium, eyestripe almost non-existent, with dark eye and rounded crown. Upperparts grey-brown, underparts wholly and characteristically silvery-white. Wings (relatively long) and tail noticeably washed yellow-green forming wing panel when fresh, but note this is weak in autumn adults prior to January-December moult. Distinctive yellowish rump contrasts with rest of body in adults, but less so in rather greyer first-winter birds, almost always difficult to see well. Legs brown. Similar species From Chiffchaff and Willow Warbler by larger size, whiter underparts usually contrasting with brighter yellow-green areas on upperparts, plain head; from even longer-winged Wood Warbler by lack of yellow on face and breast and absence of strong yellow supercilium.

HABITS Mainly associated with acacias on the sub-Sahelian plains in its winter range, but in The Gambia found in a range of lightly wooded areas, field margins, even forest edges. Usually seen in ones or twos, inclined to be shy. Does not flick tail to same extent as some other *Phylloscopus*.

VOICE Call clearly disyllabic *huu-eeet*; sings with a simple trill, reminiscent of Wood Warbler but shorter and slower; occasionally sings on migration.

STATUS AND DISTRIBUTION Uncommon but regular winter visitor, recorded from all parts of The Gambia, primarily November-March, with a few records both before and after these dates. In Senegal widespread across northern and central parts of the country, especially towards the north-west, September-April; not recorded south of The Gambia.

ORPHEAN WARBLER
Sylvia hortensis   Plate 38

Fr. Fauvette orphée

IDENTIFICATION 15cm. Sexes differ. Large warbler with dark above and pale below. Male has blackish head, merging from crown to nape into grey-brown upperparts; slightly browner wash on wings; tail duskier grey with white outer margins and narrow white tips to all but central feathers. Throat white, sharply demarcated from face below line of eye; rest of underparts are whitish, washed with pinkish-buff on breast, flanks and undertail. Eye normally bright yellowish-white, but occasionally dark brown; legs grey-brown. The female has less contrasting dark head, browner upper- and underparts and a dark eye; retains white outer tail feathers. First-winter birds comparatively uniform and dark-eyed, with dark ear-coverts and pale brown tone to outer edges of outer tail feathers. Similar species Large size is useful indicator, pale-eyed males distinctive, females and juveniles separated from Common Whitethroat by size and overall greyer tones. Barred Warbler *Sylvia nisoria* (Fr. Fauvette épervière) is a Palearctic vagrant to northern Senegal (two recent records in Djoudj). Similar in size to Orphean and larger than Garden, immature Barred is best identified by blue-grey upperparts with pale edges to wing-coverts and tertials, white in outer rectrices, barred undertail-coverts and yellowish eyes (brown in younger birds).

HABITS On migration utilises acacia scrub and broad-leaved bushland. Hops strongly within canopy or cover, feeding with hasty pecking action; characteristic pot-bellied posture, occasionally cocks tail.

VOICE Not vocal in winter; a *tut-tut* call, also a sharp *tac*, and alarm *trrrr*. Song fluid and thrush-like, lacking harsh notes.

STATUS AND DISTRIBUTION An uncommon species in The Gambia, with most records coming from well-watched coastal areas, November-March. Recent ringing results suggest probably under-recorded in coastal scrub, especially November-December. Seen on passage in northern Senegal, September-October and again March to May.

GARDEN WARBLER
Sylvia borin   Plate 38

Fr. Fauvette des jardins

IDENTIFICATION 14cm. Sexes similar. The plainest, most nondescript warbler, uniform fawn-grey tones, a large dark eye, grey legs and retiring habits.

317

Upperparts plain light grey-brown with a hint of olive. Wings and tail slightly darker brown. Hint of indistinct pale grey supercilium from bill to eye, face plain and open. Throat and belly whitish, washed brownish on flanks and undertail-coverts. Bill blackish, stubby, eye prominent and dark brown; legs grey. Immature similar, very lightly yellow-stained on the underparts. **Similar species** Nondescript appearance could be confused with Olivaceous Warbler, which has longer bill, flatter crown and is not so pot-bellied; *Acrocephalus* warblers have long undertail-coverts. From other female *Sylvia* warblers by bland face, plain tail and largish size. Pale Flycatcher is slimmer but longer, longer-billed and has very different habits.

HABITS Winter migrant to well-wooded areas in savanna and forest clearings, secondary scrub on old farmland, edges of mangroves etc. Remains within canopy of large bushes and trees much of the time. Attracted to water where available.

VOICE Mainly silent in wintering area, but uses quiet subsong all through the year, and full song recorded from January. A musical warbling at even pace, mainly in phrases of a few seconds. Contact call *chak-chak* and a harsher *charrr* alarm.

STATUS AND DISTRIBUTION Regular winter visitor, uncommon to locally frequent for short periods, recorded from all parts of the country September-April, peaking around November, with passage usually more prominent at this time than on northward movement. Passage prominent through northern Senegal October-November and again April-May, no overwintering birds in north, but present in December-January in Casamance.

BLACKCAP
Sylvia atricapilla   Plate 38

Fr. Fauvette à tête noire

IDENTIFICATION 13cm. Sexes differ. A plain, medium-sized, warbler, readily distinguished by clean-cut cap of black in the male or reddish-brown in the female. Upperparts mainly uniform grey-brown, paler grey on cheeks and nape, darker on primaries. Cap in both sexes cuts off level with the top of the eye; lores and throat pale grey. Rest of underparts greyish, tending to slightly browner in the female, the undertail-coverts are pale grey with white fringes. Legs grey. First-winter males show brown flecking on black cap. **Similar species** With good views distinctive; without view of head resembles Garden Warbler, but greyer in tone.

HABITS Most regularly encountered in riparian woodlands and bush, including mangroves, singly

or in small groups. May raise cap into short crest and flick wings and tail.

VOICE Usually quiet although full song may be heard, particularly in February-April prior to departure. A rich varied warble, with musical and harsher notes, phrases generally shorter than Garden Warbler and ending in loud rising flourish. A *tac-tac-tac* call and typical *Sylvia* churr.

STATUS AND DISTRIBUTION Winter visitor to all parts of the country from October, peaking in November, when it may be locally frequent, with a few seen thereafter each month to April/May. Numbers migrating to West Africa vary from year to year.

COMMON WHITETHROAT
Sylvia communis   Plate 40

Fr. Fauvette grisette

Other name: European Whitethroat

IDENTIFICATION 14cm. Sexes differ. Slim warbler, soft grey-browns, with rufous-brown wing edges and white throat; tends to look front-heavy, with an angular head. Breeding male pale grey over head and sides of face, whitish eye-ring; merges on nape to mid brown mantle and back, greyer rump; tail long, darkening along its length, white outer feathers. Chestnut panel on folded wing with darker feather centres. Chin and throat white, rest of underparts pale grey suffused pink, buffer on flanks to undertail-coverts. Female and first-winter lack grey tones on head, showing uniform brown from head to mantle, outer tail pale buff. Legs yellow-brown. **Similar species** From female and immature Spectacled and Subalpine Warblers by larger body size, relatively longer primaries, and heavier bill with more extensive dark tip. Does not wave tail as Spectacled. Chestnut wing-panel more or less absent on female Subalpine. Larger Orphean females greyer, also lacking wing panel.

HABITS Favours dense thickets, areas behind beach sand dunes, also orchards and areas with numerous small trees. May perch in the open briefly to view surroundings, but keeps mainly in cover.

VOICE Drawn-out *chhhh* or *chairr* call. Scratchy song occasionally heard in The Gambia.

STATUS AND DISTRIBUTION Winter visitor, mainly to coastal districts October-April, peak numbers January-March. Sporadic records inland to URD, where possibly under-recorded. Observed widely through northern Senegal on passage and wintering; less frequent south of The Gambia.

LESSER WHITETHROAT
Sylvia curruca   **Not illustrated**

Fr. Fauvette babillarde

IDENTIFICATION 13.5cm. Sexes similar. Neat, grey and white warbler with smudgy dark mask through eye and white throat; very rarely recorded. Upperparts grey-brown, crown grey-capped with blackish mask along lower edge from lores to ear-coverts; wings brown, tail dark brown with white outer fringe (sandy-brown in first-winter bird). Below, throat white, off-white underparts with strong pinkish bloom in September, wearing away; flanks buff. Legs grey. **Similar species** First-winter from Subalpine Warbler by eye mask, greyer tones and grey legs; much smaller than Orphean but see Sardinian Warbler.

HABITS Retiring, usually remaining within dense cover.

VOICE Calls with a hard *tek-tek*; song monotone rattling trill.

STATUS AND DISTRIBUTION Two records, birds netted near the coast on the north bank of the river, January and February 1996. Primarily a Palearctic migrant to East Africa; recently discovered in very small numbers in northern Senegal.

SARDINIAN WARBLER
Sylvia melanocephala   **Plate 38**

Fr. Fauvette mélanocéphale

IDENTIFICATION 13cm. Sexes differ. A dark grey, black-capped warbler with reddish eye. Adult male has black cap with red orbital ring, dark grey upperparts; tail nearly black with white edges and white outer tips. Underparts white with grey wash to flanks. Adult female browner, head dark with dull red orbital ring; tail dark brown with white edging, below off-white with grey-brown wash on flanks. First-year birds browner, tail edge grey-white. Legs yellow-brown. **Similar species** First-year from first-year Subalpine Warbler by stronger reddish eye-ring, darker upperparts and tail; much larger Orphean Warbler lacks red eye-ring.

HABITS Lively bush-loving species, moving in and around low vegetation, coming out at top to observe surroundings before diving back down or making short low flight. Tail frequently cocked and fanned.

VOICE Calls a harsh *tcha*.

STATUS AND DISTRIBUTION Not recorded from The Gambia. Bulk of population occurs in the southern Sahel region; reported several times from northern Senegal (included netted birds); one sight record from Casamance.

SUBALPINE WARBLER
Sylvia cantillans   **Plate 38**

Fr. Fauvette passerinette

IDENTIFICATION 12.5cm. Sexes differ. Adult male distinctive; uniform dark grey above, pinkish-red below fading towards white belly; prominent narrow white submoustachial stripes, reddish eye; tail as upperparts, some white on outer margins and tips. Eye yellow-brown dominated by red eye-ring, legs pale brown. Female browner above, paler below, weak pink wash limited to breast, submoustachial stripes weak, eye-ring paler. First-winter pale grey-brown above, showing broad buffish fringes to inner flight feathers; below white throat, suffused pale buff on flanks and undertail; submoustachial streak absent or very weak, eye-ring pale. **Similar species** Smaller, more compact and greyer than Common Whitethroat. First-winter birds separated from Spectacled Warbler by grey-brown upperparts; diffuse buff fringes to inner flight feathers do not create strongly contrasting wing panel; Subalpine is dumpier, shorter-tailed, with slightly longer wings. See also calls. Birds entirely orange-brown below, lacking pink tones (recorded northern Senegal) attributed to *S. c. inornata*.

HABITS Typically seen in ones or twos in dense acacia canopy, often with other small warblers, sunbirds etc. Also uses coastal mangroves.

VOICE Hard *trec trec* call.

STATUS AND DISTRIBUTION Regular winter visitor to all parts of The Gambia, present September-April, peak numbers January-February. Observed widely in season across northern Senegal, less frequent south of The Gambia.

SPECTACLED WARBLER
Sylvia conspicillata   **Plate 38**

Fr. Fauvette à lunettes

IDENTIFICATION 13cm. Sexes differ. Resembles Whitethroat, but smaller, daintier and brighter in breeding plumage. Breeding male head grey, blackish eye mask sets off white eye-ring; brown upper parts, grey-brown rump; tail dark brown, white outer margins, contrasts with upperparts. Bright rufous-brown panel in wing. Throat white, breast pale with pinkish wash, pink-buff flanks. Legs pinkish-yellow. Female brown on head and mantle, buffer wash below. First-winter females and some males lack darkening around eye, rufous wing panel present, dark centres to inner flight feathers very neat. **Similar species** Distinguished from larger Common Whitethroat by shorter wings, more solid wing panel with dark centres to inner feathers sharply pointed (rounded in Whitethroat); livelier habits with mobile, well-marked tail, rounded head shape. See also call. Subalpine Warbler lacks strong rufous-brown wing panel. Short-winged, long-tailed look may suggest non-breeding Red-winged Warbler, but bill dark-tipped and more pointed, tail broader, darker, and squarer-ended in Spectacled Warbler.

HABITS Alert and sprightly movements, usually at low level in dense hedging, mangroves, or other scrubby vegetation. Inclined to wave tail. Mainly associated with Sahelian scrub to the north of The Gambia.

VOICE Call a drawn-out rattling *trrrr, trrrrr* or *zerrr*. More elaborate song unlikely to be heard in winter.

STATUS AND DISTRIBUTION Rare winter visitor to The Gambia with a handful of sight records and one ringed bird, all December-April near the coast; once in Kiang West. Probably underrecorded. Single record from northern Senegal. Main wintering area is Mauritania.

FLYCATCHERS AND ALLIES

Muscicapidae, Platysteiridae and Monarchidae

Fourteen species, divided into three families: Muscicapidae (*Bradornis, Melaenornis, Fraseria, Muscicapa, Myioparus* and *Ficedula*); Platysteiridae (*Bias, Batis* and *Platysteira*); Monarchidae (*Elminia* and *Terpsiphone*). Generally insectivorous, they catch their prey in mid-air in short flights from a favoured, often unobtrusive perch. Most have broad flattened bills with a wide gape and prominent rictal bristles. Typically in pairs or solitary, their voices are generally unremarkable. In some species the sexes are similar, in others sexual dimorphism is extremely marked. Some are migrants, others resident.

PALE FLYCATCHER
Bradornis pallidus   Plate 41

Fr. Gobemouche pâle

Other name: Pallid Flycatcher

IDENTIFICATION 16cm. Sexes similar. Nondescript elongated flycatcher. Head and upperbody grey-brown, wings and tail slightly darker brown, with rufous edging to inner flight feathers sometimes glimpsed on the wing. Pale ring around eye weak, not always noticeable. Whitish on chin and through centre of breast, merging with pale fawn over neck, flanks and undertail-coverts. Young birds are spotted with large buff feather centres over upperparts with some narrow pale streaking on the head and darker blotchy streaking on the breast. **Similar species** Immediately recognisable as a flycatcher, but slender shape, absence of streaking, longish tail, and often subtly hunched posture all eliminate Palearctic migrant Spotted Flycatcher. Swamp Flycatcher is smaller and darker, and in different habitat.

HABITS Singly or in pairs in savanna woodland, especially areas with high tree species diversity; reported to use gallery forest and stream-side sites elsewhere. Sits low beside tracks and woodland clearings, hunched over a twig watching the ground. Periodically drops neatly to the ground. Occasionally makes aerial sallies, but mostly takes terrestrial arthropods.

VOICE Very quiet, occasional weak *churr* and a soft twittering song, very reedy quality and a throaty croaking call note.

STATUS AND DISTRIBUTION A rare to uncommon resident throughout. Least frequent in WD, with sites such as Yundum increasingly being cleared; most readily encountered in LRD and larger woodland areas to the east. In Senegal range extends only a short way north of The Gambia, otherwise thinly distributed through Casamance and the south-east.

BREEDING Feeding young in June in The Gambia and known to breed April to June further east; builds a shallow cup nest in low fork of a tree or shrub.

NORTHERN BLACK FLYCATCHER
Melaenornis edolioides   Plate 41

Fr. Gobemouche drongo

Other name: Black Flycatcher

IDENTIFICATION 20cm. Sexes similar. Distinctive

slim-bodied all-black flycatcher, fine billed with neat rounded head and long tail. Uniform dull slaty-black overall; tail narrow and square-ended, becoming rounded with wear. Females sometimes with greyer slaty tone to underside. Immature shorter-tailed, heavily spotted with tawny-brown on a dark slaty-brown background, the spots smaller on the head and larger on upper- and underparts. **Similar species** Separated from other black birds on shape, posture and actions; Fork-tailed Drongo has stiff, flared fork on tail (note: Black Flycatcher tail feathers often lie untidily divided); Square-tailed Drongo has short, stiff, indented tail, and red eye. Female White-fronted Black Chat is springier in action and short-tailed.

HABITS Usually singly or in pairs in well-wooded habitats, including thickets in open woodlands, at the edge of farmland, dense bush, forest edges and thickets along the river bank where there are emergent trees. Perches quietly, sallying out to take flying insects, or dropping to the ground to secure prey. Especially retiring in the heat of the day. Displays with flicks of half spread wings and head-low posture; pairs may combine in displaying to neighbouring pair. Especially vocal after dawn.

VOICE Persistent thin notes of squealy quality, falling and rising *tseee-weee, tsee-i-ee, i-ee*. Also more strained, grating and creaking cries, including a soft *shreei* alarm on a rising inflection. A melodious song mixed with wheezing notes.

STATUS AND DISTRIBUTION Frequent resident in savanna habitats in all parts of the country, though perhaps slightly more common in moister coastal areas and less so in URD. Widespread in Casamance and south-east Senegal; does not penetrate to the north of The Gambia save for a few observations along the coastline north to Dakar.

BREEDING A wet season breeder; nesting June-August, young birds seen July-October. Gathers twigs and grasses to construct shallow nest in sheltered location, sometimes re-lining old nest of another species.

WHITE-BROWED FOREST FLYCATCHER
Fraseria cinerascens   Plate 41

Fr. Gobemouche à sourcils blancs

IDENTIFICATION 14.5cm. Sexes similar. Dark blue-grey and white flycatcher of forested streams. Entire upperbody uniform dark bluish-grey, darker on wings and tail. Prominent short white line extends a short way above the eye. Dark cheeks cleanly separated from white chin and throat; breast

is heavily scalloped in dark grey; lower belly whitish. Heavy bill black, legs slaty-grey. Immature brown-ish, with gingery fringes and small spots on wing-coverts, chest mottled with tawny and black. **Similar species** White mark over eye, contrast between upper- and underparts; habitat distinctive. A single record of Ussher's Flycatcher *Artomyias ussheri* (Fr. Gobemouche d'Ussher) from Basse-Casamance is considered doubtful.

HABITS Spends much time perched on low twigs overlooking freshwater habitat in riverine forest. Noted for quiet and sluggish ways, and sometimes site-faithful.

VOICE Not recorded; appears to be mainly silent.

STATUS AND DISTRIBUTION No Gambian records, but suspected on one or two occasions in Abuko; might be more at home along some of the swamp forest streams of CRD, though no reports. In Senegal, locally common resident in forested areas of coastal Casamance.

BREEDING Extended breeding season, May-September in Casamance; constructs a nest of moss, leaves and rootlets in a small hole or tree stump beside water.

SPOTTED FLYCATCHER
Muscicapa striata   Plate 41

Fr. Gobemouche gris

IDENTIFICATION 14.5cm. Sexes similar. Slim, long-winged, grey-brown flycatcher with a streaky crown, characteristic sallying flights to catch flying insects. Forehead buff, finely streaked black, fading over crown to uniform grey-brown nape and upperparts. Wings brown, hint of wing bar and panel of pale edges along folded wing; longish tail plain dark brown. Breast smudgily streaked with grey-brown on dirty white background, fading to whitish lower breast and belly. Legs black. First-winter has short pale tips to wing-coverts creating stronger wing bar across top of wing. Freshly arrived immatures (October) may still show buff spotting on upperparts. **Similar species** Crown and breast streaking separates from smaller and darker Swamp Flycatcher. Pale Flycatcher is also unstreaked but shape and behavioural differences make confusion unlikely.

HABITS Perches prominently at the edge of cover overlooking a clear area, returns neatly to same perch. Found in a variety of habitats where these conditions are fulfilled. Overlaps with Swamp Flycatcher in riverbank areas, though usually not perched low over water as that species. Mainly quiet and solitary.

321

VOICE Mainly silent on migration in West Africa. May utter a quiet *tzee*, sometimes embellished to *tzee.tic.tic*, with emphasis on the first note.

STATUS AND DISTRIBUTION Uncommon to temporarily frequent passage migrant; most observations September-November and a weaker return movement March-May. Very few overwintering records. Situation similar in Senegal.

SWAMP FLYCATCHER
Muscicapa aquatica   Plate 41

Fr. Gobemouche des marais

IDENTIFICATION 13cm. Sexes similar. Small, confiding, dark grey-brown flycatcher nearly always found beside freshwater. Head brown, greyer on ear-coverts; rest of upperparts dull dark brown, often with an imprecise grey-dappled effect over the mantle. Chin and throat whitish, merging to a greyish wash over the upper breast and flanks, paling to whitish on the lower belly. Broad-based black bill, legs blackish. Juveniles spotted with buff on the upperparts and show a few dark streaks on the breast. **Similar species** Unstreaked, smaller and darker than Spotted Flycatcher. Has been confused with Cassin's Flycatcher *Muscicapa cassini* (Fr. Gobemouche de Cassin), not yet recorded closer than Guinea, which is of similar habits, but slightly larger and decidedly grey to almost blue-grey whereas Swamp Flycatcher is brown.

HABITS Generally closely associated with freshwater habitats, perching quietly, often low, on bare branches and twigs overhanging, or emerging from, still or slow-moving water. Tidal movements throughout the length of the country create an abundance of such sites inland. Sallies out in flycatcher style to capture flying insects, and take struggling insects from water surface, small water beetles being seized and beaten in kingfisher-fashion on returning to the perching post. Does not seem to be associated with reedbeds in The Gambia. Usually in pairs or family groups.

VOICE Short phrase consisting of a thin slurring three-note whistle, rising then falling, followed by a short mechanical trilling sound, *tsi-eee-uuu.trrrrrr*. Also a quiet bubbling song and a sharp *pzitt*.

STATUS AND DISTRIBUTION Frequent to common resident along freshwater courses, including banks of the main river, in CRD and URD. In Senegal an isolated population in the Senegal River delta area otherwise limited to Casamance and the south-east. Note that nominate form in The Gambia differs substantially from eastern and southern African subspecies.

BREEDING Poorly known; recorded northern Senegal in March, and feeding young in The Gambia in November. Said to build a cup-shaped nest in a crevice behind bark, or within the shell of an old weaver nest, often over water.

LEAD-COLOURED FLYCATCHER
Myioparus plumbeus   Plate 41

Fr. Gobemouche mésange

Other names: Fan-tailed Flycatcher, Grey Tit-Flycatcher, Grey Tit-Babbler

IDENTIFICATION 13-14cm. Sexes similar. Small, pale grey, slim flycatcher, with eye-catching tail-fanning habit, showing conspicuous black and white pattern. Soft pale blue-grey from head to rump; inconspicuous whitish lores. Longish pointed wings darker brown with narrow pale fringes on the flight feathers. Tail mainly black, rather long, broadens from the base, tipped with increasing graduations of pure white across the corners with pure white outer feathers. Pale grey on chin and upper breast, whitish belly, and pale buffish undertail. Legs leaden grey. Immature shows limited buff spotting on wing-coverts and pale buff-brown wash to underparts. **Similar species** Easily recognised once reasonable view obtained of greyish body, long showy black and white tail and characteristic behaviour.

HABITS Found in pairs or small family parties in a variety of habitats wherever there is reasonable tree cover, particularly dense savanna woodland, laterite bluffs, and also forest edge and thickets along the river bank. Inconspicuous initially but active, regularly flaring and fanning black and white tail, smoothly raising and spreading feathers, reminiscent of African Blue Flycatcher. May sometimes call from near the top of a small tree.

VOICE Largely quiet; emits a thin, slow *tweee* on a clear rising or slightly falling note, also a more reedy two-note call first rising from a higher pitched note, then falling from a slightly lower pitch *preeee chreee*.

STATUS AND DISTRIBUTION Uncommon local resident throughout The Gambia. Increased observations in the dry season probably reflect improved visibility caused by leaf fall in woodland habitats. Thinly scattered through Casamance and south-eastern Senegal.

BREEDING Not recorded in The Gambia. A tree-hole nester, said to line old nest of woodpecker or barbet species with a few strands of vegetation. Nesting behaviour little-known.

PIED FLYCATCHER
Ficedula hypoleuca   **Plate 41**

Fr. Gobemouche noir

IDENTIFICATION 13cm. Sexes differ. Plain, compact grey-brown flycatcher with white flash in the wing. Breeding male, rare April-May in The Gambia, unmistakable: black above, with white wing patches and white underparts. Non-breeding male: plain pale brownish head and upperparts, blackish uppertail-coverts and tail (dark brown on female). Both sexes have white outer margins to several tail feathers and bright white inner wing panel. Underparts whitish, lightly washed with diffuse grey. Short bill is black, large eye dark brown, legs black. First-winter birds show pale fringes to wing-coverts as well as tertials. **Similar species** Distinctive; Collared Flycatcher *Ficedula albicollis* (Fr. Gobemouche à collier) has once been claimed in northern Senegal. Non-breeding plumage is very similar to Pied, but confusion is likely with north-west African race of Pied *F. h. speculigera* which has larger white wing patches and has been recorded in Senegal. Red-breasted Flycatcher *Ficedula parva* (Fr. Gobemouche nain) has recently been found in northern Senegal, in November 1991 and 1992. Smaller than Pied with no white in wing and distinctive white patches at sides of tail-base. Non-breeding birds lack red breast.

HABITS A quiet and generally solitary species, in a variety of habitats, including coastal mangroves, gardens and acacia thickets, but notably associated with tall trees in forested areas, including swamp forest. Frequently flicks wings and cocks tail. Sallies in typical flycatcher fashion but readily drops to the ground to pick up small prey items.

VOICE Mainly silent in The Gambia. Commonest call a loud *wit*, or *bit*.

STATUS AND DISTRIBUTION Passage migrant seen in all divisions, first appearing late August, may become briefly frequent September-October, though generally uncommon; much weaker northward movement April-May. Sporadic records January and February suggests limited overwintering. Similar pattern in Senegal; some overwintering in Casamance.

SHRIKE-FLYCATCHER
Megabyas flammulatus   **Plate 41**

Fr. Gobemouche écorcheur

IDENTIFICATION 15cm. Sexes differ. Glossy thickset forest flycatcher, male black and white, female brown and streaky below. Male black above except white rump, cleanly divided below eye from pure white underparts. Tail all black. Female brownish on the crown, rich rufous brown on the upperparts and tail; brown centres to inner flight feathers. White underparts heavily streaked with brown from throat, over breast, fading to a white belly. Bill black, noticeably hooked and shrike-like with prominent whiskery rictal bristles; eye red; legs grey. Immature resembles female. **Similar species** From similar-sized male Northern Puffback by upright, alert posture and absence of white on primary edges. Female from female Black-and-White Flycatcher by lack of crest and streaky breast.

HABITS Forest species of the mid-canopy, moving in pairs or small parties. Tail mobile, frequently flicked or swayed side to side while perched, flight buoyant and flitting.

VOICE A tuneful disyllabic *tuick* or *chuick* reported; a series of triplets, rising in pitch through the whole sequence, each unit with emphasis on the last note *chi-ki-tik, chi-ki-tik, chiki-tik-tik*, and thought to bill snap. Also a rasping trill, *prrrt*.

STATUS AND DISTRIBUTION Vagrant. Single record of a male in Abuko in December 1988. Not recorded from Senegal or Guinea Bissau; nearest population in Sierra Leone.

BLACK-AND-WHITE FLYCATCHER
Bias musicus   **Plate 41**

Fr. Gobemouche chanteur

Other name: Vanga Flycatcher

IDENTIFICATION 16-17cm. Sexes differ. Thickset flycatcher with crest and a short tail, clean black and white in the male, rufous in the female. Male black, washed iridescent greenish, over head, crest, upperparts and tail; white primary bases make large wing patch in flight, and may form narrow wedge along the folded wing. Black throat and upper breast, fairly cleanly separated from white lower breast and belly. Female equally crested, dark brown head, grading into mid-brown mantle to coppery-rufous wings, back and tail; blackish primaries. White throat, rest of underparts washed pale rufous, heavily so on flanks. In both sexes, broad black bill surrounded by stiff rictal bristles; eye bright yellow, legs yellow. Immature resembles female. **Similar species** Crest distinguishes from other black and white forest flycatchers; male Shrike-Flycatcher lacks black throat while female Shrike-Flycatcher has heavily streaked throat.

HABITS A tree-top insectivorous species of forest clearings and adjacent farmland. Associated with tall trees, occasionally coming lower. Within normal range usually in pairs, sometimes small parties. Attracts attention through active and noisy habits. Sits upright on bare branches. Male displays in circular flight, showing off white primary patches with rapid beats in slow hovering progress around canopy.

VOICE Loud whistling song, with sharp energetic delivery, rising then falling through the scale; *wit-it-tu...wit-it-tu*, or more elaborately, *wit-tu-wit-tu-tui-tiu-tu-*, tending to accelerate through sequence. Also a harsh churring call.

STATUS AND DISTRIBUTION Vagrant. A single observation of a bird singing from a tall tree at the edge of cleared farmland in May 1976 near Seleti, WD, is the only record from Senegambia. Nearest population in Sierra Leone.

SENEGAL BATIS
Batis senegalensis   Plate 41

Fr. Batis du Sénégal ou Gobemouche soyeux du Sénégal

Other name: Senegal Puff-back Flycatcher

IDENTIFICATION 11.5cm. Sexes differ. Very small, lively black and white bird of open woodlands, neat and bright-eyed, with top-heavy upright stance. Male has slaty-black crown, bright white supercilia meet at the nape above broad jet black mask through eyes to ear-coverts. Mantle olive-washed grey grades to long soft feathering of rump; tail black with bright white outer feathers. White wing-bar forms V on the upperparts when viewed from behind. Neat blue-black breast band on otherwise pure white underparts. Female is similar pattern, but breast band tawny and a pale tawny wash to supercilium, collar and wing-bar. Bill black, eye distinctive bright pale yellow, legs black. Immature resembles female. **Similar species** Striking, but in brief view female may recall Brubru, from which separated by chestnut band across breast and yellow eye, as well as size and posture. From immature Common Wattle-eye see that species.

HABITS Alert and energetic, mainly in open woodland habitats, including comparatively mesic coastal locations, but particularly associated with dry savanna woodlands with thornier vegetation. Usually found flitting actively through small trees and shrubs, very occasionally flycatching. Usually pairs or family parties, sometimes joins savanna feeding parties; faithful to comparatively small groups of trees or bushes. Moves with rapid wing-beats and dipping flight. Aggressive interactions with similarly patterned Brubru noted.

VOICE Variety of calls, including a soft, vibrant and throaty *zit* or *zrrrt...zrrrt* and harder *chec.prrr*. A

musical disyllabic whistle described, and a more forceful *tsit.tsit.tsit* when scolding.

STATUS AND DISTRIBUTION Uncommon to locally frequent resident, mainly in dry savanna woodlands but recorded in all parts of the country where more extensive woodland occurs, including Tanji, Abuko, and Kiang West. Found throughout Senegal, possibly less common in far north and only to edge of denser southern forests.

BREEDING Builds very small cup nest of grass, lichen, bark fibre and cobwebs at no great height in fork of shrub or small tree. Breeds mainly May-July.

COMMON WATTLE-EYE
Platysteira cyanea   Plate 41

Fr. Gobemouche caronculé à collier

Other names: Brown-throated Wattle-eye, Scarlet-spectacled Wattle-eye.

IDENTIFICATION 14cm. Sexes differ. Small, chubby black and white flycatcher of denser habitats; lively, bright scarlet wattle above eye, remarkable musical calls. Male glossy blue-black over the head and cheeks, slightly greyer on the back and tail; solid white wing-bar on coverts extended by broad white edges of inner flight feathers. Below pure white except for a sharply defined glossy blue-black breast-band. Female slaty grey-black above with white wing-bar; narrow white chin with beautiful velvet-textured dark purplish-maroon from throat to breast, sharply separated from snow white belly. Bill broad and black, eye blue-grey with scarlet fleshy wattle above in male and female; legs black. Young birds buff-spotted above; immature male duller black above with a tawny breast band, not strongly defined; immature female similar with broader tawny band below; both with poorly developed pinkish-grey wattles. **Similar species** Immature from female Senegal Batis by absence of supercilium and collar, dumpier shape and habitat difference. Wattles diagnostic in adult; the only wattle-eye in Senegambian region.

HABITS Active birds of secondary forest and riverine vegetation including both *Rhizophora* and *Avicennia* mangroves, and also well established gardens. May venture to dense savanna in wet season. Usually in pairs or small parties; often in mixed species feeding parties in forest habitats. Calls intermittently throughout the day, and movements often accompanied by intense snapping and popping sounds created by bill or wing.

VOICE Remarkable ringing call, one of the characteristic sounds of Abuko. Ringing chime of five or six clear-toned notes swinging smoothly over wide intervals of pitch *do-dee-dum-di-do-do*. Female sometimes answers with a buzzing sound, but also known to seamlessly insert notes in full duet sequence when pair create elaborate stereophonic versions of main chiming song.

STATUS AND DISTRIBUTION Common resident in forest thickets of the coastal area; frequent in mangroves and dense riverine or streamside vegetation throughout the country, more restricted in the interior. Suspected of local movements in response to seasonal vegetation changes; in The Gambia recorded more frequently May-September, though regular in all months. In Senegal widespread in Casamance and the south-east, only penetrates a short way to the north of The Gambia.

BREEDING Builds very small cup nest of lichen and cobwebs low in vegetation. Young have been found in the nest over a wide period, mainly in late dry season, April-July; immatures seen in all seasons, usually in company with adults.

AFRICAN BLUE FLYCATCHER
Elminia longicauda   Plate 41

Fr. Gobemouche bleu

Other names: Blue Flycatcher, Fairy Blue Flycatcher

IDENTIFICATION 18cm. Sexes similar. Unmistakeable, small dainty and active caerulean blue flycatcher with blunt crest and long tail. Blue over entire upperparts, brightest on forehead and crown, becoming more grey-blue on tail; flight feathers dark. Small dark area on lores, and pale blue underparts, fading to whitish on the belly. Bill black, very broad and flattened. **Similar species** Unique within region.

HABITS Closely associated with water, from waterside and riverbank thickets to mangroves over brackish water. Found from water level to the high tree-tops; at falling tide active among newly exposed mangrove roots. Insect prey mainly taken from the vegetation, though sallying flights often seen. Usually in pairs or small groups, and intensely active, constantly cocking and flaring tail, drooping pointed wings below body and switching from side to side on perch with stiff sudden movements. Fond of plunge-bathing from a low twig over the water.

VOICE Clear whistling liquid and mellow song is a useful guide to its presence and location in favoured dense habitats. Also emits a sequence of scratchy notes.

STATUS AND DISTRIBUTION Uncommon species from brackish mangrove habitats (mainly *Rhizophora* spp.) in NBD, into riverine woodland and thickets along the banks of the main river all

through CRD and URD. Locally frequent in a few places; recorded in all seasons and likely to be resident. In Senegal found throughout Casamance and the south-west but not north of the Gambia River.

BREEDING Not confirmed in The Gambia or Senegal, but probable. Recorded April-May in Sierra Leone; very small shallow cup, moulded onto the upperside of a fork, decorated with lichen attached with gossamer.

RED-BELLIED PARADISE FLYCATCHER
Terpsiphone rufiventer   Plate 41

Fr. Moucherolle à ventre roux

Other name: Black-headed Paradise Flycatcher

IDENTIFICATION 16-18cm (non-breeding male and female), 30-32cm (breeding male). Slim, burnt orange-coloured bird, usually with dark head and long tail. In full breeding plumage very long central tail feathers of male trail as lax rufous ribbons. Typically glossy blue-black over entire head and neck, sharply demarcated from rufous-orange body and tail, except for white bar on coverts and flight feather edges. Wing feathers dark brown. Birds lacking white also occur, and seem not to develop such long tail ribbons. Variation in extent of black on head noted, with some having black limited to face and fore-crest only. Bill dark slaty, breeding males develop dark electric blue on upper mandible to corner of gape and around eye. Juvenile short-tailed, plain brown, prominent orange-yellow at corner of gape. **Similar species** Usually distinctive, difficulty only arising when underside appears half black merging to half rufous-orange, suggesting intermediate, possibly hybrid form with African Paradise Flycatcher. Dark-backed morphs from further east never reported in The Gambia.

HABITS Lively vociferous bird of dense vegetation, flitting with agility through thick cover, usually alone or in pairs; also joins mixed feeding parties. Alert and inquisitive while perched, seldom still for long. Compared with African Paradise Flycatcher thought to spend more time at lower levels in forest vegetation, otherwise behaviour similar. Attracted to water, visiting at intervals throughout the day, especially in hot weather to splash bathe. Displaying pairs vocal, chasing and sidling up to each other, with lively fanning of tail feathers, and crests raised.

VOICE Most frequent call a sharp, hard-edged, buzzy *zweeet, kzee.zee.zee*, articulated quickly and with emphasis on the first syllable. Sounds impatient and irritable. Also a burst of quick, fluent, pure whistled notes on a descending scale and sustained wittering in excitement.

STATUS AND DISTRIBUTION Locally common resident in patches of suitable habitat near the coast, becoming uncommon inland, though recorded in LRD and URD. Everywhere at risk from habitat destruction. Forest-dependency also evident in Senegalese distribution; mainly found in coastal Casamance, largely absent north of the Gambia River.

BREEDING Neat small cup on upperside of narrow branch, generally at a fork. Breeding season thought to be prolonged, possibly year-round.

AFRICAN PARADISE FLYCATCHER
Terpsiphone viridis   Plate 41

Fr. Moucherolle de paradis

Other name: Paradise Flycatcher

IDENTIFICATION 16-18cm, breeding male 28-35cm. Spectacular chestnut and black bird with long flowing tail ribbons. Male typically has glossy blue-black head with short thick crest; chestnut upperparts and tail streamers. Usually a white bar across wing. Underparts all slaty-black. A morph with white upperparts and white tail streamers is quite frequent elsewhere; alternatively, blackish upperparts and black streamers or streamers of two different colours possible, though none of these have been recorded in The Gambia. Females glossy black head and crest, dull chestnut mantle and tail, blackish underparts. Bill blue-grey, becoming electric blue on the upper mandible in the breeding male, eye dark brown, electric blue eye-ring in breeding season. Immature resembles female, though duller and shorter crested; juvenile initially all chocolate brown, becoming spotted with buff. **Similar species** Highly distinctive but see notes below and Red-bellied Paradise Flycatcher for discussion of forms with intermediate black and rufous on underparts.

HABITS A lively, confident bird of thick savanna woodland, using thickets, secondary forest remnants and gardens, where it may overlap with Red-bellied Paradise Flycatcher, but generally associated with slightly more open habitats. Takes large insect prey by searching vegetation and by flycatching, lax tail giving exaggerated impression of sudden twisting and undulating movements through the air. Slim and semi-upright when perched, turns quickly from side to side, looking about all the time. Often in pairs or small parties as well as with other species.

VOICE Noisy, harsh cries usually the first indication of presence; variable but not distinguishable in tone from Red-bellied Paradise Flycatcher, emphasising the close relationship of the two species. Scolding *zi-zk-zk* and a fluent rippling whistle on a falling cadence, both recognisable in Red-bellied.

STATUS AND DISTRIBUTION Frequent resident of dense habitats in all divisions though rare or absent in large tracts of open cultivated country and drier bush where trees are small and comparatively sparse. Recorded most often in LRD and CRD, where it is relatively much commoner than Red-bellied Paradise Flycatcher; the opposite situation applies in WD near the coast. Recorded in The Gambia with increased frequency May-September, implying a northward movement in anticipation of wet weather frontal systems. In Senegal found throughout Casamance, seasonally dispersing north and east of The Gambia.

BREEDING Recorded May-July, minute cup nest on the fork of a narrow branch, appears exceedingly undersized when covered by the long-tailed incubating male.

NOTE The taxonomic situation in this group is complex. Most birds can usually be assigned readily enough in the field to one of the two basic types, black-bellied or red-bellied. However both show variable morphs, and it is not known for sure the extent to which these variations signify polymorphisms or are the result of a hybridisation zone, or indeed both.

In general African Paradise Flycatcher is more associated with dense savanna woodland where various polymorphisms, notably an all-black and white plumage, are recognised. The Red-bellied Paradise Flycatcher is more closely linked to forested habitats. Over most of their ranges in West Africa, this difference appears to be sufficient to keep the two separate and they are usually regarded as separate species. However, at the western end of their range, which includes The Gambia and southern Senegal, and again in the extreme west near Lake Victoria, geographic overlap is extensive and this coincides with the occurrence of apparent hybrids. The birdwatcher in The Gambia quickly becomes aware of a great deal of variation. Tail and crest development vary with sex and season, but in addition the presence and amount of white in the wing, and the relative extent of black and rufous on the underparts or head also varies. Individuals with part grey and part rufous underparts may be taken to be hybrids. Important field information to note is the relative frequency of each type in relation to habitat type and habitat change.

BABBLERS

Timaliidae

Seven species, five rare and three of these in Senegal only. Small to small-medium rather thrush-like birds, often terrestrial with strong legs and feet, generally sombre in plumage and often untidy looking. Typically gregarious and noisy.

BROWN ILLADOPSIS
Illadopsis fulvescens   Plate 37

Fr. Grive-akalat brune

IDENTIFICATION 15cm. *I. f. gularis.* Sexes similar. Small plain brown thrush-like bird of very dense habitats; generally difficult to observe. Upperparts brown from head to mantle, slightly darker on the wings, rufous-brown tail. Cheeks grey; dirty white chin and throat merges to uniform buff-brown underparts. Legs dusky grey. Immature paler than adult, with a yellowish belly. **Similar species** Separation from other illadopses by lack of white on underparts below throat, dark legs, and voice.

HABITS Favours dense forest and adjacent thickets, clings to vertical stems close to the ground, spending less time actually on the ground than most forest floor species. Encountered in pairs or small groups, may join mixed feeding parties. Skulking. Voice an important aid to detection and recognition.

VOICE Simple melancholy song of two pure notes, each delivered slowly, with a measured pause between them, the second note lower than the first, sometimes preceded by a brief higher pitched gracenote. Most often heard early morning, may be accompanied by chirrups and clucks from a second bird. Also a testy scold in rattling monotonic triplets, *chakata, chakata, chakata.*

STATUS AND DISTRIBUTION Not recorded from The Gambia, but first collected from Basse-Casamance in 1981; now regularly heard in that region, mainly at the end of the dry season.

BREEDING Cup nest low in dense vegetation, in West Africa coincident with the wet season.

RUFOUS-WINGED ILLADOPSIS
Illadopsis rufescens   Plate 37

Fr. Grive-akalat du Libéria

IDENTIFICATION *c.*15cm. Sexes similar. Rare, little-known forest floor species. Plain brown above, merging with rufous on the uppertail-coverts and showing rufous edges to the flight feathers; tail dark chestnut-brown. Cheeks brown, throat clean white, with the rest of the underparts grey-white, diffuse grey-brown partial breast-band and grey flanks. Legs and feet whitish-flesh. **Similar species** From Brown Illadopsis by pale underparts and leg colour; from Puvel's Illadopsis by grey not peach-buff breast and flank marks, and cleaner white throat.

HABITS Confined to closed forest, where it forages primarily on the ground. Relatively solitary, but also seen in pairs, occasionally small groups and attracted to mixed species foraging parties. Maintains quiet voice contact while foraging and said to flit and hop rapidly over the ground in alarm.

VOICE Main song most prominent early morning and late evening, a sequence of evenly matched, clear whistled notes in a steady, bouncy, rhythm on a monotone (not downscale); *chit-chit-chit-tu-hoo-hoo*, repeated for long periods; beware more subdued, evenly paced and mellow sequence of White-spotted Flufftail. While feeding a quiet *whit-whit-whit.*

STATUS AND DISTRIBUTION Not recorded from The Gambia, but captured twice in Parc National de Basse-Casamance. Rare West African endemic, from southern Senegal to Ghana.

BREEDING Birds in Liberia breed in the rainy season. Nest and eggs not described.

PUVEL'S ILLADOPSIS
Illadopsis puveli   Plate 37

Fr. Grive-akalat de Puvel

IDENTIFICATION 17cm. Sexes similar. Rare forest floor species, plain rufous-brown above from crown to wings and tail, throat and belly creamy-white; diffuse peachy-buff breast-band and flanks. Legs stoutly built, pale flesh. Immature birds said to be similar, warmer-toned above, with a grey eye and bluish-white legs. **Similar species** From Brown Illadopsis by contrast between creamy-white underparts and brown upperparts; from Rufous-winged by warm peachy-orange tone (not grey) to breast-band and flanks. See also voice descriptions.

HABITS A bird of isolated forest galleries and secondary scrub. Forages mainly in leaf litter, usually singly or in pairs.

VOICE Main song a rhythmic phrase of cleanly

articulated pure-toned notes, descending evenly down a four-note scale (four central notes are bouncily repeated short semi-quavers); *wit ti-ti tu-tu too*. Repeated monotonously at short intervals for long periods.

STATUS AND DISTRIBUTION Rare resident. Voice recording from WD confirmed as this species by C. Chappuis (pers.comm.), November 1993 and again August 1996. Recently seen and mist-netted in The Gambia. In southern Senegal seen irregularly in all seasons in vicinity of Parc National de Basse-Casamance.

BREEDING Birds in Liberia in breeding condition in mid wet season. Nests on ground.

BROWN BABBLER
Turdoides plebejus   Plate 35

Fr. Cratérope brun

IDENTIFICATION 23cm. Sexes similar. *T. p. platycircus*. Plain grey and brown babbler, noticeably paler towards the head with bright orange eyes; gregarious and noisy. Brownish crown merges imperceptibly into grey sides of the head which are finely speckled with white on the ear-coverts and cheeks. Upperparts brown; plain, rounded wings and broad tail rather darker brown. Throat and neck buff finely speckled with white, merging to a dirty buff breast and belly. Eye orange, visible with binoculars at considerable range, legs black. Young birds similar to the adult. **Similar species** Distinctive within the region.

HABITS Favours a range of woodland and bush habitats; but not a regular forest bird. Almost always in tight-knit parties of 4-12, members following each other from bush to bush, swooping into cover on flatly depressed wings. Appears alert and inquisitive while foraging; noisy, contact calls frequently escalate into loud and querulous communal babbling sessions, a common background sound of the savannas. An enthusiastic mobber of small predators. Retires to dense cover to rest and roost.

VOICE Range of sounds based on a coarse and emphatical *ciaw*, either as a single cry, or, with rising excitement, repeated more rapidly at increasing volume to merge with the cries of other individuals in a rhythmic cacophony of continuous noise. Lacks the short staccato repetitions that distinguish the Blackcap Babbler; note that the voice of Green Wood-Hoopoe, is superficially similar, but distinguishable with experience.

STATUS AND DISTRIBUTION Common resident throughout the country, marginally more frequent

in the coastal region. Found throughout Senegal, less frequent towards the north.

BREEDING Cup nest placed low in a bush or tree-fork; recorded breeding over an extended season, May to December; group members cooperate in nest construction, and probably in care of young as well. A frequent host to Levaillant's Cuckoo, and may be encountered feeding cuckoo fledglings, especially December and January.

BLACKCAP BABBLER
Turdoides reinwardtii   Plate 35

Fr. Cratérope à tête noire

IDENTIFICATION 25cm. Sexes similar. Marginally larger and darker than the Brown Babbler, immediately distinguished by black head and pale eye. Entire head is black, including cheeks and ear-coverts, changing fairly cleanly to plain brown upperparts, with the wings and tail dark brown. Throat whitish, stained browner on the breast which is very lightly blotched and streaked with some brown. Eye silvery-white, legs brownish. Young birds have cap dark grey and poorly defined. **Similar species** Distinctive, but beware first-time mistakes with Common Bulbul.

HABITS Favours thickets, well-wooded areas and dense vegetation along watercourses; it overlaps with Brown Babbler in many areas, but is generally associated with denser cover. Always in parties of several individuals, staying low in vegetation or on the ground. Groups behave similarly to Brown Babbler. Maintains regular vocal contact; given to communal calling bouts, especially towards dusk.

VOICE Indulges in prolonged raspy chattering sessions, developing into loud choruses of sustained nasal calls, similar to those produced by the Brown Babbler. Blackcap Babbler punctuates this with a staccato, *cha-ka-ta.....cha-ka-ta*, both as a one-off call, or introduction to a more elaborate calling bout.

STATUS AND DISTRIBUTION Frequent to common resident in all areas, although seen slightly less often than the Brown Babbler. Found throughout southern Senegal, becoming increasingly localised to the north of The Gambia and very unusual in the extreme north; found eastward to Nigeria and Cameroon.

BREEDING Cup-shaped nest in dense vegetation, June-October. All group members bring food to nestlings and help guard fledglings. Breeding pair may start second brood (e.g. September) while group members attend to newly-mobile fledglings.

FULVOUS BABBLER
Turdoides fulvus **Not illustrated**

Fr. Cratérope fauve

IDENTIFICATION 25cm. Sexes similar. Unstreaked, long-tailed babbler; warm sandy-brown above, pale below with white throat and pale rufous flanks.

HABITS Usually in small family groups. Inhabits semi desert areas with low shrubs and bushes.

VOICE A series of 6 descending, plaintive whistles, *peeu-peeu-peeu-peeu-peeu-peeu.*

STATUS AND DISTRIBUTION No Gambian records. Small groups noted on several occasions in northern Senegal; perhaps a recent colonist there.

CAPUCHIN BABBLER
Phyllanthus atripennis **Plate 35**

Fr. Cratérope capucin

IDENTIFICATION 24cm. Sexes similar. Distinctive, thickset forest babbler with dark brown body, unusual silvered head colour giving hooded appearance, pale bill. Head and throat are silver-grey, faintly flecked with black, solid black patch from base of bill narrowly surrounds the eye. Body dark brown, enhanced by a mahogany-chestnut tone to mantle and belly. Wings brown, tail slightly darker. Bill creamy-yellow, may be tinged pale green and

usually appears ivory-white in dull forest light. Legs greenish-grey. Immature birds have not been described. **Similar species** Once seen well, unlikely to be wrongly identified.

HABITS Elusive bird of dense thicket and forest. Gregarious, groups of 6-7 normal, and indulges in raucous communal calling bouts; may venture high into trees, especially near sunrise. Maintains conversational contact notes while foraging low in the vegetation or on the forest floor, may fall silent for long periods. Moves with a bouncy hopping action among stems and branches. Several birds follow each other consecutively on short, low, flights from bush to bush; joins mixed species feeding parties. Groups seem faithful to relatively small sites. May rest in heat of day, sitting side by side along low branches in tangled thicket.

VOICE Individuals give a guttural *kiorr-kiorr-kiorr,* usually several calling simultaneously, similar to other babblers, but subtly more mellow, and lacking the machine-gun staccato notes of the Blackcap Babbler. Conversational, soft *kiok* or *quock* also given intermittently by individuals.

STATUS AND DISTRIBUTION Rare resident, known recently from only three sites in the southern parts of coastal Gambia, reported once further inland, and probably under great pressure from land clearance. Reasonably frequent in forests of coastal Casamance.

BREEDING In The Gambia collecting nest material in June; carrying food in November.

TITS, CREEPERS AND PENDULINE TITS
Paridae, Certhiidae and Remizidae

The true tits, Paridae, are represented by a single species in the genus *Parus,* as are the creepers, Certhiidae, by the genus *Salpornis.* Two penduline tit species, Remizidae, occur in the region and both are placed in the genus *Anthoscopus.* They differ from the Palearctic genus, *Remiz,* in absence of sexual dimorphism, duller coloration and short tail, but share charcteristic nest types. *Anthoscopus* and *Remiz* are sometimes regarded as congeneric (e.g. Hall & Moreau 1970).

WHITE-SHOULDERED BLACK TIT
Parus leucomelas **Plate 40**

Fr. Mésange à épaulettes blanches

Other name: White-winged Black Tit

IDENTIFICATION 14cm. Sexes similar. Striking smallish, dumpy black and white bird. Entire body uniform black, enhanced by a purple-blue gloss to the upperparts, a pristine white shoulder patch (more extensive on the male), and prominent white edges to flight feathers. Eye looks silvery-white in sunlight, in striking contrast to black body. Short

black bill, legs dark grey. **Similar species** Sooty Chat has white confined to coverts only, and is larger with longer head and bill.

HABITS Found in dry savanna woodlands and usually in small but very active parties; constantly on the move to new locations and frequently in loose association with other savanna species. Flight undulating but quick. Maintains calling while feeding.

VOICE Often first detected from calls, which are frequent but variable in content, but characterised by a buzzy, raspy toned, double *zzeuw.zzeuw* or *teer-teer.zeet.* Produces a purer-toned song when breeding.

STATUS AND DISTRIBUTION Uncommon to locally frequent resident in wooded areas throughout, WD to CRD, with main stronghold in extensive woodlands of LRD; not recently recorded in URD. Widely dispersed through central Senegal, becoming infrequent in the more forested areas of coastal Casamance and absent from the north.

BREEDING Details not well known; usually a tree-hole nester, adding a liberal lining of down; recorded in the dry season, January-May in The Gambia. In areas of low hole-availability may construct suspended nest.

SPOTTED CREEPER
Salpornis spilonotus   Plate 31

Fr. Grimpereau tachetée

IDENTIFICATION 15cm. Sexes similar. Highly distinctive, with long decurved bill, peppered and barred in black and white with woodpecker-like habits. Upperparts blackish-brown, streaked with white on the crown, heavily spotted white on the back. Wings and tail barred black and white. White supercilium and narrow dark eye-stripe. Below whitish, scalloped and barred darker brown. Long decurved slim bill, brown above and paler on the lower mandible. Immature similar, but paler and less well marked. **Similar species** Bill shape separates from all woodpeckers and Wryneck, in back view beware equally rare Little Grey Woodpecker which has scarlet rump; more frequent female Brown-backed Woodpecker has plain brown mantle.

HABITS Found climbing up trunks or branches in open dry savanna woodland. Grips bark fully with feet and does not rest on tail. Alights at base of trunk and works up in spiralling pattern, extracting insect prey from crevices with long bill. Well camouflaged, moves behind trunk in woodpecker-style when disturbed. Usually in pairs or alone, though noted to join mixed species feeding parties. Fast, undulating flight.

VOICE High pitched, hoarse, *kek.kek.kek.kek.kek* and a very high, thin and rather slow *swee-sweepy-sweep-sweep-sweep.*

STATUS AND DISTRIBUTION Rare, only four published records from The Gambia: March, May, November and December from 1965 to 1988; in WD and once in LRD near Kiang West. Recent sight records from Abuko and Sanyang, all in dry season. In Senegal, single record from April 1969, Niokolo Koba. Considered resident Guinea-Bissau and Guinea.

BREEDING Well camouflaged cup-nest over a small fork, often quite high.

YELLOW PENDULINE TIT
Anthoscopus parvulus   Plate 40

Fr. Rémiz à ventre jaune

Other name: West African Penduline Tit

IDENTIFICATION 7.5-8cm. Sexes similar. Tiny, one of the smallest birds in Africa, a truly minute yellow bird, with slightly tubby profile. Upperparts all pale yellow-green, forehead bright yellow marked by black dots (only visible at close range). Black-eyed appearance, with yellow-tinge to rump. Flight feathers pale brown, finely edged yellow, making a distinct streaked panel; tail brown, with some narrow white edging not usually visible in field. Entire underparts bright yellow. Characteristic, very finely tapered straight bill, a useful identification feature, is black; eye dark brown, legs grey. **Similar species** From small female sunbirds by bill shape; dumpier, and smaller than Yellow White-eye, which has white ring around eye. Smaller than first-winter Willow Warbler, with a more upright, round-headed stance, no supercilium, and smaller bill. See also Sennar Penduline Tit.

HABITS Most often seen searching outer branches of scrubby acacias and baobabs. Found in pairs and small groups and particularly fond of undisturbed fallow where dense but short layer of shrubs and herbs has formed, with emergent *Acacia* spp. and other small trees. Agile and active, moving upside down or hanging effortlessly by one foot.

VOICE Surprisingly loud rasping call, with stridulating quality, and a softer rapid whistle *tu-tu-tu-tu-tu*, on a monotone, regularly repeated, sometimes preceded with a rising *shree-tu-tu-tu-tu-tu.*

STATUS AND DISTRIBUTION Uncommon to locally frequent resident, recorded in all divisions except NBD, seen most regularly in LRD. Recorded widely in Senegal, including areas close to habitat of Sennar Penduline Tit in the far north.

BREEDING Single nesting record in March and a pair seen mating in June, otherwise little information. Nest typical of the family, of vegetable down woven into a strong white purse. Entrance tube closed off and protected in a slit-like structure which is forced apart by the bird on entry and exit.

SENNAR PENDULINE TIT
Anthoscopus punctifrons **Not illustrated**

Fr. Rémiz du Soudan

Other name: Sudan Penduline Tit

IDENTIFICATION 7.5-8cm. Tiny, dumpy-bodied

bird with comparatively long wings. Upperparts nondescript olive-yellow, with limited fine black dots on buff-yellow forehead only visible at close range, white on the throat and pale buff over rest of underparts. Bill appears finely pointed and black, legs black. **Similar species** Distinguished from Yellow Penduline Tit by lack of strong yellow underparts, and from *Phylloscopus* warblers by extremely small size, dumpy body and spiky bill, the dotted forehead being useful final confirmation if visible.

STATUS AND DISTRIBUTION Desert edge species found in thorny shrubs and small trees; known from a number of widely separated areas along the Senegal River valley and northern Senegal, not yet in The Gambia, which generally lacks the required open *Acacia-Balanites* habitats.

SUNBIRDS

Nectariniidae

Twelve species. Small nectar-drinking specialists, extensive plumage iridescence and many superficial parallels to New World hummingbirds but perch frequently on strong legs; energetic and mobile, frequently hanging at odd angles, even upside down, to obtain food from flowers (hence French name 'souimanga').

For this section, bill sizes are defined as:

short	= 9-13mm:	Pygmy and Mouse-brown Sunbirds
short-medium	= 13-15mm:	Collared Sunbird
medium	= 14-17mm:	Beautiful, Violet-backed, Variable, Olive-bellied Sunbirds
medium-long	= 16-20mm:	Copper Sunbird
long	= 19-25mm:	Scarlet-chested, Olive, Green-headed, Splendid Sunbirds

MOUSE-BROWN SUNBIRD
Anthreptes gabonicus   Plate 42

Fr. Souimanga brun

Other names: Brown Sunbird, Mouse-coloured Sunbird

IDENTIFICATION 10cm. Sexes similar. A dull grey-brown sunbird with bright white spectacle markings, no metallic plumage at any stage; associated with mangroves. Upperparts grey-brown from crown to tail, slightly darker on wings. Face patterned with bright white supercilium and cheek, linking around base of bill. A dark brown eyestripe just in front of eye through the ear-coverts makes dark reddish-brown eye look large. White leading edge to wing sometimes visible, pale spots on tips of all but central tail feathers may be seen in flight or from below. Underparts whitish from chin to belly, strongly suffused with grey. Bill short, slightly decurved, black; legs black. Immature ghosts pattern of adult. **Similar species** Bill shape, spectacled face pattern, uniform dull grey body and (usually) mangrove habitat all distinctive, but confirm separation from immature Collared Sunbird which has yellowish, not white, lines above and below eye and strong yellowish wash to lower body; also forest thicket habitat. From female Western Violet-backed Sunbird by greyish underparts and absence of yellow on vent.

HABITS Closely associated with both *Avicennia* and *Rhizophora* mangroves, occasionally visiting adjacent woodland. Constantly on the move with the quick energetic movements of the family, giving thin cries and searching mangrove foliage at all heights for small insects which form the bulk of its diet. Usually in pairs or small parties, making quick whirring flights, one bird shortly following behind another, over and through the vegetation; always difficult to obtain prolonged views. Dull colouring and quiet voice seem to ensure that despite energetic movements, it can be easily missed in the stillness of the tidal habitats it frequents.

VOICE Thin very high pitched *sqeee*, slightly rising in pitch, and embellishments on this in a regular conversational *wit.wit.sqeee.witter.witter.*

STATUS AND DISTRIBUTION Uncommon resident, but probably frequent within its habitat; potentially present wherever there are mangroves; recorded from sheltered creeks behind the beaches, on both sides of the main basin of the Gambia River through WD, and NBD and into the brackish creeks of LRD. Not yet recorded in freshwater bankside vegetation in CRD, though does enter such habitats in Sierra Leone. In Senegal restricted to coastal mangroves, immediately north and south of The Gambia's borders.

BREEDING Season extended; nest building in August, activity also recorded December, chicks in nest February and being fed March. Nest often in an open position, a rough woven purse of leaves and dry grass, heavily adorned with cobwebs, side entrance. Usually suspended low over water.

WESTERN VIOLET-BACKED SUNBIRD
Anthreptes longuemarei   **Plate 42**

Fr. Souimanga violet

IDENTIFICATION 13cm. Sexes differ. Unusual combination of violet upperparts (male), or strong white supercilium (female), and white underparts. Adult male dark metallic violet over entire head to chin and all upperbody to tail, metallic gloss reduced on dark tail, nearly absent on black-brown flight feathers. Cleanly divided at throat and sides of neck from pure white underparts. Bright yellow pectoral tufts sometimes show at bend of wing. Upperparts can appear all black in unfavourable light. Female upperparts and sides of face grey-brown, white supercilium; violet uppertail-coverts. Wings dark grey with lime-green-yellow wash. Below white from throat to belly; lower belly and undertail primrose yellow. Medium black bill comparatively stout, only slightly curved. Young bird resembles dark female, but is entirely pale yellow below. **Similar species** Distinctive, but if in doubt check male against *Hyliota* spp. and beware long range or glimpsed confusion with larger male Amethyst Starling. Female from Mouse-brown Sunbird by larger size, white and yellow underparts and from immature Collared Sunbird by larger size, greyer upperparts and whiter supercilium.

HABITS Favours open savanna woodlands and regenerating areas where young saplings have begun to re-establish; also noted on periphery of *Avicennia* mangroves, occasionally in forest thickets. Active, though not so frantic in its movements as some of the smaller species. Usually in pairs, sometimes associated with other species. Primarily an insectivore, but noted visiting flowers of e.g. baobab and will move into areas where e. g. ginger plum *Parinari macrophylla* is coming into bloom, when it may nevertheless be mainly catching insects.

VOICE A harsh, squeaky scold, *cha-cha-cha*, and a softer rapid chatter of whistling squeaky notes with little modulation of pitch, reminiscent of Variable Sunbird.

STATUS AND DISTRIBUTION Uncommon resident, sometimes locally frequent, in all divisions of The Gambia north and south of the river. Not easily encountered but regular, often seems transient. Most observations November-February but present in all seasons. In Senegal irregularly distributed through Casamance and the south-east, short range extension north of The Gambia restricted to coastal districts.

BREEDING Nesting in February and May, adults with fledglings in April and October. Suspended nest, a grass pouch with felted lining, elaborately decorated with dry leaves attached at all angles, creating successful illusion of debris caught in the branches.

COLLARED SUNBIRD
Anthreptes collaris   **Plate 42**

Fr. Souimanga à collier

IDENTIFICATION 9-10cm. Sexes differ, no eclipse plumage in male. Tiny, short-tailed, shiny green and bright yellow sunbird of forest thickets. Adult male bright metallic emerald-green with a golden sheen over head, mantle and throat; dull olive-brown wings; tail blue-black centre, greenish edges from above. Straight lower edge of metallic green bib narrowly fringed iridescent violet; changes crisply to bright yellow underparts, fading towards undertail-coverts (yellow pectoral tufts do not show). Adult female resembles male above, though with reduced metallic green. Below yellowish on throat, brighter yellow on breast merging to olive-yellow on flanks, belly and undertail. Black bill short, slightly decurved. Immature dull pale olive-green above and on sides of face, marked by a fine yellow supercilium above the eye and matching yellow streak below; chin greyish, rest of underparts pale yellow, bill brown above, pink below. **Similar species** Adult female distinctive; male resembles Pygmy Sunbird male lacking tail streamers; Collared separated by emerald, not coppery, green above, longer bill, dumpier body, narrow violet breast-band, absence of violet rump patch, and usually forest thicket habitat. Immature usually with adults, but see Mouse-brown Sunbird which shares similar face pattern, though in clean white tones.

HABITS Active, sociable and often noisy sunbird restricted to forest thickets and riverine forest, utilising secondary growth if sufficiently dense and tangled. Usually in pairs or parties and a frequent member of mixed species foraging parties. Primarily insectivorous, searching vegetation in warbler fashion, often high in canopy.

VOICE Song of simple hasty repetitions of a single high-pitched note, introduced and then punctuated by quick very high bat-like squeaks *tsi-tsu-tsu-tsu -tsi-tsu-tsu.tsu.tsi*; also a simple *see-suuu* double note and variations. A quavering *drree-didlee-drree-didlee* repeated hastily with a rising inflection on each component is maintained for long periods by begging young accompanying feeding adults, sometimes rising to an insistent *shreeet.shreeet*.

STATUS AND DISTRIBUTION Frequent to locally-common resident within the small patches of coastal forest in WD such as Abuko; seems to have been more often encountered in the 1980s and 1990s than by earlier observers and collectors. Found in all months, but not yet recorded inland. One or two reports from LRD not sufficiently detailed to eliminate streamerless Pygmy Sunbird. In Senegal primarily restricted to coastal Casamance.

BREEDING Breeds in the rains with young birds prominent September-December. Hanging nest of bark fibre, with downy lining and sheltered entrance.

PYGMY SUNBIRD
Anthreptes platurus   Plate 42

Fr. Souimanga pygmée

Other name: Pygmy Long-tailed Sunbird

IDENTIFICATION 9cm (breeding male 17cm). Sexes differ. A very small green and yellow sunbird of dry savanna woodlands, male with long tail streamers when breeding. Male metallic bronzy-green over head and mantle. Metallic violet rump, tail blue-black. Wing dull dark brown. Elongate narrow dark central tail feathers November-April, slightly raquet-tipped when fresh. Chin and throat metallic green, cleanly divided from dense golden-yellow on entire underside. Non-breeding male (rains) lacks streamers, resembles female above and below, sometimes retaining green gloss on coverts; moults into breeding dress with body plumage preceding development of tail streamers. Female olive-brown above, darker on wings and tail, yellow over entire underside, no streamers. Bill black, short and slightly decurved. Young birds resemble female, but young male has dark patch on throat. **Similar species** At distance check silhouetted breeding male for slim jizz and small straightish bill to separate from male Beautiful Sunbird (the only other species in The Gambia to develop elongated central tail feathers); confirm by solid yellow underparts. Shiny male lacking tail streamers, seen October-December, superficially resembles Collared Sunbird (see that species).

HABITS Found in open dry woodlands; typically in upper branches of leafless trees especially flowering *Bombax* and acacias. Usually in pairs; restless and active, flitting from flower to flower, twisting side to side on perch with tail streamers quivering, quickly moving on with bouncy flight action.

VOICE Very high pitched *tweee* and a bubbling hurried song.

STATUS AND DISTRIBUTION Reported in all seasons but with much greater frequency November-April, when males are long-tailed. Uncommon

at the coast; locally and seasonally frequent in LRD through to URD. Found throughout Senegal, though less common in the far north.

BREEDING Breeds at height of dry season, December-March. Small oval nest closely affixed to under-side of bare branch, with soft downy lining and protected side entrance, or more typical pendant sunbird-type nests.

OLIVE SUNBIRD
Nectarinia olivacea   Plate 42

Fr. Souimanga olivâtre

IDENTIFICATION *N. o. guineensis*. Sexes similar, sometimes distinguishable; no seasonal plumage changes. Distinctively large, plain dull greenish sunbird, lacking any glossy plumage. Entire upperparts dull olive, marginally darker towards the head, cleanly separated from paler grey-toned olive underparts, which are yellow washed towards the belly. Males have bright lemon yellow pectoral tufts near the bend of the wing, but these are usually hidden. Bill long and decurved, almost all black in this race **Similar species** Nondescript, leading to possible tendency to overlook as an unidentified female of more familiar species, but relatively large size and olive tones above and below should alert observer. Similarly-sized immature Green-headed Sunbird has black over head, face and upper breast. Other greenish female sunbirds much smaller.

HABITS Forest sunbird of dense thickets. Drinks nectar, but primarily hunts for insects, especially spiders, with methodical deliberation. Also noted to take seeds. Usually alone or in pairs; can be low in understorey or high in canopy.

VOICE Measured series of high-pitched but clear separate notes descending a crude scale in a slightly decelerating sequence; in the full version continues more hurriedly back up to the original starting frequency in slightly more erratic rhythm. Also a testy nasal alarm buzz.

STATUS AND DISTRIBUTION Not recorded in The Gambia. Reported from locations both to the north and south in coastal Senegal. Locally common in coastal Casamance where its forest habitat is more extensive. In most of its range typically replaced by Green-headed Sunbird as habitat diminishes.

BREEDING Pendant nest of crudely woven fibres, characteristically trailing untidy strands from the lip of the side entrance. Breeding season thought to be extended but no Senegambian details; nearest confirmed breeding Sierra Leone.

GREEN-HEADED SUNBIRD
Nectarinia verticalis   **Plate 42**

Fr. Souimanga à tête verte

Other name: Olive-backed Sunbird

IDENTIFICATION 14cm. Sexes similar but distinguishable; no seasonal plumage change. Large and long-billed sunbird, dark metallic on forequarters, yellow-olive above, grey below. Male has metallic emerald-green head, nape and face, throat and upper breast; iridescence often flashes metallic dark blue, more so on the upper breast. Plain golden-olive upperparts and fringes to brown flight and tail feathers; looks uniform in the field. Lower breast and belly dark dull grey. Pectoral tufts pale creamy-yellow, seldom visible. Female has green metallic cap down to line below eye only; rest of upperparts as male. Below very pale grey on throat darkening to mid-grey breast and belly, no pectoral tufts. Bill black, long and evenly curved. Immature non-metallic black over face, crown and upper breast; rest of underparts dull olive, yellow-washed on the breast. **Similar species** Dark metallic forequarters with olive upperparts and grey below are unique amongst Senegambian sunbirds. Olive Sunbird (unrecorded in The Gambia) lacks the metallic forequarters, and the body is more uniform plain greenish and pale grey.

HABITS Associated with well-watered dense thickets and nearby woodland. Usually in pairs working over vegetation in dense cover feeding largely on insects, also attracted to rich nectar sources such as baobab, ginger plum and paw-paw, in company with other sunbirds. Also joins mixed feeding parties. Noted in Senegal taking sap from lacerations on palm trees created by tappers collecting palm wine.

VOICE Call a distinctive *cheerick* or *tweezee*. Male also sings, but no details available.

STATUS AND DISTRIBUTION Uncommon resident, forest and thickets in WD and the swamp forests of CRD, the least-frequently recorded sunbird in The Gambia. Might be expected in dense riverside habitats of URD, but not yet recorded there. In Senegal frequent in coastal Casamance, with a few observations up the coast to Dakar; also recorded in Niokolo Koba.

BREEDING Not proven though suspected in August; likely to be wet season breeder. Nesting recorded May, August and September in Sierra Leone. Large, rough hanging nest with long pendant decorations.

SCARLET-CHESTED SUNBIRD
Nectarinia senegalensis   **Plate 42**

Fr. Souimanga à poitrine rouge

IDENTIFICATION 14cm. Sexes differ. A large black and red sunbird, with emerald crown; typically all-black silhouette intermittently glitters metallic red and green in sunlight; female dull chocolate brown. Adult male sooty-black over body; tail and, more noticeably, wing feathers contrastingly paler tawny-brown. Metallic emerald crown, chin and malar extensions; throat and extensive rounded bib on breast deep metallic scarlet, within which fine speckled bars of metallic dark blue sometimes visible at close range. Female dull brown above; brownish irregularly mottled with pale yellow and dirty white on the breast, untidily streaked yellowish on belly. Bill long, decurved, black. Immature similar to female, though more heavily mottled with yellow below; gradual development of male breeding colour creates untidy but recognisable intermediate stages. **Similar species** Adult female and immatures separated from Splendid Sunbird equivalents by browner tone, with brown underparts mottled with yellow, rather than yellowish underparts mottled with blackish, and marginally more curved, shorter and lighter bill.

HABITS Noisy sunbird of dry savanna woodlands, regenerating farmland and the edges of thicker shrubby habitats. Attracts attention by calls and male's habit of sitting at exposed point on leafless tree to sing. Feeds on insects and nectar. Much attracted to large red blossoms, of silk cotton, coral tree and scarlet blooms of the parasitic mistletoe *Tapinanthus bangwensis*. Temporary local concentrations may form where these plants are flowering.

VOICE Measured, rather slow hesitant sequence of forceful choppy notes; *chip, teuup-teupp, chip, chip, teuup-teuup* etc. varying but high pitched. Repeated over long periods with brief pauses. Chasing birds produce frequent quarrelsome chatter based on a testy, slightly hoarse, *eu-eu-eu-eu..i-i-eu-eu-eu*, similar to the quick, staccato alarm.

STATUS AND DISTRIBUTION Breeding resident, overall the most frequently recorded sunbird in The Gambia, relatively abundant inland, not common in the coastal belt. Resident in all areas of open dry woodland, uncommon to locally frequent in WD, becoming common, almost locally abundant, in LRD, CRD and URD. Found through most of Senegal, though less frequent in the extreme north.

BREEDING Wet season breeder, building nest in July, completed nest in September. Oval suspended nest with side entrance and well-defined overhanging porch, often not very far off the ground.

VARIABLE SUNBIRD
Nectarinia venusta   **Plate 42**

Fr. Souimanga à ventre jaune

Other name: Yellow-bellied Sunbird

IDENTIFICATION 10cm. Small lively sunbird, male predominately green and pale yellow, with a broad violet breast-band; female drab. Breeding male (April-December) metallic green over head and mantle to upper wings. Wings brown, metallic dark blue uppertail-coverts, blue-black tail. Mask around bill usually matt black, flashes metallic violet; breast metallic violet delimited by matt black belt. Rest of the underparts pale yellow, fading to near-white in some individuals. Yellow and orange pectoral tufts usually concealed. Adult female and non-breeding male uniform grey-brown from crown to back; thin, pale supercilium short and not noticeable; blue-black tail with whitish outer fringes may be glimpsed when hovering in front of flower. Below whitish on chin merging to pale yellow wash on breast and belly. Bill black, sharply decurved. Immature males like female, but show a dark grey patch on throat, metallic green developing from patch at bend in wing and stripe through centre of breast, otherwise like female. **Similar species** Metallic green subtly more silvery-jade in tone than other green male sunbirds. Male from Collared Sunbird by extensive violet bib, more decurved bill. Females and immatures very similar to Copper Sunbird which is more solidly yellowish below with stronger supercilium, and Beautiful Sunbird which also has more pronounced supercilium. In most situations very difficult to be confident of identity in absence of associated males.

HABITS Energetic savanna woodland and coastal scrub sunbird, attracted to small flowering bushes, flowering trees such as ginger plum and ornamental shrubs. Flies with rapid wing beats and slightly dipping flight; frequent tumbling chases. Alights at flowers and immediately commences probing for nectar, working systematically through a flower head before flitting suddenly to an adjacent blossom, intermittently looking up and calling, all with jerky movements. On leaving feeding bush often flies a considerable distance. Male gives outbursts of song from high exposed vantage point, switching body side to side with tail flared.

VOICE Main song opens with a rapid jumble of thin whistles and squeaks, closing with rapid repetition of a single note. The latter frequently used alone as the call with a variable number of units, *tew-tew-tew, tew-tew-tew-tew* etc. Clearer-toned than other sunbirds of the region.

STATUS AND DISTRIBUTION Locally common

at the coast, recorded in all months but particularly March-June, and in all divisions except CRD, though uncommon in LRD and URD. Breeding plumage males reported in all seasons. Yellow-bellied and white-bellied males recorded together, e.g. at Tanji. In Senegal main populations near the coast in Casamance north to Dakar, with an apparently disjunct population in the far south-east.

BREEDING Recorded nesting in low herbs, November, immatures seen April-May; prolonged breeding season in most of its range further east. Small oval suspended nest.

OLIVE-BELLIED SUNBIRD
Nectarinia chloropygia   **Plate 42**

Fr. Souimanga à ventre olive

IDENTIFICATION 10-11cm. *N. c. kempi*. Sexes differ. Very small green and red sunbird with a medium bill, preferring forest habitat. Adult male bright metallic green above, save for dark brown flight feathers. Chin to upper breast also metallic green, changing to a broad scarlet breast-band; belly pale olive (characteristic of this race). Pectoral tufts yellow, usually obscured. Female olive on the upperparts, with narrow supercilium and dark blue-glossed tail; whitish on the chin, paler olive-streaked and washed yellow below. Bill black. Immature males develop green gloss on upperparts first, followed by patchy scarlet on breast. **Similar species** Male from male Beautiful Sunbird lacking tail streamers by pale olive (not metallic emerald) belly, more blue-green tone to upperparts and more extensive red on breast. Proportionately longer-billed than Beautiful.

HABITS Favours forest edge and clearings, attracted to flowers, often gregarious. Particularly ready to hover in front of flowers briefly before settling to feed on nectar; also takes insects.

VOICE Faint sibilant *chwee* call; a harsh alarm note and the male has a typical thin and high-pitched sunbird twittering song ascending the scale.

STATUS AND DISTRIBUTION Unrecorded in The Gambia; could be expected in forest thicket areas of WD. Records both to the north and south of The Gambia in Senegal, where it is known from Saloum, western Casamance and from a few observations near the upper reaches of the Gambia River in Niokolo Koba, south-eastern Senegal.

BREEDING No Senegambian information; nearest confirmed breeding Sierra Leone, where builds suspended oval nest of fine grass fibres and leaves, decorated with lichen and lined with down.

COPPER SUNBIRD
Nectarinia cuprea   **Plate 42**

Fr. Souimanga cuivré

Other name: Coppery Sunbird

IDENTIFICATION 12cm. Sexes differ. Medium-sized sunbird, blackish in most initial views of breeding plumage male until metallic violet and coppery-green flashes show. Breeding male has metallic coppery over head, breast and mantle, sometimes with patches of red iridescence; metallic violet on wing-coverts to back and uppertail. Wing and tail feathers black; lower breast and belly black. Non-breeding male brown above, dull yellow below, with contrasting matt black wing, some metallic colour on shoulder and lower back; sometimes with untidy narrow stripe of dusky feathers on the belly. Female olive-brown from crown to back, black lores, narrow yellowish supercilium, pale outer web to tail. Clean division below eye and along neck between olive upper- and dull yellow underparts. Immature resembles female, male developing dusky throat. Bill black. **Similar species** Adult male distinctive in both full and eclipse plumage. Female from female Beautiful and Variable Sunbirds by darker brown tone above, tidier, with stronger yellowish supercilium, noticeably jet black bill, uniformly pale mustard below; tends to be quieter, less frenetic in manner.

HABITS Found mainly in areas of open woodland, particularly fallow fields, sometimes in thicker regenerating scrub adjacent to thickets, and hotel gardens. Perches prominently on crown of low bushes, making jerky dashing flights to flowerheads of e.g. *Combretum* and other rich nectar sources. Also takes insects. Uses dew-covered leaves to bath in the early morning.

VOICE Maintains regular calling. Sharp, rapid *chit.chit.chit* call frequent and an intermittent, thin *tsip, tsip* ; song a very rapid jumble of hoarse accelerating squeaky notes, sometimes running on from *chit* calls.

STATUS AND DISTRIBUTION Locally frequent breeding resident in WD, recorded in all seasons. Uncommon in other divisions, where most frequently encountered in the wet season. Recorded with much greater frequency May-November, but breeding plumage males only recorded Senegambia March-November; probably in eclipse November to February, which may contribute to lack of records at this time. In Senegal found to the north of The Gambia in the coastal strip only; frequent in Casamance May-November.

BREEDING Breeds in The Gambia May-October; a compact hanging nest, lined with down.

SPLENDID SUNBIRD
Nectarinia coccinigaster   **Plate 42**

Fr. Souimanga éclatant

IDENTIFICATION 15cm. Sexes differ. Most regularly encountered large dark sunbird at the coast; male heavily metallic green-purple and red, female yellowish below and streaky. Adult male glossy metallic purple head, metallic dark green mantle, inner wing and back; glossy dark blue on rump, tail blackish-edged metallic green; flight feathers black. Chin to throat metallic violet, becoming broadly mixed with metallic scarlet on the breast. Pale yellow pectoral tufts usually obscured. Belly black, undertail feathers tipped metallic dark blue. Adult female dark grey above; pale grey on throat, yellow on breast and belly with streaky grey and brown mottling, sometimes weakly tinted with additional reddish mottling. Immature male resembles darker female but develops glossy green wing-bar, and glossy violet develops progressively down chin on initially matt grey throat patch, giving rise to individuals with variably sized, often neatly demarcated, metallic patches. Immature female like adult, but blackish around face and throat. Black bill, long and broad with comparatively shallow curve. **Similar species** Female and immature from Scarlet-chested by less decurved bill; comparatively paler underparts of Splendid are mainly yellowish streakily mottled with brown and grey, not dark brown, mottled and streaked with yellow. In maturing males look for green on wings and purple on throat of Splendid.

HABITS Favours moist Guinea savanna and wooded areas with oil palm. Frequently hovers in front of leaves and flowers, alighting to dodge out of sight between leaves. Also takes insects and spiders, probing bark of palm trees, and sap leaking from incisions made in oil palms to collect palm-wine. Calls frequently.

VOICE Song phrases of methodically delivered emphatic notes, evenly paced and clear-toned *choo-tee choo-tee-tou choo-dew-dew*, each phrase tending to diminish in pitch and fade in volume.

STATUS AND DISTRIBUTION Frequent, locally common resident in WD, uncommon elsewhere. Observed with greater frequency March-June, and suspected of local migratory movements north-south in West Africa, but present in The Gambia in all months. In Senegal mainly confined to Casamance; also along the coastal strip north to Dakar.

BREEDING Mainly a wet season breeder, July-October; builds neat oval suspended nest with sheltered side entrance.

BEAUTIFUL SUNBIRD
Nectarinia pulchella   Plate 42

Fr. Souimanga à longue queue

Other names: Beautiful Long-tailed Sunbird, Long-tailed Sunbird

IDENTIFICATION 10cm (16cm breeding male). Sexes differ. Small bodied; male long-tailed, vividly metallic green with red and yellow patch on lower breast; female drab. Breeding male bright metallic emerald with golden iridescence over head, mantle and breast; wings and tail dark brown, two long central tail streamers. Bright yellow flanks either side of scarlet lower breast; undertail metallic green to black. Non-breeding plumage resembles female but retains metallic green shoulders; below sometimes shows a narrow uneven line of dark feathers down throat to breast; tail extensions show increasing wear before dropping; replacements seen in various stages of elongation. Females are plain brown to olive-brown above, with narrow yellowish supercilium; wings dark brown, tail blue-black with pale edges to outer feathers. Underparts pale washed-out yellow. Immature resembles female, though lightly mottled on chin, and males with matt grey to blackish bib. **Similar species** Silhouetted males

with streamers from Pygmy Sunbird male by longer bill; separation of females and young from equivalent stages of Variable Sunbird by more pronounced supercilium and slighter build, but often very difficult; from Copper Sunbird female by smaller size, less warmly brown above and bill not so shiny black.

HABITS Found in a wide range of habitats, including farm and fallow land, bush, coastal and savanna woodland; prominent in hotel gardens. Sprightly and much attracted to flowering trees and shrubs where it takes nectar in addition to small insects. Males sing from high exposed branches, sitting upright to expose colourful belly, flicking wings stiffly up and switching fanned tail and streamers side to side.

VOICE Frequent testy chirps, similar to Variable but fractionally reedier in quality; song begins with quick repetition of a single note before breaking into a shivering tumble of quick squeaky notes, in reverse sequence to the Variable Sunbird.

STATUS AND DISTRIBUTION Common resident in all divisions of The Gambia. Also widespread throughout Senegal, less so in the extreme north in the dry season.

BREEDING Prolonged breeding season; records from all months except at the end of the dry season. Hanging nest decorated with bark flakes and tendrils.

WHITE-EYES
Zosteropidae

Only a single species in Senegambia. White-eyes are small warbler-like birds of uncertain affinities. Their English name derives from the prominent white eye-ring (not a white eye).

YELLOW WHITE-EYE
Zosterops senegalensis   Plate 40

Fr. Zostérops jaune ou Oiseau-lunettes jaune

IDENTIFICATION 10cm. Sexes similar. Tiny green and yellow bird of the tree tops, readily recognised by the bright white eye-ring. Upperparts usually appear fairly uniform pale yellow-green. Forehead plain yellow; entire underparts bright yellow. Circle of small pure white feathers around eye creates diagnostic field mark. Fine straight black bill. Legs blue-grey. Young birds resemble adults, but darker above. **Similar species** Eye-ring diagnostic; from dumpier Yellow Penduline Tit by more horizontal posture; colours more intense than any wintering Palearctic warbler; bill shape eliminates all yellowish female Sunbirds, with which it sometimes travels. Shivery calls also useful for quick identification.

HABITS Gregarious, usually in small parties high

in leafy tree-tops. Favours woodland savanna with tree species such as the West African mahogany *Khaya senegalensis*. Active, often associated with mixed species feeding parties, constantly calling.

VOICE Sequence of short, pure, high-pitched whistles, slightly descending in pitch, with a tremolo shivering quality; maintained as an almost continuous silvery twitter by a flock of feeding birds. Forceful song *tsee-tsuu-tsuu-tsi-tsuu-tsi-tsuuu* sequence, with little variation in pitch.

STATUS AND DISTRIBUTION Frequent and widespread resident, stronghold in the woodlands of LRD; not often recorded in NBD. In Senegal found only in central and southern areas.

BREEDING An immature reported in May, nesting activity mainly May to June and a pair observed collecting nest material in August. Small cup nest slung in a narrow fork of a tree.

ORIOLES

Oriolidae

Two species. Medium-sized, starling-like, brightly coloured birds. Females duller than males, which despite vividly contrasting black and yellow plumage, often remain inconspicuous within the cover of sunlit leaves in the canopy. Best located by characteristic mellow liquid calls.

EUROPEAN GOLDEN ORIOLE
Oriolus oriolus   Plate 34

Fr. Loriot d'Europe

Other name: Golden Oriole

IDENTIFICATION 24cm. Sexes differ. Adult male's golden-yellow body contrasts with all-black fore-wing marked only by short squarish yellow bar at base of primaries. Black mask between eye and bill; tail patterned with a broad black 'Y' on yellow background. Female greenish on upperparts and folded wing showing only narrow yellow edging, faded eye mask limited to lores; off-white streaked olive below. Bill pinkish, eye red, legs grey. Immature similar to female, but duller and with blackish bill. **Similar species** Female European plainer green on mantle and shoulder area than African; separation best confirmed by dark eye mask, which extends behind the eye only in African. Note underparts of first-winter male European may show considerable yellow wash, like female African. Separation of adult male from African straightforward but requires persistence in thick foliage.

HABITS Favours forest and woodland canopies. Secretive, best located by calls.

VOICE A carrying, mellifluous whistle, with characteristic yodelled inflection, *doo-leeoo-ee*.

STATUS AND DISTRIBUTION Palearctic migrant, possibly overlooked. Observed near the coast and in CRD in March-April 1991, otherwise no recent information. Seen regularly in small groups on migration, mainly April-May, in northern Senegal.

AFRICAN GOLDEN ORIOLE
Oriolus auratus   Plate 34

Fr. Loriot doré

IDENTIFICATION 24cm. Sexes differ. Male predominantly vivid yellow, with some black streaking on the wings and clean broad black mask; duller female washed yellow above, with faded but obvious mask extending to a point behind the eye as in male. Underparts entirely yellow on the male, pale yellowish with weak olive streaking on the female; both share a yellow tail with bold black centre.

Pinkish-brown bill long and pointed, contributing to slim, streamlined jizz in flight. Eye red, legs grey. Immature resembles female with dusky bill. **Similar species** See European Golden Oriole. Yellowish female Red-shouldered Cuckoo-Shrike is barred.

HABITS Favours wooded forest tree-tops, where it moves within the canopy of large leafy trees, well camouflaged in dappled light conditions. Particularly attracted to fruiting figs. Usually shy; when disturbed, makes off quickly, with undulating flight.

VOICE Most notably clear whistling notes, rendered *fee-yoo...fee-yoo* with mellow, fluty quality typical of the genus, though not as liquid as those of the European Golden Oriole. May be preceded by a creaky, strained, squealy sounds *squee-ree-aahh, squee-reee-aahh*; pairs engage in prolonged melodic repetitions of duetted fluty whistles in the rains. Also produces a dry, hissing *kiaarr* around the nest.

STATUS AND DISTRIBUTION Frequent resident throughout The Gambia, temporarily common at local sites where ripe figs available, and similarly through southern and central Senegal, though seldom seen in northern Senegal.

BREEDING Breeds April-July in The Gambia, constructing a flimsy, hanging basket-shaped nest, often at some height and near the outer branches of the canopy.

European Golden Oriole, male

African Golden Oriole, male

338

SHRIKES AND ALLIES

Laniidae, Malaconotidae and Prionopidae

Robust predatory passerines with hooked bills, divided into three families: true shrikes, Laniidae, represented by the genera *Lanius* (four species of Palearctic migrants) and *Corvinella* (single resident species) which perch prominently to scan for prey; bush shrikes, Malaconotidae, an assortment of skulking, elusive birds (eight species), usually boldly, sometimes vividly, coloured with dramatic calls and duetting behaviour; and helmet shrikes, Prionopidae, represented by a single social species.

RED-BACKED SHRIKE
Lanius collurio   Plate 43

Fr. Pie-grièche écorcheur

IDENTIFICATION 17cm. Sexes differ. A relatively small shrike, but with typical posture and habits, adult male neat and distinctive in grey and rufous plumage. First-winter birds require care in separation from similar-age Woodchat and Isabelline Shrikes. Adult male has grey head, black mask, rufous mantle, grey rump; tail mainly black with white basal feathers at each side. Underside whitish. Female has crown grey-brown, a weak brownish mask, mantle and rump brown; underparts finely scalloped. First-winter birds similar to adult female, with increased scalloping to upperparts; tail and rump brownish sometimes with rufous-tinge reminiscent of Isabelline Shrike, but confined to uppertail only. **Similar species** First-year Isabelline usually has much lighter scalloping on the greyer-toned upperparts, and may be plain; underparts also show much less scalloping. First-winter Woodchat Shrike normally greyer overall, with line of whitish patches at shoulder and primary bases. The smaller Emin's Shrike *Lanius gubernator*, not recorded in Senegambia, is resident in Guinea: male differs in having a chestnut rump and tawnier tones to the underparts, with the black mask extending onto the forehead.

HABITS Usually seen perched on prominent lookout point. Primarliy associated with open scrub habitats.

VOICE Known to sing a subdued warble on migration in East Africa. A sharp alarm call.

STATUS AND DISTRIBUTION Not recorded with certainty from Senegambia. Previous records related to the form *isabellinus* which is now regarded as a separate species (see that species).

ISABELLINE SHRIKE
Lanius isabellinus   Plate 43

Fr. Pie-grièche isabelle

IDENTIFICATION 18cm. Plain, distinctively drab brownish shrike with rufous tail. Rufous-tinged brown crown and nape merging into a plain brown mantle, rufous rump and tail. Adult males have a well-defined blackish mask through lores and ear-coverts with a creamy supercilium above. Mask reduced to a dark smudgy patch behind the eye in females and immatures. Wings dark brown, adult males usually show a short white bar at the base of the folded primaries. Wing-bar lacking on females and first-winter birds. Bill black (pale-based in first winter), eye dark and legs grey. **Similar species** Appreciably paler than immature Woodchat which has darker crown and large whitish shoulder mark, and from immature Red-backed Shrike by absence of mantle scalloping.

HABITS As other *Lanius* shrikes, favouring semi-open habitats with readily available look-out places, including posts and wires. Usually solitary.

VOICE Soft chattering song, with mimicry, similar to Red-backed, and soft rasping call, both occasionally given on migration.

STATUS AND DISTRIBUTION Uncertain, but at best a very rare migrant; only five records in The Gambia, recorded from a handful of coastal scrub sites between 1965 and 1995. Several recent records in northern Senegal. Given the potential for confusion between immature Isabelline, Red-backed and Woodchat Shrikes, all rufous-brown migrant shrikes should be treated with care.

SOUTHERN GREY SHRIKE
Lanius meridionalis   Plate 43

Fr. Pie-grièche grise

IDENTIFICATION 24cm. Sexes similar. A large grey, black and white shrike; like others of its family, it favours perching prominently and alone. Clean powdery grey from forehead and crown to mantle and rump; sharply defined black mask on lores and ear-coverts, which in the West African races *elegans* and *leucopygos*, may lack a narrow white supercilium · above. Wings are black with a white bar on the scapulars, variable white patch at the base of the primaries and white tips to the tertials. Tail black

339

with a white margin, widest towards the tip. Underparts white, variably tinted pinkish. Bill, eye and legs blackish. Saharan *L. m. elegans* shows narrow extension of the mask to meet above the bill, and also shows more white on the scapulars and a whitish rump and uppertail-covert area, which is also a feature of *L. m. leucopygos*. (Nominate European birds have a greyish rump, thin white supercilium and pinkish underparts). The Lesser Grey Shrike, *L. minor*, has not yet been recorded in the region, but could occur; adults are smaller (20cm), show prominent extension of mask onto forehead, without a white supercilium and little white on the scapulars. First-winter Lesser Grey Shrike does not show mask projection onto forehead, and in isolation, separation from various races of Southern Grey Shrike best secured by observation of the relatively long closed wing of Lesser.

HABITS Solitary percher in shrike tradition, flying fast and low between vantage points, rapid wing beats interspersed with folded wing glides to give bounding flight action. Impales prey on thorns.

VOICE Chattering and variable song, often harsh, but with frequent mimicry and some duetting between male and female. Single *chack* call note.

STATUS AND DISTRIBUTION Rare winter visitor, with ten documented records in The Gambia in recent years, all December-April, assigned to *L. m. leucopygos*; occasional additional reports. The race *L. m. elegans* is frequent in northern Senegal, especially in winter, and is likely to occur in The Gambia.

BREEDING Saharan race breeds during the rains, in rough nest of twigs and grass placed in a bush or tree.

NOTE This species was formerly treated as conspecific with Great Grey Shrike *L. excubitor* of the Holarctic.

WOODCHAT SHRIKE
Lanius senator   Plate 43

Fr. Pie-grièche à tête rousse

IDENTIFICATION 19cm. The most regularly seen Palearctic shrike, distinctively pied body pattern of adult offset by rufous-orange crown and nape. Bold black mask from forehead through eye. Blackish mantle dominated by large oval white shoulder patch. White rump and mainly black tail with white outer feathers. Underparts white. Adult female similar but duller. First-winter birds also show ghost version of this pattern; but in early autumn may show strong traces of brown immature plumage, which is lightly barred on crown, mantle and upper breast. **Similar species** Immature more barred than

Isabelline, with whitish patch on shoulders already evident and darker face mask developing.

HABITS Favours open scrub habitats. Sits on telegraph wires.

VOICE Song involves sustained series of chattering whistles, trills and mimicry, more elaborate than other *Lanius* shrikes. Various harsh calls.

STATUS AND DISTRIBUTION Palearctic breeder, frequent to common on passage throughout The Gambia, November-April, with peak numbers at the beginning and end of this period. Found throughout Senegal at similar times, particularly in the north. Noticeably more numerous in some years than others.

YELLOW-BILLED SHRIKE
Corvinella corvina   Plate 43

Fr. Corvinelle à bec jaune

Other name: Long-tailed Shrike. Note: the name Long-tailed Shrike is also used for Magpie Shrike *Corvinella melanoleuca* of southern Africa and also refers to *Lanius schach* in Asia.

IDENTIFICATION 18cm. Sexes similar. Distinctive streaky-brown shrike with long active tail and waxy yellow bill. Upperparts mid-brown lightly streaked black. Broad, creamy supercilium above blackish-brown face mask. Tail brown, long and narrow, often untidy. Folded wings dark brown but in flight reveal prominent rufous patches in the base of the primaries, visible at long range. Underparts pale buff-white with thinly scattered pale black streaking. Prominent and diagnostic yellow bill. Eye dark with a narrow yellow eye-ring. Immature resembles adult, but shows pale buff fringes to the wing-coverts and the breast feathers are finely scalloped with dark brown. **Similar species** Unique within region.

HABITS Favours woodlands and fallow farmland with trees and shrubs to provide look-out posts. Almost always in large territorial groups of six or more, sitting on prominent perches scattered over a small area, or moving in follow-my-leader flights between a series of vantage points, telegraph wires etc., in low straight flight showing characteristic long-tailed, short-winged silhouette. Groups indulge in noisy calling bouts swinging and rotating tails in exaggerated arcs.

VOICE Harsh garrulous cries in concert are distinctive and often attract attention. A hard *squee*, *squee* in flight, also squealing notes with a guttural rattling *drreee-too, dreee-too*, developed into a communal chattering racket by a perched group. Harsh alarm, *twee-kik...wee-ti-kik*.

STATUS AND DISTRIBUTION Common resident throughout The Gambia. Similarly distributed in savannas through southern and central Senegal, but rarer in the north. Some indication of local migrations north and south with the rainy seasons in West Africa.

BREEDING Cup nest in a bush or tree, recorded breeding March-October, but mainly in the rainy season. Only a single female from the group breeds at one time, assisted by all other members, who jointly defend the territory, warn off predators and feed the young.

BRUBRU
Nilaus afer   Plate 43

Fr. Brubru

Other name: Brubru Shrike

IDENTIFICATION 14cm. Sexes differ. Small, short-tailed, compact black and white species, lacking typical shrike jizz. Male has black crown, with broad white superciliary stripe of constant width running from forehead to nape. Black eyeline broadens onto dark neck. Whitish mantle and back. Wings black, with more or less continuous buff-white wing-bar. Short tail black, edged white. Underparts white to creamy-white, save for an elongated patch of dark chestnut along either flank, sometimes obscured by the wing. Female similar but blackish-brown rather than black. Immature has white supercilium, mottled brown and buff above, barred and speckled brown below. Similar species Distinctive; but seen briefly might bring to mind female batis, which is smaller, has pale chestnut band across upper breast, contrasting yellow eye entirely set within broader black face band, upright posture and small bill, all obvious differences in reasonable views.

HABITS Unobtrusive, usually in pairs, preferring the canopy of emergent trees in dry savanna woodlands, working quietly along branches and through leaves. Often located by call, remarkably elusive.

VOICE A carrying, drawn out prrrrrreee on a reedy monotone from the male; female may join in a synchronised duet. Repeated at intervals for prolonged periods, especially in the wet season, when it becomes a feature of savanna woodlands where the Brubru is common.

STATUS AND DISTRIBUTION Frequent to locally common resident throughout, though easily missed. Appears to be most common in the relatively extensive savanna woodlands of LRD. Thinly spread throughout Senegal, though possibly absent from the moister woodlands of coastal Casamance.

BREEDING Nest is a flimsy structure in tree fork, not found in The Gambia, although adults have been seen feeding young in November.

NORTHERN PUFFBACK
Dryoscopus gambensis   Plate 43

Fr. Cubla de Gambie

Other name: Gambian Puff-back Shrike

IDENTIFICATION 18cm. Sexes differ: males black and white, females grey and apricot-buff. Usually seen in pairs. Male conspicuously capped black over head and mantle, rump (frequently puffed up – see habits) plus entire underparts pure white. Tail black; wings blackish, with strong white fringing. Heavy black bill. Eye red. Adult female lacks the puff-back, is pale grey on the head and mantle, and darker grey on tail. Entire underparts pale apricot-buff. Eye orange. Immature resembles female, but noticeably more intensely orange on underparts; developing males paler below, blackening from front of head. Similar species Female Puffbacks are usually with males, but can be distinguished from superficially similar, much rarer forest dwelling Leaflove by smaller size, orange eye, and darker throat. Two Laniarius shrikes present nearby in the region resemble male Puffback but are both substantially larger; see Tropical Boubou.

HABITS Associated with tall trees, usually in moister woodlands and thickets, occasionally in high mangroves; not normally found in dry savanna woodlands. Unobtrusive, most readily detected by voice, or sudden glimpses of the striking black and white males in flight. The puffback display of the male, dramatically alters the shape of the bird creating a soft 'woolly ball' at the rear end emphasised by drooping wings.

VOICE Loud cries, often persistent over long periods; frequent hard, rasping chiuk..chiuk, or tzzik, tzzik sounds irritable; also produces rythmic sequence of 3-5 crackly snaps and other chattering notes.

STATUS AND DISTRIBUTION Resident breeder throughout The Gambia, frequent to common in suitable habitat in all divisions; slightly more common in coastal regions. Some suggestion of seasonal movement away from URD in the dry season. Similarly distributed through southern Senegal, but range does not extend very far north of The Gambia.

BREEDING Nests in a tree or bush. Possibly extended breeding season. Adult female feeding immature birds January; collecting beakfuls of plant fibre February. Most nest records June-September, once in December. Displaying males observed as early as April, mainly July-September.

BLACK-CROWNED TCHAGRA
Tchagra senegala **Plate 43**

Fr. Tchagra à tête noire

Other names: Black-headed Bush Shrike, Black-headed Tchagra

IDENTIFICATION 22cm. Sexes similar. A medium-sized distinctive shrike, boldly striped on the head, with patterned tail tip and rufous-brown body colours. Black crown stripe, cleanly separated from broad buff-white supercilium above black eye-stripe. Upperparts mid-brown, tail mainly black, tipped white, with the two central feathers dark brown. Folded wings rich rufous-brown, primaries darker. Underparts very pale grey on the throat, grading evenly to grey on breast, belly and especially flanks. Immature dark brownish on the crown, bill horn. **Similar species** Unique coloration precludes confusion with any other species in region.

HABITS Favours scrub and woodland habitats in both dry and moister savannas. Usually remains in bush cover within one or two metres of the ground, given to flushing at close range; brief disappearing views typical, making low swift flights, looping over and between nearby bushes before dropping down. Display flight of calling male distinctive, rising steeply upward on snapping wings before either gliding down on a slanting path, or plunging more dramatically with sudden twists and turns. Also sings well from prominent perches in all seasons.

VOICE Most notably a loud burst of swooping whistles, typically careering down-scale, *cheee-cheree chee cheroo cheroo*, or rising and falling, *swooo-swooo-sweee-swee-swoo-swoo*; often preceded by shorter rattling notes; distinctive and useful for location. Prolonged dry rattle thought to come from female; duetting occurs.

STATUS AND DISTRIBUTION A frequent to common resident throughout The Gambia and Senegal, wherever there is adequate shrub cover. Nominate race in The Gambia marginally paler than isolated population, *T. s. cucullata*, in North-West Africa.

BREEDING Breeds October to December, season probably extended; flimsy nest placed low in the fork of a bush or tree.

TROPICAL BOUBOU
Laniarius aethiopicus **Not illustrated**

Fr. Gonolek à ventre blanc

IDENTIFICATION 23cm. Sexes similar. Large and elusive shrike, of uncertain status in Senegambia.

Upperparts from the head, to below eye, plus mantle and entire wings glossy black, with (in the nominate form), a bold white stripe along the folded wing, limited white spotting on the rump and a black tail. Underside white, with a pink-buff blush on the breast. Legs grey. **Similar species** From male Northern Puffback by much larger size, and a distinct white bar along the base of the folded wing, rather than diffuse whitish patch of Northern Puffback, and if heard, by flute-like whistling notes. Turat's Boubou *L. turatii* (Fr. Gonolek de Verreaux), resident in Guinea Bissau but not recorded in Senegambia, is similar, but plain black on the wing.

HABITS Usually skulking and elusive, though not necessarily shy. Favours dense thickets in well-wooded areas, and forest edges; often encountered in pairs, prominent duetting calls drawing attention to its presence.

VOICE Combines mellow flute-like whistles with harsh rattles and buzzes in duets, which either of the pair may initiate; more variable and sophisticated than those of Yellow-crowned Gonolek.

STATUS AND DISTRIBUTION No accepted records in The Gambia; occasional claims in Abuko probably mainly attributable to Northern Puffback. Two singles recorded in Senegal, at Toubakouta, north of The Gambia, and in Casamance. Otherwise nearest range is Sierra Leone.

BREEDING Flimsy cup nest in the fork of a tree or bush, recorded in May in Guinea.

YELLOW-CROWNED GONOLEK
Laniarius barbarus **Plate 43**

Fr. Gonolek de barbarie

Other names: Barbary Shrike, Gonolek

IDENTIFICATION 22cm. Sexes similar. Unmistakable vividly-coloured bush shrike, marked with bold blocks of colour in unique combination; deep scarlet below, black above including sides of face and tail; dull gold crown to nape. Undertail-coverts and thighs buff. Immature paler and duller than adult. Legs blackish. **Similar species** No other black and scarlet bird glimpsed skulking in dense cover also possesses a gold crown; male Western Bluebill otherwise has similar pattern, but is much smaller.

HABITS Usually encountered in pairs, low down in dense cover of well wooded areas, hopping down to the ground for prey as well as searching within the bushes. Also enters mangroves. Pairs maintain contact by means of regular loud calls, and frequent low-intensity confrontations between pairs in adjacent territories is routine in high density populations. Retiring, though not to the extreme extent of other bush shrikes and in The Gambia birds may be seen moving boldly in the open, for example across irrigated hotel lawns. Flight display involves deep slow stalling wing beats, with audible wing-snaps, accompanied by reverberant churring 'bed-spring' call.

VOICE A duetting bush shrike, pairs regularly producing exactly synchronised calls which superficially sound as if from one bird. Sequence initiated by a pure, liquid whistle interrupted by a neat glottle-stop, *too-lioo* and completed by a dry rasping rattle, *ch-chacha*, with the rattle commencing before the whistle is completed. A repeated, rythmic *twoo-woo* whistle from one bird also typical.

STATUS AND DISTRIBUTION A frequent to common resident breeder in all areas of The Gambia. Throughout Senegal in similar habitats, though apparently absent from drier areas of the north in the dry season.

BREEDING Builds flimsy cup nest in a tree or bush. Recorded breeding January-September in The Gambia, most activity June-August.

SULPHUR-BREASTED BUSH SHRIKE
Malaconotus sulfureopectus   Plate 43

Fr. Gladiateur soufrée

Other name: Orange-breasted Bush Shrike

IDENTIFICATION 18-19cm. Sexes similar. Neat slim shrike, greenish above, orange and yellow below, dove-grey head and dark mask. Top of head, nape and well down the back grey; rest of the upperparts and tail green, tipped yellow. Forehead and supercilium yellow contrasting with blackish mask through eye. Chin and throat bright yellow, with a deep orange-washed upper breast and yellow belly. Legs blue-grey. Immatures lack the black mask, pale around eye, and whitish malar area; two rows of pale yellowish covert spots. **Similar species** Adult readily distinguished from Grey-headed Bush Shrike by size and face markings; immature recalls that species in colour pattern of head, also showing yellow spotting on wings and tail, but size of body and bill should make separation straightforward.

HABITS Widespread in habitats offering reasonably dense shrub and tree cover; it generally remains well hidden amongst the foliage, even when its presence and location have become known through its calls. Generally seen singly or in pairs.

VOICE Variable. Most characteristically a pure mellow whistle of, usually, three short notes of even

length and emphasis, most often on a monotone, *tu-tu-tu*, but sometimes modified in pitch of opening note; a four-note *wut-wit-wit-oooo*, possibly created by duetting; a fluid pure-toned *oooit-ooi-ooooi*, rising in pitch and emphasis towards the final note. Repeats calls for long periods. Heard far more often than seen.

STATUS AND DISTRIBUTION A frequent resident in all parts of the country where there are adequate areas of trees and bushes, though not often seen if voice not recognised. Most readily encountered near the coast in the south. Found throughout southern Senegal, but range only extends a short way north of The Gambia, reaching as far as Dakar on the coast.

BREEDING Flimsy platform nest breeding not yet proven in The Gambia, but birds in immature plumage seen November-January.

GREY-HEADED BUSH SHRIKE
Malaconotus blanchoti   Plate 43

Fr. Gladiateur de Blanchot

Other name: Gladiator Bush Shrike

IDENTIFICATION 27cm. Sexes similar. Very large, green and yellow bush shrike, distinguished by a formidable hooked bill on large head. Top of head and nape to below eye, plain grey, lores white. Upperparts and tail soft green, marked by yellow spotting across the wings and yellow tip to the tail. Underparts sulphur yellow, suffused with warm orange on breast. Immature paler yellow on the underside with brownish mottling on the head. Eye yellow-orange, legs blue-black. **Similar species** From Sulphur-breasted Bush Shrike by large size, huge bill and absence of black mark through eye; note immature Sulphur-breasted also lacks black eye mark, but bill size distinguishes. See also Western Nicator.

HABITS Usually found singly or in pairs, decidedly inconspicuous if not calling. Associated with well wooded areas with tall trees throughout the savanna, seen in leafy vegetation at any height, occasionally coming to the ground, but most often quietly searching through the upper canopy. Flight across open spaces appears laboured. Bold predator, regularly killing small reptiles and birds, and may store food in larders like *Lanius* shrikes. When calling may adopt a conspicuous perch near the top of a tallest available tree, with bill pointing skywards, though may call from within canopy.

VOICE A haunting monotone whistle, slow and drawn out, with a beautiful fluty resonance, loud and far carrying when delivered from a high perch.

Occasionally terminated by an upward yodel to a stacatto finishing note. May be heard all through the year, but particularly April-July and November. Also various hoarser notes.

STATUS AND DISTRIBUTION Frequent resident throughout the country though easily overlooked. Most frequently encountered in CRD. Similarly distributed through southern Senegal, but does not penetrate far north of The Gambia.

BREEDING Flimsy and untidy cup nest, several metres up in a fork of a tree or bush, not necessarily well hidden. Breeding in The Gambia June-July, nestlings seen November.

WESTERN NICATOR
Nicator chloris   Plate 35

Fr. Nicator vert

Other name: West African Nicator

IDENTIFICATION Male 25cm, female 22cm. Sexes similar. A robust greenish bird of forest thicket with bold yellow spotting above. Head greenish shading to yellowish-green on cheeks. Back and tail pale greenish with prominent yellow spots on the wing-coverts, and fine yellowish edges to the folded primaries. Chin and underparts uniform pale grey. Long grey bill with fiercely hooked tip and curiously angular shape. **Similar species** Distinctive but typical brief glimpse may require quick separation from Grey-headed Bush Shrike: Nicator bill is more angular on a less massive head; lacks grey on the crown and nape; yellow spots on the mantle are larger and the grey underparts are also distinctive. Spotted-winged immature Sulphur-breasted Bush Shrike is smaller with a lighter bill. Nicator favours more tangled and forested habitat.

HABITS Confined to forest and thickets with dense undergrowth, where it remains extremely skulking and shy, moving singly or in pairs. The best chance of locating it is by voice.

VOICE When disturbed may give a loud *tchok* call, explosive with rich 'cork out of a bottle' hollow tone. This may be repeated several times and be accompanied by a series of clucking notes strung together. Also produces vigorous bursts of loud, clear song.

STATUS AND DISTRIBUTION Very rare in The Gambia: only four documented records, all WD, both dry and wet season. Restricted to forest and thicket. Threatened by habitat loss. In Senegal confined to coastal Casamance.

BREEDING No breeding information in The

Gambia or Senegal. Nearest recorded breeding Liberia, though resident in all countries between.

NOTE Nicator is traditionally of uncertain taxonomic status, allocated alternatively to bulbuls or shrikes. Recent DNA evidence suggests closer alliance with bulbuls, but it is retained in Malaconotidae here.

WHITE-CRESTED HELMET SHRIKE
Prionops plumatus   Plate 43

Fr. Bagadais casqué

Other name: Long-crested Helmet Shrike, White Helmet Shrike

IDENTIFICATION 20cm. Sexes similar. Striking black and white shrike usually encountered in parties, with distinctive long lax crest and yellow eye wattle. Forehead and crown plus entire underparts pure white; crown of adult drawn out into an untidy white crest, generally curling forward. Eye surrounded by bright yellow circular wattle. Black bar low down behind the eye, nape smudged with grey, collar pure white. Mantle and wings black, iridescent in bright light with broad white bar. Tail black, edged and tipped white. Immature has shorter crest, eye wattle scarcely noticeable. Legs orange-red. **Similar species** Bold patterning, crest and characteristic sociability usually preclude confusion with any species. Glimpses in forest situations may briefly suggest male Northern Puffback, which has a black head and no crest.

HABITS Sociable, travelling in parties of six or more progressing in a series of short flights on a broad front, between bouts working twigs for insect prey; rarely comes to the ground. May join mixed-species foraging parties. Favours dry lightly-wooded or woodland savanna, and moister woodlands near the coast.

VOICE Distinctive combination of dry rattles, clicks and snaps, with more conversational chattering notes, often coordinated through the party.

STATUS AND DISTRIBUTION A frequent resident breeder in woodlands throughout The Gambia with stronghold in large forest parks of LRD. Found throughout southern Senegal, but range extends only a short way north of The Gambia.

BREEDING Builds a neat cup nest on the upper side of a branch, often at a fork. Nest assistance by several members of the group frequently observed. Nesting in The Gambia May-October.

DRONGOS

Dicruridae

Two species. Medium-sized, all-black, shrike-like birds. Slim bodies and long tails, coupled with very short legs give characteristic perching posture. Mainly insectivorous, feeding by flycatching or pouncing on prey on the ground.

SQUARE-TAILED DRONGO
Dicrurus ludwigii Plate 34

Fr. Drongo de Ludwig

IDENTIFICATION 19cm. Uniformly black forest bird, with only slightly notched tail, and upright perching habit. Mantle sheen glossed dark bluish, only seen in favourable light. Tail broadens slightly towards the tip, with slight indentation in the middle, thus not completely square. Flight feathers characteristically uniform black. Female and young are less glossy than the male, otherwise similar. Heavy, hooked bill black. **Similar species** Separated from Fork-tailed Drongo by absence of pale primary panel, smaller size and proportionately shorter tail of Square-tailed. Found in different habitat to the Fork-tailed Drongo, but note that in The Gambia both species will use a common habitat edge, where savanna and forest patches abutt. From Northern Black Flycatcher by obvious postural and behavioural differences, especially tail shape. The Shining Drongo *D. atripennis*, is a bird of dense lowland forest, not recorded closer than Sierra Leone, with a long forked tail and green gloss to the feathering.

HABITS In The Gambia confined to the interior and edges of forest patch remnants, usually encountered in pairs hawking in typical swooping drongo-style. Readily joins mixed forest feeding parties. Tail flicked periodically like a Common Redstart. Periodically noisy, using strident calls, but may also be relatively quiet for long periods.

VOICE A variety of notes, many loud and harsh, variously rendered: *cherit..cherit.cheritcheritcherit, tswing..tswing. .tswing, chuit, chidder-chick.*

STATUS AND DISTRIBUTION Rare resident of isolated forests in NBD and CRD, though no modern records from well-watched and seemingly appropriate sites near the coast in WD. In Senegal common in forested areas of Basse-Casamance, but not recorded to the north or east of The Gambia.

BREEDING No confirmed breeding records from The Gambia or Senegal, though present year-round. Builds a small cup nest on a well-shaded, horizontal tree fork.

FORK-TAILED DRONGO
Dicrurus adsimilis Plate 34

Fr. Drongo brillant

Other names: Glossy-backed Drongo, Common Drongo

IDENTIFICATION 25cm. Uniform black when perched, heavy-headed, fork fish-tail silhouette, pugnacious habits. Entire body glossy black; flight feathers and tail dull black. In flight, grey-brown panel evident in primaries. Tail long, stiff and forked, but in moult looks different with an overlapped double fishtail created by new feathers. Female similar but less glossy, immature subtly spotted with buff feather tips. Bill black, heavy, broad-based and hooked; eye red. **Similar species** See Square-tailed Drongo especially in forest thicket. Black Flycatcher differs in head profile being smaller and finer billed, with rounded tail. See also male Red-shouldered Cuckoo-Shrike.

HABITS Often found in pairs, in a variety of wood and scrub habitats, including the edge of, sometimes within, forest thickets. Keeps watch like a shrike from a prominent perch. Characteristic sallying flights with vigorous wing-beats and steep, sudden twists and swoops as it pursues insect prey; attracted to bush fires. Active, noisy and loud in scolding potential predators and one of the few birds regularly seen to press aerial mobbing assaults to the point of physical contact, even with larger raptors.

VOICE A wide variety of harsh notes, usually with a rasping, ratchety quality, frequently heard up to dusk. A tinny *scrink...scrink*; also a subdued, conversational sequence of slow, nasal-toned rubbery squeaks.

STATUS AND DISTRIBUTION Common and widespread resident in savanna woodlands and farmlands throughout The Gambia. Similarly found throughout Senegal, but becoming scarce in the extreme north.

BREEDING In The Gambia, breeds in the rains, June-October, building a shallow cup nest on a narrow fork, usually fairly high, but sometimes in a surprisingly exposed location such as a leafless tree. Family parties of 3-5 birds remain together for a period post-fledging.

CROWS

Corvidae

Three species; generally large omnivorous birds with black, brown or pied plumage. Bill, legs and feet powerful. Voice loud, usually raucous.

PIAPIAC
Ptilostomus afer   Plate 24

Fr. Piac-piac

Other names: Black Magpie, African Magpie

IDENTIFICATION 46cm. Sexes similar. Medium-sized all-dark bird with deep black bill, slender body and narrow, stiffly tapered tail. Head and body black, with a slight gloss. Forehead and lores velvety, tail paler and browner than the rest of the plumage. In flight, grey-brown webbing of the primaries gives silvery-grey aspect to the open wings. Eye deep red-brown to intense mauve, long gangly legs black. Immature like adult, separable by bright pink bill, variably tipped black. **Similar species** Other gregarious, long-tailed dark birds in this size range are Green Wood-Hoopoe which has thin decurved bill and conspicuous white wing and tail spotting; Long-tailed Glossy Starling with longer, broader lax tail, iridescent blue plumage and smaller bill.

HABITS Nearly always seen in cohesive groups up to eight birds, sometimes more, with regular outbreaks of simultaneous calling. Strides through short vegetation in open woodlands and fallow cultivated areas; feeds mainly on the ground, also in the crowns of palm trees. Comes to villages and well watered lawns. Flocks often feed around grazing cattle, also perching on their backs to keep lookout. Flies with shallow, rapid wingbeats and tail trailing stiffly behind, in tight groups, often low.

VOICE Vocal and noisy, maintaining loud chattering and scolding cries with a shriller quality than those of e.g. Long-tailed Glossy Starling. Also typically short metallic explosive chirrup. Croaks when disturbed.

STATUS AND DISTRIBUTION Frequent to common resident in all divisions. Numbers remain fairly constant year-round in The Gambia. Widespread through Senegal, becoming less common in the extreme north.

BREEDING Breeds towards the end of the dry season; constructs a deep cup-shaped nest of sticks, often at the base of a palm frond. Large clutches (3-7 eggs).

PIED CROW
Corvus albus   Plate 6

Fr. Corbeau pie

IDENTIFICATION 46cm. Sexes similar. Unmistakeable large black and white crow. Upperparts at close range show steely-blue tint to the oily lustre of the black feathers. Below a clean white waistcoat on the belly extends into a complete white neck collar. Bill black and slightly bowed, eye dark brown and legs black. Immature similar, rather scraggy, lacking gloss, white parts dirtier grey. **Similar species** The only large corvid regularly encountered in The Gambia, but may be confused with small raptor in flight.

HABITS Wheeling and soaring flight often at considerable height, regularly mixing with Hooded Vultures and Black Kites. Sociable, regularly gathers in large assemblies at roost and feeding sites. Closely associated with human settlements, foraging around buildings, fields and especially rubbish dumps. Opportunistic, omnivorous feeder and nest raider. Piratical in the air, bullying terns and even skuas. Quick to identify an unusual raptor or owl in the vicinity and highly vocal group mobbing rapidly ensues. Cautious, quick to move off to safety.

VOICE A rising *kwaark*, deep pitched and resonant, and an extended nasal honk *quaaar*. Also a deep hollow sounding 'cork-out-of-a-bottle' *klok*.

STATUS AND DISTRIBUTION Found throughout all divisions; common to very abundant resident at the coast where it may be found in large numbers around the major built-up areas. Becomes progressively less common inland, though mainly associated with the bigger towns, and is seen more often in the dry season. Uncommon in URD. Distributed throughout Senegal with the exception of the extreme north, where it is seasonal in the rains.

BREEDING Breeds April-August in The Gambia, freshly-fledged birds typically appearing at the onset of the rains. Surprisingly few specific nesting records, mainly high in tall isolated trees, once on a pylon. Parasitised by the Great Spotted Cuckoo.

BROWN-NECKED RAVEN
Corvus ruficollis   **Plate 6**

Fr. Corbeau brun

Other names: African Brown-necked Raven, Desert Raven, Brown Crow

IDENTIFICATION 50cm. Sexes similar. Rare, uniformly all-dark crow; brown neck not a strong feature. Wedge-shaped tail, dark brown wash over the nape and hind neck, hint of dark brown on the underparts. The brownish nape colour is visible in favourable light, but not usually in flight. Immature all dull black. **Similar species** Separation of perched birds from slightly shorter Pied Crow straightforward: appears slimmer with longer and narrower wings than the Pied Crow in flight silhouette; wedge-shaped tail diagnostic if seen well.

HABITS Occasionally reported at rubbish tips and along beaches in The Gambia. Normally found in arid habitats and deserts where it frequents escarpments and the periphery of inhabited areas. Usually singly or in pairs, gathering in larger groups at oases and sites where food is abundant.

VOICE A harsh crow-like *karr karr karr* or *korr-korr*.

STATUS AND DISTRIBUTION A rare visitor to The Gambia from further north. A few dry season records between 1975-1995 are all from the same stretch of coastline south of Banjul, January-May. Regularly present in the dry season along the Senegal River valley in northern Senegal.

BREEDING Builds a prominent large stick nest in trees and on cliff ledges. Nearest breeding records Mauritania and Cape Verde.

STARLINGS
Sturnidae

Eleven species, including a single oxpecker. Medium-sized, generally arboreal fruit-eating birds, often colourful with metallic sheens, usually gregarious and noisy. Several are familiar Senegambian birds, both widespread and numerous.

RED-WINGED STARLING
Onychognathus morio   **Not illustrated**

Fr. Etourneau à bec robuste ou Etourneau morio

IDENTIFICATION 34-36cm. Sexes differ. A large, dark, long-tailed starling with large panels of rich chestnut in the wings, distinctive in flight. Edges of primaries show narrow chestnut margins when wing folded. Adult male larger, glossy purple-black over entire body, with a black graduated tail. Female and immature differ by greyish head and neck, streaked with purple-black. Bill black, eye dark red, legs black. **Similar species** Care required to separate from extralimital Forest Chestnut-winged Starling *O. fulgidus* (Fr. Etourneau roupenne) which is slightly smaller, has clearly demarcated green gloss to head and throat in the male, and is found in forest habitat.

HABITS Strongly associated with rocky habitats, where it is shy, and also makes use of buildings, where it may become bold. Non-migratory but thought to travel far on occasions in search of berries and fruit. Elsewhere known to use coastal habitats post-breeding. Sociable and noisy, with direct starling flight; frequently calls on the wing.

VOICE Strong whistles; *teee-jeeoooo*, chatters in flight.

STATUS AND DISTRIBUTION Not recorded in The Gambia. Rare in Senegal with a small population in extreme south-east and a few scattered records in atypical habitats elsewhere. One observation in forest in south-western Casamance when Forest Chestnut-winged Starling suspected but not confirmed.

BREEDING Builds a nest of mud and grass stems in holes in cliffs, roofs of buildings and occasionally palm trees. Nearest proven breeding in Côte d'Ivoire.

PURPLE GLOSSY STARLING
Lamprotornis purpureus   **Plate 44**

Fr. Merle métallique pourpré

IDENTIFICATION 22-23cm. Sexes similar, female slightly smaller. Broad-shouldered, bulky body with elongated and flat-crowned purple head; folded wing tips lie close to tail tip. The most distinctive of four short-tailed glossy starling species. Purple head combines uniquely with very short purple tail, but note bluer tones may dominate in some lights. Rest of upperparts shiny blue-green, violet wedge at bend of wing, two neat rows of black spots across wing in adults. Clear contrast between greenish upperparts and uniformly purplish underparts. Bill

347

black, vivid orange-yellow iris often appears large. Legs black. Immatures much duller, with dark grey underparts; wing-spots absent; iris dull grey-brown. **Similar species** Purplish head and underparts with very short tail distinguish from all other glossy starlings of the area. Head shape distinctive, giving characteristic vulturine jizz in certain postures. Overhead, short tail is emphasised by broad-based wings to give readily-recognised flight silhouette.

HABITS Highly gregarious outside breeding season; widespread, especially savannas, open bush and farmland, often with other starlings. Perches prominently on top-most branches of trees, where diagnostic silhouette is useful. Attracted to fruiting trees and water. Also forages on open ground and quick to exploit termite eruptions where it may be seen hawking flying forms. Flight swift, direct and steady.

VOICE Range of querulous squeaks, bubbling and wittering notes of typical starling character. A rising *sqeee-caree* contact note. Hoarse, persistent *shreee* scolding note.

STATUS AND DISTRIBUTION Common to abundant resident in all seasons in The Gambia. Found throughout Senegal except the extreme north.

BREEDING A hole-nester, commencing nesting in The Gambia at the end of the dry season and continuing into the rains.

BRONZE-TAILED GLOSSY STARLING
Lamprotornis chalcurus   **Plate 44**

Fr. Merle métallique à queue violette

Other name: Short-tailed Glossy Starling

IDENTIFICATION 21-22cm. Sexes similar, females slightly smaller. One of four blue glossy starling species with a relatively short tail; identified by blue or blue-green head in combination with purple-glossed tail; wing tips lie about half way along tail length. Ear-coverts blue-black, often appear black. Prominent violet patch on the lower flanks in some lights; two arcs of black spots on the shoulder (one may be hidden). Black wedge in front of the eye, creates an angry expression. Bill black, eye reddish-orange (often the first feature to attract attention in a mixed flock), legs black. Immature duller, blackish on flight feathers and flanks, eye colour progresses from grey through pale yellow before developing adult colour. **Similar species** Adult eye usually redder in tone than other species, useful but not diagnostic. From the two 'blue-eared' species by short, dark blue and purple or 'bronzy'

tail; from Purple Glossy Starling by greenish-blue upper breast and head.

HABITS Gregarious, often in mixed starling groups when it is typically the least numerous species. Attracted to fruiting trees, also forages on the ground in savannas and farmlands; comes to irrigated hotel lawns. Fond of perching on the bare branches of dead or leafless trees.

VOICE Individuals in flocks on the ground and overhead produce a simple *kwee-EE* call note with a rising pitch and inflection giving an enquiring quality.

STATUS AND DISTRIBUTION Frequent to locally common resident through most of the country; less so within a belt of 30km from the coast, readily encountered from LRD eastwards. In Senegal mostly in the south and east, rare north of 14°N.

BREEDING Hole-nester, mainly in the wet season. Fledged young in June in WD; young seen in August in Casamance.

GREATER BLUE-EARED GLOSSY STARLING
Lamprotornis chalybaeus   **Plate 44**

Fr. Merle métallique à oreillons bleus

Other name: Blue-eared Glossy Starling

IDENTIFICATION 21-23cm. Sexes similar, female slightly smaller. The largest of the four 'short-tailed' blue glossy starlings; most similar to Lesser Blue-eared. Blue-green head with blue-green tail. Entire body glossy metallic blue from crown, nape and mantle to rump and tail; usually slightly greener-blue above, bluer below on throat, breast and upper belly; area of glossy royal blue on the lower belly and thighs. The lores and ear-coverts make a broad dark violet-blue mask through eyes, often appearing black. Two bands of neat black spots across the folded wings (one may be obscured). Bill black and sturdy, adult eye usually vivid golden-yellow. Immature dull sooty blackish below, with grey eye developing through red-brown as glossy plumage emerges on wings and mantle. **Similar species** Distinguished from Purple and Short-tailed Glossy Starlings by blue or blue-green tail, folded wing tips at less than half tail-length at rest. From Lesser Blue-eared by bluer tones on upperparts, proportionately larger head and rangier build; flank patches royal blue, eye mask patch broader; usually larger than Lesser Blue-eared, but note that sex and age differences can create significant and confusing overlap in size.

HABITS Widespread in most habitats, highly

mobile. Gregarious outside breeding season, flocks may be very large especially in the dry season, when it mixes freely with other glossy starling species. Attracted to fruiting fig trees and water; also seen in numbers at termite eruptions in the early rains. Frequently feeds on the ground, working industriously forward in typical starling fashion.

VOICE A querulous, rising, *skwear* or *skwee-skwearr* with enquiring tone; a host of chattering, twittering, popping and snapping sounds, merging into a conversational background murmur in large flocks. Sharp *schwarr* alarm call.

STATUS AND DISTRIBUTION A common to abundant resident throughout The Gambia, with largest numbers in the inland divisions. More study required to establish relative abundance of this and Lesser Blue-eared, since The Gambia lies within an overlap zone between the ranges of these two. In Senegal common resident north of The Gambia, rare to the south.

BREEDING Recorded in June in The Gambia, in a hollow fence post, but typically selects a natural tree hole, lined with leaves, nesting mainly at the start of the rains.

LESSER BLUE-EARED GLOSSY STARLING
Lamprotornis chloropterus   **Plate 44**

Fr. Merle métallique de Swainson

IDENTIFICATION 19-20cm. Sexes similar, males slightly larger. Smaller than other blue glossy starlings, with a strong greenish wash on the upperparts in some lights, hence its specific name. Head, body and tail glossy blue-green; folded wings blue-green, sometimes almost bottle-green; two rows of spots across the upper wing; upper row sometimes obscured. Mask narrower than other glossy starlings, dark violet-blue, appearing blackish. Throat and breast blue-green, changing to blue with a limited magenta patch on flanks and belly. Bill and legs black; eye may be bright golden-yellow, similar to Greater Blue-eared and Purple Glossy, but often duller orange-yellow. Immature grey-brown from chin to belly, patchy green-glossed wing feathers; grey eye progresses through orange-red to pale yellow. **Similar species** Similar tail colour and proportion to Greater Blue-eared (blue-green tail, wing tips covering less than half length) eliminates Purple and Bronze-tailed Glossy. From Greater Blue-eared by neater, more compact posture; narrower, more precisely defined ear-coverts, eye usually duller orange-yellow; magenta, not deep blue, flank patch; size difference sometimes obvious and

diagnostic, but sometimes confusing (female Lesser 20% smaller than male Greater; but male Lesser averages nearly as big as female Greater with some overlap). See also voice. Immature browner on underparts than all other immature glossy starlings except much larger immature female Splendid. Emerald Starling *Lamprotornis iris* (Fr. Merle métallique vert), not recorded in our area but occurs in savanna habitats in Guinea, is even smaller, glossy bottle-green over most of the body and tail, with a purple lower breast, black central belly and dark brown eye. It is thought to be migratory, but movements are not well known.

HABITS Widespread and mobile, particularly favouring savanna woodland and farmland, but may be encountered in most habitats. Occurs in large mixed flocks with other glossy starlings in the dry season. Fond of fruiting trees, otherwise forages extensively on the ground, and attracted to water.

VOICE Sequence of cleanly clipped notes, lacking querulous quality of Greater Blue-eared Glossy; take-off call trilling *wirri-girri.*

STATUS AND DISTRIBUTION Considered widespread and common from LRD eastward; like the Greater Blue-eared Glossy Starling less frequent near the coast. Situation confused as 'blue-eared' glossy starlings are seen frequently, but seldom fully identified. In Senegal, current information suggests it is found mainly to the south of The Gambia. Thought to be migratory but regional movements not understood.

BREEDING No confirmed records from The Gambia; nearest in Sierra Leone.

SPLENDID GLOSSY STARLING
Lamprotornis splendidus   **Plate 44**

Fr. Merle métallique à oeil blanc

IDENTIFICATION 25-28cm. *L. s. chrysonotis.* Sexes similar, females smaller. Large but slim looking, with whitish eyes; particularly intense smooth metallic quality to almost entire plumage. Head and shoulders metallic greenish, changing to metallic blue on back and violet-blue rump. Broad black mask from forehead through the eyes. Breast and underparts blue-green, strongly violet-tinged on male, less so on female. Wings greenish-blue with black spots and a velvet black band across folded primary bases. Tail elongated compared to any 'yellow-eyed' glossy starlings, with diagnostic broad black band across the middle dividing a violet-blue base from a broad greenish-blue terminal band. Good light may reveal strong patches of iridescence including a coppery patch on side of neck, and a greenish reflection on ear-coverts, though these are less pronounced on females. Bill black, eye (in detail very pale yellow) appears white; legs black. Immature less glossy, underparts matt blackish (male) or brown (female). **Similar species** Large size, elongated shape, silvery eye set in a black mask unmistakeable, in combination with black-banded tail. In flight, spatulate tail, combined with a curiously pinched 'waist' at the tail base, contributes to a distinctive, hunched, silhouette. The audible whirring of the wings is more consistent and louder in this species than any other glossy starling.

HABITS A forest bird, in The Gambia strongly associated with remnant forest patches and stands of tall trees, notably oil palm; occasional in more open habitats and mangroves. Seen in pairs and groups, noisy. Frequently first noted in flight, when broad elongated tail obvious. Attracts attention from long range by its loud calls. Shy, rarely coming to water in open places as do all the other glossy starlings; not often seen on the ground.

VOICE Insistent *mi-auoow* or *qu-wow*, drawn out, nasal and cat-like. A variety of other forceful notes including a rising *kra-eek* and a hoarse, hollow, *swik-swok*.

STATUS AND DISTRIBUTION Overall uncommon. Usually seen near the coast and through WD, once LRD, and a handful of observations along swamp forest margins of the river in CRD. Recorded in all seasons. In Senegal chiefly found to the south of The Gambia, with only occasional records up the coastal strip to Dakar.

BREEDING A high tree-hole nester. Attempted breeding observed several times in The Gambia, July-August, but no successful result confirmed. Seen prospecting tree-holes in Casamance, Senegal in June, but breeding not confirmed.

LONG-TAILED GLOSSY STARLING
Lamprotornis caudatus **Plate 44**

Fr. Merle métallique à longue queue

Other name: Northern Long-tailed Starling

IDENTIFICATION 54cm, including 34cm tail. Sexes similar. The largest of the blue glossy starlings and the only one with a long, lax, tail; female slightly smaller. Face black, body metallic royal-blue, strongly tinged green on the mantle and underparts in some lights. Dark blue lower body washed violet. The long tail feathers are soft, flexing readily, especially side to side. Broad wings metallic blue, with a double arc of black spots usually hidden, but seen when the wing is extended. Pale yellow eye appears whitish, bill and legs black.

Immature tail already longer than other starlings at fledging; duller, dark grey-brown tone to the head and underparts, greener wash on the body. **Similar species** Unmistakable. In flight Piapiac distinguished even at long range by stiff tail and shallow rapid wing-beats.

HABITS Ubiquitous. A gregarious bird of woodlands, savanna and farmland, also mangroves, around villages etc. Groups of 4-10 birds normal, up to 30 or more during wet season. Wary, but noisy and easily located. Forages on the ground, standing with tail held well up, also in bushes and trees. Flight, laboured but powerful, with deep wing-beats creating an audible rush of air at close range.

VOICE Harsh and grating, measured *kchirow-kcheree*, rising in pitch, sometimes in a penetrating and garrulous chorus. An explosively hard *squip*, a grating *skaarrrr* and shrieking notes. May be vocal at night in the rainy season.

STATUS AND DISTRIBUTION Widespread and common resident throughout the country. Found throughout Senegal, though less common in the north.

BREEDING Wet season breeder, in natural tree holes. Cooperative nest care observed in captivity and several adults noted bringing food to one nest in The Gambia on several occasions.

CHESTNUT-BELLIED STARLING
Lamprotornis pulcher **Plate 44**

Fr. Etourneau à ventre roux

IDENTIFICATION 19cm. Sexes similar. Immediately recognisable by its dull grey head, dark chest cleanly separated from dull chestnut belly, and pale wing panel in flight. Upperparts progressively develop a brightening green gloss towards the short glossy green-blue tail. Pale panels in the primaries characteristic in flight. Throat and breast weakly glossed dull blue-green. Eye very pale yellow (appears silvery). Immature lacks gloss on breast, brown on head and underparts, chestnut on belly only; eye dark. **Similar species** Clean division between dark upper breast and chestnut underparts, plus pale wing panels, distinguishes from all other starlings.

HABITS Gregarious, favouring open woodland, scrub and sahel, preferring drier areas with well scattered trees. May settle near human habitation. Usually feeds on the ground in typical starling fashion. When alarmed flies directly into nearby tree tops; also perches on telegraph wires and buildings.

VOICE Softer calls than many other starlings, with

a liquid quality; call a breathy *fit-trrrr, fit-trrrr*.

STATUS AND DISTRIBUTION Generally uncommon resident in a restricted area of The Gambia scattered thinly along the north bank of the river in NBD; not recorded recently from north of the river in CRD or URD, where it might be expected. Small colonies south of the river in WD seem to fluctuate in size and may make intermittent local movements. In Senegal very common in the north, becoming less so towards the latitude of The Gambia. Absent from Casamance.

BREEDING Builds stick nest in acacia or similar tree; may build within nest of White-billed Buffalo Weaver. Nest with young in The Gambia, June; seen carrying nesting material August and December; elsewhere pre- and post-rainy season breeding periods noted.

VIOLET-BACKED STARLING
Cinnyricinclus leucogaster   Plate 44

Fr. Merle améthyste

Other names: Amethyst Starling, Plum-coloured Starling

IDENTIFICATION 18cm. Sexes differ. Adult male upperparts and tail deep glossy purple; may look blackish in some lights. Throat and upper breast also glossy purple, cleanly separated from otherwise pristine white underparts. Adult female upperparts brown, finely streaked and spotted with black; rufous-tinted sides of the head, narrow buff fringes to wing feathers. Upper breast gingery-brown wash; rest of underparts pure white, heavily streaked with dark brown teardrops, thinning out towards belly. Immature similar to female, males darker. **Similar species** Female Red-shouldered Cuckoo-Shrike, has yellow edging to feathers of upperparts, and barred underparts. Male Violet-backed Sunbird shares basic colour pattern of male starling, but is much smaller, with distinctive bill. Violet-backed Hyliota, reported once from south coastal Gambia, is also much smaller and squatter, without dark throat.

HABITS Favours light to moderately thick woodland, strongly attracted to fruiting trees such as figs. Perches in tree tops, and commonly the detection of one leads to the discovery of several more feeding unobtrusively. Appears erratically where ripe fruit trees available, highly mobile, mostly in small parties of 5-10. Flies swiftly with intermittent shallow bounding action. Displaying males sit on exposed branches giving a ringing call and flicking one or both closed wings stiffly upward, displaying their flashing metallic feathers. Often heard before seen.

VOICE Remarkable ringing whistle, a drawn-out single note with the intense quality of a tuning fork, persistently repeated at long intervals. Recalls Grey-headed Bush Shrike which is more forceful and lower pitched. Call is not particularly loud, but may carry far, heard most frequently at the end of the dry season. Also soft *sqwear* chirp, and a soft reedy flight call.

STATUS AND DISTRIBUTION Occurs in all parts of The Gambia. Numbers variable from year to year between March and November, with an obvious increase May-July. Seen in all seasons and usually most frequent near the coast. In Senegal mostly confined to the latitude of The Gambia or southward, especially the south-east April to July; also recorded in coastal regions north to Dakar.

BREEDING In The Gambia nest building recorded June and July, males seen calling and giving wing-flick display through the wet season; adults carrying food in June, immatures seen with adults in October, but young not yet found in the nest. Uses natural holes in trees at varying heights above the ground.

WATTLED STARLING
Creatophora cinerea   Not illustrated

Fr. Etourneau caronculé

IDENTIFICATION 21cm. Sexes similar. Pale grey-white starlings, with blackish wings and tail and distinctive white rump. When breeding adult male may exhibit bare black and yellow skin and several prominent black wattles on the head. At other times resembles female, with feathers of the head matching the body, small yellow patch of skin behind the eye and short streaks of bare black skin in the moustachial area. Female is browner on the body; wing and tail feathers are dark brown. Bill pinkish to pale flesh; legs flesh. Immature like female. **Similar species** Pale plumage and white rump distinguishes from other small starlings. Yellow-billed Oxpecker also shows pale rump, but is slimmer with a tapered tail.

HABITS Sociable; highly mobile according to local conditions.

VOICE A thin series of squeaking and grating notes; comparatively quiet for a starling.

STATUS AND DISTRIBUTION Accidental. Included on the basis of three records, probably of the same three birds, in WD, December-January 1975-76. Otherwise nearest regular occurrences Sudan and Gabon; widespread in central, eastern and southern Africa. Nomadic throughout this range.

BREEDING Opportunistic in times of plentiful food availability, when it creates domed stick nests in thorn trees, often placed very close to one another.

YELLOW-BILLED OXPECKER
Buphagus africanus   Plate 44

Fr. Pique-boeuf à bec jaune

IDENTIFICATION 22cm. Sexes similar. Brown to olive-brown, with swollen bright red and yellow bill, instantly recognisable by its adaptation to foraging almost exclusively on the bodies of large herbivores. Head, mantle wings and tail, brown, with hint of olive; wings darker than body. Lower back and rump pale buff-brown above the long, stiff dark brown tail. Upper breast brown, smoothly fading to a pale buff lower breast and belly. Bill bright red tipped, yellow at the base, swelling to form phlanges either side of the lower mandible; appears waxy. Eye bright orange, legs brown, sharply clawed. Immature similar to adult, with pale yellow bill. **Similar species** Attenuated bodies and tapered tails of birds in flight create unusual and distinctive silhouette,

which may confuse if seen away from large herbivores.

HABITS Gregarious when not breeding, usually seen in flocks of 6-20. Famed for specialist feeding on ectoparasites from large mammal hosts, but may also devour host body tissues, particularly of domestic cattle, sometimes excavating damaging wounds. Moves over the bodies of large animals like a woodpecker on a tree trunk. Often the majority of the flock settles on one animal at a time. Flight, often at considerable height, is accompanied by loud cries.

VOICE Chatters with a distinctive sibilant quality, mainly heard between members of a flock in flight and when squabbling for position on a host.

STATUS AND DISTRIBUTION Frequent to locally common resident throughout The Gambia giving the impression of specific local movements; sometimes temporarily uncommon. Found throughout Senegal, though less common in the south.

BREEDING Hole nester, makes use of hollow palm-log fence posts. Nest lined with hair plucked from livestock. Nest building in The Gambia completed at the end of May; fledged young late July. Adults nesting near tied cattle regularly carried beakfuls of fresh red meat to nestlings.

SPARROWS AND WEAVERS

Passeridae and Ploceidae

Twenty six species, comprising sparrows, petronias, weavers, malimbes, queleas, bishops and widowbirds. Small to medium finch-like birds, largely seed-eaters with rather rounded conical bills. Generally gregarious, sometimes forming large flocks capable of crop damage. Noted for complex nest structures of woven twigs or grasses, sometimes pendant, often colonial. Females and immatures generally drabber than males which may be vividly coloured in breeding season.

HOUSE SPARROW
Passer domesticus   Plate 45

Fr. Moineau domestique

IDENTIFICATION 15cm. Sexes differ. *P. d. indicus.* The original common streaky brown bird, and an invading coloniser in sub-Saharan Africa. Breeding male grey-capped; rich chestnut extending from behind eye to nape; neat grey-white cheeks. Mantle rich brown, streaked black, grey lower back. White bar on shoulder, brown fringing on dark flight feathers. Black streak through eye extending to throat and variable bib; rest of underparts plain grey-white. Bill black, eye dark brown, legs flesh. Non-breeding male loses chestnut nape and black bib nearly obscured; bill horn. Adult female plain grey-brown head and nape; buff supercilium from behind eye; streaky upperparts plainer tone

of brown than male; underparts plain; bill horn. Immature heavily fringed with buff above, extensive grey on body, often scruffy. **Similar species** Pale supercilium and lack of yellow tones distinguishes female from female weavers; much more heavily streaked on upperparts than female Bush Petronia which has different bill shape.

HABITS Strongly associated with human settlements, whether cities or rural. Recently appearing in between villages, e.g. Tanji Bird Reserve. Gregarious coloniser, usually nesting on buildings, occasionally in trees or tree holes. Active and alert, indulges in belligerent scuffles.

VOICE Gives a confident, throaty, *chirrup*, or *chirp*, and sings by stringing a sequence of such calls together. Insistent alarm call.

STATUS AND DISTRIBUTION An invading species, first appearing in Senegal in the 1970s and at Banjul in 1982. Now an established breeder with a growing number of colonies. By the early 1990s seen at villages down the coast (Sanyang) and appearing up river at Tendaba (peanut warehouses) and near Farafenni (ferry crossing buildings). Specimens from Senegal confirmed as *P. d. indicus*, smaller and paler-cheeked than the nominate European form. This suggests Senegambian colonisers originated on ships moving up the coast from the nearest population in South Africa, though possibility of mixing with Cape Verdian invaders of European race remains.

BREEDING Breeds year-round, lodging untidy nests in a variety of sites in roof eaves of buildings etc. Young in nest, Banjul, in most months.

GREY-HEADED SPARROW
Passer griseus   Plate 45

Fr. Moineau gris

IDENTIFICATION 15cm. Sexes similar. Slim grey and chestnut sparrow encountered in all open areas. Plain grey head, darker in front of eye; plain brown mantle grades to chestnut on rump, brown square-ended tail; single partial white wing-bar on chestnut shoulder, chestnut edging to dark brown flight feathers. White chin and throat, grey breast and flanks, lower belly pale grey. Bill black when breeding, browner at other times. Immature similar, lightly streaked on upperparts, lacking white shoulder patch. **Similar species** Distinctive once familiar; Bush Petronia smaller without contrasting grey head; Lesser Honeyguide greenish on wings, with white patches on sides of tail; Redwinged Pytilia much smaller with narrow red area on wings and flank bars.

HABITS Found in wide range of habitats, from villages and small towns, to field margins, open bush and light woodlands. Perches and feeds in trees, also in small flocks on the ground. Gathers in larger flocks in dry season. Noisy and confident, active around villages and settlements in similar way to House Sparrow, suggesting potential for direct competition with that species in domestic situations as House Sparrow expands in range. Regularly feeds alongside firefinches, indigobirds and cordon-bleus around human settlements.

VOICE Typical sparrow *chirp*, crisp and forceful; also a quick staccato *che-ke-chek*, and array of rising chirrup notes with rapid staccato articulation.

STATUS AND DISTRIBUTION Common breeding resident throughout the country. Also throughout Senegal and everywhere south of the Sahara to Tanzania, where it overlaps with paler-bellied

Southern Grey-headed Sparrow, *P. diffusus*.

BREEDING Recorded over an extended period in The Gambia, from late dry season, April, through to December, when young have been found in the nest. Primarily a hole nester, creating untidy nest of grass and feathers built in variety of opportunistic locations, from tree holes to buildings. Will take over Little Bee-eater and small kingfisher holes, occasionally evicting original owners.

SUDAN GOLDEN SPARROW
Passer luteus   Plate 45

Fr. Moineau doré

Other name: Golden Sparrow

IDENTIFICATION 12-13cm. Sexes differ. Adult male bright lemon yellow over head rump and underparts, contrasting with plain chestnut mantle and wing panel. Darker wings show two narrow pale bars. Bill black, eye dark brown, legs flesh. Adult female plain sandy over crown and nape, lightly streaked on mantle, with tendency to weak gingery wash and weak pale double wing-bars. Buff supercilium, pale sandy underparts fading to whitish belly, with variable wash of pale yellow over face and breast. Bill pale horn. Immature resembles female but greyer-toned; some younger birds may be dark legged. **Similar species** Presence of male-plumage birds precludes difficulties, but large flocks lacking such individuals do occur; note combination of supercilium, double wing-bars and general lack of streakiness to eliminate Little Weaver; from female Bush Petronia by yellowish tones, and absence of conspicuous white throat.

HABITS An arid country species primarily associated with the Sahel. In The Gambia typically seen around harvested field areas with shrubs, roadside waterpools and in dry bush. Gregarious, in flocks of several hundred in the dry season and associating freely with other small seed-eaters.

VOICE Calls similar to other sparrows, though with own recognisable quality, *schilp*. Rapid twittering flight call also noted.

STATUS AND DISTRIBUTION Dry season visitor, primarily to the north bank of the river, occurring in variable numbers from year to year; recorded November-June, with one male staying on with breeding bishops into August. Sometimes common. A handful of records on the south bank are all from the coast, notably Cape St. Mary and Tanji. Found erratically throughout central and northern Senegal in dry season; breeds in the north.

BREEDING Not recorded in The Gambia.

Colonial, with nests placed in low trees. Extended breeding season north Senegal, July-December, occasionally to March.

YELLOW-SPOTTED PETRONIA
Petronia pyrgita   Not illustrated

Fr. Moineau soulcie à point jaune

IDENTIFICATION 15cm. *P. p. pallida.* Sexes similar. Large, plain sparrow; pale grey-brown above, buffy-white below with white throat and yellow spot (smaller on female); noticeably shorter wing-tip projection than Bush Petronia. Narrow pale eye-ring and horn-coloured bill.

STATUS AND DISTRIBUTION Breeds in dry thorn scrub of northern Senegal where it is poorly known but possibly widespread.

NOTE Formerly considered conspecific with extra-limital Yellow-throated Petronia *P. xanthocollis.*

BUSH PETRONIA
Petronia dentata   Plate 45

Fr. Petit Moineau soulcie

Other name: Bush Sparrow

IDENTIFICATION 13cm. Sexes differ. Plain sparrow-like bird, recognised by details of head pattern and head shape. Male shows diagnostic chestnut superciliary stripe curling down round rear margin of ear-coverts and grey crown. Upperparts and tail plain brown, flight feathers dark brown, faint double wing-bars. Neat white throat, marked by indistinct yellow spot at lower margin when breeding (hard to see); rest of underparts plain brown fading to white belly. Bill black (pale yellowish non-breeding), longish for sparrow; eye dark brown, legs black. Adult female paler brown over crown and mantle, lightly streaked on mantle, prominent pale buff supercilium runs conventionally behind eye; bill horn colour. Immature like female, but warmer brown with heavier dark mottling on mantle and hint of malar stripe. **Similar species** Female from female House Sparrow by much less streaky upperparts. Grey-headed Sparrow is larger, with uniform grey head and chestnut back.

HABITS Strongly associated with dry woodlands in The Gambia; frequently seen perched in highest branches (often dead), of emergent trees where it may remain for long periods in characteristic upright pose. Also around cultivated areas and fallow land. Usually in pairs or small groups, exceptionally

in larger flocks e.g. at drying waterholes.

VOICE Sparrow-like *chirp*; sings with distinctive *clu-lu-lu-lu-lu*, rapid and rhythmic, on a monotone.

STATUS AND DISTRIBUTION Resident breeder, uncommon or locally frequent at the coast, becoming frequent, occasionally common, throughout the interior, where it is most readily encountered in larger woodland tracts. Widespread in Senegal, but scarce in the far north.

BREEDING Hole-nester, utilising fence posts around fields as well as natural tree holes. Prolonged breeding season: young in the nest in February and April and immatures still begging and receiving food from adults in May in The Gambia.

SPECKLE-FRONTED WEAVER
Sporopipes frontalis   Plate 45

Fr. Moineau quadrillé

Other name: Scaly-fronted Weaver

IDENTIFICATION 11-12cm. Sexes similar. Unusually marked small weaver with dull orange nape and large pale circular patches round eyes. Black forehead and narrow black moustachial stripe finely speckled with white combine with gingery nape to delineate large whitish side to head. Upperparts pale brown, lightly mottled darker; flight feathers and tail dark brown with fine buff edgings. Underparts; pearly-grey breast fading to whitish on belly. Bill conical, small, pale horn; eye dark brown, legs flesh. Immature like adult but washed out, especially on nape. **Similar species** Unmistakable in good view; Chestnut-crowned Sparrow-Weaver often found in same places, but is much larger and lacks eye mask.

HABITS Prefers dry thorn scrub and light bush; usually in small groups. Feeds mainly on the ground, using open gravel patches (and murrum roads) picking up small seeds. Long-legged stance and springy hopping action catches the eye.

VOICE Twittering or trilling song described.

STATUS AND DISTRIBUTION Possibly undergoing expansion in The Gambia. Since 1990 recorded in all seasons, and in all parts of the country north of the river from NBD through CRD to URD. Encountered along the roadside from Farafenni to Kuntaur, in April-May and July; present in the bush south of the river near Basse in August 1990 and 1991. Also at the coast near Essau. No records from south of the river. In Senegal widely distributed to the north of The Gambia, becoming increasingly common towards the north.

BREEDING Collecting nest material near Essau in

at least two years, in February on both occasions. Nest sited in acacia near village compound. Breeds September-December in northern Senegal. Nest large, globular, grass construction with side entrance.

CHESTNUT-CROWNED SPARROW-WEAVER
Plocepasser superciliosus   Plate 45

Fr. Moineau-tisserin à calotte marron

IDENTIFICATION 18cm. Sexes similar. First impression is of a large, strongly patterned, sparrow. Thick black malar stripes especially prominent. Bright chestnut crown and cheeks merge on nape to plain brown mantle, head highlighted by strong white supercilium and two heavy black malar 'bars', widening down the neck. Brown wings show double white wing-bars, narrow buff fringes to flight feathers; tail brown. Underparts dirty white. Bill pale horn, eye brown, legs brown. **Similar species** White supercilium with black malar stripes diagnostic. From any bunting by crown colour, large pale bill, larger size.

HABITS A species of well-grown dry bushland and open woodlands, typically seen searching within the canopies of large bushes and small trees; easily overlooked owing to unobtrusive ways and cryptic markings, often several together. In a few locations (mainly NBD) shares habitats and locations with Speckle-fronted Weaver. Joins loose mixed species foraging flocks in dry woodland with these and petronias, Cordon-bleus etc. On the ground walks like a pipit.

VOICE Gives a soft rapid ticking extended in a monotone reel with slight rise in pitch; recalls freewheeling bicycle, distinctive and useful for detecting and locating the bird once learnt.

Chestnut-crowned Sparrow-Weaver nest

STATUS AND DISTRIBUTION Uncommon breeding resident recorded in all areas; rare at the coast. Most likely to be found in Kiang West and associated LRD forest parks; in CRD, especially on the north bank of the river near Kaur and dry bushland of URD, where locally frequent and faithful to a few favoured areas of dry bush. Thinly scattered through central Senegal.

BREEDING Large untidy globe nest of grass and dry leaves, typically set about 2m off the ground in a shrub or small tree. Recorded in The Gambia September and November.

WHITE-BILLED BUFFALO-WEAVER
Bubalornis albirostris   Plate 45

Fr. Alecto à bec blanc

IDENTIFICATION 23-24cm. Sexes similar. Large black weaver, rangily built, gregarious and noisy. Males entirely black, save for limited diffuse white patches on back and below wing on flanks. Large conical bill swells even larger and develops chalky white colouring when breeding, otherwise black; eye dark brown, legs brown. Female resembles non-breeding male, slightly browner in strong light. Immature dark brown above, mottled below with broad, untidy, white streaking. **Similar species** Distinctive in any good view; flocks in flight from Glossy Starlings by bulkier profile of bill, head and tail proportion.

HABITS Favours open country, especially agricultural land and bush with scattered emergent trees. Gregarious, often seen working over harvested crop fields in tight-knit advancing groups, with typical scuffling movements and springy hops, surprisingly agile creeping in tree branches. Garrulous at nest colonies, attracting attention from afar with constant raucous chatter. Attracted to free standing water around cattle watering points in the dry season.

VOICE Colony members maintain constant staccato cackling chatter, loud and carrying, with dissonant squeaky hinge tone, rising and falling in intensity.

STATUS AND DISTRIBUTION Breeding resident in all areas, common inland both sides of the river. Less frequent at coast, where it seems to be increasing following relatively recent invasion, probably in association with agricultural land clearance. Found throughout Senegal except coastal Casamance.

BREEDING Birds may be seen in association with nest colonies throughout the year, but active nesting primarily in wet season, July-November, when levels of noise and excitement around the colonies much enhanced.

White-billed Buffalo-Weaver nests

NEST Massive semi-communal structures sufficiently prominent to form a characteristic landscape feature in rural areas; appear as a series of large untidy stick piles festooned through the branches of a free-standing tree, built at a range of heights from low acacia to tall kapok or baobab. Individual stick piles sometimes merged into a single enormous 'heap'. Twig and stick masses form a matrix within which several rough globular grass nests are embedded, with entrance from the underside.

GROSBEAK WEAVER
Amblyospiza albifrons   Not illustrated

Fr. Grosbec à front blanc

Other name: Thick-billed Weaver

IDENTIFICATION 17-18cm. *A. a. capitalba*. Sexes differ. Bulky dark weaver recognised by huge conical bill, upright stance. Adult male has white forehead, otherwise dull rufous-chestnut head, breast and mantle, black upperparts and tail. Clean white patch visible at base of primaries on folded wing as well as in flight. Lower belly grey, finely streaked with black feather shafts. Female dark brown above with chestnut scalloping, white below with heavy dark streaking, mainly across breast; yellowish lower mandible. Immature resembles female, but heavier brown fringing to upperbody feathers and dirty yellow bill. **Similar species** Female separated from other dark brown heavy-billed weaver species by heavy dark streaking on white underside; juvenile of common White-billed Buffalo-Weaver larger, whitish mottling on dark underside; much smaller, juvenile Crimson Seed-cracker uniform dark brown.

HABITS Gregarious colonial breeder over reed-beds near fresh water, where males chatter, flick

wings and tail restlessly from swaying stems and make circling flights over the marsh. At other seasons may disperse into wooded areas, but typically associated with giant grasses. Swooping flight.

VOICE Song a hard-toned, rapid, chattering jumble of notes. Flight call described as *tweek tweek*.

STATUS AND DISTRIBUTION Not recorded in The Gambia; included on the basis of a single sight record from Casamance in July 1977. Otherwise found Sierra Leone to Nigeria, but extensive gaps in this range.

BREEDING Weaves a globe nest around several vertical reed stems using torn strips of smaller grasses to form fine-grained, densely woven nest with open, smoothly bevelled side entrance (roosting nest) or narrower tunnel entrance (breeding nest).

LITTLE WEAVER
Ploceus luteolus   **Plate 46**

Fr. Tisserin minule

IDENTIFICATION 10-11cm. Sexes differ. Very small weaver, readily distinguished by size, male clean black and yellow when breeding, widespread but local. Breeding male developing May, full plumage June-November, partial into January: bright yellow crown, sides of neck and underparts, black mask from forecrown over ear-coverts, face and throat. Greenish-yellow mantle streaked with dark feather centers, brown wings and tail fringed with yellow. Legs blue-grey. Non-breeding male nondescript buff streaked diffusely with dusky above, white below with hint of yellow wash about face and vent. Breeding female yellowish-olive above, streaked; uniform pale yellow face and underparts, non-breeding pale streaky brown above, whitish belly. **Similar species** Small breeding male confirmed by black mask extending small way up forehead, yellow crown, absence of rusty stain tones on body, female usually accompanying. Non-breeding birds notable for small size, yellow-washed face with white belly, sometimes yellow undertail-coverts. Female and eclipse male invite confusion with similar-sized female Sudan Golden Sparrow, which is very pale tawny over crown and mantle.

HABITS Quiet and unassuming compared to typical weavers, mostly seen in pairs in open scrubland with small trees and edges of farmland; especially associated with small acacias. Perches in thorny branches or seen working methodically, almost warbler-like. Less gregarious than other weavers. Mixes with flocks of estrildids in the dry season. Site-faithful; old nest in dry season an indicator of its likely presence.

VOICE A *tsssp* call and a chattering song with grating notes.

STATUS AND DISTRIBUTION Resident breeder in The Gambia, recorded throughout the country in areas of fallow land and low bush, where it can be locally frequent, Look especially in fallow agricultural areas near the coast where young acacias persist e.g. Tanji, Yundum, Niumi-Jinnack, NBD. Recorded in all months; especially noticeable June-July shortly after acquiring fresh breeding plumage, and when at the nest.

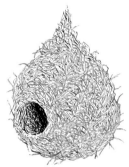

Little Weaver nest

BREEDING Nests June to October, though individual nests in varying states of decay may be found throughout the year.

NEST Densely woven globular nest of fine narrow grass strips, with short-medium tube on side entrance; soft, almost woolly texture, 3-4m above ground, in small tree. Nest clusters have been reported elsewhere, but usually isolated, typically hanging from thorny branchlets of young *Acacia albida*, often closely adjacent to hornet nest.

BLACK-NECKED WEAVER
Ploceus nigricollis   **Plate 46**

Fr. Tisserin à cou noir

Other names: Spectacled Weaver – see note below

IDENTIFICATION 16cm. Sexes differ. *P. n. brachypterus*. Mainly yellow with dull olive-green back and narrow black bandit mask through pale eye; associated with dense thickets. No sparrow-like non-breeding plumage. Adult male chestnut-stained yellow on head and nape, neat black eye mask and clean black throat bib. Mantle, wings and tail, uniform soft-toned olive. Underparts yellow. Bill black, eye stands out pale yellow in narrow mask, legs grey. Females weakly washed chestnut on breast; olive over crown, narrow yellow line between crown and black eye mask, no bib; eye may

be brown and bill horn. **Similar species** Glimpsed overhead in forest cover, female may be confused with African Golden Oriole, but much smaller and plain green upperparts quite different.

HABITS Inhabits palm thickets, relict canopied forest patches, swamp forest in freshwater margins and seasonally flooded tamarisk groves near the coast. Nests in pairs, but gathers in small flocks, especially in late dry season. Often the dominant member of mixed species forest thicket feeding parties in e.g. Abuko.

VOICE Produces sounds of unusual tone; soft wheezing quality, *prrr-it, dew, dew, twee*; several birds together produce a buzzy toned chatter.

STATUS AND DISTRIBUTION Breeding resident, recorded from all divisions. Locally frequent to common within small patches of suitable habitat at the coast, most prominent in small flocks, in late dry season; also found in swamp forest patches of CRD and locally along thickets of the riverbank in URD, elsewhere largely absent. Found through southern Senegal and a few locations north along the coast.

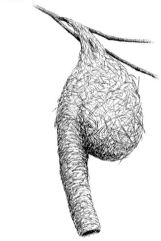

Black-necked Weaver nest

BREEDING Active nests found in The Gambia, July-November.

NEST Structure resembles Little Weaver nest in close-woven style, but coarser texture through use of larger materials such as vines and creepers, usually at low level, with 12-15cm (or longer) entrance tunnel extending down from globular chamber. Suspended, sometimes conspicuously, on acacia branch or palm frond.

NOTE The nominate race of this species (found Nigeria and eastwards) looks very different, with near-black upperparts and tail. The race seen in

The Gambia bears much closer resemblance to a separate eastern and southern African species *Ploceus ocularis* which is usually known as Spectacled Weaver. Confusingly, this name was formerly used in West Africa for *P. nigricollis*.

HEUGLIN'S MASKED WEAVER
Ploceus heuglini   Plate 46

Fr. Tisserin masqué de Heuglin

IDENTIFICATION 15cm. *P. h. neglectus*. Sexes differ. Very localised weaver, yellow and olive with restricted black face mask. Breeding male (June-December) black mask from base of bill runs just around eye, slightly pointed on throat. Crown yellow, reaching top of bill; yellow extends to nape and all underparts; forecrown and breast may be very weakly washed with faint rusty staining. Mantle pale green, faint shadow of mottling in good light, tail plain brown-olive, wings brown with yellow fringes. Bill black, eye yellow, legs brown. Female and non-breeding male olive-brown above, plain on crown, diffusely streaked on mantle, yellowish breast, white belly. Bill horn; eye pale brown. Immature similar. **Similar species** Breeding male from Vitelline Masked Weaver by diagnostic entire crown yellow on male Heuglin's (rust wash much less if present at all), pale yellow eye. Also body shape subtly chunkier, tail shorter, bill very slightly longer. Latter features may help with non-breeding plumage, which is very similar to Vitelline Masked; Heuglin's tends to more olive-toned uniform upperparts, paler eye colour.

HABITS Found in widely available habitats including secondary scrubland around cultivation, light woodland edges and into denser savanna woodland. Variably social, typically less gregarious than most weavers; single pairs may nest in isolation but colonies of nests have been reported, including mixed 'colonies' with Vitelline Masked Weaver. Males hang fluttering beneath nest in display.

VOICE Emits a soft *chaaar* and sings with mellow 'swizzling' chatter of slurry wheezes, pops and rattles.

STATUS AND DISTRIBUTION Uncommon to rare in The Gambia. Recorded locally from coastal areas, LRD and URD. Probably occurs throughout, but patchy distribution and infrequent occurrence compound tendency to overlook. All records June-November, highlighting difficulties of non-breeding recognition. In Senegal, locally recorded from a few areas in central region of the country, mainly close to north side of The Gambia, once from Casamance. Noted to be erratic in distribution throughout its range.

Heuglin's Masked Weaver nest

BREEDING Active at nest sites July-November, usually building in small acacia.

NEST Coarsely woven globular nest of wiry grass stems, resembling Village Weaver, but usually suspended from narrow branch or in tall grass, perhaps only 2-3 nests, widely spaced, not clustered like Village Weaver colony. Often quite low.

VITELLINE MASKED WEAVER
Ploceus velatus   Plate 46

Fr. Tisserin à tête rousse ou Tisserin masqué

Other name: African Masked Weaver

IDENTIFICATION 15-16cm. *P. v. vitellinus*. Sexes differ. Striking bright yellow weaver of woodland and bush, with restricted black face mask, red eye, patchy range. Breeding males (June-December) with clean black mask meeting narrowly above bill; forecrown rusty chestnut, merging evenly to bright yellow on hindcrown, extending to cheeks and entire underparts; usually some rusty staining adjacent to mask on breast. Mantle greenish-yellow, weakly mottled. Wings and tail brown cleanly edged yellow. Bill black, eye red-orange (colour retained in non-breeding plumage), legs pale pinkish-brown. Female and non-breeding male face, throat and breast buff washed with yellow, (darkens in breeding female), whiter on belly. Mantle mottled with darker feather centres. Wings and tail brown with yellow fringes. Immature similar. Female bill pinkish-brown, eye red-brown, legs pinkish-brown. **Similar species** Breeding male distinctive but check Heuglin's Masked Weaver for separation details. Non-breeding from rather similar Yellow-backed Weaver and larger Village Weaver by slimmer bill, plainer buffier appearance about face and breast. Slimmer, longer-tailed outline than dumpy Heuglin's, which has pale eye colour. Larger than Little Weaver, shares whitish belly.

358

HABITS Favours dry savanna woodland, including areas with almost closed canopy, dry scrublands with trees and out into more open areas. Gregarious, gathering in small loose colonies, which while busy, lack the intense bustle of Village and Yellow-backed Weavers. Male birds in all plumages construct nests, hanging beneath, fanning extended wings in display. Occasional mixed colonies with Heuglin's Masked Weaver. Gathers into larger flocks in dry season, spreading into agricultural areas.

VOICE Occasionally emits forceful, sharp *pink* or *zik*; sings with creaking weaver swizzling, a rapid chatter of rasps and whistles with particularly intense grating ratchet sounds mixed with bright chirps.

STATUS AND DISTRIBUTION Resident breeder, recorded in areas with more extensive woodlands. Uncommon at the coast, becoming frequent in larger protected forest parks and uncultivated areas inland. Common in some parts of LRD, notably Kiang West National Park; frequent in appropriate habitat of CRD and URD. Seasonal pattern complicated by dry season identification difficulty, when frequency of records drops dramatically, though this may not mean the birds are correspondingly less common.

Vitelline Masked Weaver nest

BREEDING Active nests found July-November, unoccupied nests throughout the year. Males may build 2-3 nests each.

NEST Usually within 2-3m of the ground. Distinctive coarse texture from use of broad-bladed grass stems, woven to produce unexpectedly smooth outer surface and characteristic onion shape. Entrance faces almost vertically down, no significant entrance tunnel. Typically affixed singly in or small clusters along, or near tip of, pendulous branch or leaf; especially favours *Acacia macrostachya*.

VIEILLOT'S BLACK WEAVER
Ploceus nigerrimus　　　　**Not illustrated**

Fr. Tisserin noir de Vieillot

IDENTIFICATION 15-16cm. *P. n. castaneofuscus.* Sexes differ. Distinctive dark weaver, may appear all black at a distance, but breeding males of the local subspecies are bicoloured in velvety black and rich bright chestnut. Black head, neck and breast; black wings and tail. Body feathers chestnut, cleanly demarcated over back, lower belly and undertail. Bill strong and black, eye bright pale yellow, legs fleshy brown. Female olive over head, with fine dark streaks; browner on mantle, more heavily streaked with dark feather centres, approaches plain rufous-brown on rump. Wings and tail very dark brown, inner flight feathers clearly fringed with pale buff. Underparts uniform pale yellow-buff, warmed by faint peachy-rufous bloom over breast and flanks. Eye pale yellow, bill brown upper mandible, fleshier lower mandible. **Similar species** Distinctive; nominate eastern race is all-black in breeding male. Female from other weavers by rufous-toned rump with plain underparts.

HABITS Typical rumbustious weaver, associated with Guinean habitats, forest clearings and adjoining savanna. Particularly noted for tendency to associate with other species, e.g. Village Weaver, Black-winged Red Bishop.

VOICE Produces typical weaver wheezing stream of chattering raspy notes.

STATUS AND DISTRIBUTION Rare in The Gambia. Five published records between 1963 and 1986 cover all seasons, with three occurrences in the Lamin-Pirang area and twice near Basse, including small nesting colony near Basse in January 1979. No recent records from Senegal, although the type specimen of the local subspecies is labelled Casamance River. Otherwise in the Guinean zone from Guinea to Nigeria.

Vieillot's Black Weaver nest

359

BREEDING Single Basse record in January, the only one for the immediate area; elsewhere breeding in most months.

NEST Similar to Village Weaver, coarsely woven from palm leaf strips, but entrance underneath lacks tunnel extension; often built in lower part of mixed species colony.

VILLAGE WEAVER
Ploceus cucullatus **Plate 46**

Fr. Tisserin gendarme

Other names: Black-headed Weaver, Spotted-backed Weaver

IDENTIFICATION 15-17cm. Sexes differ. The common weaver of The Gambia, occurring in large flocks, associating closely with human activity, nesting in huge noisy colonies in prominent locations. Breeding male (June-December) black over entire crown, face and pointed extension on throat. Broad dark chestnut wedge across nape. Centre of mantle blotchy black and yellow, forming into untidy streaks approximating a 'V' around mantle edge. Rump yellow, tail olive, wing feathers dark grey-brown, neatly fringed with yellow. Underparts solid yellow, rust stained on chest and flanks. Bill strong, black; iris red, legs pink-brown. Female and non-breeding male olive grey-green crown; grey-brown back mottled with darker central feather streaks, underparts yellow, grey-white on belly in some. Female bill fleshy-horn, eye red-brown. Juvenile similar to female, browner on the back. **Similar species** Chestnut nape on breeding male distinctive; non-breeding plumages most similar to Yellow-backed Weaver, which has pale eye, more streaked mantle, and less hefty bill.

HABITS Adaptable, found in wide range of habitats in remote as well as inhabited areas, especially near water, thrives in association with agriculture and human settlement. Giant white silk cotton, *Ceiba pentandra,* trees found at the centre of villages, hold some of the largest nesting colonies. Highly gregarious, restless and noisy. Colonies a riot of activity in season, with heavy traffic of birds coming and going and much squabbling. Moves in swift flocks using light, shallow, bounding flight action. Tears strips from e.g. oil palm leaves leaving lower fronds skeletal. A pest of seed crops but also devours large numbers of insects. Colonies frequently attract nest-robbing African Harrier-Hawk, which may be mobbed, and parasitic Diederik Cuckoo.

VOICE Produces a streaming chatter in a jumble of short, chippy, harsh notes, drawing out to wheezy, buzzy, mechanical ratchet episodes before plunging back to a babble of harder sounds. Flight call *cluck.*

STATUS AND DISTRIBUTION Abundant to very abundant breeding resident throughout The Gambia, with particularly large concentrations at the coast. Also throughout Senegal though seasonal in extreme north.

Village Weaver nest

BREEDING Extended breeding season, active from onset of rains through to November. Favours large trees and may occupy several adjacent trees.

NEST Coarsely woven, from grass, strips of plantain or palm leaf. Rounded to slightly onion-shaped with short downward extension to entrance. Lacks smooth outer layering of Vitelline Masked Weaver. Forms larger colonies than other species, with numerous nests in close proximity on, for example, hanging fronds of oil palm and large silk cotton trees. Sometimes nests are interwoven. May also build comparatively low over fresh water, close to Yellow-backed Weaver in some river bank locations.

YELLOW-BACKED WEAVER
Ploceus melanocephalus **Plate 46**

Fr. Tisserin à tête noir

Other name: Black-headed Weaver

IDENTIFICATION 15-16cm. Sexes differ, male slightly larger. Breeding male (June-December) with full black hood over head and throat, changes sharply (no chestnut wedge at hind margin) to bright clear yellow back of neck, breast and underparts. Mantle plain green, tail pale olive-brown. Wings grey-brown with yellow fringing. Bill black, eye dark brown, legs flesh-brown. Female and non-breeding male olive brown head and cheeks, streaky grey-brown mantle, plainer brown rump. Wings brown with buff and yellow fringes to inner flight feathers. Underside whitish with buff wash over breast and flanks. Bill horn, eye usually with pale grey-white iris. Immature similar. **Similar**

species Waterside habitat useful indicator. Full black hood and size separates breeding male from non-gregarious Little Weaver; non-breeding males and females usually given away by presence of pale-eyed individuals, though this is not prominent and must be consciously checked. Usually smaller than Village Weaver, but size overlap occurs; mantle also greyer, with dark feather centres more inclined to align in extended streaks; undertail-coverts usually whiter than Village, Heuglin's, and Vitelline Masked.

HABITS Gregarious, lively weaver, nesting close to water, primarily in dense vegetation along the banks of the main river. Occasionally near brackish water as on the Banjul Bund, where it mixes with Village Weavers, especially in the dry season.

VOICE Thin, squeaky-toned weaver chattering, less extended phrasing than Village Weaver; *squeee-ki-kee* mixed with hard grating and slurring creaking sounds.

STATUS AND DISTRIBUTION Frequent breeding resident, locally common to very abundant over stretches of riverbank. Recorded in all months, much increased frequency in breeding season, August-October, and from all divisions of the country. Main stronghold on riverbanks of CRD and URD, but routinely found on Banjul Bund and other localised sites at the coast, e.g. Tanji Bird Reserve. Nominate race found in an enclave from The Gambia through Casamance to Guinea-Bissau. North of The Gambia in Senegal *P. m. capitalis* patchily distributed throughout according to habitat, showing progressively more rusty staining on breast below mask.

BREEDING Recorded in The Gambia from June-September. Occurs in mixed species colonies with Village Weavers on islands in CRD, exploiting dense thorny vegetation hanging out over the water.

NEST Similar to Village Weaver but lacks extension to entrance vertically underneath. Mainly built at comparatively low level, nearly always close to or over freshwater, often in large numbers.

COMPACT WEAVER
Ploceus superciliosus   Plate 46

Fr. Tisserin gros-bec

IDENTIFICATION 13cm Sexes differ. A thickset, dark weaver, not recorded in The Gambia but sporadic in Casamance. Breeding male chestnut crown, merging to bright yellow nape. Darkish olive-brown mantle, mottled with dark feather centres, extends part way up nape; wings and tail olive-brown. Black mask over face and throat to level with top of eye,

breast and belly bright yellow, undertail buff-brown. Thickish bill black, eye brown, legs brown. Breeding female has dark olive crown extending in a wedge from point above bill continuous with upperparts; broad yellow supercilium above dark eye mask. Non-breeding similar above; pale cinnamon-buff supercilium above narrow black mask through eye, plain buff underparts both sexes. Similar species Recognised by chestnut and yellow crown with precise change to dark nape and black mask in male, strong broad supercilium in female.

HABITS In West Africa associated with northern limit of forest zone, favouring clearings and grass-land areas, and especially swamp areas with trees for breeding. Breeds in isolated pairs, gathering into flocks for the dry season.

VOICE Simple *chee* call note; song given as a repetitive *cheewery cheewery cheewery*, more musical quality than the typical weavers.

STATUS AND DISTRIBUTION No records from The Gambia. Seen on a few occasions in Casamance, near Ziguinchor and Oussouye. Perhaps a species receding from the area as a result of forest clearance; observations should be carefully documented.

Compact Weaver nest

BREEDING Elsewhere not colonial, breeding August-November.

NEST Very neat, smoothly woven nest with side entrance, built on tall grass stems resembling Grosbeak Weaver nest in both situation and appearance.

GRAY'S MALIMBE
Malimbus nitens   Plate 46

Fr. Malimbe à bec bleu

Other name: Blue-billed Malimbe

IDENTIFICATION 15cm. Sexes similar. Shy forest weaver, all black save for a neat patch of vivid red on the chest. Duller black on breast and undertail-coverts. Female less glossy, darker brown below. Strong bill is prominent blue or whitish-blue, eye reddish and legs grey. Immature resembles female, but red breast patch dull orange-tawny. **Similar species** Distinctive bird, but shares similar colours and pattern as male and immature Western Bluebill. Gray's Malimbe is slightly larger, has slimmer longer bill, no white eye-ring, and proportionately smaller red breast patch does not extend to throat. Immature Bluebill also lacks eye-ring, but has stubbier conical bill, and is smaller, usually in association with adults.

HABITS Very elusive species restricted to forest and dense thicket with freshwater, but inquisitive and approachable near nest. Found in small groups, participates in mixed species foraging parties. Fond of *Raphia* palm and swampy areas.

STATUS AND DISTRIBUTION At best very rare in The Gambia. One wet season record from Brufut area 1959. May still occur as a rare visitor to Abuko; claims of birds observed around the swampy pool margins, which fulfil classic habitat requirements, persist, but lack adequate detail to differentiate from plumages of Western Bluebill. A resident breeder in Casamance, where it is found in several forest patches in the vicinity of Oussouye and Zinguinchor. Elsewhere eastward through the forest belt to Uganda and Zaire.

BREEDING Breeds at the onset of the rains, June-July in Casamance, and over a more extended season in the heart of its range further east.

NEST Untidy globular nest, suspended from tip of palm leaf, almost always over freshwater, with entrance beneath, protected by overhanging woven wall.

RED-HEADED WEAVER
Anaplectes rubriceps　　　　　Plate 46

Fr. Tisserin écarlate ou Tisserin à ailes rouges

Synonym: *Anaplectes melanotis*

IDENTIFICATION 15-16cm. Sexes differ. Bright pink-red bill, crimson edging on flight and tail feathers, otherwise brown above and white below. Breeding male (July-December) has crimson crown, nape and upper breast surrounding neat black face mask. Upperparts brown, flight feathers edged crimson, forming weak reddish panel on folded wing, less obviously on the basal part of the tail feathers. White underparts. Bill red, eye reddish,

legs brownish-grey. Female and non-breeding male brown over head and mantle, plain whitish below. Bill bright pinkish-red in both sexes. Immature buffish over face and upper breast, hint of reddish or yellowish edging to flight feathers already present, bill dark brown. **Similar species** Bright pink to red bill immediately distinctive in all adult plumages; can appear almost 'dayglow plastic' in intensity; other red-billed seed-eaters all much smaller.

HABITS A bird of extensive dry woodlands in the Sudanian vegetation, in The Gambia most associated with large stands of mixed small trees, usually dominated by *Combretum*, with large emergent *Pterocarpus* and *Khaya*. Not particularly gregarious. Works through foliage searching for insect prey, demonstrating great agility. Flight swift and direct.

VOICE Typical weaver chattering, but transposed to much higher pitch than average, resulting in distinctive squeaky tone.

STATUS AND DISTRIBUTION Rare resident; single records from NBD, CRD and URD, together with one or two records annually from the large forest parks of LRD confirm its presence in all seasons. Not positively recorded in coastal areas. Found in the dry woodlands of southern Senegal, does not extend far north of The Gambia.

Red-headed Weaver nest

BREEDING Single nest observed in Kiang West National Park June to September but not active; like many weavers, males known to build more than one nest each season. Elsewhere usually a solitary nester.

NEST Characteristic untidy, elongated, woven nest; globular chamber suspended by a longish stem from attachment point, long pendant entrance tunnel below. Constructed mainly from leaf ribs.

RED-HEADED QUELEA
Quelea erythrops   Plate 48

Fr. Travailleur à tête rouge

Other name: Red-headed Dioch

IDENTIFICATION 11-12cm. Sexes differ. Breeding male unmistakable; crimson head and face on brown body, streaked with dark brown above, plain buff breast, paler belly. Narrow yellow fringes to inner flight feathers, blacker patch on throat. Bill blackish, eye brown, legs flesh. Female and non-breeding male streaky brown above, very finely streaked crown almost plain, fine pale yellow fringes to tertials. Yellow stained supercilium. Plain whitish below, buff stained area over breast and flanks is warmer orange-buff in breeding females. Bill fleshy-horn. **Similar species** Non-breeding plumage from Red-billed Quelea by bill colour, yellow supercilium, tawny, not grey, wash to head and breast. In this plumage very similar to non-breeding bishops; note especially relatively larger bill, also supercilium tone, plain underside and yellow fringes to tertials.

HABITS Gregarious; associated with rank grassland and agricultural land, usually close to freshwater. Rice fields are also favoured and small damp plots close to woodland. Flocks active, sometimes large, feeding in tight groups on the ground, others perched in surrounding bushes. Breeding males sit at prominent locations.

VOICE Busy, squeaky chattering from flocks; song described as churring.

STATUS AND DISTRIBUTION Uncommon wet season breeder, recorded patchily from coastal, central and inland parts of The Gambia, mainly near the river e.g. Pirang, Basse. Dry season status probably affected by non-breeding identification difficulties. Has been classed a resident in The Gambia on the basis of flocks reported on the Bund Road in April, but in Senegal, where it is limited to the south (and throughout its range through to southern Africa) usually considered an intra-African migrant. Records in The Gambia and Senegal mainly July to September.

BREEDING Recorded nesting in The Gambia in July, on large islands in the main river, CRD. Implied breeding from presence of full plumage males in August-September, at the coast and Basse. Weaves spherical nest of fine grass, with oval entrance off-centre near the top, in rank herbage.

RED-BILLED QUELEA
Quelea quelea   Plate 48

Fr. Travailleur à bec rouge

Other name: Black-faced Dioch

IDENTIFICATION 12-13cm. Sexes differ. Small streaky weaver, conical red bill with black mask in breeding season, or red bill with contrasting grey-brown head in non-breeding season. Breeding male variable; black mask surrounded by either golden-buff or rosy-pink margin over crown, neck and upper breast; occasionally the mask is missing. Mantle brown, streaked with dark brown; buff-yellow edges to wing and tail feathers. Underparts mainly plain buff, hint of dappled streaks on flanks. Relatively bulky red conical bill (breeding female may develop yellow-horn bill), eye brown with narrow pinkish-red orbital ring, legs pink. Non-breeding male and female similar, retaining orange-red bill; head grey-brown very lightly streaked, off-white supercilium, grey cheeks and short off-white mark beneath eye, creating spectacled effect; upperparts streaky brown, underparts whitish faintly dappled grey on breast. **Similar species** Bill colour shared with similar non-breeding Pin-tailed Whydah; separate by barred head pattern of the whydah, unlike the nearly unstreaked grey head of Red-billed Quelea. Non-breeding bishops have horn-coloured bills and lack yellow edging to wings.

HABITS Very gregarious, in main part of range gathering in huge flocks, notorious for demolishing agricultural crops over large areas, though this not recorded in The Gambia, where it is more likely to be encountered in small groups. Favours scrub, fallow fields and grassland. Generally associated with drier areas than Red-headed Quelea, though does occur around rice fields and swampy ground.

VOICE Chattering, with shrill high pitched whistling notes.

STATUS AND DISTRIBUTION Uncommon dry season visitor, all recent records October-May, with sightings from all divisions, mostly near grassy swamps. Noted along the coastline at Cape Point, Kotu and Tanji; in CRD at Jahali, and URD Basse. Resident in northern Senegal, breeding along the Senegal River, but noted to be declining.

BREEDING Only breeding record from The Gambia near Basse, in August in the 1960s. Light woven nests of grass, colonial. All Senegalese records in the wet season, August-November.

YELLOW-CROWNED BISHOP
Euplectes afer   **Plate 45**

Fr. Euplecte vorabé

Other name: Yellow Bishop

IDENTIFICATION 12cm. Sexes differ. Vivid, small black and yellow bishop of swampy ground, often seen plumped up in display. Breeding male (July-November) golden-yellow on crown and lower back, where elongated feathers often puffed up. Smudgily black across upper back, wing feathers and tail blackish, fringed brown. Face and underside black save for yellow flank patch at angle of wing and yellow vent. Bill black. Transitional plumages untidy but obvious. Non-breeding male and female streaky brown on crown and upperparts, broad whitish supercilium from base of bill to ear-coverts with thickish dark line marking lower margin through eye. Breeding female presents greyer overall tone than non-breeding. Grey-brown cheeks fade to whitish underparts, with variable dark streaking strongest at side of breast, fading towards centre. Bill horn, eye brown, legs brown. Immature resembles non-breeding female, possibly browner in tone. **Similar species** Separation of non-breeding plumages from slightly larger Northern Red Bishop not easy in absence of breeding male in marsh habitat; supercilium is generally more prominent though may tend to yellow-buff staining in non-breeding birds; note dark eye-line feature and extent of streaking on underparts.

HABITS Closely associated with marshes, mainly freshwater, including rice fields, but also occurs on small vegetated pools along margins of saltflats following rain. Especially fond of rushes. Polygamous breeding males indulge in characteristic bumble-bee display flights, body feathering puffed and hanging rather vertical on rapidly whirring wings, drifting forward in slow hovering progression at low level over territory, maintaining buzzing calls. Gregarious, flocking in the non-breeding season and thought to mix freely with other bishops.

VOICE Insistent, high pitched *tsrip-tsrip* breaking into short rattling buzz notes, repeated rhythmically and hurriedly *bzz-bzzp-bzzp-bzzp-bzzp*, and *tsee t-ka t-ka t-ka t-ka*; often given on the wing.

STATUS AND DISTRIBUTION Resident breeder, frequent to common within preferred habitat; recorded from all divisions. Persistent in a few favoured sites near the coast (Camaloo, Jinnack), but most readily encountered in the extensive swamps either side of the river in CRD and URD. Regional distribution dominated by association with major rivers, (Senegal River, Gambia River and Casamance), absent from dry country in between.

BREEDING Nesting recorded September-November in The Gambia. Builds loosely woven domed nest with elliptical opening near the top at one side; usually close to water level concealed within clump of vegetation, using tamarisk (Jinack) or rushes (Pakali Ba). Frequently loses nests to flooding.

BLACK-WINGED RED BISHOP
Euplectes hordeaceus   **Plate 45**

Fr. Euplecte monseigneur

Other name: Fire-crowned Bishop

IDENTIFICATION 13-15cm. Sexes differ seasonally. Brilliant scarlet and black wet season male is the only bishop in the region to show a nearly full red crown. Breeding male (late July-November) crisply delineated scarlet, sometimes orange-toned, over crown, nape and upper breast; duller orange-red on mantle. Face mask black; belly, wings and tail black; undertail and thighs fawn. Bill black. Non-breeding male upperparts heavily streaked dark on tawny-brown; whitish-buff supercilium and light streaks across buff-washed breast, plain pale belly. Retains blackish wings and tail. Bill horn, eye dark brown and legs brown. Female resembles non-breeding male, though wings not so dark. Immature resembles female with wider pale fringes to flight feathers. **Similar species** Superficial similarity of breeding male to fractionally smaller Northern Red Bishop resolved at once by red crown and black, not brown, wings. Non-breeding birds tend to present a bolder streaked pattern than paler Northern Red Bishop with greater contrast between shoulder area and underparts and heavier bill; dark wings of male, lacking yellow carpal patch of Yellow-mantled Whydah, obvious in flight. Red-collared Widowbird, not recorded for many years, also retains dark wings in streaky non-breeding plumage, and has plainer underparts. Relatively broad-based, dark wings stand out.

HABITS Most often seen in tall grasses of field edges and fallow land in the wet season, where often found alongside Northern Red Bishops. Displaying males puff feathers, sing from tall grass heads from where they dip down on whirring wings in a hesitant slow flight pattern, dropping from view. Normal flight direct and quick. In dry season small non-breeding groups sometimes encountered in dense dry scrub; singles at quiet forest pools; perhaps most often in small numbers amongst large groups of non-breeding Northern Red Bishops.

VOICE Song composed of short, sharp chirrups and twitters, not loud but high pitched and sibilant character; similar to Northern Red Bishop.

STATUS AND DISTRIBUTION Breeding resident found in all parts, locally frequent to common at the coast, marginally less so in the interior, everywhere much less common overall than Northern Red Bishop. Recorded in all months, but less frequently in the dry season when in cryptic non-breeding plumage. Found through a belt of central Senegal, east to Niokolo Koba, where it is similarly less frequent than Northern Red Bishop.

BREEDING Breeds August-November in The Gambia, constructing thinly woven globular nest of small grass stems bound onto two or three vertical stems of giant grass or small shrub.

NORTHERN RED BISHOP
Euplectes franciscanus   Plate 45

Fr. Euplecte franciscain

Other names: Red Bishop

IDENTIFICATION 11-13cm. Seasonal sex difference. Commoner black and scarlet bishop seen throughout The Gambia in the wet season. Breeding male (August-December) has black face extended to form cap over forecrown; neat black waistcoat. Broad red band (sometimes orange-toned) from throat and breast around back of head continuous with duller red mantle. Wings brown, long uppertail-coverts orange-red, covering brown tail. Bill black, eye brown, legs pink-brown. Non-breeding male resembles smaller female; brown, streaked with dark on crown to rump, pale buffy-white supercilium, underparts variably buff stained and streaked on breast and flanks, whiter on belly. Bill horn-coloured. Intermediate male plumages seen mainly July and December, untidy but obvious. Similar species Black crown and brown wings separate breeding male from Black-winged Red Bishop; non-breeding Black-winged Red Bishop has heavier bill and black wings obvious when present, though younger birds may show only dark brown wings. Non-breeding from non-breeding Yellow-crowned Bishop by tendency to warmer, browner body tone, buffer narrower supercilium, less streaking below, but not always separable in absence of accompanying breeding male plumage.

HABITS Closely associated with giant grasses and tall crops such as millet and sorghum, but found in a wide range of open habitats with rank weedy vegetation and shrubs. Breeding males perch prominently, and display with audibly whirring wings. Puffing of nape into a ruff and raising of body feathers results in substantial apparent postural and body shape changes with display patterns. Display flight slow and hovering, body near vertical with

bursts of wing-beats creating a bobbing action as they descend to nests and females hidden lower in the vegetation. Gathers in large flocks in the non-breeding season, attracted to pools of water.

VOICE An array of high pitched and emphatic short bat-squeak notes, *tsip* provide an unobtrusive but consistent background sound in the greening field areas where these birds are typically active. A wheezy, ratchety song of thin squeaks with quiet sub-text of guttural buzzes.

STATUS AND DISTRIBUTION Resident breeder throughout the country, common to abundant in all agricultural and fallow areas in the wet season. Flocks also seen through the dry season, though inevitably are less prominent. Found throughout Senegal with the exception of the most forested areas of Casamance.

BREEDING Nests in the wet season, September-November in The Gambia. Weaves a globular nest of coarse grass strips, 1-2m off the ground, often in extensive stands of tall grass or crops. Grass flower heads of nest lining project through entrance.

YELLOW-SHOULDERED WIDOWBIRD
Euplectes macrourus   Plate 45

Fr. Euplect à dos d'or

Other name: Yellow-mantled Widowbird (or Whydah)

IDENTIFICATION 18-22cm, breeding male with tail; 13-14cm, female and non-breeding male. Sexes differ. Breeding male (July-November) all black, save for bright yellow mantle and epaulettes forming single large yellow patch on upper back. Tail black, long, rounded in outline. Bill blue-black, eye dark brown, legs black. Non-breeding male retains reduced uneven yellow patches on epaulettes, otherwise brown-bodied, heavily streaked with dark brown on upperparts, yellow-buff supercilium and blackish wings; bill horn. Transitional male (July) blackish head, rump and tail, otherwise streaky brown. Female brown and well built, heavily streaked above, with pale supercilium and usually no yellow carpal patch; pale yellow wash to underparts, with light streaking on breast and flanks. Similar species Males distinctive at all times due to yellow shoulder markings; also rather rounded wing and tail profile and loose-limbed flight; slightly smaller female strongly streaked; best identified by association with male, but slightly longer-tailed than non-breeding bishops.

HABITS Especially associated with marshy habitats and rice fields in the breeding season, where males sit singing from prominent high points of vegetation, nape feathers prominently puffed. Similar behaviour seen in a few drier second growth woodland habitats as at Yundum. Indulges in halting display flight on rather stiff wings and tail. Otherwise flight direct. Gathers in loose flocks in non-breeding plumage; generally site faithful throughout the year.

VOICE Repeated, rasping, *zeep,zeep,zeep*, and crisp, hasty, twitterings, soft but penetrating.

STATUS AND DISTRIBUTION Breeding resident, most often seen at a few well-known sites near the coast e.g. Lamin, and until recently Yundum where it may have been badly affected by land use changes. Also found in swamps and marshes of CRD and URD, where likely to be under-recorded. Not recently reported from NBD though present north the river in CRD. Overall not common. In Senegal restricted to southern areas.

BREEDING Wet season breeder, mainly July-September. Weaves a loose globular nest at low level in tall vegetation, with grass inflorescences of lining material projecting from side entrance as other *Euplectes* species. Polygamous.

RED-COLLARED WIDOWBIRD
Euplectes ardens   Plate 45

Fr. Euplecte veuve-noir

Other name: Long-tailed Black Whydah

IDENTIFICATION 25-30cm, breeding male with tail; 12-13cm, non-breeding male and female. Sexes differ. *E. a. concolor*. Breeding male unmistakable, but lacks red collar; uniform black with very long tail, lax in wind, often divided; held stiff when displaying. Bill black, eye brown, legs black. Non-breeding male streaked black and tawny above; yellowish supercilium. Retains blackish wings and tail; pale throat and belly contrasts with plain buff-yellow breast. Bill horn-pink, eye brown, legs fleshy. Female similar to non-breeding male, but with brown wings. **Similar species** The only *Euplectes* lacking streaked underparts; any non-breeding bishop with this feature merits close inspection. Non-breeding Red-headed Quelea also plain below, but shows yellow edging on wings. Check for plain breasted, dark-winged males, but accompanying breeding male usually necessary for certain identification.

HABITS Gregarious, favours rank grasslands in comparatively treeless areas; sometimes also reported in association with rice and reeds. Displaying males fly slowly with tail hanging downward, wings whirring; otherwise flies swiftly with tail trailing. Feed on the ground or perched on grass heads.

VOICE Harsh *chrrt, chrrt*, forceful and short. Rapid, staccato, *chekety, chekety, chekety* in similar, reedy, hard-edged tone.

STATUS AND DISTRIBUTION Status uncertain. Few recent records, but reported persistently from The Gambia and Senegal by ornithological explorers earlier in the century, possibly even then on the basis of rather few records. Last published observations for The Gambia were two birds, at separate locations in CRD, on the same day in August 1963. In Senegal, one seen in Niokolo Koba in March 1992. Regular from Guinea eastward.

BREEDING Builds small globular woven nest of fine grass, usually 1-2m up in long grass, September-November in main range.

NOTE Patches of red in breeding males in eastern and southern Africa are the basis for the English name, otherwise inappropriate in West Africa.

PARASITIC WEAVER
Anomalospiza imberbis   **Not illustrated**

Fr. Tisserin parasite ou Tisserin coucou

Other name: Cuckoo Finch

IDENTIFICATION 13cm. Small canary-like weaver, distinguished by stubby black bill. Breeding male plain yellow over face throat and underparts; back dull olive, streaked with black, wings and tail dusky, finely edged with yellow. Short thick bill blackish, eye brown (appears dark in yellow face) and legs brown. Female browner above, also streaked blackish, with dusky line through eye; yellowish throat fading to browner wash on upper breast, lightly streaked on flanks, whitish towards belly. Bill dark horn. **Similar species** Resembles short-tailed canary; note thick base to bill and dark eye on yellow face; also direct weaver flight; more heavily streaked upperparts than any stage of Sudan Golden Sparrow.

HABITS Found in range of open habitats, including cultivation. Solitary pairs or small groups when breeding, assembling in larger flocks in non-breeding season. Feeds on the ground, and perches on flower heads of grass or herbs. Lays eggs in nests of other species, hence common name.

VOICE Bright, soft chirruping chatter in flocks is unremarkable; male song said to be raspy in tone, *tseep, krrik, krrik, krrik*.

STATUS AND DISTRIBUTION Considered a vagrant; one sight record near Banjul in September 1969, variously reported as a singleton or small flock. Not recorded from Senegal. Otherwise known from Guinea eastward but with patchy distribution and seasonal movements. Several known host species found in The Gambia and it is possible that it may occur as a very rare or sporadic wet season visitor.

BREEDING Parasite to nests of a range of *Cisticola* species and Tawny-flanked Prinia. Nearest confirmed breeding in Côte d'Ivoire.

WAXBILLS AND ALLIES

Estrildidae

Twenty three species of tiny to small, finch-like birds. Diverse and colourful, they include pytilias, seed-crackers, bluebills, firefinches, cordon-bleu, waxbills, quail-finch and mannikins. They build much smaller nests than weavers which are domed and usually made of grasses. Often gregarious.

CHESTNUT-BREASTED NEGROFINCH
Nigrita bicolor   Plate 48

Fr. Sénégali brun à ventre roux

IDENTIFICATION 11-12cm. Sexes similar. Small dark finch of forest edge and thickets, two-tone colour pattern, uniform slate above, velvety dark burgundy-chestnut below. Chestnut extends from forehead over face to all underparts; wings and tail darker than rest of body. Bill black, eye red-brown, legs dark brown. Immature similar but dull grey-brown above, orange-buff below. **Similar species** Adult distinctive; immature from immature Bronze Mannikin by darker body of negrofinch, with more contrast between upper and lower sides and smaller bill than that of the mannikin.

HABITS Forest edge species, associated with secondary growth, clearings, occasionally coming into thick bush or mangroves. Usually singly, sometimes pairs or small parties, in understorey though also moves into tree tops and creeps around canopy. Shy.

VOICE Song a low key *chi-chi-hooee* and *kiyu-kiyu-weh-weh-weh*.

STATUS AND DISTRIBUTION Rare in The Gambia; six records, all in WD, Abuko, from all seasons but none more recent than 1986. In Senegal confined to coastal Casamance. Possibly much reduced by forest clearance; continued presence in The Gambia in need of confirmation.

BREEDING Globular nest of loose dry leaves with side entrance. Up to five eggs. Nearest confirmed breeding Sierra Leone.

GREY-HEADED OLIVE-BACK
Nesocharis capistrata   Not illustrated

Fr. Sénégali vert à joues blanches

Other name: White-cheeked Olive-back

IDENTIFICATION 14cm. Sexes similar. Small estrildid, yellow-olive upperparts, strongly patterned head with white face and black chin. Crown, nape and rest of underparts grey with yellow patch along flanks. Spread wings and tail rounded in outline. Eye red, bill black, legs black. Immature similar but darker, with yellow flanks replaced by buff-brown, bill whitish.

HABITS Poorly known but mainly a canopy dwelling species of forest thickets and undergrowth, moving into dense adjacent savanna. Recorded coming to the ground to take grass seeds, also known to take snails, but typically a foliage gleaner; seen singly or in pairs.

VOICE Not vocal, but a descending song sequence has been described *chwee-chwee-chwee-chwi*.

STATUS AND DISTRIBUTION Uncertain. Included on the basis of a single very old museum specimen plus a sight record May 1968. No records from Senegal; nearest resident population Guinea-Bissau.

BREEDING Poorly known. Nearest reported breeding Sierra Leone.

RED-WINGED PYTILIA
Pytilia phoenicoptera   Plate 47

Fr. Beaumarquet aurore

IDENTIFICATION 13cm. Sexes similar. Small; red wings, rump and tail, otherwise lead-grey upperparts and dark bill. Underparts grey on breast, narrowly barred white on flanks and belly. Female duller, slightly browner-toned and less intense on red parts, more extensively barred below. Bill blackish, eye red, legs flesh-brown. Immature plain brownish above, bill pink-based; orange-red wash on wings, rump and tail, faint hint of barring below. **Similar species** Adult distinctive, much more common Lavender Firefinch always paler, without red in wings and not confusing. More likely to be overlooked as a Grey-headed Sparrow when seen perched in distant silhouette. See Green-winged Pytilia for comparison of immatures.

HABITS Favours savanna woodlands, including moister coastal areas but more typically drier savanna of the interior, especially areas with large trees and bushy thickets. Unobtrusive ground feeder, but also perches in high dead branches of emergent trees. Usually in pairs; sometimes a retiring member of mixed species parties with other waxbills, weavers and mannikins. Regular visitor to woodland and roadside waterholes, especially laterite quarries, where it may be seen well out in the open.

VOICE Various chirps and a rattling song.

STATUS AND DISTRIBUTION Uncommon resident reported from all divisions except NBD, with a particular stronghold in Kiang West and other large forest parks of LRD, where it is locally frequent. Present in all seasons but recorded with significantly increased frequency in the rains, July-October. Found in proximity to Green-winged Pytilia in LRD, but recorded only south of the river in CRD near Kudang, where Green-winged occurs mainly to the north. On both banks of the river in URD. Outnumbers Green-winged Pytilia 3:1 in overall observation frequency. In Senegal information incomplete but probably restricted to the northern border of The Gambia and the eastern interior.

BREEDING Recorded in August and September in The Gambia. Constructs globular nest from grass flower heads with side entrance at moderate level in the top of a bush. Host to the brood parasitic Exclamatory Paradise Wydah.

GREEN-WINGED PYTILIA
Pytilia melba   Plate 47

Fr. Beaumarquet melba

Other name: Melba Finch

IDENTIFICATION 13cm. Sexes differ. *P. m. citerior.* Small and colourful; vivid red face (male only) rump and tail, orange-yellow breast, soft grey and green upperparts; usually in more arid areas. Adult male scarlet face cleanly separated from grey hindcrown to nape. Underparts barred grey and white. Adult female lacks red face and yellow breast; all-grey head, red on rump and tail only. Bill scarlet (duller, darker on culmen, in non-breeding male and female), eye red, legs pale brown. Immature very dull brown-toned version of female, dull orange on rump, underparts washed plain buff, not barred, bill blackish, reddening with age. **Similar**

species Adults unmistakable in good view. Immature from browner immature Red-winged Pytilia by absence of red edging on wings and longer tail.

HABITS Unobtrusive despite bright colouring. Associated with dry thorn scrub, especially around small denser patches in open habitats, also in denser dry savanna woodlands. Usually in pairs, foraging low down or on the ground and typically first glimpsed making short dashing flight quickly into cover.

VOICE Low *wit* call; song composed of whistling and croaking notes. Lengthy trilling songs and dripping water sounds noted in races from eastern and southern Africa not reported from West African race.

STATUS AND DISTRIBUTION Uncommon, localised resident; present in all seasons; records from Kiang in LRD, CRD (notably north of Kaur and mainly on the north bank of the river) and URD (near Basse). No records but expected in NBD. In Senegal frequent in bushy plains throughout the north, down to the latitude of The Gambia; not in Casamance.

BREEDING Not proven in The Gambia, though present at same time as extended season recorded in northern Senegal, September-April. Constructs lightweight globular nest of dry grasses with side entrance. Host to the brood parasitic Sahel Paradise Wydah.

CRIMSON SEED-CRACKER
Pyrenestes sanguineus   Plate 48

Fr. Grosbec ponceau à ventre brun

IDENTIFICATION 14-15cm. Sexes differ. Massive-billed, with a crimson head; combines spectacular appearance with unusual biology, elusive habits and rarity. Adult male uniform crimson of exceptional intensity over entire head, nape, breast and rump, extending variably over flanks. Tail crimson, lightly suffused with brown. Mantle wings and belly uniform dark brown, though lit from above in dense habitat often appears lighter brown above and blackish below. Bill variable in size (though only large bills recorded to date from The Gambia, see note below), sharply-pointed thick wedge profile; appears black with gun-metal sheen and bluish mandible edges. Eye dark brown, white eyelids create bright white incomplete eye-ring, divided front and rear. Legs brown. Female similar but with restricted area of crimson on face. Juvenile uniform dark brown over head, wings and body; crimson suffusion limited to rump and tail. **Similar species** Heavy, domed head and bill distinctive; all-red crown, massive bill with straight-edged profile, and red tail, are all points of separation from any plumage of Western Bluebill.

HABITS Associated with dense, humid, swamp forest patches, especially in areas with freshwater streams or wet ground. Also comes into rank vegetation at forest margins where food plants, including hard-seeded sedges, form denser stands. Usually perches within cover and feeds on or near ground, but will fly up to trees on being disturbed and may sing from emergent tree-top. Usually seen singly or in pairs. Unobtrusive. Bulky large-headed shape also apparent in flight.

VOICE Maintains soft, hollow *tsut....tsut* call, like sound made by snapping tongue back from behind front teeth, while feeding low down in dense cover. Produces squeaky-toned six-note song phrase rising in pitch to the fourth, most emphatic note, before tailing off; *tee,ti-ti teee,ti,tee.*

STATUS AND DISTRIBUTION Rare resident breeder. All recent records in The Gambia from tidally flooded swamp forest patches on both banks of the main river in CRD between Kaur and Sapu. Threatened by habitat loss as riverbank swamp forest is replaced by farm plots and banana fields. Known from a few locations in coastal Casamance and eastward to Sierra Leone.

BREEDING Constructs a bulky globular nest up to 20cm diameter with a central side entrance. Outer layers of a nest observed in The Gambia, September, were constructed from loosely packed fern fronds, placed in screwpine one metre above wet ground which was flooded to a depth of 15cm on every tide.

Note on bill-size polymorphism in seed-crackers

Seed-crackers exhibit a range of cleanly defined bill sizes, with additional variation in brown versus black feathering and body size. All forms are sometimes treated as one variable species, others have proposed splitting them into three or even four distinct species. It has long been known that birds of differing bill size are often sympatric and mate together. Recent studies of Black-bellied Seedcrackers *P. ostrinus* in Cameroon have led to resolution of one part of the problem. Birds of different bill sizes in one population may mate at random (there is no bias towards pairings of matched bill size, or either sex preferring partners of any particular bill size) and each pair can produce offspring of two or more bill types. This state of affairs has been related to seasonal changes in relative abundance of the major food types; hard-seeded (only

accessible to large-billed birds) and soft-seeded (more efficiently handled by small-billed birds) sedge species. In addition hard-seeded sedges are relatively more abundant in northern parts of the range of *P. ostrinus*, and large-billed morphs show a comparable increase in frequency in these areas.

All old museum specimens from Senegambia, as well as all recent field observations of Seed-crackers *P. sanguineus* in The Gambia, at the northern edge of all seed-cracker ranges, have been of large-billed birds, suggesting that similar constraints may be influencing this population. It remains unknown whether the difference between brown-feathered *sanguineus* and black-feathered *ostrinus* merits full species separation. It has been suggested that each form may occur within the range of the other on occasions. This may indicate either sympatry of species, or more simply that again morphs are involved, in this case fixed at very different frequencies in western and eastern parts of the range. See Smith 1987 and Smith 1993 for detailed background.

WESTERN BLUEBILL
Spermophaga haematina   **Plate 48**

Fr. Grosbec sanguin

Other names: Blue-billed Weaver, Bluebill

IDENTIFICATION 15cm. Sexes differ. Black finch with scarlet underparts, usually seen at low level in or near forest thickets; deep conical blue bill and often stiffly fanned black tail contribute to thickset impression. Adult male glossy black over head, sides of face and all upperparts; throat, breast and flanks scarlet, belly black. Adult female duller black above, dull scarlet on rump; scarlet diffusely stained dusky on face, bright orange-toned scarlet breast and flanks, belly black heavily spotted with white. Bill shiny blue, tipped scarlet. Eye red-brown, pale lids create bright white partial eye-ring in male (duller or absent in females and young); legs blackish. Immature male dull black with reddish base to tail; red on underparts initially diffuse and weak, progressing to adult pattern; immature female similar but reddish wash on breast even weaker and emerging white spotting evident. **Similar species** Distinctive, but if in doubt check larger Gray's Malimbe, which has less conical bill, with more restricted crimson breast patch, and larger-billed Crimson Seed-cracker.

HABITS Very retiring bird of dense forest thickets and undergrowth, normally seen on or near the ground singly or in pairs. Often near damp ground, but also occurs in dry coastal palm forest and remnant forest thickets away from water. Occasionally

369

glimpsed flitting through dense creepers in mid-canopy around mixed species feeding parties. Attracted to water to drink and bathe; most easily seen from hidden positions overlooking sheltered forest pools. On the ground typically holds tail in stiff broad fan.

VOICE Typical contact call a crisp *chip...chip* or *tsik*. A singing bird in The Gambia produced a loud confident song in a continuously rising double arpeggio whistle *tsuee.tsuee.tsuee - tswee.tswee.tsuee*. Song also described as a low clucking series of tack notes rising into a trill before dying away.

STATUS AND DISTRIBUTION Uncommon resident known only from a number of forest thickets in the Guinea savanna south of the river in WD. Comes regularly to hides in Abuko Nature Reserve, Bijilo, otherwise elusive and seldom seen. In Senegal limited to forested patches in western Casamance.

BREEDING In The Gambia August-December, including nest building and observation of newly fledged young. Constructs a globular mass from ferns etc., with grass interior and side entrance, at low level in dense cover, often near water.

DYBOWSKI'S TWINSPOT
Euschistospiza dybowskii   Plate 48

Fr. Sénégali à ventre noir

IDENTIFICATION 12cm. Sexes similar but separable. A small, dusky, red and black waxbill, rare in south-east Senegal; occurrence in The Gambia would be exceptional. Male slate-grey over head and breast, crimson mantle forms saddle against plain blackish wings, crimson rump and uppertail-coverts. Tail black. Belly black marked with white spots. Bill black, eye reddish, sometimes reddish eye-ring seen, legs dark grey. Female a washed-out version of the male, some brown mottling on crimson mantle saddle, wings browner, flanks dark grey, more extensively spotted white. Juvenile dull slaty, no flank spots, reddish over mantle and rump. **Similar species** Red mantle distinctive; check more common Red-winged Pytilia where red limited to wing and tail, flanks barred, never spotted.

HABITS Poorly known. Associated with rocky or grassy hillsides, savanna or cultivation near thickets and grassy patches along forest edges. In pairs or small groups, mixes with firefinch species. A ground feeder also noted to dig in soil to feed. Male hops around female holding feather or stem, usually on the ground, in courtship display.

VOICE Soft quick *tset* or *tsit-tsit* contact calls, forceful in alarm and an elaborate variable song of trills

mixed with strong fluty and some wheezy notes, often given from low in thick cover.

STATUS AND DISTRIBUTION Not recorded from The Gambia. Two records from different locations in extreme south-eastern Senegal, 1974 and 1982, both in March; otherwise occurs patchily east across forest savanna edge as far as Sudan and Zaire.

BREEDING Globular nest of grass usually well-concealed; closest recorded breeding September-October in Sierra Leone.

BAR-BREASTED FIREFINCH
Lagonosticta rufopicta   Plate 47

Fr. Amarante pointé

Other name: Brown Firefinch

IDENTIFICATION 11cm. Sexes similar. Distinguished by carmine tint to red face and underparts, cleanly separated from olive-brown crown, nape and upperparts. Rump and tail-base crimson, darkening to blackish rounded tail tip. Upper belly and flanks richer red; white feather tips create uneven fine barring pattern across breast in good view from the front, otherwise scarcely noticeable as different from typical firefinch white speckling. Lower belly greyer and undertail dirty cream. Bill pinkish-red with black culmen, eye dark brown with narrow pale whitish ring (not prominent), legs pink-brown. Immature plain dull brown, with dull red rump and pinkish wash on breast, bill dusky. **Similar species** Wine-red tint to underparts differs from brighter scarlet tones of Red-billed Firefinch which also has all-red crown; African Firefinch has similar pattern but is darker overall with blue-black bill and black undertail.

HABITS Especially associated with rank green vegetation near damp areas and freshwater; also penetrates bushy areas in drier savanna woodland as well as irrigated gardens and cultivation. Usually feeding on the ground close to cover, in pairs or small groups, sometimes with other firefinches and waxbills. Generally shy.

VOICE High pitched *tschip* or *pik* alarm calls, sharper and harsher than Red-billed Firefinch; song a mix of metallic sounds with low pitched nasal notes repeated in a rapid and irregular pattern.

STATUS AND DISTRIBUTION Uncommon, occasionally locally frequent, in WD, CRD and URD, mainly close to freshwater sources; no recent records from NBD and LRD. Present throughout the year, recorded with marginally higher frequency in the wet season. In Senegal limited to Casamance and south-eastern areas, where generally local and scarce.

BREEDING Constructs domed nest of dry grasses, usually low in bush or tussock. Apparent family party, young birds, with soft gapes, late April. Pairs keeping very close company and on a separate occasion carrying nesting material in WD, August; otherwise no proven nest records from The Gambia or Senegal. Nest parasitised by Wilson's Indigobird.

NOTE Alternative name of Brown Firefinch (e.g. in Dowsett and Forbes-Watson 1993) derives from a decision to lump southern African *L. nitidula* with *L. rufopicta*.

RED-BILLED FIREFINCH
Lagonosticta senegala **Plate 47**

Fr. Amarante commun

Other name: Senegal Firefinch

IDENTIFICATION 10cm. Sexes differ. The common tiny red finch seen in parties on the ground around human habitation everywhere. Adult male bright red over entire head and breast, softly merging across mantle towards warm brown wings; rump red, tail red at base, darkening to blackish towards rounded tip. Some limited white spotting may be present on sides of breast (sometimes washed pink), undertail buffish-brown. Bill pink-red with black ridge down centre, eye brown with prominent yellow eye-ring, legs brownish. Adult female predominantly plain soft brown, greyer on breast, with small red patch in front of each eye and more liberal sprinkling of white spots; also has red bill. Juvenile is overall plain brown with a black bill. **Similar species** Brighter scarlet red and lack of brown wedge across crown (but some males have brown on crown) separates males from much less frequent Bar-breasted Firefinch. Females and young unlikely to be seen out of association with adult males. Males moulting into adult plumage distinguished by yellow eye-ring.

HABITS Closely associated with human settlement and activity; confiding ways much remarked upon since earliest accounts. Also found in bush and woodland, though only really frequent in areas close to water and along the river bank. Usually in small parties, often with Red-cheeked Cordon-bleu or other waxbills, and Village Indigobirds.

VOICE Soft *queet, squeet* call notes on a rising pitch are regularly given while foraging; also a testier *tzet* alarm. Song a hurried combination of rising calls, *chick-pea-pea-pea.*

STATUS AND DISTRIBUTION Common resident in all parts of the country. Similarly reported from all parts of Senegal, but locally absent far from water.

BREEDING Loose globular nest of dry grasses, lined with feathers, but readily incorporating artifacts. Entrance at the top to one side. Extended season, almost throughout the year but with lessening activity in late dry season. Usually low, often on or near buildings. Nest parasitied by Village Indigobird.

BLACK-BELLIED FIREFINCH
Lagonosticta rara **Plate 47**

Fr. Amarante à ventre noir

IDENTIFICATION 10cm. Sexes differ. *L. r. forbesi.* Adult male predominantly rich crimson over most of body, merging to brown on wings; deep red rump, tail darkening to blackish. Underparts deep red, without white flecking; belly and undertail black. Bill black with pale pink base to lower mandible; eye brown with narrow pink eye-ring; legs grey. Adult female greyish head, red patch in front of eye; pinkish-red over mantle and underparts; belly blackish to black undertail. Bill as male. Juvenile plain buff-brown body, red rump and dark tail, bill resembles adult. **Similar species** Easily overlooked; bill pattern, black belly and undertail, and lack of white flecks separates both sexes from Red-billed Firefinch, but the latter sometimes lack white flecks.

HABITS Typical firefinch, seen in pairs or small parties in grassland savannas, areas of tall herbs and thickets around cultivation, primarily feeding on the ground. Inconspicuous, not normally closely associated with human habitation.

VOICE Song *tew-tew-tew-tew*; contact call a soft whistle, quickly rising then more slowly falling back in pitch, *peeeeh* and a *chek* alarm.

STATUS AND DISTRIBUTION Not recorded in The Gambia; several observations scattered in extreme south-east Senegal suggest if may be infrequent but widespread resident in that area, and might be encountered in Niokolo Koba.

BREEDING Loose spherical nest of grasses with root or feather lining, usually low. Nearest confirmed breeding Sierra Leone. Nest parasitised by Cameroon Indigobird.

AFRICAN FIREFINCH
Lagonosticta rubricata **Not illustrated**

Fr. Amarante flambé

Other name: Blue-billed Firefinch

IDENTIFICATION 10-11cm. *L. r. polionota.* Sexes

differ. A dark firefinch, tentatively included on the basis of a single records in Senegal. Adult male recalls Bar-breasted Firefinch, but darker grey over crown and mantle. Deep red below, few white flecks on side of breast; greyish belly and black undertail; rump area deep red, most of tail black. Bill dark blue-black, eye brown with narrow pinkish eye-ring; legs grey. Adult female pinkish-grey on crown and nape, browner on back; rose-tinted underparts with white flecks at sides of breast, buffish belly, blackish undertail. Juvenile more uniform brown above, with red rump and dark tail. **Similar species** All-dark bill, darker-toned upperparts, black undertail all separate from Bar-breasted Firefinch; latter feature shared only with Black-bellied Firefinch.

HABITS Primarily a bird of woodland edge and scrub savanna; associated with remoter spots, away from human activity. General habits otherwise very similar to other firefinches.

VOICE Main call a loud *too-too-too* or *chew-chew-chew* with more staccato rattling alarm, *tchittic* or *pitpitpit*.

STATUS AND DISTRIBUTION Not recorded in The Gambia; rare in Senegal with sight records in Casamance, Niokolo Koba and eastern Senegal. A specimen of a firefinch from eastern Senegal, previously considered to be *L. rubricata*, has recently been identified as Mali Firefinch *L. virata*, the first record for Senegal.

BREEDING Typical firefinch globe-shaped nest with side entrance and feather lining. Nest parasitised by Cameroon Indigobird.

BLACK-FACED FIREFINCH
Lagonosticta larvata   Plate 47

Fr. Amarante masqué

Other name: Vinaceous Firefinch

NOTE The form in Senegambia is sometimes considered to be a separate species, *L. vinacea*.

IDENTIFICATION 11.5cm. Sexes differ. *L. l. vinacea*. Pale grey crown merging into increasingly vinous-pink wash over grey-brown body, dull crimson rump and tail; distinctive pale pink firefinch, even before diagnostic large but neat matt black mask of adult male noted. Male underparts pink, speckled with small white dots on flanks; black undertail. Adult female soft pale grey-brown on head, buff on throat, increasingly pink-washed on body with white flank spotting and pale buff undertail. Eye dark reddish, bill dark grey, legs grey-brown. Immature pale grey-brown with dull red at base of tail. **Similar species** All ages distinguished

from other firefinches by pale vinous-pink and grey tones with dark bill.

HABITS Retiring species generally found in grassland patches and bamboo thickets in savanna woodland, though recent observations in The Gambia more associated with thickets and woodland close to the banks of the main river. Associates with other firefinches and waxbills, usually in pairs or small groups. May perch in small trees or shrubs, but most likely to be seen feeding on the ground near dense cover at a secluded spot.

VOICE Two-note whistle, first low-pitched and descending, second higher pitched and rising, *whee-hew, whee-hew*. A sharp *dwit-it-it* alarm and *seesee* contact call. All calls imitated by Baka Indigobird.

STATUS AND DISTRIBUTION Has always been uncommon, and long thought to be in decline, linked to reduction in bamboo thickets. Lack of recent sightings in comparatively well-watched areas where it was formerly present in WD and LRD suggest this trend continues. The most recent observations are all from vicinity of Bansang through to Basse in CRD and URD in November-May where it is rare, and pair in WD in April. Information insufficient to determine possible seasonal changes; formerly regarded as resident and may still be so. In Senegal found close to the northern borders of The Gambia and in the east, with a relative stronghold in the extreme south-east. Regionally mainly confined to the drainage basin of the Gambia River.

BREEDING No record from The Gambia or Senegal, though a young male developing face mask observed in CRD in April. Constructs a loose globular nest from grasses, with side entrance. Nest parasitised by Baka Indigobird.

RED-CHEEKED CORDON-BLEU
Uraeginthus bengalus   Plate 47

Fr. Cordonbleu à joues rouges

IDENTIFICATION 12cm. Sexes differ. The common, very small, vivid pale blue and brown finch seen around human habitation everywhere. Pale brown above, intensely pale blue underparts with deep red ear-patch on male, longish pointed tail pale blue, and pinkish bill. Adult female lacks the red ear-patch, blue parts more restricted and paler in tone; both sexes pale brown in centre of belly. Eye reddish-brown, bill chalky-pink, darker towards tip, legs yellow-brown. Immature like female, but blue restricted to face and throat, with duskier bill. **Similar species** Unmistakable.

HABITS Confiding species found in a very wide

range of habitats, but especially notable for freely using human habitation, including village compounds and cultivation as well as lush hotel gardens. Widespread in both moist and dry savanna woodlands and thornscrub; commonly encountered along bush tracks. Primarily a ground feeder in small groups, often with Red-billed Firefinch. Attractive nuptial display, hopping on perch beside mate while holding grass stem in bill.

VOICE Thin-toned piping call, similar in quality to several other small waxbills, *tsee-tsee*. Song a more elaborate *wit-sit-diddly-diddly-ee-ee*.

STATUS AND DISTRIBUTION Abundant breeding resident, widespread throughout. Similarly distributed throughout Senegal.

BREEDING Constructs a woven globe nest of flowering grass heads with a side entrance at low level in small bushes. Will also utilise old nest of e.g. Little Weaver. Breeding commences at onset of rains in July, extending well into the early dry season up to February.

LAVENDER WAXBILL
Estrilda caerulescens   Plate 47

Fr. Astrild gris-bleu

Other names: Lavender Finch, Lavender Firetail

IDENTIFICATION 10cm. Sexes similar. Uniform grey body, paler towards the head, contrasting with deep crimson rump, tail and undertail-coverts. Narrow dark streak through eye; grey of belly merges with small area of dark slaty-grey on lower flanks, marked by a few white spots. Deep conical bill blackish, pinker towards base, eye dark brown with black orbital, legs black-brown. Immature paler with duller red parts and ill-defined eyestreak.

HABITS Found in a wide variety of habitats from thickets and thicket edge, extending into woodland, occasionally rank patches in more open land, e.g. at the margins of cultivation. Often encountered in larger groups than other small waxbills, with which it often associates. Feeds on the ground but also more arboreal than other species.

VOICE Maintains high piping calls at constant pitch, thin in tone, *see-see-squee-see*. Female call *tseeht-tseeht* answered by a measured, thin whistle *seet-tyoo* from male, accent and highest pitch at beginning of second note.

STATUS AND DISTRIBUTION A widespread resident, locally common in many areas, though not generally penetrating drier bush country. Found in equal abundance in all divisions of the country. In Senegal similarly widespread at the latitude of

The Gambia and southward, with a handful of records from further north.

BREEDING Recorded breeding in The Gambia in mid rainy season, August-September; builds comparatively bulky globular nest; one described from Senegal built from grasses with long entrance tunnel at top, hidden in green creepers well off the ground; notably secretive around the nest.

ORANGE-CHEEKED WAXBILL
Estrilda melpoda   Plate 47

Fr. Astrild à joues oranges

IDENTIFICATION 10cm. Sexes similar. Tiny waxbill; orange cheeks, head and nape soft grey, warm brown upperparts with bright scarlet rump, black tail. Underparts pale grey with yellow wash in centre of belly. Bill red, eye brown, legs brown. Immature similar but orange cheeks paler and less well defined; bill black. **Similar species** Unmistakable.

HABITS Found in a range of habitats, including grassland, cultivation, forest and stream edges, oil palm thickets; usually close to shady rank vegetation. Primarily a ground feeder. Gregarious, flying up in close packed flocks to nearby low shrubs, several perching side by side along the same grass stem. Often apart from other waxbills, though sometimes joins mixed species flocks.

VOICE Soft high-pitched *tsee-tsee* calls, a testy *chi-dee-chi* and a thin song *tsee-reee-ree, tsee-ree-reee*.

STATUS AND DISTRIBUTION Frequent to locally common resident, widespread and most readily encountered in WD near the coast, also present in NBD, CRD and URD though no recent records from LRD. In Senegal mainly found to the south in Casamance and the south-east; also in oil palm woodlands northward near the coast.

BREEDING Nest building in September; young birds seen in October-November. Constructs an elongated grass nest at low level in a grass tussock or on the ground under a bush, with short tunnel entrance to one side.

BLACK-RUMPED WAXBILL
Estrilda troglodytes   Plate 47

Fr. Bec-de-corail cendré

IDENTIFICATION 10cm. Sexes similar. Tiny pale grey-brown waxbill, crimson streak through eye, black tail and rump. Underparts pale, flushed with pink on breeding male; back and flanks barred with fine dark striations (not a clear field mark); outer tail feather margins white. Bill red, eye red-brown, legs dark brown. Immature black-billed with dark smudge in place of crimson eye-streak. **Similar species** Combination of all-pale body tone with black tail unlike any other waxbill in region; Common Waxbill *Estrilda astrild* (Fr. Astrild bec-de-corail) superficially similar, but is darker, brown tailed with heavier fine barring and not recorded closer than Guinea. African Silverbill also has pale body with dark tail, but lacks eye-streak and has a blue-grey bill..

HABITS Favours comparatively open dry bush and clearer areas in dry savanna woodland. In pairs or more often in small tight flocks; mainly at low level in grasses and bushes, dropping to pick up small seeds from the ground. Flocks and feeds together with other waxbill species, but less likely to come close to villages. Flicks tail side to side.

VOICE Repetitive loud *heu-cheu* and in flight maintains *tiup,tiup, tiup* call. Song an emphatic *tch-tcheer. cheeer*, rising on the second syllable.

STATUS AND DISTRIBUTION Resident; uncommon in WD, locally frequent further inland, especially in LRD and CRD. Overall the least commonly recorded of three *Estrilda* waxbills found in The Gambia. Widely but patchily distributed in northern and southern Senegal.

BREEDING Nests at ground level, constructing an elongated oval grass nest chamber with extra side compartment used by non-incubating bird. Breeding season not well documented in The Gambia, but seen carrying food in September, and an older December breeding record.

ZEBRA WAXBILL
Amandava subflava   Plate 47

Fr. Bengali zébré, Ventre-Orange, Astrild à flancs rayés

Other names: Orange-breasted Waxbill, Goldbreast

IDENTIFICATION 10cm. Sexes differ. Tiny colourful waxbill, olive above, orange below, with splashes of red. Adult male olive-brown over upperparts, greyer on crown. Vivid red eyestripe above narrow black lores; red rump. Short, rounded tail blackish, feathers narrowly tipped white. Below yellow on throat to deep orange on breast, extending variably to belly; olive barring along flanks. Conical bill red, highlighted by black margins above and below; eye red, legs flesh. Adult female lacks red eyestripe (has narrow black lores), rump duller red, underparts pale yellow with weaker flank barring.

Juvenile plain brown, no flank barring, all-dark bill, hint of weak double wing-bars. **Similar species** Distinctive; Quail-Finch lacks red plumage, much finer paler flank bars, all-dark face.

HABITS Found in small parties near rank herbs and thickets, mainly near waterside swamps; will use margins of neglected rice fields. Confiding, but not typically seen near habitation.

VOICE Calls mainly short, including *chit chit* and *ink churr churr*; also a *trip* or *zink zink* flight call.

STATUS AND DISTRIBUTION Formerly locally common, but today rare in The Gambia; in the last thirty years only a handful of records from the Banjul–Cape St. Mary area, all December-January, with three more recent inland records of small groups near rice fields and swamps in February-March (CRD along north bank of river, east of Kaur) and July (CRD on south bank near Sapu). In Senegal widely but very patchily and thinly recorded from northern, western and south-eastern areas, and likewise noted to have diminished as a result of rice cultivation reducing its habitat.

BREEDING May build own barrel-shaped nest, but also makes use of nests of other species, a habit perhaps under-recorded in other small waxbills. No breeding records from The Gambia; in northern Senegal, January and November. Parasitised by Goldbreast Indigobird.

QUAIL-FINCH
Ortygospiza atricollis   Plate 48

Fr. Astrild-caille

Other name: West African Quail-Finch

IDENTIFICATION 9.5-10cm. Sexes similar. Tiny; almost always first seen after flushing from underfoot, springing steeply up from rice stubble or swamp grassland, showing dumpy, bobbing silhouettes, circling against the sky uttering small tinny cries. Adult male: black face mask merges to grey-brown upperparts. Breast and flanks striated with vertical black and brown bars, smudgily interrupted mid-belly by a clear chestnut patch. Tail blackish with white outer corners. Female similar, but lacks dark face mask, and lateral barring less fully developed. Bill red in breeding male, brown above, red below in female; eye brown, legs fleshy-brown. Juvenile almost plain brown, warmer tone on breast, faint barring on flanks, bill dark, pinkish at base. **Similar species** Unmistakable in clear view; flying Bronze Mannikins may look and sound superficially similar, but produce a less tinny note.

HABITS Closely associated with treeless areas of

flat grassland floodplains adjacent to water and swamps, including dry rice fields where post-harvest stubble is a particularly favoured habitat. Almost exclusively terrestrial. Elusive; after flushing flocks circle and drop back to ground level, usually nearby, to freeze near a tussock. Camouflage remarkably good. Usually in small groups, though flocks up to 30 reported.

VOICE Characteristic hard tinny chirrups in flight are useful aid to location: compressed squeaky repeated *trink-trink* or *trillink, trillink*. Song given on the ground a rapid *click, clack, cloik* repeated in syncopated sequence.

STATUS AND DISTRIBUTION Breeding resident throughout The Gambia; locally common within the constraint of its habitat preference on the vegetated margins of sandflats near the coast, or the freshwater floodplains inland; especially prominent on the flood plain swamps of URD. Rice cultivation is likely to have reduced its breeding habitat. Present year-round but observed most frequently in the early dry season. Widespread along the major river basins of Senegal in north and south.

BREEDING Ground nester, constructing globular or pear-shaped grass nest with side entrance and rootlet lining, in base of tussock. Observed in The Gambia in November, recorded in Senegal in September-December, with one record February. Parasitised by Quail-Finch Indigobird.

AFRICAN SILVERBILL
Lonchura cantans   Plate 47

Fr. Spermète bec-d'argent

Other name: Warbling Silverbill

IDENTIFICATION 10cm. Sexes similar. A plain pale waxbill, distinguished only by bill colour and blackish wings and tail. Body overall weak milky-tea buff; very lightly mottled and scalloped on paler head. Wings and mantle greyer-brown, fine transverse vermiculations across upperparts seen in good view. Flight feathers, rump and tail blackish; central tail feathers oddly elongated and pointed. Breast pale buff, belly and undertail white. Stout conical bill pale blue-grey, eye black with pale blue eye-ring, legs pale grey. Juvenile similar, lacking vermiculations. **Similar species** Relatively featureless compared to other waxbills; immature Black-rumped Waxbill is similar but paler with a dark bill.

HABITS Gregarious; less inclined to mix with other waxbill species; subtly quieter manner, sitting shoulder to shoulder for long periods. Generally associated with drier acacia scrub habitats, but also present in thorn scrub behind beaches. Will use beach shades and village roof thatching for perches and nest sites, but not closely associated with habitation.

VOICE Bright silvery contact calls *cheep* or *chink* and maintains a soft but sustained trilling song, rising and falling in pitch.

STATUS AND DISTRIBUTION Uncommon resident, persistent in locally favoured sites from the coast through to CRD. Well-known from Kotu beaches, otherwise mainly at scattered locations on the north bank of the river in NBD and CRD east to Kaur. One record from Bansang. Present year-round, but seen most frequently through the dry season. In Senegal widely distributed in the northwest, becoming rarer down the coast; rare in Casamance.

BREEDING Globular nest of dry grass stems lined with feathers and hair, entrance spout pointing upward, usually in small trees, sometimes thatched eaves; will also take-over old weaver nests. Extended season from late rains through most of the dry season, September-May.

BRONZE MANNIKIN
Lonchura cucullata   Plate 47

Fr. Spermète nonnette

IDENTIFICATION 9-10cm. Sexes similar. Common tiny black, brown and white mannikin seen widely in fields, villages and bush. Face, head and throat black glossed purple, warm brown suffusion on nape fades to cooler grey-brown upperparts, flight feathers black; white rump barred with dark brown, tail black. Metallic bronzy-green wedge on shoulder. Underparts white; smudgy black and green-glossed flank patch breaks into even black barring towards tail. Bill, dark grey upper mandible, pale blue-grey lower mandible; eye brown, legs dark grey. Juvenile plain brown above, rump barred buff-brown, tail and wings dark brown; bill all-dark grey. **Similar species** Adults distinctive; immatures nondescript and confusing at first; deeper brown and dumpier build with shorter tail than plain juvenile indigobirds, African Silverbill, or immature Black-rumped Waxbill.

HABITS Lively and highly social birds, moving in small family groups or larger flocks, perching in rows on low arching stems side by side. Found in a range of scrub and field margin habitats, floodplain swamps and grasslands around sandflats. Attracted to water and will come into dense thickets to find it. Noted for rapid construction of communal roosting 'nests' holding small groups of birds overnight and which may be dismantled and remade daily.

VOICE A short, fluid trilling *rreep..trreep* sustained by all members of flock in flight; various *tsek* and *chik,chik, chirra* calls; maintains twittering while perching in groups.

STATUS AND DISTRIBUTION Breeding resident throughout, abundant in the coastal area, common to frequent inland, seen year-round. In Senegal widespread across the south, extending northward over a broad belt up the western half of the country.

BREEDING Bulky globular grass nest with side entrance, usually in a small tree, occasionally on buildings, often several in proximity. Mainly recorded September-December, young birds in good numbers throughout this period.

BLACK-AND-WHITE MANNIKIN
Lonchura bicolor   **Plate 47**

Fr. Spermète à bec bleu

Other name: Blue-billed Mannikin

IDENTIFICATION 10cm. Sexes similar. Only Mannikin entirely lacking brown colouring; associated with forest; included on basis of one poorly documented sight record from Casamance. Entire upperparts, throat and upper breast black, weakly green-glossed. Unbarred rump and tail black. A few tiny white dots seen on folded inner wing feathers. Lower breast and underparts contrasting pure white, with short smudgy black barring along flanks. Large conical bill pale blue, eye brown, legs dark grey. Juvenile dusky to sooty-brown above, wings and tail very dark; below paling through buffish-grey to whitish. **Similar species** Plain black and white pattern of adult with pale blue bill distinctive, slightly more robust than Bronze Mannikin.

HABITS As other Mannikins but much the most forest associated of the trio, so unlikely to be persistent in our area.

VOICE A short whistling *seet-seet* and a *kip* call.

STATUS AND DISTRIBUTION No Gambian records. In Senegal possibly vagrant; in need of substantiation. One observation of 10 birds in a Casamance hotel garden for three days January 1977. Resident through forest belt from Guinea Bissau eastward.

BREEDING Globular nest with side entrance, in small tree, among bunches of bananas etc.; will sometimes use old nest of another species. Nearest recorded breeding Sierra Leone.

MAGPIE MANNIKIN
Lonchura fringilloides   **Plate 47**

Fr. Spermète pie

Other name: Pied Mannikin

IDENTIFICATION 13cm. Sexes similar. The largest of the three mannikins in the region, the comparatively bulky head and bill lending a weaver-like appearance. Entire head glossy blue-black down to nape and throat. Upperparts warm dark brown, subtly mottled with black and pale shaft streaks. Wings brown, rump and tail black. Below white, fading into creamy undertail; a black smudge at side of breast emphasises projection of white at lower side of neck; narrow strip of blotchy alternating black and chestnut bars along flanks. Bill, blackish upper mandible, blue-grey lower mandible, eye brown, legs dark grey to black. Juvenile plain brown above, buffer on underparts but shows same dark tail as adult; bill dark grey. **Similar species** Size and shape draw attention; confirm by noting more restricted distribution of black on throat, combination of brown back with absence of barring on rump and smudgy flank-bar pattern.

HABITS Found around well vegetated thickets, often near cultivation, especially dry rice fields. Usually in small feeding groups, assembles in larger evening roosts in swamp grasses; frequently associates with Bronze or Black-and-White Mannikins. Less inclined than these to feed on the ground, taking favoured rice and seeds from standing vegetation. Nomadic in south of range, possibly also in Senegambia.

VOICE Simple *tsek* or *cheep* calls, and a repetitive *pee-oo-pee-oo* song.

STATUS AND DISTRIBUTION Today rare, probably vagrant, in The Gambia; considered common at the start of the century, e.g. when present in thousands near Banjul in June. Disappeared almost entirely around 1918; since then two records at Bakau in December-January 1974 and Bijilo June 1989. Similarly rare and sporadic records in southern Senegal from Casamance to Niokolo Koba, where it may be resident.

BREEDING Globular nest of grasses with side entrance, placed well up in a small tree, sometimes communally with Bronze Mannikin. Nearest confirmed breeding Sierra Leone.

CUT-THROAT FINCH
Amadina fasciata   **Plate 48**

Fr. Amadine cou-coupé

Other name: Cut-throat

IDENTIFICATION 11-12cm. Sexes differ. Small; crimson gash across throat of male on otherwise sandy-buff scaly-patterned body unmistakable. Finely barred with black on head, face and rump, more scalloped on back and flanks; tail dark brown with white tip. Plain sandy-buff breast, small chestnut patch on belly. Female similar but lacks crimson throat gash. Large conical bill grey-white, eye dark brown, legs fleshy. Immature similar to adult female; partial red throat gash apparent on young male. **Similar species** Male unmistakable. Both sexes are uniquely barred on the head.

HABITS Primarily associated with dry woodland savanna, occasionally open woodland habitats nearer the coast. Frequently seen on bare branches perched high in tree-tops, often in pairs, but also assembles in flocks, especially in the dry season when it may mix in larger groups with other small estrildids. Feeds mainly on the ground. Attracted to water.

VOICE Various clear sparrow-like chirps and wheezier notes; gives thin *ee-ee-ee* in flight. Sings with sustained low-pitched buzzy notes mixed with richer warbles.

STATUS AND DISTRIBUTION Mainly a dry season visitor, recorded in all parts of The Gambia; rare at the coast, becoming locally frequent in LRD (Kiang) though to URD (Basse area). Majority of observations November-May, peaking December-March. One record from URD in August. Very widespread in Senegal, but seems not to be generally common.

BREEDING Not proven in The Gambia. Globular nest of dry grasses lined with feathers low in a bush or in a tree hole. Also commonly utilises other species' nests, particularly old weaver nests. Adult seen emerging from Village Weaver nest in CRD, November. Breeds northern Senegal in December-March and August.

INDIGOBIRDS AND WHYDAHS
Viduidae

This confusing group, comprising indigobirds and whydahs (sometimes known as widows), presents enormous taxonomic problems. As a result of research by R. B. Payne, who provided assistance with the preparation of the texts, six species of indigobirds and three species of whydah are recognised here.

Indigobirds

Identification of these birds presents unusual difficulties. They are brood parasites, laying their eggs in the nest of an array of small estrildid finches, growing up alongside, not killing, the host nestlings. Adult males are black, variously glossed with green, blue or purple, but all look very similar; females are all streaky brown with longitudinally barred crowns. Both sexes have white patches on the flanks, occasionally glimpsed on displaying males.

The key to indigobird identification has been the understanding that as adults, these birds mimic the calls and songs of their natal hosts and that in most cases there has been sufficient genetic selection and differentiation to ensure that those parasitising one estrildid species also mimic the unique fledgling gape patterns of that species (Nicolai 1964, Payne 1982). Thus there is a tendency for a one-to-one relationship between indigobird and host. It is known that matching of species to host is not always perfect, and that the genetic differences between Indigobird species are small, suggesting a complex evolutionary background. From the field ornithologist's angle the main point is that they behave as different species. Song mimicry is dominant in identification.

Points to bear in mind in the field

i) Relative abundance of host species. In The Gambia there are at least five species that are now considered to have their own indigobird parasite. By far the most abundant is Red-billed Firefinch followed by Quail-Finch, Bar-breasted Firefinch, Black-faced Firefinch and Zebra Waxbill or Goldbreast. In addition, three rarely recorded waxbills from Senegal only, African Firefinch, Black-bellied Firefinch and Dybowski's Twinspot, are currently thought to share a possible sixth indigobird parasite, subject to further investigation (see below for details).

ii) Leg, bill and wing colour are important features of males, but all species show variation in intensity. The wings may at times show bleaching on a narrow margin of the outer primary feathers, which should not be confused with browner primary colour.

iii) In addition to mimetic calls and songs, singing indigobird males also produce their own rather chattering and complex song sequences; mimicry indicating 'parentage' typically takes up around 20% of their vocal output.

iv) In nearly all situations more than one of the known hosts of indigobirds can be found in close proximity, so simple presence of an indigobird near one of the rarer host species is not in itself evidence for one of the lesser known indigobird species.

v) New records of unusual indigobirds should seek as far as possible to link indigobird to host species, and accurate knowledge of song is clearly the prime factor in establishing this.

vi) In practice the Village Indigobird seems to be by far the most frequently encountered in The Gambia, and it is to be expected that a high proportion of individuals seen will turn out to be this species.

Summary of indigobird field identification pointers (adult males), organised in descending order of relative abundance of all known host species found in The Gambia and Senegal (hence expected rarity of indigobird species increases down the table). Based on Payne (1982, 1985, 1996) and Payne & Payne (1994).

Name	English name	Male body gloss	Male wing (most show pale primary edging)	Breeding male bill (B) and leg (L) colour	Host (and song mimicry)
Vidua chalybeata	Village Indigobird	blue-black or blue-green	blackish	B: whitish L: orange	Red-billed Firefinch
Vidua nigeriae	Quail-Finch Indigobird	dull green	brown	B: whitish L: light purplish	Quail-Finch
Vidua wilsoni	Wilson's Indigobird	purple	dark brown	B: whitish L: light purplish	Bar-breasted Firefinch
Vidua larvaticola	Baka Indigobird	green	dark brown	B: whitish L: light purplish	Black-faced Firefinch
Vidua raricola	Goldbreast Indigobird	green	brown	B: whitish L: light purplish	Zebra Waxbill
Vidua camerunensis	Cameroon Indigobird	blue	brown	B: whitish L: light purplish	African & Black-bellied Firefinches, Dybowski's Twinspot

Paradise whydahs

The paradise whydahs (or widows), now united with indigobirds in one genus, *Vidua,* also have a turbulent taxonomic history. They lay their eggs in the nests of *Pytilia* spp.

The background situation in The Gambia is not complicated. There are two *Pytilia* species in our area, the Green-winged *melba* and the Red-winged *phoenicoptera*. Their ranges overlap in The Gambia; overall Red-winged observations outnumber Green-winged observations 3:1, but the Red-winged Pytilia does not penetrate far north into Senegal, where only the Green-winged Pytilia is found. In areas such as

Kiang West National Park both *Pytilia* species share the same woodlands. Traditionally one form of paradise whydah has been recognised in both The Gambia and Senegal as Broad-tailed Paradise Whydah *Vidua orientalis aucupum*.

Recent field work in other parts of West and southern Africa suggests that some revision of this situation is necessary. First the English common name 'Broad-tailed Paradise Whydah' has now been assigned to *Vidua obtusa*, a specialist mimic and brood parasite of the Orange-winged Pytilia *Pytilia afra*, both central and southern African species. So the most widely published English name for paradise whydahs in Senegambia is no longer available and must be changed. Secondly, the two pytilias may be parasitised by different wydah species.

It is proposed that the brood parasite of Green-winged Pytilia should be distinguished as the Sahel Paradise Whydah *Vidua orientalis*, while Red-winged Pytilia is parasitised by the Exclamatory Paradise Whydah *Vidua interjecta*. Birds observed in The Gambia appear to be attributable to the latter species (Payne pers. comm.) in line with relative abundance of hosts.

These two whydahs are very similar in appearance; *interjecta* develops a longer tail up to 3x, rather than 2.5x, the body length in full development, but this is of limited use if completion of tail development cannot be ascertained. *Vidua interjecta* also shows reddish bill and legs in non-breeding plumages. It should be borne in mind that preliminary comparison of genetic material (from as yet very few birds) suggests that differences between them at this level are extremely small (less than can be observed within some single wide ranging species) and may not even be consistent (Klein *et al.* 1993). Nevertheless, like indigobirds, it may be that a behavioural mechanism of breeding population isolation (by song mimicry) is running ahead of genetic differentiation.

Difficulties in understanding the breeding behaviour of these birds are not new. Bannerman, correctly speculating 50 years ago that the failure of one of his field contacts to locate a nest was circumstantial evidence that it might be a brood parasite, penned the memorable footnote: 'a bird that can hide its nest from Sir Charles Belcher must be no ordinary bird.'

VILLAGE INDIGOBIRD
Vidua chalybeata   Plate 48

Fr. Combassou du Sénégal

Other name: Senegal Indigo Finch

IDENTIFICATION 11-12cm. Sexes differ seasonally. Breeding male (July-February) head and body uniform black, glossed bluish, occasionally blue-green; wings and tail dark brown to blackish, with little contrast to body. Axillaries streaky black and white. Bill whitish, eye dark brown, legs orange to pinkish. White flank spots on both sexes are only occasionally glimpsed. Non-breeding male and female distinguished by bold head striping; two broad blackish lateral crown stripes either side of pale tawny median streak, thin dark eye and malar stripes. Upperparts grey-brown, streaked with blackish feather centres. Underparts buffy, fading to whiter on belly. Bill dull horn, legs orange-tinted. Immature: upperparts indistinctly streaked grey-brown, plain grey-brown crown with weak pale supercilium; below plain buff. **Similar species** Breeding male from other indigobird males by combination of dark wings and bright leg colour, though latter variable and fades late season. Females from Pin-tailed Whydah by bill colour, but not safely distinguishable from other indigobird females. Broad head bars separates from streaky-crowned non-breeding bishops. Juvenile also very similar to Pin-tailed juvenile: greyer, paler-throated than Bronze Mannakin immature.

HABITS Host species and song mimicry: Red-billed Firefinch. Found in farmland, light woodlands and around the edges of swamps etc., but most readily encountered in and around villages, perched on projecting hut roof poles or feeding on the ground near grain stores etc. Will enter compounds, especially after grain pounding. Males perch and sing prominently in the breeding season; females come and go from these song posts. Various flight displays, especially hovering just above perched female, circular loops etc. In the dry season males in transitional plumage often present.

VOICE Male Village Indigobirds produce rapid, sputtering churring song, within which is embedded distinct phrases mimicking songs and calls of Red-billed Firefinch, especially clean whistled notes *chick-pea-pea-pea* with rising inflection to the latter notes.

STATUS AND DISTRIBUTION Widely distributed breeding resident, locally frequent at the coast, becoming frequent to common from LRD through CRD and URD. Recorded with diminished frequency in late dry season, when male breeding plumage not seen. Found throughout Senegal.

BREEDING Males hold territories where they attract females with song and display. Brood parasite of Red-billed Firefinch; young may be seen being fed by firefinches in all seasons.

GOLDBREAST INDIGOBIRD
Vidua raricola   Not illustrated

Fr. Combassou jambandu

IDENTIFICATION 11-12cm. Sexes differ seasonally. Breeding male black, brightly glossed with green, wings brown. Bill whitish, eye black, feet pale purplish. Female and immatures as Village Indigobird. **Similar species** Brighter green gloss than much more probable Quail-Finch Indigobird but confirmation of Zebra Waxbill voice mimicry essential for acceptable field identification in The Gambia.

HABITS Host species and song mimicry Zebra Waxbill (or Goldbreast), which was formerly common, but recorded only three times in recent years, on the north bank of the river.

VOICE Singing males particularly incorporate reeling juvenile begging call of Zebra Waxbill as well as simpler *chit, chit* and *ink, churr, churr*.

STATUS AND DISTRIBUTION Not yet identified in The Gambia or Senegal; nearest known populations in Sierra Leone; host species a rare resident in the region and the least frequently recorded indigobird host in The Gambia.

WILSON'S INDIGOBIRD
Vidua wilsoni   Plate 48

Fr. Combassou noir

Other names: Pale-winged Indigobird, Bar-breasted Firefinch Indigobird

IDENTIFICATION 11cm. Sexes differ seasonally. Breeding male uniform black, with purple gloss in appropriate light; wings and tail dark brown. Bill whitish, eye dark brown, legs pale grey with purple or pinkish cast. Females and immatures as Village Indigobird. **Similar species** Legs separate male from Village Indigobird. Contrasting brown wing and distinctly purplish gloss thought to be unique to this species in the region.

HABITS Host species and song mimicry: Bar-breasted Firefinch, which is found patchily throughout The Gambia, usually in undergrowth close to freshwater.

VOICE Mimics the erratic chinking song of Bar-breasted Firefinch, a jingling mix of short nasal and high metallic notes. Also a sharp *pik* alarm.

STATUS AND DISTRIBUTION A small number of sight records of purple-glossed indigobirds probably referable to this species. The host species is

thinly scattered throughout The Gambia, but conclusive proof of Bar-breasted Firefinch mimicry and parasitism by purple-glossed indigobird not yet obtained. Known from southern Senegal and patchily eastward to Sudan and Zaire.

CAMEROON INDIGOBIRD
Vidua camerunensis   Not illustrated

Fr. Combassou du Cameroun

IDENTIFICATION 11-12cm. Sexes differ seasonally. Breeding male black, glossed with blue; pale brown wings. Bill whitish, eye dark, legs pale purplish. Females and immatures as Village Indigobird. **Similar species** Very similar to Baka Indigobird and confirmation of song mimicry essential for acceptable field identification.

HABITS Host species and song mimicry African Firefinch, Black-bellied Firefinch, and Dybowski's Twinspot.

VOICE Mimics a rapid *pitpitpit* call from African Firefinch, a low whistled, *tew-tew-tew-tew* from Black-bellied Firefinch; more elaborate song from Dybowski's Twinspot.

STATUS AND DISTRIBUTION None of the host species are recorded from The Gambia; three of them occur rarely in Senegal, mainly in the southeast. Museum specimens from Senegal have recently been attributed to this species, but a single sight record from February 1986 in The Gambia (Gore 1990) made before details of breeding associations discovered, now seems doubtful.

NOTE The status of this indigobird is likely to be subject to revision, in view of the multiple host species currently associated with it.

BAKA INDIGOBIRD
Vidua larvaticola   Not illustrated

Fr. Combassou bakra

Other names: Black-faced Firefinch Indigobird

IDENTIFICATION 11-12cm. Sexes differ seasonally. Breeding male body feathers black, glossed greenish-blue. Flight feathers darkish brown. Bill whitish, legs light grey-brown, sometimes with purplish cast. Females and immatures as Village Indigobird. **Similar species** Wing colour browner than Village Indigobird, body gloss if seen eliminates purplish Wilson's Indigobird; Quail-Finch Indigobird said to be duller green, but unlikely to be distinguishable without definite confirmation of song mimicry; similarly with very remote possibility of Goldbreast Indigobird.

HABITS Host species and song mimicry: Black-faced Firefinch, which is found along dry woodland areas along north and south banks of the river in CRD and URD.

VOICE Mimicry includes slurred whistles of host, *whee-hew, whee-hew* and sharper *dwit-it-it* alarm.

STATUS AND DISTRIBUTION Considered likely to occur in The Gambia on basis of distribution of host, which is present though infrequent; formal demonstration of birds mimicking and parasitising Black-faced Firefinch in The Gambia not yet established. Like Village Indigobirds, Baka Indigobirds from further east are more bluish-glossed in males.

QUAIL-FINCH INDIGOBIRD
Vidua nigeriae   Not illustrated

Fr. Combassou du Nigéria

IDENTIFICATION 11-12cm. Sexes differ seasonally. Breeding male uniform black, rather dull greenish in appropriate light but seldom easily discernable; wings and tail contrasting browner than body. Bill whitish, eye dark brown, legs, pale purplish. Females and immatures as Village Indigobird.
Similar species Similar to Baka Indigobird; song mimicry or other significant association with Quail-Finch the best means of identification; note that simultaneous presence with Quail-Finch not enough, since other indigobird host species may also be present in such areas.

HABITS Host species and song mimicry: Quail-Finch, which favours open grassland around saline flats, open swamp and rice fields. Any indigobirds singing from isolated acacias or emergent shrubs in such places merit scrutiny.

VOICE Particularly mimics the rapid *click, clack, cloik* song of Quail-Finch.

STATUS AND DISTRIBUTION Three old museum specimens from The Gambia near Kaur attributed to this species; Quail-Finch host found throughout and is the second most frequently recorded of the indigobird host species found in the country. Like other indigobirds with browner wings, confirmation of mimicry and parasitism of Quail-Finch not yet obtained in The Gambia. Elsewhere presently known only from Nigeria, Cameroon and the Sudan.

PIN-TAILED WHYDAH
Vidua macroura   Plate 48

Fr. Veuve dominicaine

Other name: Pin-tailed Widow

IDENTIFICATION 12-13cm; 26-34cm with tail in breeding male. Sexes differ. Breeding males (June-December) black and white, elongated, lax, central four tail feathers and red bill. Cap, upperparts and tail black; pure white wing patches, collar and underparts; small extensions of black either side of breast. Bill bright coral or orange-red, eye brown, legs blackish. Female and non-breeding male (slightly larger and more boldly marked) sparrow-like streaked body with boldly marked head; pale tawny median crown stripe bordered by broad black lateral crown stripes, buff supercilium, dark line through eye and on cheek. Wings and tail dark brown edged buff with extensive white on edges of inner tail feathers. Bill pinkish. Juvenile, plain grey-brown above, plain buff below with dusky bill.
Similar species Longitudinally barred head pattern distinguishes non-breeding birds from all weavers, bishops and queleas; look for red or pinkish tones in bill to separate from streaky phases of smaller greyer indigobirds; note larger bill of Exclamatory Paradise Whydah which is also reddish; head darker. Plain juveniles greyer than all brown immature Mannakins.

HABITS Found in a range of farmland and woodlands, mostly in small groups with one male present. Abrupt twitchy movements; dancing flight display of breeding male. Feeds on the ground, skitters jerkily backwards on tarsi (like other *Vidua*), quickly pecking at freshly disturbed ground. Sits prominently at song posts including telegraph wires, trees or bushes. Displays low over grass. Often drinks at roadside pools. Brood parasite of various waxbills; regular at sites frequented by e.g. Orange-cheeked Waxbill; 2-3 males often seen close together.

VOICE Energetic delivery of hasty but hesitant short squeaky notes within a narrow range of pitch *tsip, tsit, tsit, see, see, churr, chrrt* etc. Flight call *tsip tseep*. Does not mimic songs of brood hosts.

STATUS AND DISTRIBUTION Resident breeder, widely and evenly distributed throughout, can be locally frequent, seldom common. Probably under-recorded in non-breeding season, but increased prominence September and October, well after males have developed distinctive plumage, may indicate local immigration to breed. Widely distributed in Senegal and throughout sub-Saharan Africa.

BREEDING Males hold territories and attract females with song and display. Brood parasite of estrildid waxbills. Five different hosts recorded throughout the species' range; these show a broadly similar gape pattern in their nestlings and this is mimicked by the Pin-tailed Whydah nestling (see also indigobirds). Black-rumped and Orange-cheeked Waxbill occur in The Gambia, both recorded as hosts in Senegal. Does not kill host species' nestlings.

EXCLAMATORY PARADISE WHYDAH
Vidua interjecta   Plate 48

Fr. Veuve d'Uelle

Other name: Uelle Paradise Whydah

IDENTIFICATION 12-13cm, plus 27-30cm tail in full breeding male. Sexes differ seasonally. Breeding males unmistakable, seen developing long tail feathers from July; most noticeable in full plumage November-December. Black cap, face and throat; chestnut and golden-yellow collar. Black upperparts, wings and tail which includes two elongated and two very elongated trailing broad black plumes to 3x body length at full development, twisted vertically. Dark chestnut breast continuous with collar, fades across lower breast through golden-yellow to pale buff belly. Bill black, eye brown, legs brownish. Non-breeding male blackish on the head with pale median crown streak, face and throat mottled whitish, narrow dark eye-streak; tawny mantle streaked black, variably black in wings and tail; below buffish breast, fading to whiter belly. In transition (June-July) a strange, elongated, harlequin-patterned finch. Adult female streaky brown on upperparts, pale median crown streak separated by broad black lateral head stripes, whitish supercilium; below plain grey fading to whiter belly. In non-breeding plumage males and females have reddish-tinted bill and legs. Juvenile plain brown above and over breast, paler grey on belly. **Similar species** Very similar to Sahel Paradise Whydah; see note above. Immature heavier-billed, bulkier than similar plumage of Pin-tailed Whydah, bigger than dark-billed immature indigobirds; juvenile stubbier-billed and plainer than petronias. Whydahs in transitional plumages may be confused with buntings and Chestnut-crowned Sparrow-Weaver.

HABITS Found primarily in larger tracts of dry savanna woodlands. Single breeding males perch in topmost twigs of leafless tree, often high, where they may be visited by females. Flies off with tail trailing high, developing see-saw rocking action with bursts of wing beats, flying in a circle 50m above the ground and moving from treetop to tree-top. Non-breeding birds more likely to be encountered in small groups at low level in wooded undergrowth.

VOICE Simple *chip* calls. Sings from bare treetop perch, but more subdued style than indigobird; a steady chattering that incorporates mimicry of Red-winged Pytilia.

STATUS AND DISTRIBUTION Resident, locally frequent in the woodlands and forest parks inland, especially LRD and CRD; uncommon at the coast. Recorded with increased frequency from the onset of the rains in July, when males develop distinctive plumage; records peak November-December. Local movement has been inferred from absence of records through the dry season, but plumage characteristics must also influence this result.

BREEDING Males hold territories where they attract females with song and display. Eggs laid in nest of Red-winged Pytilia. See notes on Paradise Whydah breeding biology above.

SAHEL PARADISE WHYDAH
Vidua orientalis   Not illustrated

Fr. Veuve de paradis

IDENTIFICATION Breeding male c.30cm; non-breeding male and female 12-13cm. *V. o. aucupum.* Breeding male (July to December), more or less identical to Exclamatory Paradise Whydah save that when tail plumes fully developed, they extend to 2.5x, rather than 3.0x, the body length. Non-breeding birds generally show horn bill and fleshy legs. **Similar species** See Exclamatory Paradise Whydah.

HABITS As for Exclamatory Paradise Whydah, except this species parasitises Green-winged Pytilia.

VOICE Similar to Exclamatory Paradise Whydah, but should mimic Green-winged Pytilia.

STATUS AND DISTRIBUTION Not known. Formerly regarded as the common Paradise Whydah of the region. However, this will not be the case if it is accepted that this is a consistent specialist parasite of the Green-winged Pytilia since Red-winged Pytilia, parasitised by *Vidua interjecta,* is much more common. By the same argument, this is likely to be the only Paradise Whydah throughout northern Senegal. A single tail feather from this species from Belel Forest Park (CRD) constitutes the first proof of this whydah in The Gambia.

BREEDING A brood parasite of Green-winged Pytilia. See above.

FINCHES

Fringillidae

Three species, two resident and one vagrant. Linnet *Acanthis cannabina* (Fr. Linotte mélodieuse) is known from a single vagrant captured in northern Senegal in March 1971. Finches are small, conical-billed seed-eaters, noted for song.

WHITE-RUMPED SEEDEATER
Serinus leucopygius   Plate 45

Fr. Serin à croupion blanc ou Chanteur d'Afrique

Other names: Grey Canary, White-rumped Serin

IDENTIFICATION 10-12cm. Sexes similar. Small streaky grey finch, with a stubby bill. Grey brown head to mantle lightly and evenly streaked, browner wings (with slight pale wing-bar), brown tail. White rump may be glimpsed in flight, though not always prominent. Underparts whitish, with greyer area over breast marked by distinct streaking. Juvenile similar but browner and more heavily streaked. **Similar species** Distinctive in good view, but can be overlooked among immature and female indigo-birds. Uniform streakiness without barring around head plus pale bill distinguishes without view of white rump.

HABITS Associated with agricultural land and small rural settlements in the drier regions of The Gambia. Usually seen in pairs or small groups, often with waxbills, Indigo Finch and Yellow-fronted Canary, forming small mixed flocks in or near villages. Perch high in baobabs, rooftop thatch poles of Fula villages; descend to the ground to raid dropped millet etc. Well camouflaged on grey dusty soils; generally undemonstrative ways.

VOICE Song subdued, resembling Yellow-fronted Canary, though opinion divided on whether it is a richer songster. Call not described.

STATUS AND DISTRIBUTION Resident, found in all regions and seasons; uncommon at the coast. Can occasionally be seen around villages and even towns (Soma) from LRD eastward; perhaps more reliable around small villages north of the river in CRD and URD. Throughout Senegal north of The Gambia, but not in Casamance.

BREEDING Not yet proven in The Gambia; most records from Senegal in dry season, September-March. Compact cup nest in extremities of tree branches.

YELLOW-FRONTED CANARY
Serinus mozambicus   Plate 45

Fr. Serin du Mozambique

Other name: Yellow-eyed Canary

IDENTIFICATION 11-13cm. *S. m. caniceps*. Sexes differ slightly. Compact small treetop finch with yellow underparts. Adult male bright yellow forehead and face, set off by neat black eyeline and malar stripe; crown to nape pale grey; lightly streaked pale grey-green mantle. Yellow rump may show when flying; folded wing shows single bar in fresh plumage. Mainly brown tail feathers narrowly fringed greenish and tipped with white. Underparts bright yellow. Female similar but head striping weaker, yellow less extensive and intense on underparts. Juvenile greyer, with only faintest yellow wash mainly about head, and echo of adult face pattern; wing-bar may be more prominent. **Similar species** Immature from White-rumped Seedeater by pale face with echo of adult pattern, usually some pale yellow wash and less heavy streaking.

HABITS Favours savanna woodlands, also coming into field areas and secondary scrub where there are taller trees, in small parties or large flocks. Males are prominent songsters. Erratic bounding flight.

VOICE Short but piercing whistled and chirruped warbling phrases, *zeee-zeree-chereeo*, quickly repeated, sometimes organised into repetitive rhythmic phrases *tew-to-teee, tew-to-teee*. Simple *tsssp* call.

STATUS AND DISTRIBUTION Common resident breeder found throughout the country, especially abundant in the large forest reserves of LRD and particularly prominent June-October when breeding. Thought to have expanded from inland strongholds since the 1970s and still less frequent in coastal regions than elsewhere, though readily found. Throughout Senegal except the extreme north.

BREEDING Wet season breeder constructing compact cup nest of woven fibre.

BUNTINGS

Emberizidae

Six species, four rare. Corn Bunting *Miliaria calandra* (Fr. Bruant proyer) is known from a single record of three individuals in northern Senegal in February 1974. Buntings are rather elongated, long-tailed, finch-like birds with conical bills, upper mandible smaller than lower. Sexes may differ. Generally terrestrial feeders.

ORTOLAN BUNTING
Emberiza hortulana   Plate 46

Fr. Bruant ortolan

IDENTIFICATION 16-17cm. Sexes similar but distinguishable. Unobtrusive, and uncommon. Dumpy body, with bright eye-ring, pink bill, droopy moustachial streaks. Male head and breast pale grey with stong olive wash; prominent but narrow pale eye-ring, pale yellow throat. Brown mantle streaked with dark feather centres, flight feathers dark brown fringed chestnut-brown; tail dark brown, with narrow rufous edging, white outer fringes. Rufous cinammon tone to belly. Female and non-breeding male similar, but streakier on crown and breast, paler. Bill uniform pink-orange at all ages, legs flesh. First-year birds start heavily streaked brown, with reddish bill, pale eye-ring, shadow of adult head pattern detectable. **Similar species** Eye-ring plus white in tail separates from House Bunting and Cinammon-breasted Bunting; characteristically uniform reddish-pink bill always distinctive.

HABITS Appears to prefer highland areas (not available in The Gambia) in its main wintering range further south in West Africa. Feeds on the ground in short vegetation, flushing suddenly with a hesitant bounding flight action before returning to the ground.

VOICE On migration gives a short *ie.....plit* flight call.

STATUS AND DISTRIBUTION Rare winter migrant in The Gambia; only two documented records, Banjul November 1955 and Tendaba April 1981; perhaps under-recorded. Scattered records through Senegal, mainly to the north of The Gambia, suggest both passage, and overwintering birds.

HOUSE BUNTING
Emberiza striolata   Plate 46

Fr. Bruant striolé

IDENTIFICATION 14cm. *E. s. sahari*. Sexes similar. Nondescript, inconspicuous bunting. Breeding male smoky-grey over head, nape, cheeks and breast, unevenly streaked black on crown, more broadly on breast. Untidy and variable pale super-

cilium and moustachial streaks. Upperparts, wing-coverts to uppertail-coverts cinnamon-rufous, only faintly streaked. Inner flight feathers dark brown broadly fringed rufous, tail dark brown with rufous-buff fringes to outer edges. Underparts paler cinnamon-rufous. Bill, dark brown upper mandible, flesh-yellow lower mandible, legs yellow-brown. Female similar to male, but duller; young birds similar with all-dark bill. **Similar species** From Cinnamon-breasted Bunting by diffusely streaked head pattern, often appears uniform grey and rufous at a distance; greater extent of grey on breast; upperparts much less streaked if at all. Immature Ortolan has pinkish bill and pale eye-ring.

HABITS Associated with human settlements and rocky outcrops, confiding and approachable.

VOICE Short scratchy bunting phrase, rendered *witch witch a wee* with emphasis on last notes.

STATUS AND DISTRIBUTION Vagrant to The Gambia; recorded Albreda in 1926, and at Barra nearby fifty years later in 1976 and 1982. Single record from near Dakar 1983. Nearest regular breeding range Mauritania, Mali and northward.

BREEDING Not recorded in Senegambia, elsewhere twig nest with grass lining in crevice or amongst loose stones, mostly in wet season .

CINNAMON-BREASTED BUNTING
Emberiza tahapisi   Plate 46

Fr. Bruant cannelle

Other names: Cinnamon-breasted Rock Bunting, African Rock Bunting

IDENTIFICATION 12.5-13.5cm. *E. t. goslingi*. Sexes similar. Unobtrusive dark bunting. Breeding male black and white striped head; white supercilium especially prominent; mantle rich brown mottled with dark feather centres; wings dark brown, extensive chestnut on inner flight feathers may create bright panel on folded wing, and shows in flight; tail dark brown. Small plain smoky blue-grey bib distinguishes West African race. Rest of underside uniform rich cinnamon. Bill dark brown above, lower mandible prominent pale orange; legs yellow-brown. Female and young have tawny, not white, head streaks. **Similar species** From similar House

Bunting by brighter head striping, especially stronger median crown stripe; smaller bib plain or mottled with white; more intensely streaked upper-parts including rump. House Bunting has less chestnut in wings, stronger buff fringes to outer tail feathers. Ortolan shows prominent pale eye-ring, no supercilium, reddish bill, white outertail.

HABITS Strongly associated with eroded lateritic bluffs, but also found on bare ground in drier savannas, attracted to freshwater. Most often in pairs.

VOICE Song a quick short jangle of 4-5 notes, opening with a trill and accelerating to a hastily stifled twittering close, *zrr-ze-ze-zewiliwi.* Wheezy *sweee* call.

STATUS AND DISTRIBUTION Generally uncommon, patchily distributed even within favoured habitat; locally frequent in preferred sites along the laterite ridges and bluffs in LRD, CRD and URD. Rarely at the coast.

BREEDING Breeds at the end of the rains, nest found in The Gambia in November and males singing prominently through to at least late February. Nest a low shallow cup of grass and roots.

GOLDEN-BREASTED BUNTING
Emberiza flaviventris   Not illustrated

Fr. Bruant à poitrine dorée

IDENTIFICATION 14-15cm. Sexes distinguishable with care. Black head with broad white central crown stripe and white lines running above and below eye. Mantle chestnut in male, with grey rump and brown tail showing white distal third of outer feathers in flight. White patch on wing-coverts (larger on male than female), with secondary smaller white wing bar. Throat and belly yellow; central breast tinged orange-yellow. Female similar but drabber. Bill, dark brown upper mandible, pale lower mandible. Immature shows buff in head streaking and more washed-out coloration. **Similar species** Look for white wing patches to separate from smaller Brown-rumped Bunting, also broader black cheek patch, contrast between grey rump and brown mantle in flight; note that Brown-rumped does show a narrow grey band at base of tail.

HABITS Typical bunting, perching near tops of trees without much moving around, but dropping to the ground to feed.

VOICE Contact notes, *zizi-zizi;* song of quick, emphatic and cleanly whistled notes, each with emphasis and lacking typical bunting jangle, so quite distinct from Brown-rumped.

STATUS AND DISTRIBUTION Not recorded in The Gambia; only one published in eastern-central

Senegal. West African, *E. f. flavigastra* is associated with the arid acacia belt of the southern Sahel from Mauritania, through Mali and eastward to the Sudan. A species that might be expected to appear in The Gambia if regional climatic and vegetation trends continue towards increasing aridity.

BREEDING Elsewhere, nests at the onset of the rains, June-July.

BROWN-RUMPED BUNTING
Emberiza affinis   Plate 46

Fr. Bruant à ventre jaune

IDENTIFICATION 12.5-14.5cm. *E. a. nigeriae.* Sexes similar. Colourful bunting of drier woodland, inconspicuous unless singing. Distinctive black and white striped head; mantle mottled chestnut, sometimes cinnamon; upper rump pale cinammon-brown with a narrow grey patch at base of tail; tail dark brown with white corners visible in flight. Wings dark, inner feathers broadly fringed cinnamon-brown. Underparts mainly bright sulphur-yellow; chin and undertail white. Bill, brown upper mandible, pale lower mandible. Juvenile head sandy-buff dark streaked; chestnut-streaked upperparts, little white in tail; yellowish below, tinged rusty on the breast. **Similar species** Cinnamon-breasted Buntings lack yellow underparts; from Golden-breasted Bunting by absence of white wing-covert patch, whitish lores, smaller size and song.

HABITS Found primarily in dry savanna woodlands of the lateritic soil zones, less frequently near where shrubs are scattered. Intermittent song often draws attention. Inconspicuous, usually approachable.

VOICE Song a variable but distinctive jangle, repeated regularly, a rapid delivery of silvery notes organised in two neat phrases, ending with a modestly emphasised trill, - *diddly du-du . didi-chiddly trrrr,* or *rijidi-durrr-ridgi-widgeri.* Short *chip* in flight.

STATUS AND DISTRIBUTION Uncommon to locally frequent member of the dry woodland community in LRD, through CRD to URD; stronghold in Kiang West where it can be frequent to common. Recorded in all months. Increase in observations June-September perhaps due to prominent singing behaviour. In Senegal moderately widespread through eastern and cental regions; largely absent from the far north.

BREEDING Singing males are prominent in the early rains. Adults with well grown fledgling seen in December 1996.

USEFUL ORGANISATIONS

Tanji Birders

Tanji Birders was formed in 1993 by a group of British birdwatchers with the object of raising funds to support the then newly gazetted Bird Reserve at Tanji. Since that time the aims of the group have broadened. The aims now include support for conservation projects and the promotion of ecotourism to The Gambia. They also help with and provide funding for scientific research into the Gambian avifauna. Their newsletter *Tanji Talk*, published twice a year, contains articles both about Gambian conservation and also of more general ornithological interest.

In all their activities the organisation has worked with and received the cooperation of the Gambian Department of Parks and Wildlife Management and the Ministry of Tourism.

Membership of Tanji Birders is open to all who have an interest in The Gambia and its wildlife. Tanji Birders may be contacted by writing c/o The Gambian National Tourist Office, The Gambia High Commission Building, 57 Kensington Court Road, London W8 5DG, U.K.

African Bird Club

Founded in 1994, the African Bird Club aims to provide a worldwide focus for African ornithology and to encourage an interest in the conservation of the birds of the region. A twice-yearly colour Bulletin is published which contains much material relevant to West Africa.

To join ABC, or for further information, contact: African Bird Club, c/o BirdLife International, Wellbrook Court, Girton Road, Cambridge CB3 0NA, U.K.

West African Ornithological Society

The West African Ornithological Society was formed in 1978 and exists to promote West African ornithology. One of its aims is to support ornithologists on the ground in West Africa by means of providing grants. Its journal, *Malimbus*, is published twice per year.

For details of membership, or further information, contact: R. E. Sharland, 1 Fisher's Heron, East Mills, Fordingbridge, Hampshire SP6 2JR, U.K.

BIBLIOGRAPHY

Aidely, D.J. & Wilkinson, R. 1987. The annual cycle of six *Acrocephalus* warblers in a Nigerian reed bed. *Bird Study* 34: 226-234.

Allport, G.A., Ausden, M.J., Fishpool, L.D.C., Hayman, P.V., Robertson, P.A. & Wood, P. 1996. Identification of Illadopsises *Illadopsis* spp. in the Upper Guinea forest. *Bull. African Bird Club* 3: 26-30.

Anon. 1993. What's that wader and why's it in The Gambia? *Birdwatch* 16: 74.

Anon. 1994. Documentation now and for posterity. *Brit. Birds* 87: 583-584.

Ash, J.S., Pearson, D.J., Nikolaus, G. & Colston, P.R. 1989. The Mangrove Reed Warblers of the Red Sea and Gulf of Aden coasts, with description of a new race of the African Reed Warbler, *Acrocephalus baeticatus. Bull. Brit. Orn. Club* 109: 36-43.

Baillon, F. 1992a. *Streptopelia hypopyrrha*, nouvelle espèce de tourterelle pour le Sénégal. *Oiseaux Rev. Fr. Orn.* 62: 320-334.

Baillon, F. 1992b. The nest and eggs of *Estrilda caerulescens. Bull. Brit. Orn. Club* 112: 274-275.

Baillon F. & Dubois P.J. 1992. Nearctic Gull species in Senegal and The Gambia. *Dutch Birding* 14: 49-50.

Bannerman, D.A. 1930-1951. *The Birds of Tropical West Africa.* 8 vols. Crown Agents, London.

Bannerman, D.A. 1953. *The Birds of West and Equatorial Africa.* 2 vols. Oliver & Boyd, London.

Bates, G.L. 1930. *Handbook of the Birds of West Africa.* John Bale, Sons & Danielson, London.

Blasdale, P. 1984. Nestling-feeding behaviour of Lesser Wood Hoopoes. *Malimbus* 6: 91-92.

Blasdale, P. 1984. Some observations on Black-crowned and White-backed Night Herons. *Malimbus* 6: 85-89.

Borrow, N. 1997. Red-crested Bustard *Eupodotis ruficrista* and Adamawa Turtle Dove *Streptopelia hypopyrrha*, new to The Gambia, and sightings of Great Snipe *Gallinago media. Malimbus* 19: 36-38.

Bouet, G. 1955. *Oiseaux de l'Afrique Tropicale*, première partie. In *Faune de l'Union Française* XVI, ORSTOM, Paris.

Brooke, R.K. 1971. Taxonomic and distributional notes on the Africa Chaeturini. *Bull. Brit. Orn. Club* 91: 76-79

Brown, L.H., Urban E.K & Newman, K. 1982 *The Birds of Africa*, Vol.I. Academic Press, London.

Bruggers, R.L. & Bortoli, L. 1979. Notes on breeding, parasitism and association with wasps by Heuglin's Weaver nesting on telephone wires in Mali. *Malimbus* 1: 135-144.

Budgett, J.S. 1901. On the ornithology of the Gambia River. *Ibis* 8: 481-497.

Cade, T.J. & Digby, R.D. 1982. *The Falcons of the World.* Collins, London.

Cawkell, E.M. & Moreau, R.E. 1965. Notes on birds in The Gambia. *Ibis* 105: 56-178

Chantler, P. & Driessens, G. 1995. *Swifts.* Pica Press, Sussex.

Chappuis, C. 1974-1985. Illustration sonore de problèmes bioacoustiques posés par les oiseaux de la zone éthiopienne. *Alauda* 42-53. Supplement sonores.

Chappuis, C. & Erard, C.1991. A new cisticola from west-central Africa. *Bull. Brit. Orn. Club* 111: 59-70.

Chappuis, C., Erard, C. & Morel, G.J. 1989. Type specimens of *Prinia subflava* Gmelin and *Prinia fluviatilis* Chappuis. *Bull. Brit. Orn. Club* 109: 108-110.

Chappuis, C., Erard, C. & Morel, G.J.1992. Morphology, habitat, vocalisations and distribution of the River Prinia *Prinia fluviatilis* Chappuis. Proc.VII Pan-Afr. Orn. Congr. pp. 481-488.

Cheke, R.A. & Walsh, J.F. 1989. Westward range extension into Togo of the Adamawa Turtle Dove *Streptopelia hypopyrrha. Bull. Brit. Orn. Club* 109: 47-48.

Cheke, R.A. & Walsh, J.F. 1997. *The Birds of Togo.* B.O.U. Check-list No. 14. British Ornithologists' Union, Tring.

Clancy, P.A. 1987. Subspeciation in the Afrotropical Gabar Goshawk *Micronisus gabar. Bull. Brit. Orn. Club* 107: 173-177.

Clark, W. S. 1992. The taxonomy of Steppe and Tawny Eagles, with criteria for separation of museum specimens and live eagles. *Bull. Brit. Orn. Club* 112: 150-157.

Clement, P. 1987. Field identification of West Palearctic wheatears. *Brit. Birds* 30: 137-157; 187-238.

Clement, P., Harris, A. & Davis, J. 1993. *Finches and Sparrows: An Identification Guide.* Christopher Helm, London.

Collar, N.J. & Stuart, S.N. 1985. *Threatened Bird of Africa and related islands.* ICPB, Cambridge.

Collar, N.J., Crosby, M.J. & Stattersfield, A.J. 1994. *Birds to Watch 2: The World List of Threatened Birds.* BirdLife International, Cambridge.

Colston, P.R. & Curry-Lindahl, K. 1986. *The birds of Mount Nimba, Liberia.* British Museum (Natural History), London.

Colston, P.R., & Morel, G.J. 1985. A new subspecies of the Rufous Swamp Warbler *Acrocephalus rufescens* from Senegal. *Malimbus* 7: 61-62.

Colston, P.R., & Morel, G.J. 1984. A new subspecies of African Reed Warbler *Acrocephalus baeticatus* from Senegal. *Bull. Brit. Orn. Club* 104: 3-5.

Condamin, M. 1987. Le Pluvier de Leschenault *Charadrius leschenaultii* espèce nouvelle pour le Sénégal. *Malimbus* 9: 131-133.

Craig, A. J. F. K. 1992. The identification of *Euplectes* species in non-breeding plumage. *Bull. Brit. Orn. Club* 112: 102-108.

Cramp, S. *et al.* (eds.) 1977-1994. *The Birds of the Western*

Palearctic. Vols. I-IX. Oxford Univ. Press, Oxford.

Curry-Lindahl, K. 1981. *Bird Migration in Africa: movements between six continents.* Academic Press, London.

de Naurois, R. & Morel, G.J. 1995. Description des oeufs et du nid de la Prinia aquatique *Prinia fluviatilis. Malimbus* 17: 28-31.

del Nevo, A.J., Rodwell, S., Sim, I.M.W., Saunders, C.R., & Wacher, T. 1994. Audouin's Gulls *Larus audouinii* in Senegambia. *Seabird* 16: 57-61.

Demey, R. 1995. Notes on the birds of coastal and Kindia areas, Guinea. *Malimbus* 17: 85-99.

Devisse, R. 1992. Première observation au Sénégal du Martinet marbre. *Malimbus* 14: 16.

Dickerman, R.W. 1989. Notes on the Malachite Kingfisher *Corythornis (Alcedo) cristata. Bull. Brit. Orn. Club* 109: 158-159.

Dowsett, R.J. & Dowsett-Lemaire, F. 1993. *A contribution to the Distribution and Taxonomy of Afrotropical and Malagasy Birds.* Tauraco Research Report No. 5. Tauraco Press, Liège.

Dowsett, R.J. & Forbes-Watson, A.D. 1993. *Checklist of Birds of the Afrotropical and Malagasy Regions.* Tauraco Press, Liège.

Dowsett-Lemaire, F. & Dowsett, R.J.1987. European and African Reed Warblers, *Acrocephalus scirpaceus* and *A. baeticatus*: vocal and other evidence for a single species. *Bull. Brit. Orn. Club* 107: 74-85.

Dupuy, A.R. 1984. Synthèse sur les oiseaux de mer observés au Sénégal. *Malimbus* 6: 79-84.

Dupuy, A.R. 1985. Découverte d'un nouvel aigle forestier au Sénégal, l'Aigle d'Ayres *Hieraaetus dubius. Malimbus* 7: 114.

Edberg, E. 1982. *A Naturalist's Guide to The Gambia.* J.G.Sanders, St.Anne.

Elgood, J.H. 1982. The case for the retention of *Anaplectes* as a separate genus. *Bull. Brit. Orn. Club* 102: 70-75.

Elgood, J.H. *et al.*1994. *The Birds of Nigeria.* B.O.U. Check-list No.4, Second Edition. British Ornithologists' Union, Tring.

Ellenberg. H., Galat-Luong, A., von Maydell, H., Muhlenberg, M., Panzer, K.F., Schmidt-Lorenz, R., Sumser, M., & Szolnoki, T.W. 1988. *Pirang -Ecological investigations in a forest island in The Gambia.* Stiftung Walderhaltung in Afrika, Hamburg.

Erard, C. & Morel, G.J. 1994. La sous-espèce du Chevis modeste, *Galerida modesta*, en Sénégambie. *Malimbus* 16: 56-57.

Ericsson, S. 1989. Notes on birds observed in Gambia and Senegal in November 1984. *Malimbus* 11: 88-94.

Farmer, R. 1984. Ear tufts in a *Glaucidium* owl. *Malimbus* 6: 67-69.

Field, G.D. 1973. Ortolan and Blue Rock Thrush in Sierra Leone. *Bull. Brit. Orn. Club* 93: 81-82.

Field, G.D. 1979. Barn Owl movement. *Malimbus* 1: 67-68.

Fishpool, L.D.C., Demey, R., Allport, G. & Hayman, P.V. 1994. Notes on the field identification of

bulbuls Pycnonotidae of Upper Guinea. Part 1: The genera *Criniger, Bleda* and *Andropadus. Bull. African Bird Club* 1: 32-38.

Fishpool, L.D.C., Demey, R., Allport, G. & Hayman, P.V. 1994. Notes on the field identification of bulbuls Pycnonotidae of Upper Guinea. Part II: *Phyllastrephus* and remaining genera. *Bull. African Bird Club* 1: 90-95.

Fishpool, L.D.C., van Rompaey, R. & Demey, R. 1989. Call of White-crested Tiger Heron *Tigriornis leucolophus* attributed to the Rufous Fishing Owl *Scotopelia ussheri. Malimbus* 11: 96-97.

Forshaw, J.M & Cooper, W.T. 1989. *Parrots of the World.* 3rd revised edition. Blandford.

Friedmann, H. 1955. *The Honeyguides.* United States National Museum Bulletin 208, Smithsonian Institution, Washington D.C.

Fry, C.H. 1981a. The diet of Large Green Bee-eaters *Merops superciliosus* supersp. and the question of bee-eaters fishing. *Malimbus* 3: 31-38.

Fry, C.H. 1981b. On the breeding season of *Merops persicus* in West Africa. *Malimbus* 3: 52.

Fry, C.H. 1982a. Spanish Black Kites in West Africa. *Malimbus* 4: 48.

Fry, C.H. 1982b. The Moreau ecological overview. *Ibis* 134 suppl.1: 3-6.

Fry, C.H. 1983. Red mandibles in the Woodland Kingfisher superspecies. *Malimbus* 5: 91-93.

Fry, C.H. 1984. *The Bee-eaters.* T. & A.D. Poyser, Calton.

Fry, C.H., Keith, S. & Urban, E.K. 1988. *The Birds of Africa,* Vol.III. Academic Press, London.

Fry, C.H., Fry, K, & Harris, A. 1992. *Kingfishers, Bee-eaters and Rollers.* Christopher Helm, London.

Geeson, J. & Geeson, J. 1990. First Red Kite record for The Gambia. *Malimbus* 11:144.

Génsbol, B. 1992. *Birds of Prey of Britain and Europe.* Collins, London.

Gibbon, G. 1991. *Southern African Bird Sounds.* 6 tapes. Southern African Birding, Hilary.

Goodwin, D. 1982 *Estrildid Finches of the World.* Oxford Univ. Press, Oxford.

Gore, M.E.J. 1980. Millions of Turtle Doves. *Malimbus* 2: 78.

Gore, M.E.J. 1990. *Birds of The Gambia.* B.O.U. Check-list No.3. Second edition. British Ornithologists' Union, Tring.

Grimes, L.G. 1980. Observations of group behaviour and breeding biology of the Yellow-billed Shrike *Corvinella corvina. Ibis* 122: 166-192.

Grimes, L.G. 1987. *The Birds of Ghana.* B.O.U. Check-list No.9. British Ornithologists' Union, Tring.

Hall, B.P. 1963. The Francolins, a Study in Speciation. *Bull. Brit. Mus. Nat. Hist. Zool.* 102: 105-204.

Hall, B.P. & Moreau, R.E. 1970. *An Atlas of Speciation in African Passerine Birds.* British Museum (Natural History), London.

Harris, A., Tucker, L. & Vinicombe, K. 1990. *The Macmillan Field Guide to Bird Identification.*

Macmillan, London and Basingstoke.

Harris, A., Shirihai, H. & Christie, D. 1996. *The Macmillan Birder's Guide to European and Middle Eastern Birds*. Macmillan, London and Basingstoke.

Harrison, P. 1983. *Seabirds: An identification guide*. Croom Helm, Beckenham.

Hayman, P., Marchant, J. & Prater, T. 1986. *Shorebirds: An identification guide to waders of the world*. Croom Helm, London.

Hazevoet, C.J. 1995. *The Birds of the Cape Verde Islands*. B.O.U. Check-list No. 13. British Ornithologists' Union, Tring.

Herremans, M. & Stevens, J. 1983. Moult of the Long-tailed Nightjar *Caprimulgus climacurus* Vieillot. *Malimbus* 5: 5-16.

Hirschfeld, E. & Stawarczyk, T. 1994. Leg colour of Kentish Plovers *Charadrius alexandrinus* in Bahrain. *Bull. Orn. Soc. Middle East* 33: 10-11.

Hollom, P.A.D., Porter, R.F. & Christensen, S. 1988. *Birds of the Middle East and North Africa*. T. & A. D. Poyser, Calton.

Huff, J.B. & Auta, J. 1977. Nests of White-fronted Black Chats. *Bull. Niger. Orn. Soc.* 13 (44): 148.

Irwin, M.P.S. 1987. Geographical variation in the plumage of female Klaas's Cuckoo. *Malimbus* 9: 43-46.

Jackson, H.D. 1984. Key to the Nightjar species of Africa and its Islands, Aves: Caprimulgidae. *Smithersia* 4: 1-55.

Jackson, H.D. 1985. Mouth size in *Macrodipteryx* and other African nightjars. *Bull. Brit. Orn. Club* 105: 51-54.

James, P.C. & Robertson, H.A. 1985. Soft-plumaged Petrels *Pterodroma mollis* at Great Salvage Island. *Bull. Brit. Orn. Club* 105: 25-26.

Jensen J.V. & Kirkeby, J. 1980. *The Birds of The Gambia*. Aros Nature Guides, Aarhus.

Jensen J.V. & Kirkeby, J. 1987. Records of Rock Thrush *Monticola saxatilis* in The Gambia. *Malimbus* 9: 123-124.

Jobling, J.A. 1991. *A Dictionary of Bird Names*. Oxford Univ. Press, Oxford.

Johnsgard, P.A. 1988. *The Quails, Partridges and Francolins of the World*. Oxford Univ. Press, Oxford.

Johnsgard, P.A. 1991. *Bustards, Hemipodes and Sandgrouse: Birds of Dry Places*. Oxford Univ. Press, Oxford.

Johnson, D.N. & Horner R.F, 1986. Identifying Widows, Bishops and Queleas in female plumage. *Bokmakierie* 38: 13-17.

Jones, R.M., 1991 The status of larks in The Gambia, including first records of the Sunlark *Galerida modesta*. *Malimbus* 13: 67-73.

Jones, S. 1992. The Gambia and Senegal. Pp. 175-182 *in* Sayer, J.A., Harcourt, C.S. & Collins, N.M. (eds.) *The Conservation Atlas of Tropical Forests: Africa*. IUCN World Conservation Monitoring Centre, Cambridge.

Jonsson, L. 1992. *Birds of Europe with North Africa and the Middle East*. Christopher Helm, London.

Kasper, P. 1993. *Some Common Flora of The Gambia*. Stiftung Walderhaltung in Afrika, Hamburg.

Keith, S. & Gunn, W.W.H. 1971. *Birds of the African Rainforests*. Sounds of Nature No. 9. Federation of Ontario Naturalists, Ontario.

Keith, S., Urban, E.K. & Fry, C.H. 1992. *The Birds of Africa*, Vol.IV. Academic Press, London.

Kemp, A. 1995. *The Hornbills*. Oxford Univ. Press, Oxford.

Klein, N.K., Payne, R.B., & Nhlane M.E.D. 1993. A molecular genetic perspective on speciation in the brood parasitic *Vidua* finches. *In* Wilson, R.T. (ed.) Birds and the African Environment: Proc. VIII Pan-Afr. Orn-Congr. *Annales Musee Royal de l'Afrique Centrale Zoologie* 268: 29-39.

Knox, A. 1994. Lumping and splitting of species. *Brit. Birds* 87: 149-159.

Lang, J.R. 1969a. A Spotted Eagle Owl's nest. *Bull. Niger. Orn. Soc.* 6 (23):101-103.

Lang, J.R. 1969b. The nest and eggs of the Moho or Oriole Babbler. *Bull. Niger. Orn. Soc.* 6 (24): 127-128.

Lewington, I., Alström, P. & Colston, P. 1991. *A Field Guide to the Rare Birds of Britain and Europe*. HarperCollins, London.

Louette, M. 1992. The identification of forest *Accipiters* in central Africa. *Bull. Brit. Orn. Club* 112: 50-53.

Lynes, H. 1930. Review of the genus *Cisticola*. Ibis Ser.12 (6). Suppl.

Mackworth-Praed, C.W. & Grant, C.H.B. 1970-73. *Birds of West Central and Western Africa*, 2 vols. Longman, London.

Maclean, G.L.1988. *Roberts' Birds of South Africa*. New Holland, London.

Madge, S. & Burn, H. 1988. *Wildfowl: An identification guide to the ducks, geese and swans of the world*. Christopher Helm, London.

McGregor, I.A. & Thomson, A.L. 1965. Blue Rock-Thrush *Monticola solitaria* in The Gambia. *Ibis* 107: 410.

Mild, K. & Shirihai, H. 1994. Field identification of Pied, Collared and Semi-collared Flycatchers. Part 2: females in breeding plumage. *Birding World* 76: 231-240.

Moore, A. 1980. Some observations on a brood of Grey Woodpeckers in The Gambia. *Malimbus* 2: 159.

Moore, A. 1983. On the nesting of the Lavender Fire-Finch. *Malimbus* 5: 56.

Moore, A. 1984. Levaillant's Cuckoo *Clamator levaillanti* fed by Brown Babblers *Turdoides plebejus*. *Malimbus* 6: 94-95.

Morel, G.J. 1981. Réponse à la pluie de *Mirafra javanica*. *Malimbus* 3: 57.

Morel, G.J. & Browne, P.W.P. 1981. Les *Buteo* paléarctiques en Mauritanie et au Sénégal. *Malimbus* 3: 2-6.

Morel, G.J., Monnet, C. & Rouchouse, C. 1983. Données nouvelles sur *Monticola solitaria* et *Monticola saxatilis* en Sénégambie. *Malimbus* 5: 1-4

Morel, G.J. & Morel, M.-Y. 1979. La Tourterelle des bois dans l'extrême Ouest-Africain. *Malimbus* 1: 66-67.

Morel, G.J. & Morel, M.-Y. 1990. *Les Oiseaux de Sénégambie*. ORSTOM, Paris.

Morel, G.J. & Morel, M.-Y. 1992. Habitat use by Palearctic migrant passerine birds in West Africa. *Ibis* 134 suppl: 83-88.

Morel, M.-Y. 1987. *Acrocephalus scirpaceus* and *Acrocephalus baeticatus* dans la region de Richard-Toll, Sénégal. *Malimbus* 9: 47-55.

Moynihan, M. 1987. Notes on the behaviour of Giant Kingfishers. *Malimbus* 9: 97-104.

Mullié, W.C. & Keith, J.O. 1991. Notes on the breeding biology, food and weight of the Singing Bush-Lark *Mirafra javanica* in Northern Senegal. *Malimbus* 13: 24-39.

Nelson, J.B. 1978. *The Sulidae: Gannets and Boobies*. Oxford Univ. Press, Oxford.

Newell, D. Porter, R. & Marr, T. 1997. South Polar Skua - an overlooked bird in the eastern Atlantic. *Birding World* 10: 229-235.

Newman, K.1983. *Birds of Southern Africa*. Southern Book Publishers, Johannesburg.

Newman, K. & Solomon, D, 1994. *Look-alike Birds*. Southern Book Publishers, Johannesburg.

Olsen, K.M. & Larsson, H. 1995. *Terns of Europe and North America*. Christopher Helm, London.

Olsen, K.M. & Larsson, H. 1997. *Skuas and Jaegers*. Pica Press, Sussex.

Parmenter, T. & Byers, C. 1991. *A Guide to the Warblers of the Western Palearctic*. Bruce Coleman Books, Uxbridge.

Parrott, J. 1979. Kaffir Rail, *Rallus caerulescens,* in West Africa. *Malimbus* 1:145-146.

Payne, R.B. 1982. Species limits in the Indigobirds Ploceidae, *Vidua* of West Africa: Mouth Mimicry, Song Mimicry and Description of New Species. *Misc. Publ. Mus. Zool.* Univ. of Michigan, 162: 1-96.

Payne, R.B. 1985. The species of parasitic finches in West Africa. *Malimbus* 7: 103-113.

Payne, R.B. 1991. Female and first-year male plumages of paradise wydahs *Vidua interjecta*. *Bull. Brit. Orn. Club* 111: 95-100.

Payne, R.B. 1996. Field identification of the Indigobirds. *Bull. African Bird Club* 3: 14-25.

Payne, R.B. 1997a. Field identification of the brood-parasitic whydahs *Vidua* and Cuckoo Finch *Anomalospiza imberbis*. *Bull. African Bird Club* 4: 18-28.

Payne, R.B. 1997b. The Mali Firefinch *Lagonosticta virata* in Senegal. *Malimbus* 19: 39-41.

Payne, R.B. & Payne, L.L. 1994. Song mimicry and species associations of West African Indigobirds *Vidua* with Quail-finch *Ortygospiza atricollis,* Goldbreast *Amandava subflava* and Brown Twinspot *Clytospiza monteiri*. *Ibis* 136: 291-304.

Payne R.B. & Payne, L.L. 1995. Song mimicry and association of brood-parasitic Indigobirds *Vidua* with Dybowski's Twinspot *Euschistospiza dybowskii*.

The Auk 112: 649-658.

Perrins, C.M., Lebreton, J-D. & Hirons, G.J.M. 1991. *Bird Population Studies*. Oxford Univ. Press, Oxford.

Peske, L., Bobek, M., Pojer, F., Simek, J. & Mriek, V. 1996. Satellite and VHF radio-tracking of Black Storks, migrating from Europe to Africa. *Argos Newsletter* 51.

Porter, R., Newell, D., Marr, T. & Jolliffe, R. 1997. Identification of Cape Verde Shearwater. *Birding World* 10: 222-228.

Potti, J., & Merino, S. 1995. Some male Pied Flycatchers *Ficedula hypoleuca* in Iberia become collared with age. *Ibis* 137: 405-409.

Rand, A.L. 1951. Birds from Liberia, with a discussion on barriers between Upper and Lower Guinea subspecies. *Fieldiana, Zool.* 32: 561-653.

Randall, R.D. 1995. Greater Swamp Warbler *Acrocephalus rufescens* on the Chobe River. *Babbler* 29-30.

Reyer, H U. 1980. Sexual dimorphism and cooperative breeding in the Striped Kingfisher. *Ostrich* 51: 117-118.

Richards, M.W. & Boswell, J. 1985. Some Egyptian Plover nests in Senegal. *Malimbus* 7: 128.

Riddiford, N. 1990. Collared Flycatcher *Ficedula albicollis* in Senegal. *Malimbus* 11: 149-150.

Ripley, S. D. 1977. *Rails of the World: A monograph of the family Rallidae*. M. F. Fehely, Toronto.

Rodwell, S.P., Sauvage, A., Rumsey, S.J.R. & Braünlich, A. 1996. An annotated checklist of birds occurring at the Parc National des Oiseaux du Djoudj in Senegal, 1984-1994. *Malimbus* 18: 74-111.

Rouchouse, C. 1985. Sedentarisation de *Monticola solitarius* au Cap de Naze, Sénégal. *Malimbus* 7: 91-92.

Rowan, M.K. 1983. *The Doves, Parrots, Louries and Cuckoos of Southern Africa*. Croom Helm, London.

Rumsey, S.J.R. & Rodwell, S.P. 1992. *A Field Checklist for the Birds of Senegambia*. The Wetland Trust, Icklesham.

Ryall, C. & Stoorvogel, J.J. 1995. Observations on nesting and associated behaviour of the Shrike Flycatcher *Megabyas flammulatus* in Tai National Park, Ivory Coast. *Malimbus* 17: 19-24.

Safford, R.J., Ash, J.S., Duckworth, J.W., Telfer, M.G. & Zewdie, C. 1995. A new species of nightjar from Ethiopia. *Ibis* 137: 301-397.

Sauvage, A. & Rodwell, S.P. 1998. Notable observations of birds in Senegal (excluding Parc National des Oiseaux du Djoudj), 1984-1994. *Malimbus* 20: 75-122.

Savalli, U.M. 1995. The evolution of tail-length in widow-birds Ploceidae: tests of alternatives to sexual selection. *Ibis* 137: 389-395.

Serle, W. & Morel, G.J. 1977. *A Field Guide to the Birds of West Africa*. Collins, London.

Shirihai, H., Harris, A. & Cottridge, D. 1991. Identification of Spectacled Warbler. *Brit. Birds* 81: 423-430.

Sibley, C.G. & Monroe, B.L. Jnr. 1990. *Distribution and Taxonomy of Birds of The World.* Yale Univ. Press, New Haven.

Sinclair, I., Hockey, P. & Tarboton, W. 1993. *Illustrated Guide to the Birds of Southern Africa.* New Holland, London.

Skead, C.J. 1967. *The Sunbirds of Southern Africa.* South African Bird Book Fund, A.A. Balkema, Cape Town.

Skinner, N.J. 1969. Notes on the breeding of the Pygmy Long-tailed Sunbird. *Bull. Niger. Orn. Soc.* 6 (24):124-126.

Small, B. 1996. Identification of male Spectacled Warbler. *Brit. Birds* 89: 275-280.

Smalley, M.E. 1979a. Dowitcher in The Gambia. *Malimbus* 1: 68.

Smalley, M.E. 1979b. Cattle Egret feeding on flies attracted to mangos. *Malimbus* 1: 114-117.

Smalley, M.E. 1983a. The Marsh Owl *Asio capensis:* a wet season migrant to The Gambia. *Malimbus* 5: 31-33.

Smalley, M.E. 1983b. Abyssinian Rollers *Coracias abyssinica* and European Rollers *C. garrulus* in The Gambia. *Malimbus* 5: 34-36.

Smalley, M.E. 1984. Predation by Pied Crows *Corvus albus* on Gambian Epauletted Fruit Bats *Epomophorus gambianus. Bull. Brit. Orn. Club* 104: 77-79.

Smith T.B. 1987. Bill size polymorphism and interspecific niche utilisation in an African finch. *Nature* 329: 717-719.

Smith, T.B. 1993. Ecological and evolutionary significance of a third bill form in the polymorphic finch *Pyrenestes ostrinus.* Proc. VIII Pan-Afr. Orn. Congr pp. 61-66.

Snow, D.W. 1978. *An Atlas of Speciation in African Non-passerine Birds.* British Museum (Natural History), London.

Stacey, P.B. & Koenig, W.D. 1990. *Co-operative Breeding in Birds.* Cambridge Univ. Press, Cambridge.

Steyn, P. 1982. *Birds of Prey of Southern Africa.* Croom Helm, Beckenham.

Steyn, P. 1984. *A Delight of Owls.* David Phillip, Cape Town & Johannesburg.

Stickley, J. 1966. Nesting of the Cut-throat Weaver. *Bull. Niger. Orn. Soc.* 3 (11): 34-36.

Stoate, C. 1995. The impact of Desert Locust *Schistocerca gregaria* swarms on pre-migratory fattening of Whitethroats *Sylvia communis* in the western Sahel. *Ibis* 137: 420-422.

Summers-Smith, J.D. 1988. *The Sparrows.* T. & A.D. Poyser, Calton.

Svensson, L. 1992. *Identification Guide to European Passerines.* Fourth edition. Stockholm.

Tarboton, W.R. 1981. Co-operative breeding and group territoriality in the Black Tit. *Ostrich* 52: 215-225.

Thiollay, J-M. 1977. Distribution saisonnière des rapaces diurnes en Afrique occidentale. *L'Oiseau* 47: 25-85.

Turner, A. & Rose, C. 1989. *A Handbook to the Swallows and Martins of the World.* Christopher Helm, London.

Tye, A. 1984. Long-tailed nightjar drinking in flight. *Malimbus* 6: 4.

Urban, E.K., Fry, C.H. & Keith, S. 1986. *The Birds of Africa,* Vol.II. Academic Press, London.

Urban, E.K., Fry, C.H. & Keith, S. 1997. *The Birds of Africa,* Vol.V. Academic Press, London.

van Perlo, B. 1995. *Birds of Eastern Africa.* HarperCollins, London.

Wacher, T.J. 1993. Some new observations of forest birds in The Gambia. *Malimbus* 15: 24-37.

Walsh, J.F. 1985. Extension of known range of the African Black Duck *Anas sparsa* in West Africa. *Bull. Brit. Orn. Club* 105: 117.

Walsh, J.F. 1991. On the occurrence of the Black Stork *Ciconia nigra* in West Africa. *Bull. Brit. Orn. Club* 111: 209-215.

Weick, F. & Brown, L.H. 1980. *Birds of Prey of the World.* Collins, London.

White, C.M.N. 1960-1963. *A Revised Checklist of African Passerine Birds.* Govt. Printer, Lusaka.

Wilkinson, R. 1978a. Co-operative breeding in the Chestnut-bellied Starling *Spreo pulcher. Bull. Niger. Orn. Soc.* 14 (46): 71-72.

Wilkinson, R. 1978b. Behaviour of Grey-headed Bush Shrikes at their nest. *Bull. Niger. Orn. Soc.* 14: 87.

Wilkinson, R. 1982. A colour variant of *Spreo pulcher* at Kano, Nigeria. *Malimbus* 4: 89.

Wilkinson, R. 1982. Seasonal movements of the Pygmy Kingfisher, *Ceyx picta,* in West Africa. *Malimbus* 4: 53-54 & 108.

Wilkinson, R. 1983. Biannual breeding and moult-breeding overlap in the Chestnut-bellied Starling *Spreo pulcher. Ibis* 125: 353-361.

Wilkinson, R. 1984. Variation in eye colour of Blue-eared Glossy Starlings. *Malimbus* 6: 2-4.

Wilkinson, R. 1988. Long-tailed Glossy Starlings in field and aviary with observations on co-operative breeding in captivity. *Avic. Mag.* 94: 143-154.

Williams, J.G. & Arlott, N. 1980. *A Field Guide to the Birds of East Africa.* Collins, London.

Wink, M. 1981. On the diets of warblers, weavers and other Ghanaian Birds. *Malimbus* 3: 114-115.

Winkler, H., Christie, D.A. & Nurney, D. 1995. *Woodpeckers.* Pica Press, Sussex.

Wood, B. 1975. Observations on the Adamawa Turtle Dove. *Bull. Brit. Orn. Club* 95: 68-73.

Zimmerman, D.A., Turner, D.A. & Pearson, D.J. 1996. *Birds of Kenya and Northern Tanzania.* Christopher Helm, London.

INDEX

Species are listed by their English vernacular name, under their family or group name (e.g. Bulbul, Common), together with alternative names where relevant. Scientific names are listed under their respective genera. Subspecific names are not given, but synonyms are included. Numbers in roman type refer to the first page of the systematic entry. Numbers in *italic* type refer to alternative English names or synonyms. Numbers in **bold** type refer to plate numbers.